Theory of Atomic Nuclei:
Quasiparticles and Phonons

Theory of Atomic Nuclei: Quasiparticles and Phonons

Vadim G Soloviev

Joint Institute for Nuclear Research, Dubna, Russia

Translated from the Russian by

Vitaly I Kisin

Institute of Physics Publishing
Bristol and Philadelphia

British Library Cataloguing-in-Publication Data. A catalogue record for this book is available from the British Library.

ISBN 0-7503-0131-7

Library of Congress Cataloging-in-Publication Data are available.

Published by IOP Publishing Ltd, a company wholly owned by the Institute of Physics, London

IOP Publishing Ltd
Techno House, Redcliffe Way, Bristol BS1 6NX, UK

US Editorial Office: IOP Publishing Inc., The Public Ledger Building, Suite 1035, Independence Square, Philadelphia, PA 19106

Typeset in TeX by IOP Publishing Ltd

Printed in Great Britain by Galliard (Printers) Ltd, Great Yarmouth

Contents

Preface

This book deals with a microscopic description of the structure of complex nuclei, at low and intermediate excitation energies, in terms of quasiparticle and phonon operators. It also compares the results of calculations with the available experimental data. The book continues an exposition of the theoretical physics of nuclei, which was earlier presented in a number of textbooks and monographs, e.g., [1–12].

The present book belongs to a series of monographs on the theory of nuclei that I am writing. It is not, however, connected directly to the opening volume, *'Theory of Atomic Nuclei: Nuclear Models'* [12]. The reader need not refer to this volume to be able to understand the present one since the information on the mean field of the nucleus and on superconductivity-type pair correlations is briefly outlined here as well. I hope that the macroscopic description of neutron resonances, analogue states and various types of electrical and magnetic resonances will be given in the next volume of the series, *'Theory of Atomic Nuclei: High-Excitation States'*.

The progress in the theory of atomic nuclei has been very impressive recently, bringing out quite clearly the internal unity in our understanding of nuclear structures. As a result, it is possible to select the deductive method of exposition. This monograph gives the solution of the nuclear many-body problem in the framework of the Hartree–Fock–Bogoliubov method. The theory of nuclear vibrations is given in the general form, on the basis of the Bogoliubov self-consistent field method. The mean field of the nucleus, the interaction resulting in pairing, and interactions in the particle–hole and particle–particle channels are singled out. The phonon operators, including the neutron–proton ones, are defined, and the random-phase approximation for the electric- and magnetic-type states is presented exhaustively for the first time.

The book gives the general scheme of the boson representation of fermion operators. The collective Hamiltonian for quadrupole excitations is constructed in the SU(6) approximation, and the efficiency of the approximation is demonstrated for nuclei of the transitional regions. The mathematical apparatus of the interacting bosons model and of its modifications is presented. The application of this model to spherical and deformed nuclei is described, and its relation to the Bohr–Mottelson model is discussed.

The monograph gives, for the first time, a formulation of the quasiparticle–phonon nuclear model, suggested to describe the few-quasiparticles components of wave functions at low, medium and high excitation energies. The general calculation scheme is given, and also the advantages of the model, which follow from the phonon basis, the quasiparticle–phonon interaction, the approximation based on the expansion of the wave function in the number of phonons, and the application of the strength function method.

The structure of the states of the deformed and spherical nuclei with energy below the neutron binding energy is described at a higher level. Strength functions are derived for the description of the fragmentation of single-quasiparticle, one-phonon configurations and the quasiparticle ⊗ phonon configurations. The book presents, for the first time, a systematic exposition of the fragmentation of one- and two-particle states in spherical nuclei. The structure of low non-rotational states of even–even deformed nuclei is described. It is clearly demonstrated that contradictions between three models are typical for these states: the Bohr–Mottelson model and its microscopic analog, the quasiparticle–phonon nuclear model and the interacting bosons model.

This monograph does not offer a complete bibliography on the aspects of nuclear physics involved. I cite only the most important papers and books, and those papers whose results I have used. I also cite those papers which make it possible to learn the details of the work presented in the book and those in which this work was further developed.

In some sections of the monograph, the presentation follows the courses of lectures on the theory of nuclei that I gave to students of the physics department of Moscow State University and of its Dubna affiliate. The book also includes some material presented earlier in lectures at the JINR Schools in Alushta in 1972, 1974, 1980, 1983 and 1985, at the USSR Schools in Voronezh, Bukhara and Khumsan, at the international schools in Italy, Poland, Czechoslovakia and Yugoslavia, and also in talks delivered at a number of institutes in the Bulgaria, Czechoslovakia, Denmark, France, Germany, Hungary, Japan, Mongolia, Poland, USA and the former USSR.

I assume that the reader is familiar with the foundations of quantum mechanics, electrodynamics, statistical physics and nuclear physics.

I am grateful to N N Bogoliubov, A I Vdovin, V V Voronov, R V Jolos, V A Kuzmin, G Kyrchev, L A Malov, V O Nesterenko, P Nikolaeva, V Yu Ponomarev, Ch Stoyanov, A V Sushkov and N Yu Shirikova for useful discussions and help in improving the presentation of some aspects.

Notes on terminology

This monograph generally employs the notation and terminology recommended by the Symbols, Units and Nomenclature Commission of the International Union of Pure and Applied Physics (IUPAP) [13], in which $\hbar = c = 1$. The phases of wave functions are chosen as in [14], unless otherwise specified. The Clebsch–Gordan coefficients are denoted by $\langle I_1 M_1 I_2 M_2 | I_3 M_3 \rangle$. The Wigner–Eckart theorem is given as formulated in [14]. For transformations, the quantum theory of angular momentum is given in accordance with [14]. Matrix elements are written in the form $(\Psi_f^* \mathfrak{M}(\lambda) \Psi_j)$, where Ψ_i, Ψ_f are the wave functions of the initial and finite states, and $\mathfrak{M}(\lambda)$ is an operator of multipolarity λ. In cases of possible confusion, a hat is used above an operator, for example, $\widehat{N}$.

Even–even nuclei are nuclei with even numbers of neutrons and protons; *odd nuclei* are nuclei in which either the number of neutrons or that of protons is odd; and *odd–odd nuclei* are those with odd numbers of both neutrons and protons. The term *complex nuclei* (it covers medium and heavy nuclei) is used for nuclei with $A > 40$; the term *light nuclei* stands for nuclei with $A \leqslant 40$.

The set of quantum numbers characterizing single-particle states is denoted by f. If the number $\sigma = \pm 1$ is specified (σ relates the states that are conjugate with respect to the time reversal operation), we use the notation $q\sigma$, $s\sigma$ and $r\sigma$, with q characterizing the states of the neutron and the proton systems, s characterizing only the neutron system, and r, only the proton system in deformed nuclei. In spherical nuclei, single-particle states are denoted by $nljm$, where the total momentum j and its projection m, the orbital moment l and the number n indicate that a given value lj appears in a sequence for the nth time; a contracted notation jm is often used. The notation $j_{\rm n} m_{\rm n}$ for the neutron systems and $j_{\rm p} m_{\rm p}$ for the proton systems is also employed.

Notation

A	– mass number
$a_{q\sigma}, a_{jm}, a^+_{q\sigma}, a^+_{jm}$	– nucleon absorption and creation operators
a^{qk}_{nlj}	– coefficients of expansion of a single-particle wave function in the spherical basis
$\mathcal{A}(q_1q_2;\mu\sigma)$, $\mathcal{A}(j_1j_2;\lambda\mu)$, $\mathcal{A}^+(q_1q_2;\mu\sigma)$, $\mathcal{A}^+(j_1j_2;\lambda\mu)$	– two-quasiparticle operators entering into interactions of the magnetic type
$A(q_1q_2;\mu\sigma)$, $A(j_1j_2;\lambda\mu)$ $A^+(q_1q_2;\mu\sigma)$, $A^+(j_1j_2;\lambda\mu)$	– two-quasiparticle state operators entering into electric-type interactions
$B(E\lambda)$,$B(M\lambda)$	– reduced probabilities of $E\lambda$ and $M\lambda$ transitions
$B(E\lambda)_{\text{s.p.u.}}$, $B(M\lambda)_{\text{s.p.u.}}$	– reduced probabilities in single-particle units
$B(qq';\mu\sigma)$, $B(jj';\lambda\mu)$	– operators containing $\alpha^+\alpha$ and entering into electric-type interactions
$\mathcal{B}(qq';\mu\sigma)$, $\mathcal{B}(jj';\lambda\mu)$	– operators containing $\alpha^+\alpha$ and entering into interactions of the magnetic type
C	– pairing function
C_p, C_n	– proton and neutron pairing functions
$C^2_J(\eta)$	– strength function describing the fragmentation of the nlj sub-shell
C^2_JS	– spectroscopic factors in spherical nuclei
D^J_{MK}	– coefficients of transformation from one rotating reference frame to another
d^+_μ, d_μ	– creation and absorption operators for quadrupole bosons, $\mu = -2, \ldots, +2$
$\bar{E}$	– energy centroid
$E\lambda$	– electric transition of multipolarity λ
$E(q)$, $E(j)$	– energy of the mean field single-particle levels
$\mathcal{E}_0$	– ground-state energy of even–even nucleus
$e^{(\lambda)}_{\text{eff}}(\tau)$	– effective charge
$f^{\lambda\mu}(qq')$, $f^\lambda(jj')$	– matrix elements of multipole forces
$f^{\lambda LK}(qq')$, $f^{\lambda L}(jj')$	– matrix elements of spin–multipole forces
G	– pairing constant

G_Z, G_N	– proton and neutron pairing constants
$g_s^{\text{eff}}(\tau)$, $g_s(\tau)$, $g_l^{\text{eff}}(\tau)$, $g_l(\tau)$	– effective and free gyromagnetic ratios for the spin and the orbital angular moment
H	– system's Hamiltonian
H_{av}	– Hamiltonian of single-particle motion
H_{pair}	– Hamiltonian of the interaction leading to pairing
H_{M}^{ph}, $H_{\text{CM}}^{\text{ph}}$	– Hamiltonians of multipole interaction in particle–hole channel
H_S^{ph}, H_{CS}^{ph}	– Hamiltonians of spin–multipole interaction in particle–hole channel
H_T^{ph}, H_{CT}^{ph}	– Hamiltonians of tensor interaction in particle–hole channel
H_{QP}	– Hamiltonian in the quasiparticle–phonon nuclear model
H_{vq}	– Hamiltonian of quasiparticle–phonon interaction
$\boldsymbol{I}$	– total angular momentum ($I_\pm = I_x \pm \mathrm{i}I_y$)
i	– number of the root of the RPA secular equation
$\boldsymbol{J}$	– angular momentum of internal motion
K	– projection of $\boldsymbol{I}$ on the nucleus' symmetry axis
$\mathcal{K}$	– a function describing the effect of the Pauli exclusion principle on two-phonon states
$\mathcal{L}$	– a function describing the effect of the Pauli exclusion principle on quasiparticle ⊗ phonon states
$M\lambda$	– magnetic transition of multipolarity λ
$\mathcal{M}$	– matrix element
$\mathfrak{M}(E;\lambda\mu)$	– operator of the electric $E\lambda$ transition
$\mathfrak{M}(M;\lambda\mu)$	– operator of the magnetic $M\lambda$ transition
m	– nucleon mass
N_d	– the number of d-bosons
N_s	– the number of s-bosons
N	– the number of neutrons; the principal oscillator number; the total number of bosons
$Nn_z\Lambda$	– asymptotic quantum numbers
n_γ	– the number of γ-bosons
n_β	– the number of β-bosons
$\boldsymbol{p}$	– particle's momentum
$Q_{\lambda\mu i\sigma}$, $Q_{\lambda\mu i}$, $Q_{\lambda\mu i\sigma}^+$, $Q_{\lambda\mu i}^+$	– phonon absorption and creation operators
R, $R(\theta,\varphi)$	– radius of a nucleus
$R_\lambda(r)$	– radial function of residual separable interactions
$\boldsymbol{r}$	– radius-vector of a particle
$S_{nlj}^{q_0K_0}(\eta)$	– strength functions describing the fragmentation of a sub-shell nlj, which is comprised in a single-particle state q_0

$\widetilde{S}^{q_0K_0}_{nlj}(\eta)$, $\approx{S}^{q_0K_0}_{nlj}(\eta)$	– spectroscopic srength functions of (dp) and (dt) reactions
s, s^+	– boson absorption and creation operators for $J = 0$
$\mathfrak{S}$	– seniority
T	– nuclear isospin
T_0	– nuclear isospin in ground state
$\mathfrak{T}$	– time reflection operator
u_{fg}, v_{fg}, u_q, v_q, u_j, v_j	– coefficients of the Bogoliubov transformation $u^{(\pm)}_{qq'} = u_q v_{q'} \pm u_{q'} v_q$; $v^{(\pm)}_{qq'} = u_q u_{q'} \pm v_q v_{q'}$
$u_{nlj}(r)$	– radial part of the single-particle wave function of a spherical nucleus
$V(r)$, $V_{ls}(r)$, $V_c(r)$	– central, spin-orbital and the Coulomb parts of the Saxon–Woods potential
$V_0^{N,Z}$	– depth of potential well
W	– transition probability
x	– spatial coordinates
$Y_{lm}(\theta, \varphi)$	– spherical functions
Z	– the number of protons
α	– spreading constant
$\alpha_{q\sigma}$, α_{jm}, $\alpha^+_{q\sigma}$, α^+_{jm}	– quasiparticle absorption and creation operators
β_2, β_4	– parameters of quadrupole and hexadecapole deformations
$\Gamma\downarrow$	– spreading width
γ	– deformation parameter
Δ	– averaging interval
ϵ_q, ϵ_j	– quasiparticle energy $(\epsilon_{qq'} = \epsilon_q + \epsilon_{q'})$
η_ν	– energies of excited states
θ	– axial angle
$\kappa_0^{(\lambda\mu)}$, $\kappa_0^{(\lambda)}$, $\kappa_1^{(\lambda\mu)}$, $\kappa_1^{(\lambda)}$	– isoscalar and isovector constants of multipole particle–hole interactions
$\kappa_0^{(\lambda LK)}$, $\kappa_0^{(\lambda L)}$, $\kappa_1^{(\lambda LK)}$, $\kappa_1^{(\lambda L)}$	– isoscalar and isovector constants of spin-multipole particle–hole interactions
$\kappa_T^{(LK)}$, $\kappa_T^{(L)}$	– isovector constant of tensor ph interactions
Λ	– projection of L on the symmetry axis; asymptotic quantum number
λ	– multipolarity; chemical potential
λ_p, λ_n	– proton and neutron chemical potentials
μ	– projection of multipolarity λ
μ_j, μ_0	– magnetic moment and nuclear magneton
μ_p, μ_n	– proton and neutron magnetic moments
ν	– number of roots of the QPNM secular equation
π	– parity

$\rho(f, f')$	– density function
$\rho_0(q)$	– diagonal density function for quasiparticle vacuum
$\rho_{\lambda i}(r)$	– transition charge density
$\rho^{c,m}_{\lambda Li}(r)$	– transition convectional and magnetic current densities
Σ	– projection of σ on the symmetry axis
σ	– spin matrix, $\sigma = \pm 1$
τ	– isospin matrix
φ	– azimuthal angle
$\varphi_K(q)$	– single-particle wave function of a deformed nucleus
φ_{nljm}	– single-particle spherically-symmetric wave function
Ψ^I_{MK}	– wave function of an even–even deformed nucleus
Ψ_{00}	– particle vacuum
Ψ^0_0	– quasiparticle vacuum
Ψ_0	– phonon vacuum
$\Psi_\nu(K_0^{\pi_0}\sigma_0)$, $\Psi_\nu(JM)$	– wave functions in the quasiparticle–phonon nuclear model
$\psi^{\lambda\mu i}_{qq'}$, $\psi^{\lambda i}_{jj'}$, $\phi^{\lambda\mu i}_{qq'}$, $\phi^{\lambda i}_{jj'}$	– direct and inverse phonon operator amplitudes
ω_g, $\omega_{\lambda i}$	– one-phonon energies
$\Delta\omega(Jj\lambda i)$, $\Delta\omega(qg)$	– shifts of quasiparticle $\otimes$ phonon poles
$\Delta\omega(g_1, g_2)$, $\Delta\omega(\lambda_1 i_1, \lambda_2 i_2)$	– shifts of two-phonon poles
$\Omega_{\lambda\mu i\sigma}$, $\Omega_{\lambda\mu i}$, $\Omega^+_{\lambda\mu i\sigma}$, $\Omega^+_{\lambda\mu i}$	– neutron–proton absorption and creation operators

Introduction

An approximation which treats a nucleus with mass number A as consisting of N neutrons and Z protons, $A = N + Z$, is considered to be good for describing the structure of atomic nuclei. The explicit effects of non-nucleon degrees of freedom at non-relativistic energies and low momentum transfer are weak. The fundamental properties of nuclear matter, such as its density and binding energy, are implied by the laws of quantum chromodynamics. The non-nucleon degrees of freedom must play the decisive role in processes with high energy transfer, especially in regions that are kinematically forbidden for nucleons. Multi-quark configurations are likely to play an important role at very high excitation energies [15,16].

The analysis of the effect of non-nucleon degrees of freedom on the structure of nuclei has only just begun. This is a time- and labour-consuming process. To be successful, it must be based on a sufficiently consistent theory of nuclei at the nucleon level. We are to expect a clear-cut display of non-nucleon degrees of freedom precisely when the nucleon interpretation meets with insurmountable difficulties. We may divide the nuclei–nuclei interaction, with certain qualifications, into three energy levels. The first level is the interaction of bare nuclei at very low energies, when nuclei can be treated as elementary fermions or bosons regardless of their internal structure. The second level is the interaction of nuclei at energies from several MeV to several GeV, with low momentum transfer. This level corresponds to the neutron–proton interpretation, with the Pauli exclusion principle taken into account for the nucleons. It is at this level that the theory of nuclear structure has been developed. Finally, the third — the quark–gluon — level covers the interactions of nuclei at high momentum transfer; these interactions result in high excitation energies and high local densities of nuclear matter. The analysis in this book will be limited to the nucleon level, on which the structure of nuclei is described in terms of quasiparticle and phonon operators.

The first step in solving the nuclear many-body problem is to find the effective interactions in nuclei. In doing this, one begins with the interaction of free nucleons and employs the quark–gluon interpretation and the exchange of various mesons. Phenomenological nucleon–nucleon potentials are often used. It is essential in the process of solving the nuclear many-body problem that one obtains the form of effective interactions and singles out such properties of nuclei as the mean field, rotation and some others that allow phenomenological interpretation. The object of the theory of nuclei is not so much the strictest possible solution of the many-body problem as the most accurate description of those characteristics of nuclei that have been measured in experiments or are expected to be measured in the near future. We need to find such an approximation of the problem that would enable us to obtain a highly accurate microscopic description of a large ensemble of the experimental nuclear characteristics and to make predictions.

The introduction of the nuclear mean field is a reflection of the fundamental properties of atomic nuclei. The possibility of defining this field is implied, first, by the Pauli exclusion principle and, second, by the ratio of the Fermi momentum to the momentum of the repulsive core of the nucleon–nucleon potential. It is the mean field of the nucleus which is responsible for the diversity of properties of atomic nuclei. Hence, the theory of nuclei must determine an approximate solution of the $A = N + Z$ problem which would correctly describe the differences in the properties of isobars and of nuclei containing $A \pm 1$, $A \pm 2$, etc nucleons.

The microscopic theory of nuclei was greatly influenced by the mathematical methods of the theories of superfluidity [17], superconductivity [18] and Fermi liquid [19]. In their turn, the mathematical methods of the microscopic theory of atomic nuclei may prove fruitful for other fields of theoretical physics.

A considerable number of modifications of the descriptions of complex nuclei were developed. For example, collective states are described on the basis of the boson representation of fermion operators. The solutions in the general form are best found by the Green's function and self-consistent field method. Algebraic methods found widespread application to theories of nuclei. The microscopic theory of collective excitations of atomic nuclei operates in terms of the method of generalized hyperspherical functions [20].

The maximum attention is paid in this book to describing the properties of nuclear states in terms of quasiparticle and phonon operators. The wave functions of the excitation states are represented by expansions in numbers of quasiparticles and phonons. The utilization of wave functions makes it possible to describe various processes in which a given state is excited. Few-quasiparticle components of wave functions are described with the quasiparticle–phonon interaction taken into account at low, intermediate and high excitation energies.

It is essential that relative, not absolute values are mostly calculated: the differences between the energies of the ground and excitation states; matrix elements reflecting the changes in the wave functions of the final states in comparison with the initial ones, and so forth. The calculation of the relative values enables one to avoid difficulties inherent in the many-body problem and to improve the accuracy of the results obtained.

1

Nuclear many-body problem

1.1 HARTREE–FOCK–BOGOLIUBOV METHOD

1.1.1

The Hartree–Fock–Bogoliubov method (HFB) is one of the basic and most often used methods of solving the many-body problem. Bogoliubov [21] suggested a generalization of the Hartree–Fock method, in which the minimum of the energy of a system is sought on a broader class of functions than in the Hartree–Fock method. In addition to the wave functions of individual particles, the method takes into account the wave functions of pairs with zero angular momentum. The HFB approach is widely used to analyse the properties of atomic nuclei. Let us formulate the method.

We write the Hamiltonian of the system in the general form:

$$H = \sum_{ff'} T'(f, f') a_f^+ a_{f'} - \frac{1}{4} \sum_{f_1 f_2 f_2' f_1'} G(f_1, f_2; f_2', f_1') \times a_{f_1}^+ a_{f_2}^+ a_{f_2'} a_{f_1'}. \tag{1.1}$$

Here f is the set of quantum numbers characterizing the state of a nucleon; a_f and a_f^+ are the nucleon absorption and creation operators satisfying the commutation relations

$$a_f^+ a_{f'} + a_{f'} a_f^+ = \delta_{ff'} \qquad a_f a_{f'} + a_{f'} a_f = 0 \tag{1.2}$$

$$T'(f, f') = T(f, f') - \lambda \delta_{ff'} \tag{1.3}$$

where λ is the chemical potential, introduced to satisfy the condition of conservation of average particle number.

The functions $T(f, f')$ and $G(f_1, f_2; f_2', f_1')$ satisfy the following relations:

$$T(f,f') = T^*(f',f)$$
$$G(f_1,f_2;f_2',f_1') = -G(f_1,f_2;f_1',f_2') = G^*(f_1',f_2';f_2,f_1) \tag{1.4}$$

Among the set of quantum numbers f, we single out the numbers $\sigma = \pm 1$, so that the states $f = q\sigma$ that differ in the sign of q are conjugate with respect to the time reversal operation. For example, σ can be the sign of the projection of the angular momentum on the symmetry axis of the nucleus. The time reversal operation transforms the operator $a_{q\sigma}^+$:

$$\mathfrak{T}^{-1} a_{q\sigma}^+ \mathfrak{T} = \gamma_\sigma a_{q-\sigma} \tag{1.5}$$

where $\mathfrak{T}$ is the time reversal operator, and γ_σ are coefficients satisfying the relations

$$\gamma_\sigma \gamma_{-\sigma} = -1 \qquad \gamma_\sigma^2 = 1 \qquad \gamma_{-\sigma} = -\gamma_\sigma \tag{1.5'}$$

(as a particular case, we can assume $\gamma_\sigma = \sigma$). The invariance of Hamiltonian (1.1) with respect to the time reversal operation implies

$$T(q\sigma, q'\sigma') = \gamma_\sigma \gamma_{\sigma'} T^*(q-\sigma, q'-\sigma')$$
$$G(q_1\sigma_1, q_2\sigma_2; q_2'\sigma_2', q_1'\sigma_1') = \gamma_{\sigma_1}\gamma_{\sigma_2}\gamma_{\sigma_2'}\gamma_{\sigma_1'}$$
$$\times G(q_1-\sigma_1, q_2-\sigma_2; q_2'-\sigma_2', q_1'-\sigma_1'). \tag{1.6}$$

For the sake of simplification, we assume hereafter that the functions T and G are real.

Let us define two functions,

$$\Phi(f_1,f_2) = \langle | a_{f_1} a_{f_2} | \rangle \qquad \Phi^*(f_1,f_2) = \langle | a_{f_2}^+ a_{f_1}^+ | \rangle \tag{1.7}$$

and the density function

$$\rho(f_1,f_2) = \langle | a_{f_1}^+ a_{f_2} | \rangle \tag{1.8}$$

so that

$$\Phi(f_1,f_2) = -\Phi(f_2,f_1) \qquad \rho^*(f_2,f_1) = \rho(f_1,f_2).$$

The average value is taken here over an arbitrary state $|\rangle$.

Let us consider the amplitudes a_f in the Heisenberg representation where $a_f(t)$ are explicit functions of time, that is, $a_f(t) = \exp(\mathrm{i}Ht)\, a_f \exp(-\mathrm{i}Ht)$. Let us introduce time-dependent functions $\rho_t(f_1,f_2)$, $\Phi_t(f_1,f_2)$. The equations of motion then become

$$\mathrm{i}\frac{\partial}{\partial t}\rho_t(f_1,f_2) = \langle | [a_{f_1}^+(t) a_{f_2}(t), H] | \rangle \tag{1.9}$$

$$\mathrm{i}\frac{\partial}{\partial t}\Phi_t(f_1,f_2) = \langle | [a_{f_1}(t) a_{f_2}(t), H] | \rangle. \tag{1.10}$$

We need to rewrite them in expanded form for the case when the Hamiltonian H is given by (1.1):

$$\frac{\partial}{\partial t}\rho_t(f_1,f_2) = \sum_f \{T'(f_2,f)\rho_t(f_1,f) - T'(f,f_1)\rho_t(f,f_2)\} + \frac{1}{2}\sum_{ff_2'f_1'} \{G(f_1',f_2';f,f_1)\langle|a^+_{f_1'}a^+_{f_2'}a_f a_{f_2}|\rangle - G(f_2,f;f_2'f_1')\langle|a^+_{f_1}a^+_f a_{f_2'}a_{f_1'}|\rangle\} \qquad (1.9')$$

$$\mathrm{i}\frac{\partial}{\partial t}\Phi_t(f_1,f_2) = \sum_f \{T'(f_1,f)\Phi_t(f,f_2) + \Phi_t(f_1,f)T'(f_2,f)\} + \frac{1}{2}\sum_{f_1'f_2'} G(f_1,f_2;f_2',f_1') \times \Phi_t(f_1',f_2') - \frac{1}{2}\sum_{ff_2'f_1'}\{G(f_1,f;f_2',f_1') \times \langle|a^+_f a_{f_2}a_{f_2'}a_{f_1'}|\rangle + G(f,f_2;f_2',f_1') \langle|a^+_f a_{f_1}a_{f_2'}a_{f_1'}|\rangle\}. \qquad (1.10')$$

Now we need to write equations for

$$\langle|a^+_{f_1}a^+_{f_2}a_{f_2'}a_{f_1'}|\rangle \quad \text{and} \quad \langle|a^+_{f_1}a_{f_2}a_{f_2'}a_{f_1'}|\rangle.$$

The first of them has the form

$$\begin{aligned}\mathrm{i}\frac{\partial}{\partial t}\langle|a^+_{f_1}a^+_{f_2}a_{f_3}a_{f_4}|\rangle &= \langle|[a^+_{f_1}a^+_{f_2}a_{f_3}a_{f_4},H]|\rangle\\ &= \sum_f\{-T'(f,f_1)\langle|a^+_f a^+_{f_2}a_{f_3}a_{f_4}|\rangle - T'(f,f_2)\langle|a^+_{f_1}a^+_f a_{f_3}a_{f_4}|\rangle\\ &\quad + T'(f_3,f)\langle|a^+_{f_1}a^+_{f_2}a_f a_{f_4}|\rangle + T'(f_4,f)\langle|a^+_{f_1}a^+_{f_2}a_{f_3}a_f|\rangle\}\\ &\quad - \frac{1}{2}\sum_{ff'f_2'}\{G(f,f';f_2',f_1)\langle|a^+_f a^+_{f'}a^+_{f_1}a_{f_2'}a_{f_3}a_{f_4}|\rangle\\ &\quad - G(f,f';f_2',f_2)\langle|a^+_f a^+_{f'}a^+_{f_1}a_{f_2'}a_{f_3}a_{f_4}|\rangle\\ &\quad + G(f_3,f;f_2',f')\langle|a^+_{f_1}a^+_{f_2}a^+_f a_{f_4}a_{f_2'}a_{f'}|\rangle\\ &\quad - G(f_4,f;f_2',f')\langle|a^+_{f_1}a^+_{f_2}a_f a_{f_3}a_{f_2'}a_{f_1'}|\rangle\}\\ &\quad + \sum_{ff'}\{G(f,f';f_2,f_1)\langle|a^+_f a^+_{f'}a_{f_3}a_{f_4}|\rangle\\ &\quad + G(f_3,f_4;f',f)\langle|a^+_{f_1}a^+_{f_2}a_{f'}a_f|\rangle\}.\end{aligned} \qquad (1.11)$$

The equation for $\langle|a^+_{f_1}a_{f_2}a_{f_2'}a_{f_1'}|\rangle$ has a similar structure. As a result, these two distribution functions are found to be related to distribution functions of still higher order. These latter functions are expressed in terms of distribution functions of even higher order, and so on. In the theory of nuclei, this hierarchy is finite, owing to the finite number of nucleons in nuclei. Alas, this finiteness does not facilitate the solution of the nuclear many-body problem.

One of the approaches to an approximate solution of the nuclear many-body problem is to represent higher-order distribution functions via lower-order ones. The following approximation is postulated in the HFB method:

$$\begin{aligned}\langle| a^+_{f_1} a^+_{f_2} a_{f'_2} a_{f'_1} |\rangle &= \rho(f_1, f'_1)\rho(f_2, f'_2) \\ &\quad - \rho(f_1, f'_2)\rho(f_2, f'_1) + \Phi^*(f_2, f_1)\Phi(f'_2, f'_1) \\ \langle| a^+_{f_1} a_{f_2} a_{f'_2} a_{f'_1} |\rangle &= \rho(f_1, f_2)\Phi(f'_2, f'_1) \\ &\quad - \rho(f_1, f'_2)\Phi(f_2, f'_1) + \rho(f_1, f'_1)\Phi(f_2, f'_2).\end{aligned} \tag{1.12}$$

It results in a closed set of equations that we can recast to the form [21, 22]

$$\mathrm{i}\frac{\partial}{\partial t}\rho_t(f_1, f_2) = \mathfrak{B}(f_1, f_2) \tag{1.13}$$

$$\mathrm{i}\frac{\partial}{\partial t}\Phi_t(f_1, f_2) = \mathfrak{A}(f_1, f_2). \tag{1.14}$$

Let us define a function

$$C_{f_1 f_2} = \frac{1}{2}\sum_{f'_1 f'_2} G(f_1, f_2; f'_2, f'_1)\Phi(f'_2, f'_1) \tag{1.15}$$

and write the following expressions for the functions $\mathfrak{A}(f_1, f_2)$ and $\mathfrak{B}(f_1, f_2)$ [22]:

$$\begin{aligned}\mathfrak{A}(f_1, f_2) &= \sum_f \{\xi(f_1, f)\Phi(f, f_2) + \Phi(f_1, f)\xi(f_2, f)\} \\ &\quad - \sum_f \{C_{f_1 f}\rho(f, f_2) + C_{f f_2}\rho(f, f_1)\} + C_{f_1 f_2}\end{aligned} \tag{1.16}$$

$$\begin{aligned}\mathfrak{B}(f_1, f_2) &= \sum_f \{\xi(f_2, f)\rho(f_1, f) - \xi(f, f_1)\rho(f, f_2)\} \\ &\quad + \sum_f \{C_{f_2 f}\Phi^*(f_1, f) - C^*_{f_1 f}\Phi(f_2, f)\}\end{aligned} \tag{1.17}$$

where

$$\xi(f, f') = T'(f, f') - \sum_{f_1 f_2} G(f, f_1; f_2, f')\rho(f_1, f_2). \tag{1.18}$$

As a result, we arrive at time-independent equations of the HFB method.

$$\mathfrak{B}(f_1, f_2) = 0 \qquad \mathfrak{A}(f_1, f_2) = 0. \tag{1.19}$$

It must be mentioned that the HFB approximation actually omits a large number of equations. The importance of the influence of the neglected equations on the calculation results has not yet been quantified. A very satisfactory description of a number of properties of complex nuclei in the framework of the HFB method indicates that the role played by these equations is often very minor. It is possible that when this method is used, the effect of neglected equations could be partially compensated for by selecting a form of the residual equations and by finding the constants on the basis of experimental data. In any case, the application of the HFB techniques to the nuclear many-body problem ought not to result in our forgetting about the discarded equations.

In some papers (see, e.g., [23]), the distribution functions were expressed via lower-order distribution functions more accurately than in the HFB method, by adding a number of excitation states as intermediates, in addition to the ground state. This work is based on an exact relation of the type

$$\langle|a^+_{f_1}a^+_{f_2}a_{f'_2}a_{f'_1}|\rangle = \sum_n \{\langle|a^+_{f_1}a_{f'_1}|n\rangle\langle n|a^+_{f_2}a_{f'_1}|\rangle - \langle|a^+_{f_1}a_{f'_2}|n\rangle\langle n|a^+_{f_2}a_{f'_1}|\rangle + \langle|a^+_{f_1}a^+_{f_2}|n\rangle\langle n|a_{f'_2}a_{f'_1}|\rangle\} \quad (1.20)$$

where $|n\rangle$ is a complete set of states; note that in order not to set the last terms to zero, $|n\rangle$ must not be the eigenfunctions of the particle number operator. Instead of a complete system of states $|n\rangle$, one selects those excitation states of the nucleus in which the corresponding non-diagonal matrix elements are of the same order as the diagonal ones. For example, collective vibrational states or the states of one rotational band can be selected among the intermediate states [24, 25]. The choice of the intermediate states is dictated by the form of the interaction $G(f_1, f_2; f'_2, f'_1)$ and by its relation to certain single-particle operators. A description of low-energy vibrational and rotational states of complex nuclei has been obtained in this manner.

1.1.2

In order to express the distribution function in terms of lower-order distribution functions, such as ρ and Φ, the Bogoliubov canonical transformation is used:

$$a_f = \sum_g (u_{fg}\alpha_g + v_{fg}\alpha^+_g). \quad (1.21)$$

In order for the quasiparticle operators α_g and α^+_g to satisfy the fermion commutation relations, the functions u_{fg} and v_{fg} must satisfy the conditions

$$\zeta_0(f,f') \equiv \sum_g (u_{fg}u^*_{f'g} + v_{fg}v_{f'g}) - \delta_{gg'} = 0$$
$$\zeta_1(f,f') \equiv \sum_g (u_{fg}v_{f'g} + u'_{f'g}v_{fg}) = 0$$
$$\zeta^*_1(f,f') = 0 \tag{1.22}$$

such that $\zeta_0(f,f') = \zeta^*_0(f',f)$, $\zeta_1(f,f') = \zeta_1(f',f)$.

Using relations (1.22), it is not difficult to find α_g:

$$\alpha_g = \sum_f (u^*_{fg}a_f + v_{fg}a^+_f). \tag{1.21'}$$

The functions u_{fg} and v_{fg} also satisfy the relations

$$\sum_f (u^*_{fg}u_{fg'} + v_{fg}v^*_{fg'}) = \delta_{gg'}$$
$$\sum_f (v^*_{fg}u_{fg'} + v^*_{fg'}u_{fg}) = 0 \tag{1.22'}$$
$$\sum_f (v_{fg}u^*_{fg'} + v_{fg'}u_{fg}) = 0.$$

If we single out of the set of quantum numbers the number $\sigma = \pm 1$, that is, $f = q\sigma$, $g = p\sigma$, then the canonical transformation described by formulas (1.21), (1.21') changes to

$$a_{q\sigma} = \sum_p (u_{qp}\alpha_{p\sigma} + \sigma v_{qp}\alpha^+_{p-\sigma})$$
$$\alpha_{p\sigma} = \sum_q (u_{qp}a_{q\sigma} - \sigma v_{qp}a^+_{q-\sigma}). \tag{1.21''}$$

The functions u_{qp}, v_{qp} are real and relations (1.22) become

$$z_0(q,q') \equiv \sum_p (u_{qp}u_{q'p} + v_{qp}v_{q'p}) - \delta_{qq'} = 0$$
$$z_1(q,q') \equiv \sum_p (u_{qp}v_{q'p} - v_{qp}u_{q'p}) = 0.$$

In the case of canonical transformation (1.21), functions (1.7) and (1.8) take the form

$$\rho(f_1,f_2) = \sum_g v^*_{f_1g}v_{f_2g} + \sum_{g_1g_2} \{u^*_{f_1g_1}u_{f_2g_2}\langle|\alpha^+_{g_1}\alpha_{g_2}|\rangle$$

$$- v^*_{f_1 g_1} v_{f_2 g_2} \langle |\alpha^+_{g_2} \alpha_{g_1}|\rangle + u^*_{f_1 g_1} v_{f_2 g_2} \langle |\alpha^+_{g_1} \alpha^+_{g_2}|\rangle$$
$$+ v^*_{f_1 g_1} u_{f_2 g_2} \langle |\alpha_{g_1} \alpha_{g_2}|\rangle \} \tag{1.23}$$

$$\Phi(f_1, f_2) = \sum_g u_{f_1 g} v_{f_2 g} + \sum_{g_1 g_2} \{ v_{f_1 g_1} u_{f_2 g_2} \langle |\alpha^+_{g_1} \alpha_{g_2}|\rangle$$
$$- u_{f_1 g_1} v_{f_2 g_2} \langle |\alpha^+_{g_2} \alpha_{g_1}|\rangle + u_{f_1 g_1} u_{f_2 g_2} \langle |\alpha_{g_1} \alpha_{g_2}|\rangle$$
$$+ v_{f_1 g_1} u_{f_2 g_2} \langle |\alpha^+_{g_1} \alpha^+_{g_2}|\rangle \}. \tag{1.23'}$$

We define the ground state of a system consisting of an even number of particles as the vacuum with respect to the operators α_g, that is

$$\alpha_g |\rangle_0 = 0. \tag{1.24}$$

As shown in [6], the wave function $|\rangle_0$ has the form

$$|\rangle_o = \mathfrak{N} \exp\Big\{ - \frac{1}{2} \sum_{ff'g} (u^*_{fg})^{-1} v_{f'g} a^+_f a^+_{f'} \Big\} \Psi_{00} \tag{1.24'}$$

where $\mathfrak{N}$ is a normalization factor and Ψ_{00} is the vacuum for particles, that is

$$a_f \Psi_{00} = 0. \tag{1.25}$$

If averaging is carried out over the quasiparticle vacuum, functions (1.7) and (1.8) transform to

$$\Phi_0(f_1, f_2) = \sum_g u_{f_1 g} v_{f_2 g} \qquad \Phi^*_0(f_1, f_2) = \sum_g u^*_{f_1 g} v^*_{f_2 g}$$
$$\rho_0(f_1, f_2) = \sum_g v^*_{f_1 g} v_{f_2 g}. \tag{1.26}$$

Following [26], we find the relation between the functions $\Phi_0(f_1, f_2)$ and $\rho_0(f_1, f_2)$. Let us introduce combined indices $\zeta = (f, \vartheta)$, $\eta = (g, \vartheta)$, $\vartheta = 0$ and 1, and set

$$\varphi_{g\,0}(f, 0) = v^*_{fg} \qquad \varphi_{g\,0}(f, 1) = u_{fg}$$
$$\varphi_{g1}(f, 0) = u^*_{fg} \qquad \varphi_{g1}(f, 1) = v_{fg}.$$

In this notation, relations (1.22) change to

$$\sum_\eta \varphi^*_\eta(\zeta) \varphi_\eta(\zeta') = \delta_{\zeta\zeta'}. \tag{1.22''}$$

Let us define a matrix

$$F(\zeta, \zeta') = \sum_\eta \varphi^*_\eta(\zeta) \varphi_\eta(\zeta') \bar{n}_\eta$$

in which $\bar{n}_{g,0} = 1$, $\bar{n}_{g,1} = 0$. Having used (1.22″), we obtain

$$F(\zeta, \zeta') = \begin{vmatrix} \rho_0(f', f) & -\Phi_0(f, f') \\ \Phi_0^*(f, f') & \delta_{ff'} - \rho_0(f, f') \end{vmatrix}.$$

The definition of the operator F shows that $\varphi_\eta(\zeta)$ are the eigenvectors and $\bar{n}_\eta$ are the eigenvalues of this operator. Since these eigenvalues equal zero or unity, we find that $F^2 = F$. Having expanded this expression, we find the following conditions relating the functions $\rho_0(f, f')$ and $\Phi_0(f, f')$:

$$\begin{aligned} \rho_0(f, f') &= \sum_{f_2} [\rho_0(f, f_2)\rho_0(f_2, f') + \Phi_0^*(f_2, f)\Phi_0(f_2, f')] \\ \sum_{f_2} [\rho_0(f, f_2)&\Phi_0(f_2, f') + \rho_0(f', f_2)\Phi_0(f_2, f)] = 0. \end{aligned} \tag{1.27}$$

Using canonical transformation (1.21) and quasiparticle vacuum (1.24), we arrive at equations (1.13) and (1.14), in which $\Phi(f, f')$ and $\rho(f, f')$ are replaced by $\Phi_0(f, f')$ and $\rho_0(f, f')$. These equations can be used for the calculation of u_{fg} and v_{fg}. It was shown (see [26]) that the functions $\mathfrak{A}_0(f, f')$ and $\mathfrak{B}_0(f, f')$ are not independent; the relation between them is

$$\begin{aligned} \sum_{ff'} [u_{fg} v_{f'g'} \mathfrak{A}_0^*(f, f') + u_{f'g}^* v_{fg'}^* \mathfrak{A}_0(f, f') \\ + (u_{fg} u_{f'g'}^* - v_{f'g}^* v_{fg'}) \mathfrak{B}_0(f, f')] = 0 \end{aligned} \tag{1.28}$$

Hence, if $\mathfrak{A}_0(f, f') = 0$ then $\mathfrak{B}_0(f, f') = 0$ and only one of equations (1.19) need be analysed.

Equations (1.19) can be obtained from the variational HFB principle. To do this, we need to find the mean value of the Hamiltonian H (1.1) over the state (1.24) and then determine u_{fg} and v_{fg} from the condition of minimum of $\langle|H|\rangle_0$, which we write in the form

$$\begin{aligned} \delta\{\langle|H|\rangle_0 + \sum_{ff'} [\lambda_0(f', f)\zeta_0(f.f') \\ + \lambda_1^*(f', f)\zeta_1(f, f') + \lambda_1(f', f)\zeta_1^*(f, f')]\} = 0. \end{aligned} \tag{1.29}$$

Here $\lambda(f', f)$ are Lagrange multipliers, and the variations δu_{fg}^*, δu_{fg}, δv_{fg} and δv_{fg}^* can be treated as independent. The chemical potential λ, which also plays the role of a Lagrange multiplier, is found from the condition of conservation of the average number of particles N, that is

$$N = \sum_f \langle|a_f^+ a_f|\rangle_0 = \sum_{fg} v_{fg}^* v_{fg}. \tag{1.30}$$

As a result of transformations, we arrive at equations (1.19), in which $\Phi_0(f, f')$ and $\rho_0(f, f')$ are written in the form (1.26).

In order to find the minimum energy of a system of interacting nucleons when solving the stationary equations on the basis of the variational principle, it is necessary for the second variation to be positive for solutions (1.19). This condition can be rewritten in the form

$$\sum_{fg}(\mathfrak{E}_1 \delta v^*_{fg} \delta v_{fg} + \mathfrak{E}_2 \delta u^*_{fg} \delta u_{fg}) > 0. \tag{1.31}$$

It is satisfied for the positive eigenvalues $\mathfrak{E}_1$ and $\mathfrak{E}_2$, with $\mathfrak{E}_1$, $\mathfrak{E}_2$, δu_{fg}, δv_{fg} determined from the equations of the eigenvalue problem.

The Hartree–Fock–Bogoliubov variational principle is a generalization of the familiar Hartree–Fock method. The solutions of the HFB method always contain those of the Hartree–Fock method. It is of interest, therefore, to formulate a condition under which the application of the Hartree–Fock method does not result in minimizing the energy of a system of interacting particles. This condition can be written in the following form:

$$\sum_{fg}(\mathfrak{E}_1 \delta v^*_{fg} \delta v_{fg} + \mathfrak{E}_2 \delta u^*_{fg} \delta u_{fg}) < 0$$

where $\mathfrak{E}_1$, $\mathfrak{E}_2$, δu_{fg}, δv_{fg} are found from the solutions of the Hartree–Fock method. In this case, the energy minimum must be sought on a broader class of solutions, which also take into account the pairing correlations of particles. Hence, if the Hartree–Fock method does not find the energy minimum for a system on a class of functions, it can be regarded as the condition of existence of pairing correlations (see [27]).

1.1.3

Let us find the mean nuclear field and the interactions which lead to superfluid pairing correlations. It was shown in [28] that for practically any type of interaction between nucleons, we can choose a diagonal function $\rho_0(f, f')$ and transform the function $\Phi_0(f, f')$ to the canonical form, that is,

$$\begin{aligned} \rho_0(f.f') &= \rho_0(f)\delta_{ff'} = \rho_0(q)\delta_{qq'}\delta_{\sigma\sigma'} \\ \Phi_0(f.f') &= \Phi_0(f)\delta_{f,-f'} = \Phi_0(\sigma q)\delta_{qq'}\delta_{\sigma,-\sigma'}. \end{aligned} \tag{1.32}$$

Equation (1.27) then takes the form

$$\rho_0(q) = \rho_0^2(q) + \Phi_0^*(q)\Phi_0(q). \tag{1.33}$$

Equality $\rho_0^*(f_2, f_1) = \rho(f_1, f_2)$ implies that $\rho_0(q) = \rho_0^*(q)$. If $\Phi_0(q) = \Phi_0^*(q) = 0$, then $\rho_0(q) = \rho_0^2(q)$, which implies either $\rho_0 = 1$ or $\rho_0 = 0$.

Let us derive an expression for energy and an explicit form of one of the equations (1.19) when the functions Φ_0 and ρ_0 are reduced to the form (1.32). The average value of the energy operator over state (1.24) is

$$\langle|H|\rangle_0 = \sum_f \{T'(f) - \frac{1}{2}\sum_{f'} G(f,f';f',f)\rho_0(f')\}\rho_0(f)$$
$$- \sum_{qq'} G(q+,q-;q'-,q'+)\Phi_0^*(q)\Phi_0(q'). \qquad (1.34)$$

The basic equations are

$$2\xi(q)\Phi_0(q) - [1 - 2\rho_0(q)] \times \sum_{q'} G(q+,q-;q'-,q'+)\Phi_0(q') = 0 \qquad (1.35)$$

$$N = 2\sum_q \rho_0(q) \qquad (1.36)$$

where

$$\xi(q) = T'(q) - \sum_{q'\sigma'} G(q+,q'\sigma';q'\sigma',q+)\rho_0(q') \equiv E(q) - \lambda. \qquad (1.18')$$

Here $E(q)$ are the mean field single-particle energies.

The most general form of the two-particle interaction between nucleons thus makes it possible to single out the mean nuclear field and the interactions of nucleon pairs in states that are conjugate with respect to the time reversal operation. This procedure of determining the mean nuclear field can be regarded as one of the steps to justifying the model of independent particles. Finding the mean field is not connected, therefore, either with the value or with the form of the potential that describes the interaction between nucleons.

It is necessary to mention that the method of deriving basic equations (1.19) is an approximate one since correlation functions like $\langle|a^+_{f_1}a^+_{f_2}a_{f'_2}a_{f'_1}|\rangle$ are given approximately as functions of ρ and Φ. Hence, the results obtained must be considered as a first approximation. The convergence of this method, which depends on the type of interaction, has not been analysed yet. The only exception is the particular case of the model Hamiltonian in the superconductivity theory, for which Bogoliubov derived asymptotic estimates [29].

The theory of nuclei uses the following postulate: the mean nuclear field corresponds to a representation for which the density matrix $\rho_0(f,f')$ is diagonal for the ground states of some even–even nuclei [28]. In this representation, the residual interactions completely reduce to interactions that result in superfluid pairing correlations. Consequently, it is not necessary to take into account any other residual interactions.

1.1.4

The Hamiltonian of a system in which superfluid pairing correlations have been taken into consideration can be written in the form

$$H = \sum_{q\sigma}[E_0(q) - \lambda]a^+_{q\sigma}a_{q\sigma} - \sum_{qq'} G(q+, q-; q'-, q'+)a^+_{q+}a^+_{q-}a_{q'_-}a_{q'_+}.$$

Let us introduce new symbols:

$$\Phi_0(q) = u_q v_q = \Phi^*(q) \qquad \rho_0(q) = v_q^2 \tag{1.37}$$

where u_q and v_q are real functions. Condition (1.33) becomes

$$u_q^2 + v_q^2 = 1. \tag{1.38}$$

Equation (1.35) can be transformed to

$$2\xi(q)u_q v_q - (u_q^2 - v_q^2)\sum_{q'} G(q+, q-; q'-, q'+)u_{q'}v_{q'} = 0$$

where $\xi(q) = E(q) - \lambda$.

The function $C_{f,-f}$ will be denoted by C_q and written in the form

$$C_q = \sum_{q'} G(q+, q-; q'-, q'+)u_{q'}v_{q'}. \tag{1.39}$$

With u_q^2 and v_q^2 transformed to

$$u_q^2 = \frac{1}{2}\left(1 + \frac{\xi(q)}{\epsilon_q}\right) \qquad v_q^2 = \frac{1}{2}\left(1 - \frac{\xi(q)}{\epsilon_q}\right) \tag{1.40}$$

we obtain

$$u_q v_q = (1/2)C_q/\epsilon_q. \tag{1.40$'$}$$

Setting $u_q^2 v_q^2$ of (1.40) equal to (1.40′) squared, we find

$$\epsilon_q = \sqrt{C_q^2 + \xi^2(q)} \tag{1.41}$$

This gives us the following set of equations:

$$C_q = \frac{1}{2}\sum_{q'} G(q+, q-; q'-, q'+)\frac{C_{q'}}{\sqrt{C_{q'}^2 + \xi^2(q')}} \tag{1.42}$$

$$N = \sum_q \left[1 - \frac{\xi(q)}{\epsilon_q}\right] \tag{1.43}$$

(these equations were derived in [30, 31]).

Equation (1.42) has two solutions: $C_q \neq 0$, which corresponds to the superfluid state of the system, and $C_q = 0$, which corresponds to the normal state:

$$u_q = 1 - \Theta_F(q) \qquad v_q = \Theta_F(q).$$

Here $\theta_F(q) = 1$ if the energy $E(q)$ of a state q is less than the Fermi energy E_F, and $\theta_F(q) = 0$ if the energy $E(q) > E_F$. Of the two solutions $C_q \neq 0$ and $C_q = 0$, the real system is described by the one which minimizes the energy of the system.

1.1.5

Let us look at the formulas of the theory of superfluid pairing correlations in atomic nuclei. Since the forces producing pairing correlations are short-range forces, it has been assumed in [32] that $G(q+, q-; q'-, q'+)$ are independent of q and q', that is,

$$G(q+, q-; q'-, q'+) = G_{N,Z} = g_{N,Z}/A. \tag{1.44}$$

Arguments supporting this assumption can be found in [6, 7, 12, 28, 33]. This approximation was proved to be surprisingly good. Since there is no neutron–proton pairing in complex nuclei, the neutron and the proton systems are described independently of each other and have different constants G_N and G_Z. The Hamiltonian of the system then splits into two components:

$$H_0 = H_0(N) + H_0(Z) \tag{1.45}$$

where

$$H_0(N) = \sum_{q\sigma}[E_0(q) - \lambda]a^+_{q\sigma}a_{q\sigma} - G_N \sum_{qq'} a^+_{q+}a^+_{q-}a_{q'-}a_{q'+}.$$

The summation in the second sum is carried over single-particle levels of the neutron (proton) system. In approximation (1.44), the function C_q of (1.39) is independent of q and is denoted by C. In this approximation, equations (1.42) and (1.43) for finding C and the chemical potential λ for the twice-degenerate single-particle levels become

$$\begin{aligned} 1 &= \frac{G}{2}\sum_q \frac{1}{\sqrt{C^2 + [E(q) - \lambda]^2}} \\ N &= \sum_q \left\{1 - \frac{E(q) - \lambda}{\sqrt{C^2 + [E(q) - \lambda]^2}}\right\} \end{aligned} \tag{1.46}$$

To describe pairing, one usually employs the canonical Bogoliubov transformation

$$a_{q\sigma} = u_q \alpha_{q\sigma} + \sigma v_q \alpha^+_{q-\sigma}. \tag{1.47}$$

The ground state is a quasiparticle vacuum; it is described by the wave function

$$\Psi_0^0 = \prod_q (u_q + v_q a^+_{q+} a^+_{q-}) \Psi_{00} \tag{1.48}$$

where $a_{q\sigma} \Psi_{00} = 0$. Its energy is

$$\mathcal{E}_0 = 2 \sum_q v_q^2 - \frac{C^2}{G}. \tag{1.49}$$

Let us specify the main equations (1.46) for quasiparticle excitation states, by rigorously taking the Pauli exclusion principle into account and demanding the conservation of the average number of particles. It is then necessary to find for each quasiparticle state an average of $H_0(N)$ or $H_0(Z)$ and obtain, using the variational principle, the main equations, that is, to solve the variational problem for each excitation state of the system.

The effect of unpaired particles on the superfluid properties of the system in each state is known as the blocking effect. An atomic nucleus becomes superfluid as a result of nucleon interactions described by Hamiltonian (1.45). The characteristic of this Hamiltonian is that nucleons transfer from one level to another in pairs. Hence, if a doubly degenerate level of the mean field is occupied by one nucleon, then this level is forbidden by the Pauli exclusion principle to be occupied by a pair. Blocking then means that when superfluid properties of certain states are to be calculated, the twice-degenerate levels of the mean field occupied by quasiparticles are ignored. This changes the values of the functions C and of the chemical potentials in comparison with their values in the states without quasiparticles, or with a different number of quasiparticles, or with quasiparticles on other mean field levels. The blocking effect was introduced in [32, 34] and was analysed in detail in [6, 12, 28, 35–37].

The lowest-excitation states of even–even nuclei are those with one broken pair, that is, two-quasiparticle states. The wave function of a two-quasiparticle state has the form

$$\alpha^+_{q_1\sigma_1} \alpha^+_{q_2\sigma_2} \Psi_0^0. \tag{1.50}$$

We now introduce

$$C(q_1, q_2) = G_N \sum_{q \neq q_1, q_2} u_q(q_1 q_2) v_q(q_1 q_2)$$

and rewrite the basic equations and the energy as follows:

$$\begin{aligned}1 &= \frac{G_N}{2} \sum_{q \neq q_1, q_2} \frac{1}{\sqrt{C^2(q_1, q_2) + [E(q) - \lambda(q_1, q_2)]^2}} \\ N &= 2 + \sum_{q \neq q_1, q_2} \left\{1 - \frac{E(q) - \lambda(q_1, q_2)}{\sqrt{C^2(q_1, q_2) + [E(q) - \lambda(q_1, q_2)]^2}}\right\}.\end{aligned} \tag{1.51}$$

$$\begin{aligned}\mathcal{E}(q_1, q_2) = E(q_1) + E(q_2) + 2 \sum_{q \neq q_1, q_2} E(q) v_q^2(q_1, q_2) \\ - C^2(q_1, q_2)/G_N.\end{aligned} \tag{1.52}$$

Let us consider a system of an odd number of nucleons. The wave function, the basic equations and the energy of the system with a quasiparticle at a single-particle level q_2 are:

$$\alpha^+_{q_2\sigma_2} \Psi^0_0 \tag{1.50'}$$

$$\begin{aligned}1 &= \frac{G_N}{2} \sum_{q \neq q_2} \frac{1}{\sqrt{C^2(q_2) + [E(q) - \lambda(q_2)]^2}} \\ N &= 1 + \sum_{q \neq q_2} \left\{1 - \frac{E(q) - \lambda(q_2)}{\sqrt{C^2(q_2) + [E(q) - \lambda(q_2)]^2}}\right\}\end{aligned} \tag{1.53}$$

$$\mathcal{E}(q_2) = E(q_2) + 2 \sum_{q \neq q_2} 2E(q) v_q^2(q_2) - C^2(q_2)/G_N. \tag{1.54}$$

Each single-quasiparticle state is characterized by the quantities $C(q_2)$ and $\lambda(q_2)$; the excitation energies are determined by the difference $\mathcal{E}_0(q_2) - \mathcal{E}_0(q_F)$, where q_F is the single-particle Fermi level. The quasiparticle creation operator is

$$\alpha^+_{q\sigma} = u_q a^+_{q\sigma} + \sigma v_q a_{q-\sigma}. \tag{1.55}$$

For states q with energy much greater than the Fermi energy, the quasiparticle operator coincides with that of a particle; for states q with energy much lower than the Fermi energy, it is a hole operator; and for q close to the Fermi energy, the quasiparticle operator is a superposition of the particle and hole operators.

An important role of the blocking effect has been mentioned recently in the description of a number of properties of nuclei, such as quasi-crossing of high-spin bands. The time has now come to calculate blocking effect with higher accuracy; the results have been reported in a number of papers (see, e.g., [38–41]).

Formulas (1.45)–(1.55) describe superfluid pairing correlations in deformed nuclei. In what follows, we will need the correspondent formulas for spherical nuclei. We will give them here (see [12]):

$$H_0 = \sum_{jm}[E_0(j) - \lambda]a^+_{jm}a_{jm} - \frac{G}{4}\sum_{jmj'm'}(-1)^{j-m}$$
$$\times (-1)^{j'-m'}a^+_{jm}a^+_{j-m}a_{j'-m'}a_{j'm'} \tag{1.56}$$
$$a_{jm} = u_j\alpha_{jm} + (-1)^{j-m}v_j\alpha^+_{j-m} \tag{1.57}$$
$$\Psi^0_0 = \prod_{j,m>0}[u_j + (-1)^{j-m}v_j a^+_{jm}a^+_{j-m}]\Psi_{00}. \tag{1.58}$$

The quantities u_j, v_j and $C = G\sum_j(j+\frac{1}{2})u_jv_j$ are independent of m; the basic equations for C and λ are now

$$1 = \frac{G}{2}\sum_j \frac{j+1/2}{\sqrt{C^2 + [E(j) - \lambda]^2}} \tag{1.59}$$
$$N = \sum_j \left(j + \frac{1}{2}\right)\left(1 - \frac{E(j) - \lambda}{\sqrt{C^2 + [E(j) - \lambda]^2}}\right). \tag{1.60}$$

Formulas for two- and single-quasiparticle states are similar to (1.50)–(1.54). They can be found, for example, in [12].

1.2 HARTREE–FOCK METHOD AND MEAN NUCLEAR FIELD

1.2.1

The existence of the mean field is a fundamental property of atomic nuclei, responsible for the diversity of nuclear characteristics. The mean nuclear field is singled out in the Hartree–Fock approximation when solving the many-body problem. The condition for a successful application of the Hartree–Fock method in the theory of nuclei is the demand that the nucleon–nucleon potential be smooth. We know [42] that this potential is repulsive at short distances, which is an obstacle for Hartree–Fock applications. It is essential that forces in nuclei are effective forces. The difference between an effective and a vacuum force is caused by the influence of other nucleons in nuclei. This influence depends on the number of nucleons in the region where nuclear forces are strong, that is, on the density of nuclear matter.

A considerable number of papers have been published on the microscopic calculation of the main characteristics of nuclei in the Brueckner–Hartree–Fock approximation. Their authors start with realistic nucleon–nucleon potentials, sum up a chain of certain diagrams and thus arrive at effective interactions. The calculated effective interactions are found to be smooth even if hard-core potentials are chosen as the original interaction. The smoothing of the effective interaction potential occurs because the mean distance between nucleons in nuclei considerably exceeds the core radius. The applicability of the Hartree–Fock method to description of nuclei is justified by the smoothness of the potentials that describe the interactions between nucleons in nuclei. Note that calculations in the Brueckner–Hartree–Fock approximation do not provide a correct description of nuclear radii and simultaneously of binding energies.

A considerable advance in the application of the Hartree–Fock method to the theory of nuclei was connected with the introduction of the density-dependent Skyrme potential. The interaction in the form of a δ function greatly simplifies the calculations. Indeed, the Hartree–Fock single-particle potential consists of two parts, one being determined by the direct term of the two-particle potential and another, by the exchange term. If the two-particle potential is local, then the direct term of the Hartree–Fock potential is found to be local while the exchange term is non-local. The exchange term of the potential in nuclei is very important, and calculations with a non-local potential are extremely difficult. If the Skyrme potential is used, the Hartree–Fock exchange potential is also local, owing to the δ functions. The Schrödinger equation is then also local, and efficient methods of solving it are known.

1.2.2

The Hartree–Fock method allows us to reduce the problem of many interacting nucleons to the problem of motion of a particle in an external field. This gives us Hartree–Fock equations using the secondary quantization techniques. They can also be obtained as a particular case of the Hartree–Fock–Bogoliubov method.

The Hamiltonian of the system will be taken in the form (1.1). The set of quantum numbers will be denoted by g and f, with g standing for the states occupied by particles and f, for the free and occupied states. Now we apply the linear canonical transformation:

$$a_g = \sum_f v_{gf} a_f^+ . \tag{1.61}$$

For the new operators a_f, a_f^+ to satisfy commutation relations (1.2), the following conditions must be met:

$$\eta(g,g') = \sum_f v^*_{gf} v_{g'f} - \delta_{gg'} = 0 \qquad \sum_g v^*_{gf} v_{gf'} = \delta_{ff'}. \tag{1.62}$$

The ground state of a system composed of an even number of particles can be defined as a vacuum with respect to the operators a_f, that is

$$a_f |\rangle_g = \sum_g v_{gf} a_g^+ |\rangle_g = 0. \tag{1.63}$$

Hence,

$$|\rangle_g = \prod_g a_g^+ \Psi_{00} = \prod_{\tilde{g}} a^+_{\tilde{g}+} a^+_{\tilde{g}-} \Psi_{00} \tag{1.64}$$

where $g = \tilde{g}\sigma$, $\sigma = \pm 1$, since all the g states are occupied by particles. The density function and the mean value of the particle number operator over state (1.64) are given by

$$\begin{aligned} \rho(g,g') &= {}_g\langle | a_g^+ a_{g'} |\rangle_g = \sum_f v^*_{gf} v_{g'f} \\ N &= {}_g\langle | \sum_q a_q^+ a_g |\rangle_g = \sum_f v^*_{gf} v_{gf}. \end{aligned} \tag{1.65}$$

Let us find the mean value of the system's Hamiltonian in state (1.64):

$$\begin{aligned} {}_g\langle |H|\rangle_g = \sum_{gg'} T(g,g')\rho(g,g') - \frac{1}{4} \sum_{g_1 g_2 g_2' g_1'} G(g_1 g_2; g_2', g_1') \\ \times [\rho(g_1, g_1')\rho(g_2, g_2') - \rho(g_1, g_2')\rho(g_2, g_1')]. \end{aligned} \tag{1.66}$$

This expression coincides with the formula for the mean energy, derived by Fock. Let us determine v_{gf} and v^*_{gf} from the minimum condition, (1.66) which we can rewrite as

$$\delta\{{}_g\langle |H|\rangle_g + \sum_{gg'} \eta(g,g')\lambda(g,g')\} = 0 \tag{1.67}$$

where $\lambda(g,g')$ are Lagrange multipliers, and δv_{gf}, δv^*_{gf} are treated as independent variations. As a result, we obtain

$$\begin{aligned} \sum_{gg'} \{T(g,g') - \frac{1}{2} \sum_{g_2 g_2'} [G(g', g_2; g_2', g) \\ - G(g', g_2; g, g_2')]\rho(g_2, g_2') + \lambda(g', g)\} v^*_{g'f} = 0 \end{aligned} \tag{1.68}$$

which is the Schrödinger equation describing the motion of a particle in a field.

Let us determine the self-consistent Hamiltonian by the formula

$$H^{\text{s.c.}} = \sum_{ff'}\{T(f,f') - \frac{1}{2}\sum_{g_2 g_2'}[G(f,g_2;g_2',f') - G(f,g_2;f',g_2')]\rho(g_2,g_2')\}a_f^+ a_{f'}. \tag{1.69}$$

It is now possible to change to a representation in which $H^{\text{s.c.}}$ is diagonal. In this approximation, the new occupied levels are linear combinations of the former occupied levels. The Hamiltonian $H^{\text{s.c.}}$ is so constructed that it is invariant with respect to such transformations: indeed, the sum in (1.69) over g_2 equals the trace in the subspace of the occupied levels. In the new representation,

$$T(f,f') - \frac{1}{2}\sum_{gg'}[G(f,g;g',f') - G(f,g;f',g')]\rho(g,g') = E(f)\delta_{ff'} \tag{1.70}$$

where $E(f)$ are the eigenvalues of the Hamiltonian $H^{\text{s.c.}}$. In the representation where $H^{\text{s.c.}}$ is diagonal, this operator is treated as the mean field Hamiltonian and is denoted by H_{av}. The eigenvalues $E(f)$ can be interpreted as approximate energy values of a particle at a level f.

Let us rewrite the Hartree–Fock equation in a familiar form. To do this, we use the representation in which $H^{\text{s.c.}}$ is diagonal, introduce the direct and exchange potentials,

$$V_d(g) = \frac{1}{2}\sum_{g_2 g_2'} G(g.g_2;g_2',g)\rho(g_2,g_2')$$

$$V_{\text{ex}}(g,g') = -\frac{1}{2}\sum_{g_2} G(g,g_2;g',g)\rho(g_2,g')$$

denote $\lambda(g,g) = -E(g)$, and rewrite equation (1.68) in the form

$$[T(g) - V_d(g)]v_{gf}^* - \sum_{g'} V_{\text{ex}}(g,g')v_{g'f}^* = E(g)v_{gf}^*. \tag{1.71}$$

The energies and wave functions of single-particle states are found by solving equations (1.71).

For an effective density-dependent interaction, we have to modify the Hartree–Fock equations. If G is a function of density, then additional terms, proportional to $\delta G/\delta v_{gf}$, appear in the self-consistent field when δv_{gf} is varied. The corresponding equations are given in many papers (see, e.g., [43, 44]). The calculations use the effective nucleon mass, which is then found with greater accuracy by the self-consistency procedure.

In describing non-stationary processes, for example, nucleus–nucleus interactions, the time-dependent Hartree–Fock theory is used. It is based on the assumption that a self-consistent field exists at all times. The time-dependent self-consistent field is determined by the position of all particles in the system. Hence, the coherent motion of several nucleons results in a change in the mean field, which affects the motion of all other nucleons. If this coupling is strong, the process can be described by introducing a time-dependent mean field. The appropriate equations are given, for instance, in [44].

Distributions of charge density and nuclear matter have been calculated in the last 10–15 years for many, especially magic, nuclei using the Hartree–Fock method; the phenomenological Skyrme potential containing density-dependent terms has been used. The Skyrme potential parameters in [43, 44] and in other papers were chosen so as to obtain the values of binding energies and the root-mean-square radii of magic nuclei in agreement with experimental data. It is not always possible to use the same parameters of the Skyrme potential, so that a new paper often introduces a new set of parameters. The results of Hartree–Fock calculations are in good agreement with the experimental data on the scattering of fast electrons and protons. This means that we obtained a correct distribution of charge density and nuclear matter in atomic nuclei, especially in the surface region. The Hartree–Fock method also yielded the position of single-particle levels close to the Fermi energy. Further elaboration of the parameters of the Skyrme potential was needed to improve the agreement between the calculated and the experimental energy levels of near-magic nuclei. It can be stated that microscopic Hartree–Fock calculations give a sufficiently good description of single-particle level density in near-magic nuclei close to the Fermi energy. The results of these calculations can be treated as a justification of the single-particle shell model of nuclei.

1.2.3

Single-particle energies and wave functions are calculated using the Hartree–Fock method or by solving the Schrödinger equation with a phenomenological mean field potential of the atomic nucleus. Hartree–Fock calculations represent an earlier stage of parametrization as compared to calculations with phenomenological potentials. The Hartree–Fock method uses a certain freedom in the choice of the effective interaction parameters. At the same time, the freedom in choosing the parameters of the phenomenological mean field potentials is very much restricted.

We know that the finite surface potential, which reproduces the dependence of the nuclear matter density on radius, provides the best description of the nuclear mean field. The main parameters of the nuclear potential are found from the real part of the optical potential. The parameters of the optical potential are calculated using the rich ensemble of experimental data on scattering of nucleons by nuclei. Typically, the Saxon–Woods potential is chosen as the finite surface mean field nuclear potential. This is the potential of a spherically symmetric finite-depth potential well whose value on the surface $r = R_0$ equals one half of its value at the centre of the nucleus.

The nuclear part of the Saxon–Woods potential contains two terms: the central term

$$V(r) = -\frac{V_0^{N,Z}}{1+\exp[\alpha(r-R_0)]} \tag{1.72}$$

and the spin-orbital term

$$V_{\text{ls}} = -\kappa\,\frac{1}{r}\,\frac{\mathrm{d}V(r)}{\mathrm{d}r}(\boldsymbol{l}\,\boldsymbol{s}) \tag{1.73}$$

where V_0^N, V_0^Z are the the neutron and proton potential well depths, α is the diffusivity parameter, κ is the spin-orbital interaction constant and $R_0 = r_0 A^{1/3}$.

In calculating the levels of the proton system, the mean field potential, consisting of (1.72) and (1.73), must be complemented by a term for the Coulomb interaction:

$$V_c(r) = \frac{(Z-1)\mathrm{e}^2}{r}\begin{cases}(3/2)r/R_0 - (1/2)(r/R_0)^3 & \text{if } r \leqslant R_0\\ 1 & \text{if } r > R_0.\end{cases} \tag{1.74}$$

The advantages of the Saxon–Woods potential also include the correct $r \to \infty$ asymptotics of single-particle wave functions.

The shape of a deformed nucleus is described by the function

$$R(\Theta,\varphi) = R_0[1 + \tilde{\beta} + \beta_2 Y_{20}(\Theta,\varphi) + \beta_4 Y_{40}(\Theta,\varphi)] \tag{1.75}$$

where $R_0 = r_0 A^{1/3}$ is the radius of the spherical nucleus of the same volume; the constant $\tilde{\beta}$ is often introduced to improve the conservation of the nucleus' volume; β_2 and β_4 are the parameters of the quadrupole ($\lambda = 2$) and the hexadecapole ($\lambda = 4$) deformations. In some papers, a term $R_0\beta_6 Y_{60}(\theta,\varphi)$ is added to formula (1.75). The nuclear part of the Saxon–Woods potential consists of a central and a spin-orbital terms:

$$V_{\text{nuc}}(r) = V(r) + V_{\text{ls}}(r) \tag{1.76}$$

$$V(r) = \frac{-V_0^{N,Z}}{1+\exp\{\alpha[r - R(\Theta,\varphi)]\}} \tag{1.76$'$}$$

$$V_{\text{ls}}(r) = -\kappa[\boldsymbol{p}\times\boldsymbol{\sigma}]\cdot\boldsymbol{\nabla}V(r) \tag{1.76$''$}$$

where $\boldsymbol{p}$ is the momentum of a nucleon and $\boldsymbol{\sigma}$ are the Pauli matrices. For the proton system, we have to add a Coulomb term

$$V_c(r) = \frac{3}{4\pi}\,\frac{(Z-1)\mathrm{e}^2}{R_0^3}\int\frac{n(\boldsymbol{r}')\,\mathrm{d}\boldsymbol{r}'}{|\boldsymbol{r}-\boldsymbol{r}'|} \tag{1.77}$$

where the charge distribution density in the nucleus is

$$n(r) = \{1 + \exp[a(r - R(\theta, \varphi))]\}^{-1}.$$

The problem of determining the energy eigenvalues and the eigenfunctions of the Schrödinger equation with a non-spherical Saxon–Woods potential was first solved in [45]. To solve the Schrödinger equation, the set of differential equations was solved numerically to obtain the single-particle energies and wave functions for the neutron and proton systems. Other methods of solving the Schrödinger equation with the Saxon–Woods potential were suggested later (see [46–48]). A modification of the Saxon–Woods potential may be used, in which different values of the parameters a and r_0 are chosen for the central and the spin-orbital parts.

1.3 THEORY OF NUCLEAR VIBRATIONS

1.3.1

Let us derive the general expressions for nuclear vibrations, using the self-consistent method in Bogoliubov's formulation [26]. We will make use of equations (1.13), (1.14), rewriting them in the form

$$\begin{aligned} \mathrm{i}\frac{\partial}{\partial t}\rho(f, f') &= \mathfrak{B}(f, f') \\ \mathrm{i}\frac{\partial}{\partial t}\Phi(f, f') &= \mathfrak{A}(f.f') \\ -\mathrm{i}\frac{\partial}{\partial t}\Phi^*(f, f') &= \mathfrak{A}^*(f.f'). \end{aligned} \qquad (1.78)$$

The functions $\mathfrak{A}(f, f')$, $\mathfrak{B}(f, f')$ are given by formulas (1.16), (1.17); note that $\mathfrak{B}^*(f, f') = -\mathfrak{B}(f', f)$. The functions $\Phi(f, f')$, $\Phi^*(f, f')$ and $\rho(f, f')$, defined by (1.7) and (1.8), are related by the formulas

$$\begin{aligned} \rho(f, f') &= \sum_{f_2}\{\rho(f, f_2)\rho(f_2, f') + \Phi^*(f_2, f)\Phi(f_2, f')\} \\ 0 &= \sum_{f_2}\{\rho(f, f_2)\Phi(f_2, f') + \rho(f', f_2)\Phi(f_2, f)\} \\ 0 &= \sum_{f_2}\{\rho(f_2, f)\Phi^*(f_2, f') + \rho(f_2, f')\Phi^*(f_2, f)\}. \end{aligned} \qquad (1.79)$$

Formula (1.27) is a particular case of these relations.

If interactions between quasiparticles are ignored, the ground state of an even–even nucleus is interpreted as a quasiparticle vacuum. If averaging is carried out over the quasiparticle vacuum, then the functions $\rho(f, f')$ and $\Phi(f, f')$ stand for $\rho_0(f, f')$ and $\Phi_0(f, f')$. If the nuclear mean field is chosen in the most optimal way, the function $\rho_0(f, f')$ is diagonal while $\Phi_0(f, f')$ is transformed to the canonical form. In order to take into account the interactions between quasiparticles that produce low-amplitude vibrations, small deviations must be added to ρ_0, Φ_0 and Φ_0^*:

$$
\begin{aligned}
\rho(f, f') &= \rho_0(f, f') + \delta\rho(f, f') \\
\Phi(f, f') &= \Phi_0(f, f') + \delta\Phi(f, f') \\
\Phi^*(f, f') &= \Phi_0(f, f') + \delta\Phi^*(f, f')
\end{aligned}
\tag{1.80}
$$

where $\delta\rho(f, f') = \delta\rho^*(f', f)$. These increments satisfy the equations

$$
\begin{aligned}
\mathrm{i}\frac{\partial}{\partial t}\delta\rho(f, f') &= \delta\mathfrak{B}(f, f') \\
\mathrm{i}\frac{\partial}{\partial t}\delta\Phi(f, f') &= \delta\mathfrak{A}(f, f') \\
-\mathrm{i}\frac{\partial}{\partial t}\delta\Phi^*(f, f') &= \delta\mathfrak{A}(f, f')
\end{aligned}
\tag{1.81}
$$

$$
\begin{aligned}
\delta\mathfrak{B}(f, f') = \sum_{f_2} &[\delta\xi(f', f_2)\rho_0(f, f_2) + \xi_0(f', f_2)\delta\rho(f, f_2) \\
&- \delta\xi(f_2, f)\rho_0(f_2, f') - \xi_0(f_2, f)\delta\rho(f_2, f')] \\
&+ \sum_{f_2}[\delta C_{f'f_2}\Phi_0^*(f, f_2) + C^0_{f'f_2}\delta\Phi^*(f, f_2) \\
&- \delta C^*_{ff_2}\Phi_0(f', f_2) - C^{*0}_{ff_2}\delta\Phi(f', f_2)]
\end{aligned}
\tag{1.82}
$$

$$
\begin{aligned}
\delta\mathfrak{A}(f, f') = \sum_{f_2} &[\delta\xi(f, f_2)\Phi_0(f_2, f') + \xi_0(f, f_2)\delta\Phi(f_2, f') \\
&+ \delta\Phi(f, f_2)\xi_0(f', f_2) + \Phi_0(f, f_2)\delta\xi(f', f_2)] \\
&- \sum_{f_2}[\delta C_{ff_2}\rho_0(f_2, f') + C^0_{ff_2}\delta\rho(f_2, f') + \delta C_{f_2f'}\rho_0(f, f_2) \\
&+ C^0_{f_1f'}\delta\rho(f, f_2)] + \delta C_{ff'}
\end{aligned}
\tag{1.83}
$$

where

$$
\xi_0(f, f') = T(f, f') - \lambda\delta_{ff'} - \sum_{f_1'f_2'} G(f, f_1'; f_2', f')\rho_0(f_1', f_2')
$$

$$
C^0_{ff'} = \frac{1}{2}\sum_{f_1'f_2'} G(f, f'; f_2', f_1')\Phi_0(f_2', f_1').
$$

It is not difficult to derive the expressions for $\delta\mathfrak{A}^*(f,f')$. Furthermore, $\delta\rho$, $\delta\Phi$ and $\delta\Phi^*$ are related by expressions implied by (1.80). Owing to this relationship, it will be convenient to introduce new functions that automatically satisfy these relations.

As a result of interactions between quasiparticles, the wave function of the ground state of an even–even nucleus ceases to be a quasiparticle vacuum. Assuming that the average number of quasiparticles in the ground state is small, we can use the approximation

$$\langle|\alpha^+_{g_1}\alpha_{g_2}|\rangle = 0. \tag{1.84}$$

To characterize the deviation of the wave function from the quasiparticle vacuum, we introduce the functions

$$\mu(g_1,g_2) = \langle|a_{g_1}a_{g_2}|\rangle \qquad \mu^*(g_2,g_1) = \langle|a^+_{g_1}a^+_{g_2}|\rangle \tag{1.85}$$

which satisfy the conditions

$$\mu(g_1,g_2) = -\mu(g_2,g_1) \qquad \mu^*(g_2,g_1) = -\mu^*(g_1,g_2).$$

Making use of approximation (1.84) and substituting functions (1.85) into (1.23) and (1.23′), we arrive at the following expressions:

$$\begin{aligned}\delta\rho(f_1,f_2) &= \sum_{g_1g_2}[v^*_{f_1g_1}u_{f_2g_2}\mu(g_1,g_2) \\ &\quad + u^*_{f_1g_1}v_{f_2g_2}\mu^*(g_2,g_1)] \\ &= \delta\rho^*(f_2,f_1)\end{aligned} \tag{1.86}$$

$$\begin{aligned}\delta\Phi(f_1,f_2) &= \sum_{g_1g_2}[u_{f_1g_1}u_{f_2g_2}\mu(g_1,g_2) \\ &\quad + v_{f_1g_1}v_{f_2g_2}\mu^*(g_2,g_1)]\end{aligned} \tag{1.86′}$$

$$\begin{aligned}\delta\Phi^*(f_1,f_2) &= \sum_{g_1g_2}[u^*_{f_1g_1}u^*_{f_2g_2}\mu^*(g_1,g_2) \\ &\quad + v^*_{f_1g_1}v^*_{f_2g_2}\mu(g_2,g_1)].\end{aligned} \tag{1.86″}$$

Let us derive equations for $\mu(g_1,g_2)$ and $\mu^*(g_1,g_2)$; we begin with giving them in terms of $\delta\rho$, $\delta\Phi$ and $\delta\Phi^*$. We multiply (1.86) by v_{f_1g} and (1.86′) by $u^*_{f_1g}$, add them up and carry out summation over f_1. Using orthonormalization (1.22′), we obtain

$$\begin{aligned}&\sum_{f_1}[v_{f_1g}\delta\rho(f_1,f_2) + u^*_{f_1g}\delta\Phi(f_1,f_2)] \\ &\qquad = \sum_{f_1g_1g_2}\{[v_{f_1g}v^*_{f_1g_1} + u^*_{f_1g}u_{f_1g_1}]u_{f_2g_2}\mu(g_1,g_2) \\ &\qquad\quad + [v_{f_1g}u^*_{f_1g_1} + u^*_{f_1g}v_{f_1g_1}]v_{f_2g_2}\mu^*(g_2,g_1)\} \\ &\qquad = \sum_{g_2}u_{f_2g_2}\mu(g,g_2)\end{aligned} \tag{1.87}$$

Now we multiply $\delta\rho^*(f_1, f_2)$ by $u^*_{f_1 g}$ and $\delta\Phi^*(f_1, f_2)$ by $v_{f_1 g}$, add them up and carry out summation over f_1:

$$\sum_{f_1}[u^*_{f_1 g}\delta\rho^*(f_1, f_2) + v_{f_1 g}\delta\Phi^*(f_1, f_2)] = -\sum_{g_2} v^*_{f_2 g_2}\mu(g, g_2). \qquad (1.87')$$

We will now resort to this technique again, multiplying (1.87) by $u^*_{f_2 g'}$ and (1.87′) by $v_{f_2 g'}$, subtracting (1.87′) from (1.87) and carrying out summation over f_2. This gives

$$\mu(g, g') = \sum_{f_1 f_2}[v_{f_1 g}u^*_{f_2 g'}\delta\rho(f_1, f_2) + u^*_{f_1 g}u^*_{f_2 g'}\,\delta\Phi(f_1, f_2)$$
$$- u^*_{f_1 g}v_{f_2 g}\delta\rho(f_2, f_1) - v_{f_1 g}v_{f_2 g'}\,\delta\Phi^*(f_1, f_2)].$$

Let us differentiate this expression with respect to t, taking into account equations (1.81); this produces an equation for $\mu(g_1, g_2)$:

$$\mathrm{i}\frac{\partial}{\partial t}\mu(g_1, g_2) = \sum_{f_1 f_2}[u^*_{f_1 g_1}u^*_{f_2 g_2}\,\delta\mathfrak{A}(f_1, f_2) + v_{f_1 g_1}v_{f_2 g_2}\,\delta\mathfrak{A}^*(f_1, f_2)$$
$$+ (u^*_{f_2 g_2}v_{f_1 g_1} - u^*_{f_2 g_1}v_{f_1 g_2})\,\delta\mathfrak{B}(f_1, f_2)]. \qquad (1.88)$$

The equation for $\mu^*(g_1, g_2)$ is found in a similar manner.

Let us substitute the expressions for $\delta\mathfrak{A}$, $\delta\mathfrak{A}^*$ and $\delta\mathfrak{B}$ into (1.88), make use of formulas (1.86), (1.86′) and (1.86″), and express $\delta\rho$ and $\delta\Phi$ in terms of μ and μ^*. Collecting the corresponding terms, we make use of orthonormalization conditions (1.22′) and obtain, after cumbersome manipulation

$$\mathrm{i}\frac{\partial}{\partial t}\mu(g_1, g_2) = \sum_{g'}[\Omega(g_2, g')\mu(g_1, g') - \Omega(g_1, g')\mu(g_2, g')]$$
$$+ \sum_{g'_1 g'_2}[X(g_1, g_2; g'_1, g'_2)\mu(g'_1, g'_2)$$
$$+ Y(g_1, g_2; g'_1, g'_2)\mu^*(g'_2, g'_1)]. \qquad (1.89)$$

Here

$$\Omega(g, g') = \sum_{f f'}[\xi_0(f, f')(u^*_{fg}u_{f'g'} - v^*_{fg}v_{f'g'})$$
$$- C^0_{ff'}u^*_{fg}v^*_{f'g'} - C^{0*}_{ff'}v_{f'g}u_{fg'}] \qquad (1.90)$$

$$X(g_1, g_2; g'_1, g'_2) = -\frac{1}{2}\sum_{f_1 f_2 f'_1 f'_2} G(f_1, f_2; f'_1, f'_2)$$

$$\times \left[u_{f_1g}u_{f_2g_1}u_{f_1'g_2'}u_{f_2'g_1'} + v_{f_1'g_1}v_{f_2'g_2}v_{f_1g_1'}v_{f_2g_2'} + (v_{f_1'g_1}u_{f_1g_2} - u_{f_1g_1}v_{f_1'g_2})(v_{f_2g_1'}u_{f_2'g_2'} - v_{f_2g_2'}u_{f_2'g_1})\right] \tag{1.91}$$

$$Y(g_1, g_2; g_1', g_2') = -\frac{1}{2} \sum_{f_1f_2f_2'f_1'} G(f_1, f_2; f_2', f_1') \times \left[u_{f_2g_1}u_{f_1g_2}v_{f_2'g_1'}v_{f_1'g_2'} + v_{f_1'g_1}v_{f_2'g_2}u_{f_2g_2'}u_{f_1g_1'} + (v_{f'g_1}u_{f_1g_2} - v_{f_1'g_2}u_{f_1g_1}) \times (u_{f_2g_1'}v_{f_2'g_2'} - u_{f_2g_2'}v_{f_2'g_1'})\right]. \tag{1.91'}$$

The second equation is obtained from (1.89) by taking its complex conjugate.

Equations (1.89) were derived by Bogoliubov [26]. They were used as a basis for obtaining the time-dependent adiabatic Hartree–Fock–Bogoliubov approximation, which was later elaborated in [49, 50] and in other publications. A particular case of this approximation, which ignores pairing, was analysed in [51–53] and in some other papers; it is now known as the time-dependent adiabatic Hartree–Fock approximation. Fast vibrations and slow high-amplitude collective processes, often used to describe nucleus–nucleus interactions, have been studied in the framework of these approximations.

Let us consider now the stationary case and seek solutions of homogeneous equations (1.89), and of those complex-conjugate to them, in the form

$$\begin{aligned} \mu(g_1, g_2) &= \sum_i \exp(-i\omega_i t)\zeta_i(g_1, g_2) \\ \mu^*(g_1, g_2) &= \sum_i \exp(-i\omega_i t)\eta_i(g_1, g_2) \end{aligned} \tag{1.92}$$

demanding that

$$\begin{aligned} &\eta_i(g_1, g_2) = \zeta_i^*(g_1, g_2; -\omega_i) \quad \zeta_i(g_1, g_2) = -\zeta_i(g_2, g_1) \\ &\eta_i(g_1, g_2) = -\eta_i(g_2, g_1). \end{aligned} \tag{1.93}$$

As a result, we obtain

$$\begin{aligned} \omega_i\zeta_i(g_1, g_2) = &\sum_{g'}[\Omega(g_2, g')\zeta_i(g_1, g') - \Omega(g_1, g')\zeta_i(g_2, g')] \\ &+ \sum_{g_1'g_2'}[X(g_1, g_2; g_1', g_2')\zeta_i(g_1'g_2') \\ &- Y(g_1, g_2; g_1', g_2')\zeta_i(g_1', g_2')] \end{aligned} \tag{1.94}$$

$$
\begin{aligned}
-\omega_i\eta_i(g_1,g_2) = \sum_{g'}&[\Omega^*(g_2,g')\eta_i(g_1,g') - \Omega^*(g_1,g')\eta_i(g_2,g')] \\
&+ \sum_{g_1'g_2'}[X^*(g_1,g_2;g_1',g_2')\eta_i(g_1'g_2') \\
&- Y(g_1,g_2;g_1',g_2')\zeta_i(g_1',g_2')].
\end{aligned} \tag{1.95}
$$

Note that if ω_i, ζ_i, η_i are solutions of (1.94) and (1.95), then the transformation

$$\omega_i \to -\omega_i \qquad \xi_i \to \xi_i^* \qquad \eta_i \to \eta_i^*$$

again yields the solution of the same set of equations.

1.3.2

Let us consider a representation in which the function $\rho_0(f,f')$ is diagonal while $\Phi_0(f,f')$ has canonical form. It is given by formulas (1.32) in which $f = q\sigma$, $\sigma = \pm 1$. In this representation,

$$
\begin{aligned}
u_{fg} &= u_f\delta_{fg} = u_q\delta_{fg} & u_{q\sigma} &= u_q & & \\
v_{fg} &= v_f\delta_{f,-g} = v_q\sigma\delta_{\sigma,-\sigma'}\delta_{qq'} & v_{q\sigma} &= \sigma v_q & g &= q'\sigma' \\
u_q &= u_q^* \quad v_q = v_q^* & & & &
\end{aligned} \tag{1.96}
$$

and the function is diagonal

$$\xi_0(f,f') = \xi_0(q)\delta_{ff'} = [E(q) - \lambda]\delta_{qq'}\delta_{\sigma\sigma'} \tag{1.97}$$

where $E(q)$ are single-particle mean field energies, and

$$C_{ff'}^0 = C_{ff'}^{0*} = C_q\delta_{f,-f'} \qquad C_q = \sum_{q'} G(q+,q-;q'-,q'+)u_{q'}v_{q'}. \tag{1.98}$$

Furthermore, in this representation (see [6])

$$\Omega(g,g') = \delta_{gg'}\epsilon_q \quad X = X^* \quad Y = Y^*. \tag{1.99}$$

We will now introduce new functions

$$\nu_i^{(\pm)}(g_1,g_2) = \frac{1}{2}[\zeta_i(g_1,g_2) \pm \eta_i(g_1,g_2)]$$

make use of relations (1.6) and notation

$$
\begin{aligned}
u_{g_1g_2}^{(\pm)} &\equiv u_{q_1\sigma_1,q_2\sigma_2}^{(\pm)} = \sigma u_{q_1q_2}^{(\pm)} & u_{q_1q_2}^{(\pm)} &= u_{q_1}v_{q_2} \pm u_{q_2}v_{q_1} \\
u_{g_1,-g_2}^{(\pm)} &= -\sigma u_{q_1q_2}^{(\mp)} & & \\
v_{g_1g_2}^{(\pm)} &\equiv v_{q_1\sigma,q_2\sigma}^{(\pm)} = v_{q_1q_2}^{(\pm)} & v_{q_1q_2}^{(\pm)} &= u_{q_1g_2} \pm v_{q_1}v_{q_2} \\
v_{g_1,-g_2}^{(\pm)} &= v_{q_1q_2}^{(\mp)} & &
\end{aligned} \tag{1.100}
$$

and, after some manipulation (see [6, 54]), transform equations (1.94) and (1.95) into

$$\begin{aligned}
&\epsilon_{q_1 q_2}\nu_i^{(\pm)}(g_1, g_2) - \omega_i \nu_i^{(\mp)}(g_1, g_2) \\
&\quad - \frac{1}{4}\sum_{g_1' g_2'} G(g_1, g_2; g_2', g_1')(1 + \sigma_1\sigma_2\sigma_2'\sigma_1') v^{(\mp)}_{g_1 g_2} v^{(\mp)}_{g_1' g_2'} \nu_i^{(\pm)}(g_1', g_2') \\
&\quad - \frac{1}{4}\sum_{g_1' g_2'} G(g_1, -g_2'; g_1', -g_2) \mp G(g_1, -g'; g_2', -g_2)(1 + \sigma_1\sigma_2\sigma_2'\sigma_1') \\
&\quad \times u^{(\pm)}_{g_1 g_2} u^{(\pm)}_{g_1' g_2'} \nu_i^{(\pm)}(g_1', g_2') - \frac{1}{4}\sum_{g_1' g_2'} G(g_1, g_2; g_2', g_1')(1 - \sigma_1\sigma_2\sigma_2'\sigma_1') \\
&\quad \times v^{(\mp)}_{g_1 g_2} v^{(\pm)}_{g_1' g_2'} \nu_i^{(\mp)}(g_1', g_2') - \frac{1}{4}\sum_{g_1' g_2'} [G(g_1, -g_2'; g_1', -g_2) \\
&\quad \pm G(g_1, -g_1'; g_2', -g_2)(1 - \sigma_1\sigma_2\sigma_2'\sigma_1') u^{(\pm)}_{g_1 g_2} u^{(\mp)}_{g_1' g_2'} \nu_i^{(\mp)}(g_1', g_2') = 0. \quad (1.101)
\end{aligned}$$

where $\epsilon_{q_1 q_2} = \epsilon_{q_1} + \epsilon_{q_2}$.

We now introduce a new function

$$Z_i^{(\pm)}(q_1, q_2) = \sum_{\sigma} [\nu_i^{(\pm)}(q_1\sigma, q_2\sigma) + \nu_i^{(\mp)}(q_1\sigma, q_2 - \sigma)]$$

and denote the interactions in the particle–hole and particle–particle channels by $G^{\mathrm{ph}}(q_1, q_2'; q', q)$ and $G^{\mathrm{pp}}(q_1, q_2; q_2', q_1')$. These functions are expressed in [54] in terms of the sums $G(q_1\sigma_1, q_2\sigma_2; q_2'\sigma_2', q_1'\sigma_1')$ with the appropriate values of σ_1, σ_2, σ_2', σ_1'. Let us make use of notation (1.100) and find, after a number of transformations of (1.101),

$$\begin{aligned}
&\epsilon_{q_1 q_2} Z_i^{(\pm)}(q_1, q_2) - \omega_i Z^{(\mp)}(q_1, q_2) \\
&\quad - \sum_{q_1' q_2'} G^{\mathrm{pp}}(q_1, q_2; q_2', q_1') v^{(\mp)}_{q_1 q_2} v^{(\mp)}_{q_1' q_2'} Z^{(\pm)}(q_1' q_2') \\
&\quad - 2\sum_{q_1' q_2'} G^{\mathrm{ph}}(q_1, q_2'; q_1', q_2) u^{(\pm)}_{q_1 q_2} u_{q_1' q_2'} Z^{(\pm)}(q_1', q_2') = 0. \quad (1.102)
\end{aligned}$$

These are the basic equations for finding the energies ω_i of nuclear vibrations. In deriving them, the interactions between quasiparticles must be taken in the general form. The terms in equations (1.102), containing $u^{(\pm)}_{q_1 q_2}$, describe interactions between quasiparticles in the particle–hole channel. The terms containing $v^{(\pm)}_{q_1 q_2}$ describe interactions in the particle–particle channel.

1.3.3

Let us generalize equations (1.102) to the case of a weak external field, and transform them to a form in which they can be compared with the equations of the theory of finite Fermi systems [55]. When introducing the mean field, we add to the Hamiltonian (1.1) a term

$$\sum_{ff'} I(f, f') a_f^+ a_{f'} \tag{1.103}$$

where the function $I(f, f') = I^*(f', f)$ characterizes the external field. The introduction of term (1.103) makes it necessary to add $I(f, f')$ to the terms containing $T'(f, f')$ in the functions $\mathfrak{A}(f_1, f_2)$ and $\mathfrak{B}(f_1, f_2)$. The expressions $\delta\mathfrak{A}(f_1, f_2)$ and $\delta\mathfrak{B}(f_1, f_2)$, defined by formulas (1.83) and (1.82), must be supplemented by the terms

$$\delta\mathfrak{A}_{\text{ex}}(f_1, f_2) = \sum_f [\delta I(f_1, f)\Phi(f, f_2) + \delta I(f_2, f)\Phi(f_1, f)]$$

$$\delta\mathfrak{B}_{\text{ex}}(f_1, f_2) = \sum_f [\delta I(f_2, f)\rho(f_1, f) - \delta I(f, f_1)\rho(f, f_2)]$$

where

$$\delta I(f, f') = \sum_i \exp(-\mathrm{i}\omega t)\delta I_\omega^i(f, f').$$

We will now carry out the same transformations as in deriving equations (1.102) (see [6, 26, 54]) and obtain

$$\begin{aligned} &\epsilon_{q_1 q_2} Z_i^{(\pm)}(q_1, q_2) - \omega_i Z_i^{(\mp)}(q_1, q_2) \\ &\quad - \sum_{q_1' q_2'} G^{\text{pp}}(q_1, q_2; q_1', q_2') v_{q_1 q_2}^{(\mp)} v_{q_1' q_2'}^{(\mp)} Z_i^{(\pm)}(q_1', q_2') \\ &\quad - 2\sum_{q_1' q_2'} G^{\text{ph}}(q_1, q_2'; q_1', q_2) u_{q_1 q_2}^{(\pm)} u_{q_1' q_2'}^{(\pm)} Z^{(\pm)}(q_1', q_2') \\ &\quad = \frac{1}{2} u_{q_1 q_2}^{(\pm)} [\delta I_\omega^i(q_1, q_2) \pm \delta I_{-\omega}^i(q_1, q_2)] \end{aligned} \tag{1.104}$$

where

$$\delta I_\omega^i(q_1 q_2) = \sum_\sigma [\delta I_\omega^i(q_1\sigma, q_2\sigma) - \sigma\delta I_\omega^i(q_1\sigma, q_2 - \sigma)]$$

$$\delta I_{-\omega}^i(q_1 q_2) = \sum_\sigma [\delta I_{-\omega}^i(q_1\sigma; q_2\sigma) - \sigma\delta I_{-\omega}^{*i}(q_1\sigma, q_2 - \sigma)].$$

Now we can transform equations (1.104) to a form not very different from that given in the theory of finite Fermi systems. The following new functions will be needed for that:

$$d^{(\pm)}(q_1, q_2) = \sum_{q_1' q_2'} G^{\text{pp}}(q_1, q_2; q_2', q_1') v_{q_1' q_2'}^{(\mp)} Z_i^{(\pm)}(q_1', q_2') \tag{1.105}$$

$$V^{(\pm)}(q_1,q_2) = 2\sum_{q_1'q_2'} G^{\mathrm{ph}}(q_1,q_2';q_1',q_2)u^{(\pm)}_{q_1'q_2'}Z_i^{(\pm)}(q_1',q_2')$$
$$+ V_0^{(\pm)}(q_1,q_2) \qquad (1.106)$$
$$V_0^{(\pm)}(q_1,q_2) = \frac{1}{2}[\delta I^i_\omega(q_1,q_2) \pm \delta I^{*i}_{-\omega}(q_1,q_2)].$$

Substituting them into (1.104) and carrying out the appropriate transformations (see [6, 54]), we arrive at the following equations for the functions $V^{(\pm)}$ and $d^{(\pm)}$:

$$V^{(\pm)}(q_1,q_2) = V_0(q_1,q_2) + 2\sum_{q_1'q_2'} G^{\mathrm{ph}}(q_1,q_2;q_1'q_2')$$
$$\times [u^{(\pm)}_{q_1'q_2'}/(\epsilon^2_{q_1'q_2'} - \omega_i^2)]\{\epsilon_{q_1'q_2'}[u^{(\pm)}_{q_1'q_2'}V^{(\pm)}(q_1',q_2')$$
$$+ v^{(\mp)}_{q_1'q_2'}d^{(\pm)}(q_1',q_2')] + \omega_i[u^{(\mp)}_{q_1'q_2'}V^{(\mp)}(q_1',q_2')$$
$$+ v^{(\pm)}_{q_1'q_2'}d^{(\mp)}(q_1',q_2')]\} \qquad (1.105')$$
$$d^{(\pm)}(q_1,q_2) = \sum_{q_1'q_2'} G^{\mathrm{pp}}(q_1,q_2;q_2'q_1')$$
$$\times [v^{(\mp)}_{q_1'q_2'}/(\epsilon^2_{q_1'q_2'} - \omega_i^2)]\{\epsilon_{q_1'q_2'}[u^{(\pm)}_{q_1'q_2'}V^{(\pm)}(q_1',q_2')$$
$$+ v^{(\mp)}_{q_1'q_2'}d^{(\pm)}(q_1',q_2')] + \omega_i[u^{(\mp)}_{q_1'q_2'}V^{(\mp)}(q_1',q_2')$$
$$+ v^{(\pm)}_{q_1'q_2'}d^{(\mp)}(q_1',q_2')]\}. \qquad (1.106')$$

If we redefine G^{ph} by $\mathcal{F}^\omega$ and G^{pp} by $\mathcal{F}^\epsilon$, then (1.105) and (1.106) take the form of equations in the theory of finite Fermi systems [55]. Note that only two of the four equations (1.105) and (1.106) are independent. Indeed, if we solve equations (1.104), find $Z_i^{(\pm)}(q_1,q_2)$ and substitute them into (1.105) and (1.106), we obtain four functions $d_i^{(\pm)}(q_1,q_2)$ and $V_i^{(\pm)}(q_1,q_2)$.

The local interactions between quasiparticles are not calculated in the theory of finite Fermi systems; they are found by fitting the results of calculations to experimental data. It is assumed that these parameters are universal, that is, they take on the same values in all complex nuclei. Their radial dependence is taken in the form of a δ function. It is then assumed that

$$\mathcal{F}^\omega(r) = \mathcal{F}^\omega \frac{\mathcal{F}^\omega_{\mathrm{ex}} - \mathcal{F}^\omega}{1 + \exp[-\alpha(r - R_0)]}$$

that is, that outside of the nucleus, the interaction $\mathcal{F}^\omega$ is allowed to take on values different from those inside the nucleus. Since $\mathcal{F}^\omega$ is used to describe processes characterized by low momentum transfer, it is assumed that $\mathcal{F}^\omega$ is independent of the momentum transferred; hence, it can be written as:

$$\mathcal{F}^\omega = \frac{\pi^2}{m^* p_F}\{f + g(\boldsymbol{\sigma}^{(1)}\boldsymbol{\sigma}^{(2)}) + [f' + g'(\boldsymbol{\sigma}^{(1)}\boldsymbol{\sigma}^{(2)})](\boldsymbol{\tau}^{(1)}\boldsymbol{\tau}^{(2)})\}. \qquad (1.107)$$

The quantities f, f', g and g' are functions of the momentum of quasiparticles; in fact, the transitions occur close to the Fermi surface, so that all momenta may be replaced by the Fermi momentum $p_{\rm F}$. The interaction $\mathcal{F}^{\omega}$ is a function of only the angle between input momenta, so that it can be expanded in a series in Legendre polynomials. The interaction $\mathcal{F}^{\omega}$ is written as a δ function. The theory of finite Fermi systems is widely applied in calculations of the characteristics of atomic nuclei (see [9, 56–59]).

1.3.4

Let us consider how equations (1.102) are simplified in two particular cases.

Case One. Vibrational states are known to result from particle–hole interactions. The greatest coherent contribution comes from the interactions proportional to $u^{(+)}_{qq'}$. We will first consider a particular case of a negligible effect of the interaction in the particle–particle channel on the properties of vibrational states. Correspondingly, we retain in equations (1.102) only the terms containing $u^{(+)}_{qq'}$ and obtain

$$\begin{aligned} &\epsilon_{qq'} Z_i^{(+)}(q,q') - \omega_i Z_i^{(-)}(q,q') \\ &\qquad - 2\sum_{q_1'q_2'} G^{\rm ph}(q,q_2';q_1',q') u^{(+)}_{qq'} u^{(+)}_{q_1'q_2'} Z_i^{(+)}(q_1',q_2') = 0 \\ &\epsilon_{qq'} Z_i^{(-)}(q,q') - \omega_i Z_i^{(+)}(q,q') = 0. \end{aligned}$$

Finding $Z_i^{(-)}(q,q')$ from the second equation, we substitute it into the first. This gives

$$\begin{aligned} (\epsilon^2_{qq'} - \omega_i^2) Z_i^{(+)}(q,q') &= \epsilon_{qq'} u^{(+)}_{qq'} D_i(q,q') \\ D_i(q,q') &= 2\sum_{q_2q_2'} G^{\rm ph}(q,q_2';q_2,q') u^{(+)}_{q_2q_2'} Z_i^{(+)}(q_2,q_2'). \end{aligned}$$

The first equation now gives $Z^{(+)}(q,q')$; we substitute it into the second equation and obtain the following homogeneous equation:

$$D_i(q,q') = 2\sum_{q_2q_2'} G^{\rm ph}(q,q_2';q_2,q') \frac{(u^{(+)}_{q_2q_2'})^2 \epsilon_{q_2q_2'}}{\epsilon^2_{q_2q_2'} - \omega_i^2} D_i(q_2,q_2'). \tag{1.108}$$

The energies ω_i are found from the secular equation in the space of two-quasiparticle states:

$$\det \left\| \delta_{qq',q_2q_2'} - 2G^{\rm ph}(q,q_2';q_2,q') \frac{(u^{(+)}_{q_2q_2'})^2 \epsilon_{q_2q_2'}}{\epsilon^2_{q_2q_2'} - \omega_i^2} \right\| = 0. \tag{1.109}$$

The rank of this determinant equals the number of two-quasiparticle states. We have thus obtained a random phase approximation (RPA) equation for the case of non-separable interactions. Equation (1.109) was solved in a number of papers, such as [60], and the energies ω_i of vibrational states were calculated.

We choose $G^{\rm ph}$ in the separable form which corresponds to the case of the multipole–multipole interaction:

$$G^{\rm ph}(q,q_2';q_2,q') = \frac{\kappa}{2} f(q,q') f(q_2,q_2'). \tag{1.110}$$

Correspondingly,

$$\begin{aligned} D_i(q,q') &= \kappa f(q,q') \sum_{q_2 q_2'} f(q_2,q_2') u^{(+)}_{q_2 q_2'} Z_i^{(+)}(q_2,q_2') \\ &\equiv \kappa f(q,q') D_0 . \end{aligned}$$

Substituting it into equation (1.108), we come to the secular equation in the form

$$1 = \kappa \sum_{qq'} \frac{[f(qq')u^{(+)}_{qq'}]^2 \epsilon_{qq'}}{\epsilon^2_{qq'} - \omega_i^2}. \tag{1.111}$$

We have thus obtained the familiar RPA equation that is widely used to calculate vibrational states of spherical and deformed nuclei ([61–63]).

Case Two. Let us consider isoscalar and isovector multipole–multipole interactions in the particle–hole channel, containing $u^{(+)}_{qq'}$. In this case,

$$\begin{aligned} G^{\rm ph}(q,q_2';q_2,q') = {} & \frac{1}{2}(\kappa_0+\kappa_1)[f(ss')f(s_2s_2') + f(rr')f(r_2r_2')] \\ & + (\kappa_0 - \kappa_1) f(ss') f(rr') \end{aligned} \tag{1.112}$$

where s, s' and r, r' refer to the neutron and the proton systems, respectively. Equations (1.102) change to

$$\begin{aligned} &\epsilon_{s,s'} Z_i^{(+)}(ss') - \omega_i Z_i^{(-)}(ss') - (\kappa_0+\kappa_1) f(ss') u^{(+)}_{ss'} D_i(n) \\ &\quad - (\kappa_0-\kappa_1) f(ss') u^{(+)}_{ss'} D_i(p) = 0 \\ &\epsilon_{rr'} Z_i^{(+)}(r,r') - \omega_i Z_i^{(-)}(r,r') - (\kappa_0+\kappa_1) f(rr') u^{(+)}_{rr'} D_i(p) \\ &\quad - (\kappa_0-\kappa_1) f(rr') u^{(+)}_{rr'} D_i(n) = 0 \\ &Z_i^{(-)}(q,q') = \frac{\omega_i}{\epsilon_{qq'}} Z_i^{(+)}(q,q') \end{aligned}$$

where $D_i(\tau) = \sum^{\tau}_{qq'} f(qq') u^{(+)}_{qq'} Z^{(+)}(q,q')$ and $\tau = n$ or $\tau = p$. Let us find the functions $Z^{(+)}(s,s')$, $Z^{(+)}(r,r')$ and substitute them into the expressions for $D_i(\tau)$. We will use the functions $X^i(n)$ and $X^i(p)$ in which the summation is carried out over the neutron and proton states, respectively. After appropriate transformations, we obtain the following secular equation:

$$\begin{Vmatrix} (\kappa_0 + \kappa_1) X_i^M(n) - 1 & (\kappa_0 - \kappa_1) X_i^M(n) \\ (\kappa_0 - \kappa_1) X_i^M(p) & (\kappa_0 + \kappa_1) X_i^M(p) - 1 \end{Vmatrix} = 0 \tag{1.113}$$

This equation is frequently used to calculate the energies of collective quadrupole and octupole low-lying states and electric-type giant resonances (see [6, 61, 64–66]).

Equations (1.102) have thus proved to be sufficiently general and complete. Different choices of interactions in the particle–particle and particle–hole channels lead to particular cases of these equations, applicable to the analysis of the collective vibrational states of complex nuclei.

2

Secondary quantization method

2.1 ONE-PHONON STATES IN DEFORMED NUCLEI

2.1.1

One of the important problems is to select the collective vibrational degrees of freedom. Several methods are available for solving this problem. In a phenomenological description, the collective motions are the vibrations of the nuclear surface and the vibrations of neutrons with respect to protons. The microscopic theory analyses, in addition to vibrations having a classical analogue, the collective motions connected with charge exchange, spin flip and pairing vibrations.

The microscopic description of nuclear vibrations is based on the method of approximate secondary quantization. Its most frequently used and efficient formulation is known as the random phase approximation(RPA). Among the advantages of the RPA are, *first*, a sufficiently accurate and physically clear description of low-lying vibrations and high-lying giant-resonance-type states, *second*, a unified description of collective, weakly collective and two-quasiparticle states, and *third*, the possibility of taking into account in a relatively simple way the coupling to quasiparticle states. The RPA allows the calculation of the energy and wave functions of one-phonon states with the interaction in the particle–hole channel taken into consideration. If necessary, the particle–particle channel is included in the calculations. It is possible to produce a description which simultaneously covers the isoscalar and isovector multipole and spin-multipole interactions, and so forth.

The present chapter describes one-phonon states in deformed and spherical nuclei, including, among other things, neutron–proton phonons. The material presented here is a basis for further description of the structure of complex nuclei. We will begin with one-phonon states in deformed nuclei.

The nuclear Hamiltonian will be

$$H = H_{\mathrm{av}} + H_{\mathrm{pair}} + H_M^{\mathrm{ph}} + H_S^{\mathrm{ph}} + H_T^{\mathrm{ph}} . \tag{2.1}$$

We have shown in Chapter 1 how to identify in nucleon–nucleon interactions the mean field H_{av} and the interactions H_{pair} leading to superfluid pairing correlations. In (2.1), H_{av} is the mean field in the neutron and proton systems and H_M^{ph} and H_S^{ph} are the sums of isoscalar and isovector multipole and spin-multipole interactions in the particle–hole channel. If necessary, tensor interactions H_T^{ph} are introduced. Hamiltonian (2.1) is sometimes supplemented by terms which correspond to the interactions in the particle–particle channel. The analysis below is limited to non-rotational states, so that H does not include either the kinetic energy of rotation or the Coriolis interaction which describes the coupling of the internal motion and rotation.

Since the nuclear mean field is considered separately for the neutron and the proton systems and since pairing acts only between neutrons and protons, the terms $H_{\text{av}} + H_{\text{pair}}$ are written separately for the two systems. For deformed nuclei,

$$H_{\text{av}} + H_{\text{pair}} = H_0(\text{n}) + H_0(\text{p}0) \tag{2.2}$$

$$H_0 = \sum_{q\sigma}[E_0'(q) - \lambda]a_{q\sigma}^+ a_{q\sigma} - G\sum_{qq'} a_{q+}^+ a_{q-}^+ a_{q'-} a_{q'+}\,. \tag{2.3}$$

Here $E'(q)$ denotes single-particle energies that are not completely renormalized; G denotes G_N and G_Z for the neutron and the proton systems, respectively; $q\sigma$, $\sigma = \pm 1$, denotes the set of quantum numbers that characterizes single-particle states of a deformed nucleus in the neutron and proton systems. The chemical potentials $\lambda(\lambda_{\text{n}}, \lambda_{\text{p}})$ are found from the condition of conservation of the average number of protons and neutrons.

Let us transform H_0, carrying out Bogoliubov canonical transformation (1.47) and making use of equations (1.46). Correspondingly, we introduce the operators

$$A^+(q) = 2\alpha_{q+}^+\alpha_{q-}^+ \qquad A(q) = 2\alpha_{q-}\alpha_{q+} \qquad B(q) = \sum_\sigma \alpha_{q\sigma}^+\alpha_{q\sigma} \tag{2.4}$$

drop the terms which are independent of operators, and rewrite H_0 in the form (see [6])

$$H_0 = \sum_q \epsilon_q B(q) + H_0^\beta + H_0' + H_0'' \tag{2.5}$$

where

$$\epsilon_q = \sqrt{C^2 + [E(q) - \lambda]^2}$$

$$H_0^\beta = -\frac{G}{4}\sum_{qq'}[u_q^2 A^+(q) - v_q^2 A(q)][u_{q'}^2 A(q') - v_{q'}^2 A^+(q')]. \tag{2.6}$$

$$H_0' = -G\Big\{\sum_q u_q v_q B(q)\Big\}^2$$
$$H_0'' = -\frac{G}{2}\sum_{qq'}(u_q^2 - v_q^2)u_{q'}v_{q'}[A^+(q)B(q') + B(q')A(q)]. \tag{2.7}$$

When considering one-phonon states with $K^\pi \neq 0^+$ in (2.5), we take into account only one term:

$$\sum_q \epsilon_q B(q). \tag{2.8}$$

In order to describe non-diagonal parts of the density matrix, we will introduce an effective interaction between quasiparticles. Its central part will be

$$V(|\boldsymbol{r}_1 - \boldsymbol{r}_2|) + V_\sigma(|\boldsymbol{r}_1 - \boldsymbol{r}_2|)(\boldsymbol{\sigma}^{(1)}\boldsymbol{\sigma}^{(2)}) + \{V_\tau(|\boldsymbol{r}_1 - \boldsymbol{r}_2|) + V_{\tau\sigma}(|\boldsymbol{r}_1 - \boldsymbol{r}_2|)(\boldsymbol{\sigma}^{(1)}\boldsymbol{\sigma}^{(2)})\}(\boldsymbol{\tau}^{(1)}\boldsymbol{\tau}^{(2)}). \tag{2.9}$$

We can expand it in a series in multipoles and spin-multipoles, that is,

$$V(|\boldsymbol{r}_1 - \boldsymbol{r}_2|) = \sum_\lambda R^\lambda(r_1, r_2)\frac{4\pi}{2\lambda+1} \times \sum_{\mu=-\lambda}^{\lambda}(-)^\mu Y_{\lambda\mu}(\Theta_1,\varphi_1)Y_{\lambda-\mu}(\Theta_2,\varphi_2) \tag{2.10}$$

$$V_\sigma(|\boldsymbol{r}_1 - \boldsymbol{r}_2|)((\boldsymbol{\sigma}^{(1)}\boldsymbol{\sigma}^{(2)}) = \sum_L \sum_{\lambda=L,L\pm1} R_\sigma^{\lambda L}(r_1, r_2) \times \frac{4\pi}{2\lambda+1}(-)^{\lambda-L+L}\sum_{M=-L}^{L}(-)^M\{\boldsymbol{\sigma}^{(1)} \times \mathbf{Y}_\lambda(\Theta_1,\varphi_1)\}_{LM}\{\boldsymbol{\sigma}^{(2)}\mathbf{Y}_\lambda(\Theta_2,\varphi_2)\}_{L-M} \tag{2.11}$$

where

$$\{\boldsymbol{\sigma}\mathbf{Y}_\lambda(\Theta,\varphi)\}_{LM} = \sum_{\rho=0,\pm1}\sum_{\mu=-\lambda}^{\lambda}\langle 1\rho\lambda\mu|LM\rangle\sigma_\rho Y_{\lambda\mu}(\Theta,\varphi). \tag{2.12}$$

The radial part of the interaction can be chosen in a different manner. In many papers, it was taken in a separable form which makes it possible to give a sufficiently good representation of practically any effective interaction. The separable representation is successfully used to analyse few-nucleons systems, in which the sensitivity to the radial dependence is considerably stronger than in complex nuclei. The application of a separable radial component allows one to avoid the diagonalization of high-order matrices and to obtain a simple secular equation. Correspondingly, we assume in expansions (2.10) and (2.11) that

$$R^{\lambda}(r_1, r_2) = \kappa^{(\lambda)} R_{\lambda}(r_1) R_{\lambda}(r_2)$$
$$R_{\sigma}^{\lambda L}(r_1, r_2) = \kappa^{(\lambda L)} R_{\lambda}(r_1) R_{\lambda}(r_2). \tag{2.13}$$

If we choose a separable interaction of finite rank $n_{\max} > 1$, then

$$R^{\lambda}(r_1, r_2) = \kappa^{(\lambda)} \sum_{n=1}^{n_{\max}} R_n^{\lambda}(r_1) R_n^{\lambda}(r_2)$$
$$R^{\lambda L}(r_1, r_2) = \kappa^{(\lambda L)} \sum_{n=1}^{n_{\max}} R_n^{\lambda L}(r_1) R_n^{\lambda L}(r_2). \tag{2.13'}$$

If necessary, we can supplement (2.9) by tensor interactions. In the case of deformed nuclei, isoscalar and isovector constants of particle–hole multipole $(\kappa_0^{(\lambda\mu)}, \kappa_1^{(\lambda\mu)})$ and spin-multipole $(\kappa_0^{(\lambda LK)}, \kappa_1^{(\lambda LK)})$ interactions depend on the projections μ and K of the moments $\boldsymbol{\lambda}$ and $\boldsymbol{L}$ onto the symmetry axis of the nucleus.

Taking into consideration formulas (2.10) and (2.11) and also recalling that the Hamiltonian is Hermitian and invariant under time reversal operation, we can write the Hamiltonian for multipole, spin-multipole and tensor forces in the particle–hole channel in the form

$$H_M^{\mathrm{ph}} = -\frac{1}{2} \sum_{\lambda\mu} [\kappa_0^{(\lambda\mu)} + \kappa_1^{(\lambda\mu)} (\boldsymbol{\tau}^{(1)} \boldsymbol{\tau}^{(2)})] \sum_{\sigma} M_{\lambda\sigma\mu} M_{\lambda-\sigma\mu} \tag{2.14}$$

$$H_S^{\mathrm{ph}} = -\frac{1}{2} \sum_{LK} \sum_{\lambda=L,L\pm1} [\kappa_0^{(\lambda LK)} + \kappa_1^{(\lambda LK)} \times (\boldsymbol{\tau}^{(1)} \boldsymbol{\tau}^{(2)})] \sum_{\sigma} S_{L\sigma K}^{\lambda} S_{L-\sigma K}^{\lambda} \tag{2.15}$$

$$H_T^{\mathrm{ph}} = -\frac{1}{2} (\boldsymbol{\tau}^{(1)} \boldsymbol{\tau}^{(2)}) \sum_{LK} \kappa_T^{(LK)} \times [(S_{LK}^{L-1})^{+} S_{LK}^{L+1} + (S_{LK}^{L+1})^{+} S_{LK}^{L-1}] \tag{2.16}$$

$$M_{\lambda\sigma\mu} = \sum_{q_1\sigma_1 q_2\sigma_2} \langle q_1\sigma_1 | R_{\lambda}(r) Y_{\lambda\sigma\mu}(\Theta, \varphi) | q_2\sigma_2 \rangle a_{q_2\sigma_1}^{+} a_{q_2\sigma_2}$$
$$M_{\lambda-\sigma\mu}^{+} = M_{\lambda\sigma\mu} \tag{2.17}$$

$$S_{L\sigma K}^{\lambda} = \sum_{q_1\sigma_1 q_2\sigma_2} \langle q_1\sigma_1 | R_{\lambda}(r) \{\sigma Y_{\lambda}(\Theta, \varphi)\}_{L\sigma K} | q_2\sigma_2 \rangle a_{q_1\sigma_1}^{+} a_{q_2\sigma_2}. \tag{2.18}$$

The matrix elements are taken over the single-particle wave functions of the Saxon–Woods potential of a deformed nucleus using formulas (1.76) and (1.77). The single-particle states are characterized by the quantum numbers $q\sigma$, $q = K^\pi[Nn_z\Lambda]$, where K is the projection of the angular momentum on the symmetry axis of the nucleus; the asymptotic quantum numbers $Nn_z\Lambda$ are defined in [6, 12]. Now we use the notation $Nn_z\Lambda\uparrow$ if $K = \Lambda + 1/2$ and $Nn_z\Lambda\downarrow$ if $K = \Lambda - 1/2$. Introducing $\sigma = \pm 1$, we assume that $K > 0$ and $\mu > 0$. The dependence of matrix elements on σ will be explicitly introduced. The matrix element of a multipole operator is

$$\begin{aligned}
\langle q_1\sigma_1|R_\lambda(r)[Y_{\lambda\sigma\mu}(\Theta,\varphi)+(-)^\mu Y_{\lambda-\sigma\mu}(\Theta,\varphi)]\,|q_2\sigma_2\rangle \\
&= \delta_{\sigma_1K_1-\sigma_2K_2,\sigma\mu}\langle q_1\sigma_1|f^{\lambda\mu}|q_2\sigma_2\rangle \\
\langle q_1\sigma|f^{\lambda\mu}|q_2\sigma\rangle &= \langle q_1-\sigma|f^{\lambda\mu}|q_2-\sigma\rangle \equiv \tilde{f}^{\lambda\mu}(q_1q_2) = \tilde{f}^{\lambda\mu}(q_2q_1) \\
\langle q_1\sigma|f^{\lambda\mu}|q_2-\sigma\rangle &= -\langle q_1-\sigma|f^{\lambda\mu}|q_2\sigma\rangle \\
&\equiv \sigma\bar{f}^{\lambda\mu}(q_1q_2) = -\sigma\bar{f}^{\lambda\mu}(q_2q_1).
\end{aligned}$$

We denote it by

$$f^{\lambda\mu}(q_1q_2) = \begin{cases} \tilde{f}^{\lambda\mu}(q_1q_2) & \mu = |K_1 - K_2| \\ \bar{f}^{\lambda\mu}(q_1,q_2) & \mu = K_1 + K_2. \end{cases} \tag{2.19}$$

The symmetry of the matrix elements of the spin-multipole operators

$$\begin{aligned}
&\langle q_1\sigma_1|R_\lambda(r)[\{\sigma Y_\lambda(\Theta,\varphi)\}_{L\sigma K} + (-)^K\{\sigma Y_\lambda(\Theta,\varphi)_{L-\sigma K}\}]|q_2\sigma_2\rangle \\
&= \delta_{\sigma_1K_1-\sigma_2K_2,\sigma K}\langle q_1\sigma_1|f^{\lambda LK}|q_2\sigma_2\rangle
\end{aligned}$$

changes with changing λ. If $\lambda = L$, they have:

$$\begin{aligned}
\langle q_1\sigma|f^{\lambda\lambda K}|q_2\sigma\rangle &= \langle q_1-\sigma|f^{\lambda\lambda K}|q_2-\sigma\rangle \\
&\equiv \tilde{f}^{\lambda\lambda K}(q_1q_2) = -\tilde{f}^{\lambda\lambda K}(q_2q_1) \\
\langle q_1\sigma|f^{\lambda\lambda K}|q_2-\sigma\rangle &= -\langle q_1-\sigma|f^{\lambda\lambda K}|q_2\sigma\rangle \\
&\equiv \sigma\bar{f}^{\lambda\lambda K}(q_1q_2) = \sigma\bar{f}^{\lambda\lambda K}(q_2q_1).
\end{aligned}$$

If $\lambda = L \pm 1$, their symmetry is identical to that of the matrix elements of the magnetic transitions and

$$\begin{aligned}
\langle q_1\sigma|f^{L\pm1\,K}|q_2\sigma\rangle &= -\langle q_1-\sigma|f^{L\pm1\,L-K}|q_2-\sigma\rangle \\
&\equiv \sigma\tilde{f}^{L\pm1\,LK}(q_1q_2) = \sigma\tilde{f}^{L\pm1\,LK}(q_2q_1) \\
\langle q_1\sigma|f^{L\pm1\,LK}|q_2-\sigma\rangle &= \langle q_1-\sigma|f^{L\pm1\,LK}|q_2\sigma\rangle \\
&\equiv \bar{f}^{L\pm1\,LK}(q_1q_2) = \bar{f}^{L\pm1\,LK}(q_2q_1).
\end{aligned}$$

We denote them by

$$f^{\lambda LK}(q_1q_2) = \begin{cases} \tilde{f}^{\lambda LK}(q_1q_2) & K = |K_1 - K_2| \\ \bar{f}^{\lambda lK}(q_1,q_2) & K = K_1 + K_2. \end{cases} \tag{2.19$'$}$$

We will need new operators in order to express operators $M_{\lambda\sigma\mu}$ and $S^{\lambda}_{L\sigma K}$ in terms of the operators of quasiparticles [67]:

$$A^+(q_1q_2;\mu\sigma) = \begin{cases} \widetilde{A}^+(q_1q_2;\mu\sigma) = \sum_{\sigma'} \delta_{\sigma'(K_1-K_2),\sigma\mu}\sigma'\alpha^+_{q_1\sigma'}\alpha^+_{q_2-\sigma'} \\ \bar{A}^+(q_1q_2;\mu\sigma) = \sum_{\sigma'} \delta_{\sigma'(K_1+K_2),\sigma\mu}\alpha^+_{q_2\sigma'}\alpha^+_{q_1\sigma'} \end{cases} \tag{2.20}$$

$$A(q_1q_2;\mu\sigma) = \begin{cases} \widetilde{A}^+(q_1q_2;\mu\sigma) = \sum_{\sigma'} \delta_{\sigma'(K_1-K_2),\sigma\mu}\sigma'\alpha_{q_2-\sigma'}\alpha_{q_1\sigma'} \\ \qquad = \widetilde{A}(q_2q_1;\mu\sigma) \\ \bar{A}(q_1q_2;\mu\sigma) = \sum_{\sigma'} \delta_{\sigma'(K_1+K_2),\sigma\mu}\alpha_{q_1\sigma'}\alpha_{q_2\sigma'} \\ \qquad = -\bar{A}(q_2q_1;\mu\sigma) \end{cases}$$

$$\mathcal{A}^+(q_1q_2;\mu\sigma) = \begin{cases} \widetilde{\mathcal{A}}^+(q_1q_2;\mu\sigma) = \sum_{\sigma'} \delta_{\sigma'(K_1-K_2),\sigma\mu}\alpha^+_{q_1\sigma'}\alpha^+_{q_2-\sigma'} \\ \bar{\mathcal{A}}^+(q_1q_2;\mu\sigma) = \sum_{\sigma'} \delta_{\sigma'(K_1+K_2),\sigma\mu}\sigma'\alpha^+_{q_2\sigma'}\alpha^+_{q_1\sigma'} \end{cases} \tag{2.20$'$}$$

$$\mathcal{A}^+(q_1q_2;\mu\sigma) = -\mathcal{A}^+(q_2q_1;\mu\sigma)$$

$$B(q_1q_2;\mu\sigma) = \begin{cases} \tilde{B}(q_1q_2;\mu\sigma) = \sum_{\sigma'} \delta_{\sigma'(K_1-K_2),\sigma\mu}\alpha^+_{q_1\sigma'}\alpha_{q_2\sigma'} \\ \bar{B}(q_1q_2;\mu\sigma) = \sum_{\sigma'} \delta_{\sigma'(K_1+K_2),\sigma\mu}\sigma'\alpha^+_{q_1\sigma'}\alpha_{q_2-\sigma'} \end{cases} \tag{2.21}$$

$$\tilde{B}^+(q_1q_2;\mu\sigma) = \tilde{B}(q_2q_1;\mu-\sigma) \qquad \bar{B}^+(q_1q_2;\mu\sigma) = -\bar{B}(q_2q_1;\mu-\sigma)$$

$$\mathfrak{B}(q_1q_2;\mu\sigma) = \begin{cases} \tilde{\mathfrak{B}}(q_1q_2;\mu\sigma) = \sum_{\sigma'} \delta_{\sigma'(K_1-K_2),\sigma\mu}\sigma'\alpha^+_{q_1\sigma'}\alpha_{q_2\sigma'} \\ \bar{\mathfrak{B}}(q_1q_2;\mu\sigma) = \sum_{\sigma'} \delta_{\sigma'(K_1+K_2),\sigma\mu}\alpha^+_{q_1\sigma'}\alpha_{q_2-\sigma'} \end{cases}$$

$$\mathfrak{B}^+(q_1q_2;\mu\sigma) = \mathfrak{B}(q_2q_1;\mu-\sigma). \tag{2.21$'$}$$

At $\mu \neq 0$, the operators satisfy the following commutation relations:

$$[\widetilde{A}(q_1'q_2';\mu'\sigma'), \widetilde{A}^+(q_1q_2;\mu\sigma)]$$
$$= \delta_{\mu\mu'}\delta_{\sigma\sigma'}\delta_{|K_1-K_2|,\mu}(\delta_{q_1q_1'}\delta_{q_2q_2'} + \delta_{q_1q_2'}\delta_{q_2q_1'})$$

$$
\begin{aligned}
&-\sum_{\sigma_3}\delta_{\sigma_3(K_1-K_2),\sigma\mu}[\delta_{\sigma_3(K_1'-K_2'),\sigma'\mu'}(\delta_{q_1q_1'}\\
&\times \alpha^+_{q_2-\sigma_3}\alpha_{q_2'-\sigma_3} + \delta_{q_2q_2'}\alpha^+_{q_1\sigma_3}\alpha_{q_1'\sigma_3}) + \delta_{\sigma_3(K_2'-K_1'),\sigma'\mu'}\\
&\times (\delta_{q_1q_2'}\alpha^+_{q_2-\sigma_3}\alpha_{q_1'-\sigma_3} + \delta_{q_2q'}\alpha^+_{q_1\sigma_3}\alpha_{q_2'\sigma_3})]
\end{aligned}
\qquad (2.22)
$$

$$
\begin{aligned}
&[\bar{A}(q_1'q_2';\mu'\sigma'), \bar{A}^+(q_1q_2;\mu\sigma)]\\
&\quad = \delta_{\mu\mu'}\delta_{\sigma\sigma'}\delta_{(K_1+K_2),\mu}(\delta_{q_1q_1'}\delta_{q_2q_2'} - \delta_{q_1q_2'}\delta_{q_2q_1'})\\
&\quad - \delta_{\sigma\sigma'}\delta_{(K_1-K_2),\mu}\delta_{(K_1'+K_2'),\mu'}(\delta_{q_1q_1'}\alpha^+_{q_2\sigma}\alpha_{q_2'\sigma}\\
&\quad + \delta_{q_2q_2'}\alpha^+_{q_1\sigma}\alpha_{q_1'\sigma} - \delta_{q_1q_2'}\alpha^+_{q_2\sigma}\alpha_{q_1'\sigma} - \delta_{q_2q_1}\alpha^+_{q_1\sigma}\alpha_{q_2'\sigma})
\end{aligned}
\qquad (2.22')
$$

$$
\begin{aligned}
&[A(q_1'q_2';\mu'\sigma'), A(q_1q_2;\mu\sigma)]\\
&\quad = [A^+(q_1'q_2;\mu'\sigma'), A^+(q_1q_2;\mu\sigma)] = 0
\end{aligned}
\qquad (2.22'')
$$

$$
\begin{aligned}
&[\tilde{\mathcal{A}}(q_1'q_2';\mu'\sigma'), \tilde{\mathcal{A}}^+(q_1q_2;\mu\sigma)]\\
&\quad = \delta_{|K_1-K_2|,\mu}\delta_{\mu\mu'}\delta_{\sigma\sigma'}(\delta_{q_1q_1'}\delta_{q_2q_2'} - \delta_{q_1q_2'}\delta_{q_2q_1'})\\
&\quad - \sum_{\sigma_3}\delta_{\sigma_3(K_1-K_2),\sigma\mu}[\delta_{\sigma_3(K_1'-K_2'),\sigma'\mu'}(\delta_{q_2q_2'}\alpha^+_{q_1\sigma_3}\\
&\quad \times \alpha_{q_1'\sigma_3} + \delta_{q_1q_1'}\alpha^+_{q_2-\sigma_3}\alpha_{q_2'-\sigma_3}) - \delta_{\sigma_3(K_2'-K_1'),\sigma'\mu'}\\
&\quad \times (\delta_{q_1'q}\alpha^+_{q_1\sigma_3}\alpha_{q_2'\sigma_3} + \delta_{q_1q_2'}\alpha^+_{q_2-\sigma_3}\alpha_{q_1'-\sigma_3})]
\end{aligned}
\qquad (2.23)
$$

$$
\begin{aligned}
&[\bar{\mathcal{A}}(q_1'q_2';\mu'\sigma'), \bar{\mathcal{A}}^+(q_1q_2;\mu\sigma)]\\
&\quad = \delta_{\mu\mu'}\delta_{\sigma\sigma'}\delta_{(K_1+K_2),\mu}(\delta_{q_1q_1'}\delta_{q_2q_2'}\\
&\quad - \delta_{q_1q_2'}\delta_{q_2q_1'}) - \delta_{\sigma\sigma'}\delta_{(K_1'+K_2'),\mu'}\delta_{(K_1+K_2),\mu}\\
&\quad \times (\delta_{q_1q_1'}\alpha^+_{q_2\sigma}\alpha_{q_2'\sigma} + \delta_{q_2q_2'}\alpha^+_{q_1\sigma}\alpha_{q_1'\sigma}\\
&\quad - \delta_{q_1q_2'}\alpha^+_{q_2\sigma}\alpha_{q_1'\sigma} - \delta_{q_2q_1'}\alpha^+_{q_1\sigma}\alpha_{q_2'\sigma})
\end{aligned}
\qquad (2.23')
$$

$$
\begin{aligned}
&[\tilde{\mathcal{A}}(q_1'q_2';\mu'\sigma'), \mathcal{A}(q_1q_2;\mu\sigma)]\\
&\quad = [\mathcal{A}^+(q_1'q_2';\mu'\sigma'), \mathcal{A}^+(q_1q_2;\mu\sigma)] = 0.
\end{aligned}
\qquad (2.23'')
$$

Taking into account the above formulas, we rewrite the operators $M_{\lambda\sigma\mu}$ and $S^{\lambda}_{L\sigma K}$ in the form

$$
\begin{aligned}
M_{\lambda\sigma\mu} &= \sum_{qq'\sigma'}[\tilde{f}^{\lambda\mu}(qq')\delta_{\sigma'(K-K'),\sigma\mu}a^+_{q\sigma'}a_{q'\sigma'}\\
&\quad + \bar{f}^{\lambda\mu}(qq')\sigma'\delta_{\sigma'(K+K'),\sigma\mu}a^+_{q\sigma'}q_{q'-\sigma'}]\\
&= \frac{1}{2}\sum_{qq'} f^{\lambda\mu}(qq')\{u^{(+)}_{qq'}[A^+(qq';\mu\sigma)\\
&\quad + A(qq';\mu-\sigma)] + 2v^{(-)}_{qq'}B(qq';\mu\sigma)\}
\end{aligned}
\qquad (2.24)
$$

$$
S^L_{L\sigma K} = \sum_{q_2q_2'\sigma'}[\tilde{f}^{LLK}(q_2q_2')\delta_{\sigma'(K_2-K_2'),\sigma K}a^+_{q_2\sigma'}a_{q_2'\sigma'}
$$

$$+ \bar{f}^{LLK}(q_2q_2')\sigma'\delta_{\sigma'(K_2+K_2'),\sigma K} a^+_{q_2\sigma'} a_{q_2'-\sigma'}]$$

$$= \frac{1}{2}\sum_{q_2q_2'} f^{LLK}(q_2q_2')\{u^{(-)}_{q_2q_2'}[A^+(q_2q_2';K\sigma)$$

$$- A(q_2q_2';K-\sigma)] + 2v^+_{q_2q_2'} B(q_2q_2';K\sigma\rangle\} \tag{2.25}$$

$$S^{L\pm1}_{L\sigma K} = \sum_{q_2q_2'\sigma'} [\tilde{f}^{L\pm1\,LK}(q_2q_2')\sigma'\delta_{\sigma'(K_2-K_2'),\sigma K} a^+_{q_2\sigma'} a_{q_2'\sigma'}$$

$$+ \bar{f}^{L\pm1\,LK}(q_2q_2')\delta_{\sigma'(K_2+K_2'),\sigma K} a^+_{q_2\sigma'} a_{q_2'-\sigma'}]$$

$$= \frac{1}{2}\sum_{q_2q_2'} f^{L\pm1\,LK}(q_2q_2')\{u^{(-)}_{q_2q_2'}[\mathcal{A}^+(q_2q_2';K\sigma\rangle$$

$$+ \mathcal{A}(q_2q_2';K-\sigma)] + 2v^{(+)}_{q_2q_2'}\mathfrak{B}(q_2q_2';K\sigma)\}. \tag{2.25'}$$

Here we have dropped all operator-independent terms and used the notation

$$u^{(\pm)}_{qq'} = u_q v_{q'} \pm u_{q'} v_q \qquad v^{(\pm)}_{qq'} = u_q u_{q'} \pm v_q v_{q'}. \tag{2.26}$$

The multipole and spin-multipole effective particle–hole interactions will be written in the following form:

$$H^{\mathrm{ph}}_M = H^{\mathrm{ph}}_{M1} + H^{\mathrm{ph}}_{M2} + H^{\mathrm{ph}}_{M3} \tag{2.27}$$

$$H^{\mathrm{ph}}_{M1} = \frac{1}{8}\sum_{\lambda\mu}[\kappa_0^{(\lambda\mu)} + \kappa_1^{(\lambda\mu)}(\boldsymbol{\tau}^{(1)}\boldsymbol{\tau}^{(2)})]$$

$$\times \sum_{q_1q_2q_1'q_2'\sigma} f^{\lambda\mu}(q_1q_2)u^{(+)}_{q_1q_2} f^{\lambda\mu}(q_1'q_2')$$

$$\times u^{(+)}_{q_1'q_2'}[A^+(q_1q_2;\mu\sigma) + A(q_1q_2;\mu-\sigma)]$$

$$\times [A^+(q_1'q_2';\mu-\sigma) + A(q_1'q_2';\mu\sigma)] \tag{2.28}$$

$$H^{\mathrm{ph}}_{M2} = -\frac{1}{4}\sum_{\lambda\mu}[\kappa_0^{(\lambda\mu)} + \kappa_1^{(\lambda\mu)}(\boldsymbol{\tau}^{(1)}\boldsymbol{\tau}^{(2)})]$$

$$\times \sum_{q_1q_2q_1'q_2'\sigma} f^{\lambda\mu}(q_1q_2)u^{(+)}_{q_1q_2} f^{\lambda\mu}(q_1'q_2')v^{(-)}_{q_1'q_2'}$$

$$\times \{[A^+(q_1q_2;\mu\sigma) + A(q_1q_2;\mu-\sigma)]$$

$$\times B(q_1'q_2';\mu-\sigma) + \mathrm{h.c.}\} \tag{2.28'}$$

$$H^{\mathrm{ph}}_{M3} = -\frac{1}{2}\sum_{\lambda\mu}[\kappa_0^{(\lambda\mu)} + \kappa_1^{(\lambda\mu)}(\boldsymbol{\tau}^{(1)}\boldsymbol{\tau}^{(2)})]$$

$$\times \sum_{q_1q_2q_1'q_2'\sigma} f^{\lambda\mu}(q_1q_2)v^{(-)}_{q_1q_2} f^{\lambda\mu}(q_1'q_2')v^{(-)}_{q_1'q_2'}$$

$$\times B(q_1q_2;\mu\sigma)B(q_1'q_2';\mu-\sigma) \tag{2.28''}$$

$$H_S^{\rm ph} = H_{SE}^{\rm ph} + H_{SM}^{\rm ph} \tag{2.29}$$

$$\begin{aligned} H_{SE}^{\rm ph} &= H_{SE1}^{\rm ph} + H_{SE2}^{\rm ph} + H_{SE3}^{\rm ph} \\ H_{SM}^{\rm ph} &= H_{SM1}^{\rm ph} + H_{SM2}^{\rm ph} + H_{SM3}^{\rm ph} \end{aligned} \tag{2.29'}$$

$$\begin{aligned} H_{SE1}^{\rm ph} = -\frac{1}{8}\sum_{LK} & [\kappa_0^{(LLK)} + \kappa_1^{(LLK)}(\boldsymbol{\tau}^{(1)}\boldsymbol{\tau}^{(2)})] \\ & \times \sum_{q_1 q_2 q_1' q_2' \sigma} f^{LLK}(q_1 q_2) u_{q_1 q_2}^{(-)} f^{LLK}(q_1' q_2') \\ & \times u_{q_1' q_2'}^{(-)} [A^+(q_1 q_2; K\sigma) - A(q_1 q_2; K-\sigma)] \\ & \times [A^+(q_1' q_2'; K-\sigma) - A(q_1' q_2'; K\sigma)] \end{aligned} \tag{2.30}$$

$$\begin{aligned} H_{SE2}^{\rm ph} = -\frac{1}{4}\sum_{LK} & [\kappa_0^{(LLK)} + \kappa_1^{(LLK)}(\boldsymbol{\tau}^{(1)}\boldsymbol{\tau}^{(2)})] \\ & \times \sum_{q_1 q_2 q_1' q_2' \sigma} f^{LLK}(q_1 q_2) u_{q_1 q_2}^{(-)} f^{LLK}(q_1' q_2') v_{q_1' q_2'}^{(+)} \\ & \times \{[A^+(q_1 q_2; K\sigma) - A(q_1 q_2; K-\sigma)] \\ & \times B(q_1' q_2'; K-\sigma) + \text{h.c.}\} \end{aligned} \tag{2.30'}$$

$$\begin{aligned} H_{SM1}^{\rm ph} = -\frac{1}{8}\sum_{LK\lambda=L\pm1} & [\kappa_0^{(\lambda LK)} + \kappa_1^{(\lambda LK)}(\boldsymbol{\tau}^{(1)}\boldsymbol{\tau}^{(2)})] \\ & \times \sum_{q_1 q_2 q_1' q_2' \sigma} f^{\lambda LK}(q_1 q_2) u_{q_1 q_2}^{(-)} f^{\lambda LK}(q_1' q_2') u_{q_1' q_2'}^{(-)} \\ & \times [\mathcal{A}^+(q_1 q_2; K\sigma) + \mathcal{A}(q_1 q_2; K-\sigma)] \\ & \times [\mathcal{A}^+(q_1' q_2'; K-\sigma) + \mathcal{A}(q_1' q_2'; K\sigma)] \end{aligned} \tag{2.31}$$

$$\begin{aligned} H_{SM2}^{\rm ph} = -\frac{1}{4}\sum_{LK\lambda=L\pm1} & [\kappa_0^{(\lambda LK)} + \kappa_1^{(\lambda LK)}(\boldsymbol{\tau}^{(1)}\boldsymbol{\tau}^{(2)})] \\ & \times \sum_{q_1 q_2 q_1' q_2' \sigma} f^{\lambda LK}(q_1 q_2) u_{q_1 q_2}^{(-)} f^{\lambda LK}(q_1' q_2') v_{q_1' q_2'}^{(+)} \\ & \times \{[\mathcal{A}^+(q_1 q_2; K\sigma) + \mathcal{A}(q_1 q_2; K-\sigma)] \\ & \times \mathfrak{B}(q_1' q_2'; K-\sigma)] + \text{h.c.}\} \end{aligned} \tag{2.32}$$

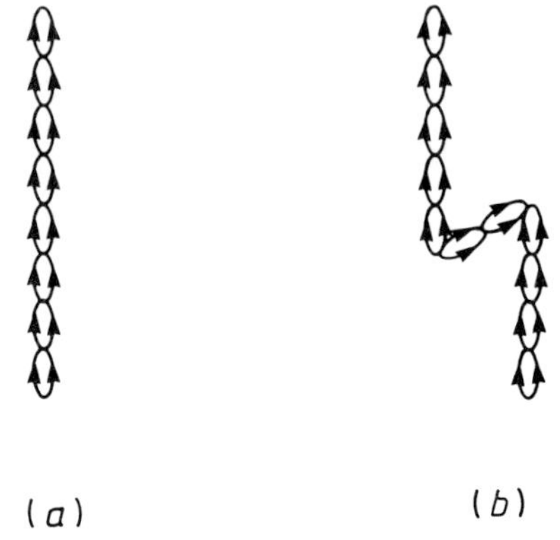

Figure 2.1 Diagrams covered by the RPA (a, b). Diagram (a) is taken into account in the Tamm–Dankoff method.

2.1.2

Let us derive equations for the RPA. This approximation is widely used to calculate the energies and wave functions of collective vibrational states [6, 61, 68]. The RPA takes into account the interactions between quasiparticles in the ground and the excitation states; the diagrams shown in figure 2.1 (a) and (b) are covered by the RPA.

The interaction between quasiparticles leads to changes in the ground state of an even–even nucleus. The wave function of the ground state contains, in addition to quasiparticle-free terms, four-quasiparticle, eight-quasiparticle, and other similar terms. We will limit the discussion only to those cases in which the mean number of quasiparticles in the ground state is small. The main RPA constraint,

$$\langle \alpha^+_{q\sigma}\alpha_{q\sigma}\rangle = 0 \tag{2.33}$$

coincides with condition (1.84) which is used in the self-consistent field method. Approximation (2.33) makes it possible to drop $\alpha^+_{q\sigma}\alpha_{q\sigma}$-containing terms in the commutation relations (2.22)–(2.23″). The operators $A(qq';\mu\sigma)$, $A^+(qq';\mu\sigma)$ are treated as boson operators and the approximation itself is sometimes called the *quasi-boson approximation*. The commutation relations (2.22)–(2.23) are replaced by the formulas

$$\begin{aligned}
&[\widetilde{A}(q_1'q_2';\mu'\sigma'), \widetilde{A}^+(q_1q_2;\mu\sigma)] \\
&\quad = \delta_{\mu\mu'}\delta_{\sigma\sigma'}\delta_{|K_1-K_2|,\mu}(\delta_{q_1q_1'}\delta_{q_2q_2'} + \delta_{q_1q_2'}\delta_{q_2q_1'}) \\
&[\bar{A}(q_1'q_2';\mu'\sigma'), \bar{A}^+(q_1q_2;\mu\sigma)] \\
&\quad = \delta_{\mu\mu'}\delta_{\sigma\sigma'}\delta_{|K_1+K_2|,\mu}(\delta_{q_1q_1'}\delta_{q_2q_2'} - \delta_{q_1q_2'}\delta_{q_2q_1'})
\end{aligned} \tag{2.34}$$

$$\begin{aligned}
&[\widetilde{A}(q_1'q_2';\mu'\sigma'),\widetilde{A}^+(q_1q_2;\mu\sigma)] \\
&\quad = \delta_{\mu\mu'}\delta_{\sigma\sigma'}\delta_{|K_1-K_2|,\mu}(\delta_{q_1q_1'}\delta_{q_2q_2'} - \delta_{q_1q_2'}\delta_{q_2q_1'}) \\
&[\bar{A}(q_1'q_2';\mu'\sigma'),\bar{A}^+(q_1q_2;\mu\sigma)] \\
&\quad = \delta_{\mu\mu'}\delta_{\sigma\sigma'}\delta_{|K_1+K_2|,\mu}(\delta_{q_1q_1'}\delta_{q_2q_2'} - \delta_{q_1q_2'}\delta_{q_2q_1'}).
\end{aligned} \tag{2.34$'$}$$

Now we introduce a phonon absorption operator which depends explicitly on σ :

$$Q_{\lambda\mu i\sigma} = \frac{1}{2}\sum_{qq'}[\psi_{qq'}^{\lambda\mu i}A(qq';\mu\sigma) - \varphi_{qq'}^{\lambda\mu i}A^+(qq';\mu-\sigma)]. \tag{2.35}$$

The wave functions of one-phonon states will be written in the form

$$Q^+_{\lambda\mu i\sigma}\Psi_0 \tag{2.36}$$

where Ψ_0 is the wave function of the ground state of an even–even nucleus, that we define now as the phonon vacuum. Here $i = 1,2,3\ldots$ is the number of one-phonon states. If the operators $A(qq';\mu\sigma)$, $A^+(qq';\mu\sigma)$ satisfy commutation relations (2.24), then phonon operators satisfy the following commutation relations:

$$\begin{aligned}
&[Q_{\lambda'\mu'i'\sigma'},Q^+_{\lambda\mu i\sigma}] = \delta_{\lambda\lambda'}\delta_{\mu\mu'}\delta_{\sigma\sigma'}\delta_{ii'} \\
&[Q_{\lambda'\mu'i'\sigma'},Q_{\lambda\mu i\sigma}] = [Q^+_{\lambda\mu i\sigma},Q^+_{\lambda'\mu'i'\sigma'}] = 0.
\end{aligned} \tag{2.37}$$

These commutation relations ensure that the conditions of orthonormalization of the wave functions of the ground and excitation states are met. They impose the following constraints on the functions $\psi_{qq'}^{\lambda\mu i}$ and $\varphi_{qq'}^{\lambda\mu i}$:

$$\frac{1}{2}\sum_{qq'}(\psi_{qq'}^{\lambda\mu i}\psi_{qq'}^{\lambda'\mu'i'} - \varphi_{qq'}^{\lambda\mu i}\varphi_{qq'}^{\lambda'\mu'i'}) = \delta_{\lambda\lambda'}\delta_{\mu\mu'}\delta_{ii'} \tag{2.38}$$

$$\sum_{qq'}(\psi_{qq'}^{\lambda\mu i}\varphi_{qq'}^{\lambda'\mu'i'} - \psi_{qq'}^{\lambda'\mu'i'}\varphi_{qq'}^{\lambda\mu i}) = 0. \tag{2.38$'$}$$

Using commutation relations and conditions (2.38) and (2.38′), we obtain

$$A(qq';\mu\sigma) = \sum_i(\psi_{qq'}^{\lambda\mu i}Q_{\lambda\mu i\sigma} + \varphi_{qq'}^{\lambda\mu i}Q^+_{\lambda\mu i-\sigma}) \tag{2.39}$$

Commutation relations (2.34) imply the equations

$$\begin{aligned}
&\sum_i(\tilde{\psi}_{qq'}^{\lambda\mu i}\tilde{\psi}_{q_2q_2'}^{\lambda\mu i} - \tilde{\varphi}_{qq'}^{\lambda\mu i}\tilde{\varphi}_{q_2q_2'}^{\lambda\mu i}) = \delta_{qq_2}\delta_{q'q_2'} + \delta_{qq_2'}\delta_{q_2q'} \\
&\sum_i(\bar{\psi}_{qq'}^{\lambda\mu i}\bar{\psi}_{q_2q_2'}^{\lambda\mu i} - \bar{\varphi}_{qq'}^{\lambda\mu i}\bar{\varphi}_{q_2q_2'}^{\lambda\mu i}) = \delta_{qq_2}\delta_{q'q_2'} - \delta_{qq_2'}\delta_{q'q_2} \\
&\sum_i(\psi_{qq'}^{\lambda\mu i}\varphi_{q_2q_2'}^{\lambda\mu i} - \psi_{q_2q_2'}^{\lambda\mu i}\varphi_{qq'}^{\lambda\mu i}) = 0
\end{aligned} \tag{2.39$'$}$$

where $\mu = |K - K'|$ for $\widetilde{\psi}_{qq'}^{\lambda\mu i}$ and $\mu = K + K'$ for $\bar{\psi}_{qq'}^{\lambda\mu i}$.

The ground state wave function is defined as the phonon vacuum

$$Q_{\lambda\mu i\sigma}\Psi_0 = 0.$$

It was shown in [6] that the wave function Ψ_0 equals, up to a normalization factor,

$$\Psi_0 \propto \exp\left\{\frac{1}{4}\sum_{q_2 q_2'}(\psi_{qq'}^{\lambda\mu i})^{-1}\varphi_{q_2 q_2'}^{\lambda\mu i}A^+(qq';\mu\sigma)A^+(q_2q_2';\mu-\sigma)\right\}\Psi_{00}.$$

It contains a quasiparticle-free, four-quasiparticle, and some other terms but does not include two-quasiparticle, six-quasiparticle and other similar terms.

We can now find the secular equation for determining the energies of one-phonon electric-type states. We will neglect the contribution of spin-multipole interactions, that is, we neglect H_{SE}^{ph}. We will consider now the states with $K^\pi \neq 0^+$; correspondingly, we take from (2.5) only one term $\sum_q \epsilon_q B(q)$ and from (2.27), only the term H_{M1}^{ph}, since the contribution $H_{\mathrm{M2}}^{\mathrm{ph}}$ in the one-phonon approximation is zero while H_{M3}^{ph} gives a non-coherent contribution which we will evaluate later. We thus make use of (2.39) and rewrite the Hamiltonian in the form

$$\begin{aligned}
H_M^{(1)} = \sum_q \epsilon_q B(q) &- \frac{1}{8}\sum_{\lambda\mu}[\kappa_0^{(\lambda\mu)} + \kappa_1^{(\lambda\mu)}(\boldsymbol{\tau}^{(1)}\boldsymbol{\tau}^{(2)})] \\
&\times \sum_{ii'\sigma}\sum_{q_1q_2q_1'q_2'} f^{\lambda\mu}(q_1q_2)u_{q_1q_2}^{(+)}f^{\lambda\mu}(q_1'q_2') \\
&\times u_{q_1'q_2'}^{(+)}g_{q_1q_2}^{\lambda\mu i}g_{q_1'q_2'}^{\lambda\mu i'}(Q_{\lambda\mu i\sigma}^+ + Q_{\lambda\mu i-\sigma}) \\
&\times (Q_{\lambda\mu i'-\sigma}^+ + Q_{\lambda\mu i'\sigma})
\end{aligned} \tag{2.40}$$

where $g_{qq'}^{\lambda\mu i} = \psi_{qq'}^{\lambda\mu i} + \varphi_{qq'}^{\lambda\mu i}$; $w_{qq'}^{\lambda\mu i} = \psi_{qq'}^{\lambda\mu i} - \varphi_{qq'}^{\lambda\mu i}$.

In what follows, we deal with the summation $\sum_\tau$ over the neutron (n) and the proton (p) levels, so that it is convenient to introduce the notation $\tau = \{n, p\}$. The replacement $\tau \to -\tau$ is equivalent to the replacement $n \to p$:

$$\sum_\tau A(\tau) = A(n) + A(p)$$

$$\sum_\tau A(\tau)B(-\tau) = A(p)B(n) + A(n)B(p)$$

$$\sum_{\tau\rho\pm 1} A(\tau)B(\rho\tau) = \sum_\tau [A(\tau)B(\tau) + A(\tau)B(-\tau)].$$

Since

$$(\boldsymbol{\tau}^{(1)}\boldsymbol{\tau}^{(2)}) = \tau_z^{(1)}\tau_z^{(2)} + \frac{1}{2}(\tau_+^{(1)}\tau_-^{(2)} + \tau_-^{(1)}\tau_+^{(2)})$$

we can retain only the term $\tau_z^{(1)}\tau_z^{(2)}$ (neglecting the charge-exchange phonons). Let us introduce the function

$$D_{\rho\tau}^{\lambda\mu i} = \sum_{qq'}{}^{\rho\tau} f^{\lambda\mu}(qq')u_{qq'}^{(+)}g_{qq'}^{\lambda\mu i} \tag{2.41}$$

and rewrite (2.40) in the form

$$H_M^{(1)} = \sum_q \epsilon_q B(q) - \frac{1}{8}\sum_{\lambda\mu i i'}\sum_{\sigma\tau\rho\pm 1}\left(\kappa_0^{(\lambda\mu)} + \rho\kappa_1^{(\lambda\mu)}\right)D_\tau^{\lambda\mu i}$$
$$\times D_{\rho\tau}^{\lambda\mu i'}(Q_{\lambda\mu i\sigma}^+ + Q_{\lambda\mu i-\sigma})(Q_{\lambda\mu i'-\sigma}^+ + Q_{\lambda\mu i'\sigma}). \tag{2.40$'$}$$

We can calculate the mean value $H_M^{(1)}$ over one-phonon state (2.36):

$$\langle Q_{\lambda\mu i\sigma}H_M^{(1)}Q_{\lambda\mu i\sigma}^+\rangle = \frac{1}{2}\sum_{qq'}\epsilon_{qq'}[(\psi_{qq'}^{\lambda\mu i})^2 + (\varphi_{qq'}^{\lambda\mu i})^2]$$
$$- \frac{1}{4}\sum_{\tau\rho=\pm 1}(\kappa_0^{(\lambda\mu)} + \rho\kappa_1^{(\lambda\mu)})D_\tau^{\lambda\mu i}D_{\rho\tau}^{\lambda\mu i}.$$

In order to calculate the energies $\omega_{\lambda\mu i}$, we will use the variational principle in the form

$$\delta\left[\langle Q_{\lambda\mu i\sigma}H_M^{(1)}Q_{\lambda\mu i\sigma}^+\rangle - \frac{\omega_{\lambda\mu i}}{2}\left(\sum_{qq'}g_{qq'}^{\lambda\mu i}w_{qq'}^{\lambda\mu i} - 2\right)\right] = 0$$

in which the normalization condition (2.38) is expressed though $g_{qq'}^{\lambda\mu i}$ and $w_{qq'}^{\lambda\mu i}$; $\omega_{\lambda\mu i}$ acts as a Lagrange factor and the variations $\delta g_{qq'}^{\lambda\mu i}$ and $\delta w_{qq'}^{\lambda\mu i}$ are treated as independent variations. Let us carry out the variations over $\delta g_{qq'}^{\lambda\mu i}$ and $\delta w_{qq'}^{\lambda\mu i}$; this gives us the equations

$$\epsilon_{qq'}g_{qq'}^{\lambda\mu i} - \omega_{\lambda\mu i}w_{qq'}^{\lambda\mu i} - f^{\lambda\mu}(qq')u_{qq'}^{(+)}[(\kappa_0^{\lambda\mu} + \kappa_1^{\lambda\mu})D_\tau^{\lambda\mu i}$$
$$+ (\kappa_0^{(\lambda\mu)} - \kappa_1^{(\lambda\mu)})D_{-\tau}^{\lambda\mu i}] = 0$$
$$w_{qq'}^{\lambda\mu i} = (\frac{\omega_{\lambda\mu i}}{\epsilon_{qq'}})g_{qq'}^{\lambda\mu i}.$$

Having found $g_{qq'}^{\lambda\mu i}$ from this expression and substituting it into (2.41), we arrive at the secular equation

$$\mathcal{F}^{M}_{\lambda\mu}(\omega_i) = \begin{vmatrix} (\kappa_0^{(\lambda\mu)} + \kappa_1^{(\lambda\mu)})X^{\lambda\mu i}(\mathrm{n}) - 1 & (\kappa_0^{(\lambda\mu)} - \kappa_1^{(\lambda\mu)})X^{\lambda\mu i}(n) \\ (\kappa_0^{(\lambda\mu)} - \kappa_1^{(\lambda\mu)})X^{\lambda\mu i}(p) & (\kappa_0^{(\lambda\mu)} + \kappa_1^{(\lambda\mu)})X^{\lambda\mu i}(p) - 1 \end{vmatrix} = 0 \tag{2.42}$$

where

$$X^{\lambda\mu i}(\tau) = (1+\delta_{\mu 0}) \sum_{qq'}{}^{\tau} \frac{[f^{\lambda\mu}(qq')u^{(+)}_{qq'}]^2 \epsilon_{qq'}}{\epsilon^2_{qq'} - \omega^2_{\lambda\mu i}}. \tag{2.42'}$$

The factor $1+\delta_{\mu 0}$ is introduced to describe the states $K^{\pi} = 0^-$.

In order to find $\psi^{\lambda\mu i}_{qq'}$ and $\varphi^{\lambda\mu i}_{qq'}$, we make use of normalization condition (2.38); calculations now give (see [64, 65]):

$$\begin{aligned} \psi^{\lambda\mu i}_{qq'}(\tau) &= \frac{1}{(2\mathcal{Y}^{\lambda\mu i}_{\tau})^{\frac{1}{2}}} \frac{f^{\lambda\mu}(qq')u^{(+)}_{qq'}}{\epsilon_{qq'} - \omega_{\lambda\mu i}} \\ \varphi^{\lambda\mu i}_{qq'}(\tau) &= \frac{1}{(2\mathcal{Y}^{\lambda\mu i}_{\tau})^{\frac{1}{2}}} \frac{f^{\lambda\mu}(qq')u^{(+)}_{qq'}}{\epsilon_{qq} + \omega_{\lambda\mu i}} \end{aligned} \tag{2.43}$$

or

$$\begin{aligned} g^{\lambda\mu i}_{qq'}(\tau) &= \left(\frac{2}{\mathcal{Y}^{\lambda\mu i}_{\tau}}\right)^{\frac{1}{2}} \frac{f^{\lambda\mu}(qq')u^{(+)}_{qq'}\epsilon_{qq'}}{\epsilon^2_{qq'} - \omega^2_{\lambda\mu i}} \\ w^{\lambda\mu i}_{qq'}(\tau) &= \left(\frac{2}{\mathcal{Y}^{\lambda\mu i}_{\tau}}\right)^{\frac{1}{2}} \frac{f^{\lambda\mu}(qq')u^{(+)}_{qq'}\omega_{\lambda\mu i}}{\epsilon^2_{qq} - \omega^2_{\lambda\mu i}} \end{aligned} \tag{2.43'}$$

$$\begin{aligned} \mathcal{Y}^{\lambda\mu i}_{\tau} &= Y^{\lambda\mu i}_{\tau} + (\mathbf{Y}^{\lambda\mu i}_{-\tau})^2 Y^{\lambda\mu i}_{-\tau} \\ \mathbf{Y}^{\lambda\mu i}_{\mathrm{p}} &= \frac{(\kappa_0^{(\lambda\mu)} - \kappa_1^{(\lambda\mu)})X^{\lambda\mu i}(n)}{1 - (\kappa_0^{(\lambda\mu)} + \kappa_1^{(\lambda\mu)})X^{\lambda\mu i}(p)} = \frac{1}{\mathbf{Y}^{\lambda\mu i}_{\mathrm{n}}} \\ Y^{\lambda\mu i}_{\tau} &= (1+\delta_{\mu o}) \sum_{qq'}{}^{\tau} \frac{(f^{\lambda\mu}(qq')u^{(+)}_{qq'})^2 \epsilon_{qq'}\omega_{\lambda\mu i}}{(\epsilon^2_{qq'} - \omega^2_{\lambda\mu i})^2} \\ &= \frac{1}{2} \frac{\partial}{\partial\omega_{\lambda\mu}} X^{\lambda\mu}(\tau)\Big|_{\omega_{\lambda\mu} = \omega_{\lambda\mu i}} \\ \epsilon_{qq'} &\equiv \epsilon_q + \epsilon_{q'} \end{aligned} \tag{2.44}$$

When calculating the energies and wave functions of one-phonon states of electric type, we can simultaneously take into account the multipole and spin-multipole interactions. In this case, the Hamiltonian is written in the form

$$H^{(1)}_{MS} = \sum_q \epsilon_q B(q) + H^{\rm ph}_{M1} + H^{\rm ph}_{SE1}. \tag{2.45}$$

The phonon operator now becomes (2.35), relations (2.39) are used, and the Hamiltonian $H^{(1)}_{MS}$ is found in terms of the phonon operators. A variational procedure is then used, and we find the secular equation and the wave function of such one-phonon states. If only isoscalar forces for $\lambda\mu = 22$ are taken into account, we can find the corresponding formulas in [6].

If we consider only isoscalar forces and set $\kappa_1^{(\lambda\mu)} = 0$, equation (2.42) simplifies to

$$\frac{1}{\kappa_0^{(\lambda\mu)}} = X^{\lambda\mu i} = \sum_{qq'} \frac{[f^{\lambda\mu}(qq')u^{(+)}_{qq'}]^2 \epsilon_{qq'}}{\epsilon^2_{qq'} - \omega^2_{\lambda\mu i}} \tag{2.46}$$

where the summation is carried out over the neutron and proton states. Let us analyse the specific features of this secular equation. The function $X^{22}(\omega)$ in ^{164}Dy is plotted in figure 2.2 for states with $K^\pi = 2^+$. For the two-quasiparticle state energies $\omega_{\lambda\mu i} = \epsilon_{qq'}$, the function $X^{22}(\omega)$ has a number of poles. The points of intersection of the straight line $1/\kappa_0^{(22)}$ with $X^{22}(\omega)$ are the roots of equation (2.46). There is a single root below the first pole if

$$\kappa_0^{(\lambda\mu)} < \kappa^{\lambda\mu}_{0\,\rm crit} = \left\{ \sum_{qq'} \frac{[f^{\lambda\mu}(qq')u^{(+)}_{qq'}]^2}{\epsilon_{qq'}} \right\}^{-1}.$$

Further on, the corresponding roots are located in between the neighbouring poles.

If the energy of a one-phonon state is close to the value of the corresponding pole, the horizontal line $1/\kappa_0^{(\lambda\mu)}$ intersects the function $X^{(\lambda\mu)}(\omega)$ at nearly right angles. This corresponds to the third and fourth roots shown in figure 2.2, for a value of $1/\kappa_0^{(\lambda\mu)}$ three times greater than shown in the figure. The one-phonons are weakly collective in such cases. When the value of the first root of equation (2.46) is considerably lower than that of the first pole, the first one-phonon state is strongly collective. The line $1/\kappa_0^{(\lambda\mu)}$ then intersects the curve $X^{(\lambda\mu)}(\omega)$ at a small angle. This is exactly the case that figure 2.2 illustrates for the first root. This figure shows that a small change in the value of $1/\kappa_0^{(\lambda\mu)}$ in the vicinity of $1/\kappa^{(\lambda\mu)}_{0,\rm crit}$ substantially changes the energy ω_{221} of the first one-phonon state. If the state is very strongly collectivized and the energy ω_{221} is much less than the value corresponding to the first pole, the accuracy of the quasi-boson approximation becomes insufficient because the quasiparticle number density $\langle \alpha^+_{q\sigma}\alpha_{q\sigma}\rangle$ ceases to be a small quantity.

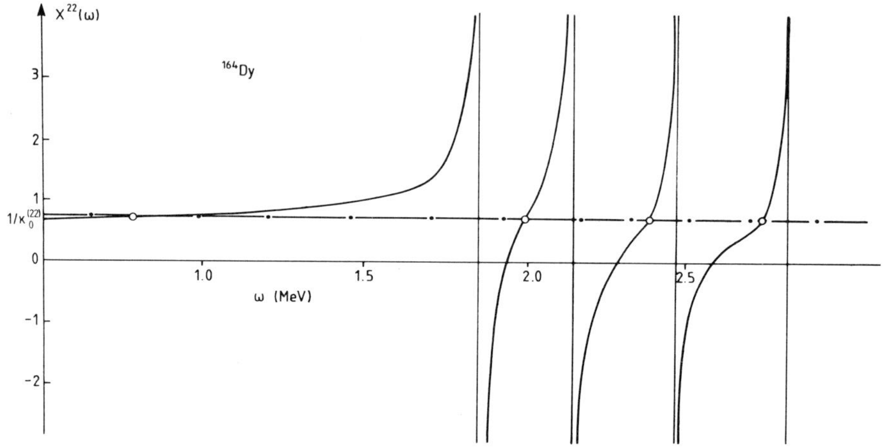

Figure 2.2 The behaviour of the right-hand side of equation (2.46) as a function of energy ω for $\lambda\mu = 22$ in ^{164}Dy: vertical lines give the poles of (2.46); open circles — the roots of (2.46) for a fixed value of $1/\kappa_0^{(22)}$.

The wave function of a one-phonon state $Q^+_{\lambda\mu i\sigma}\Psi_0$ is a sum of two-quasiparticle components having unequal weights. Table 2.1 lists components with values above 1%. The wave function of a collective state contains a large number of two-quasiparticle components. Such states are the first 2^+_1 states in deformed nuclei. As the degree of collective behaviour diminishes, the contribution of one component grows, as was the case for $\lambda\mu i = 222$ in ^{156}Gd. As we go from $\lambda\mu i = 221$ to $\lambda\mu i = 222$, $\lambda\mu i = 223$, ..., the space of two-quasiparticle states expands. The total number of one-phonon states equals the number of two-quasiparticle states.

Let us analyse the behaviour of the wave function of a one-phonon state in two limiting cases: when the root $\omega_{\lambda\mu i\sigma}$ approaches a fixed pole, and when the root $\omega_{\lambda\mu 1}$ is much smaller than the value of the first pole. Let us find the limit to which the function $Q^+_{\lambda_2\mu_2 i_2\sigma}\Psi_0$ tends when $\omega_{\lambda_2\mu_2 i_2}$ tends to the neutron pole $\epsilon_{q_1 q_2}$ (its matrix element is $f^{\lambda_2\mu_2}(q_1 q_2)$):

$$Q^+_{\lambda_2\mu_2 i_2\sigma}\Psi_0\Big|_{\epsilon_{q_1q_2}-\omega_{\lambda_2\mu_2 i_2}\to 0}$$

Table 2.1 Contribution of two-quasiparticle components to the wave functions of the states with $\lambda\mu i = 221$ and $\lambda\mu i = 222$ in ^{156}Gd and ^{176}Hf

Nucleus	Two-quasiparticle configuration	State with $K_i^\pi = 2_1^+$ $\lambda\mu i = 221$	State with $K_i^\pi = 2_2^+$ $\lambda\mu i = 222$
^{156}Gd	nn 642↑ − 660↑	0.231	0.723
	nn 402↓ + 400↑	0.150	—
	nn 651↑ + 660↑	0.132	0.046
	nn 521↑ + 521↓	0.097	0.067
	nn 633↑ − 651↑	—	0.024
	pp 411↑ + 411↓	0.085	0.045
	pp 413↓ − 411↓	0.071	0.038
^{176}Hf	nn 512↑ − 510↑	0.420	0.010
	nn 512↑ − 521↓	—	0.980
	nn 514↓ − 512↓	0.130	—
	nn 624↑ − 642↑	0.090	—

$$= \frac{1}{2}\left\{\frac{1}{\sqrt{2\mathcal{Y}_n^{\lambda_2\mu_2 i_2}}}\sum_{qq'}^{n}\left[\frac{f^{\lambda_2\mu_2}(qq')u_{qq'}^{(+)}}{\epsilon_{qq'}-\omega_{\lambda_2\mu_2 i_2}}A^+(qq';\mu_2\sigma)\right.\right.$$
$$\left.\left. - \frac{f^{\lambda_2\mu_2}(qq')u_{qq'}^{(+)}}{\epsilon_{qq'}+\omega_{\lambda_2\mu_2 i_2}}A(qq';\mu_2-\sigma)\right]\Psi_0\right\}_{\epsilon_{q_1q_2}-\omega_{\lambda_2\mu_2 i_2}\to 0}$$
$$= \frac{1}{2}\left\{\left[\frac{(f^{\lambda_2\mu_2}(qq')u_{qq'}^{(+)})^2\omega_{\lambda_2\mu_2 i_2}\epsilon_{q_1q_2}}{(\epsilon_{q_1q_2}^2-\omega_{\lambda_2\mu_2 i_2}^2)^2}\right]^{-\frac{1}{2}}\right.$$
$$\left.\times\left[\frac{f^{\lambda_2\mu_2}(q_1q_2)u_{q_1q_2}^{(+)}}{\epsilon_{q_1q_2}-\omega_{\lambda_2\mu_2 i_2}}A^+(q_1q_2;\mu_2\sigma)\right]\Psi_0\right\}_{\epsilon_{q_1q_2}-\omega_{\lambda_2\mu_2 i_2}\to 0}$$
$$= \frac{1}{2}\frac{\epsilon_{q_1q_2}+\omega_{\lambda_2\mu_2 i_2}}{(\omega_{\lambda_2\mu_2 i_2}\epsilon_{q_1q_2})^{\frac{1}{2}}}A^+(q_1q_2;\mu\sigma)\Psi_0 = A^+(q_1q_2;\mu\sigma)\Psi_{00}. \tag{2.47}$$

These calculations demonstrate that as the root approaches the pole, the wave function of a one-phonon state transforms to the wave function of a two-quasiparticle state. This is the behaviour of a state with $\lambda\mu i = 222$ in ^{176}Hf (see table 2.1). In this interpretation, collective states are considered together with two-quasiparticle states. The wave functions of one-phonon states are superpositions of two-quasiparticle states.

Let us consider another limiting case: assume the values of the first root, $\omega_{\lambda\mu 1}$, to be much less than the first pole, that is

$$\omega_{\lambda\mu 1} \ll \min \epsilon_{qq'}. \tag{2.48}$$

This case corresponds to the adiabatic limit which is the basis of the phenomenological description. Secular equation (2.50) is rewritten in the form

$$\frac{1}{\kappa_0^{(\lambda\mu)}} = \sum_{qq'} \frac{[f^{\lambda\mu}(qq')u_{qq'}^{(+)}]^2}{\epsilon_{qq'}} \frac{1}{1-\omega_{\lambda\mu 1}^2/\epsilon_{qq'}^2}$$
$$= \sum_{qq'} \frac{[f^{\lambda\mu}(qq')u_{qq'}^{(+)}]^2}{\epsilon_{qq'}} + \omega_{\lambda\mu 1}^2 \sum_{qq'} \frac{[f^{\lambda\mu}(qq')u_{qq'}^{(+)}]^2}{\epsilon_{qq'}^3} + \ldots .$$

If we only retain here the term proportional to $\omega_{\lambda\mu 1}^2$, we obtain

$$\omega_{\lambda\mu 1}^2 = \frac{(1/\kappa_0^{(\lambda\mu)}) - \sum_{qq'}(1/\epsilon_{qq'})[f^{\lambda\mu}(qq')u_{qq'}^{(+)}]^2}{\sum_{qq'}(1/\epsilon_{qq'}^3)[f^{\lambda\mu}(qq')u_{qq'}^{(+)}]^2} \tag{2.48'}$$

that is, we arrive at the adiabatic approximation in which $\omega_1 = (C/B)^{1/2}$.

The multipole and spin-multipole particle–particle (pp) interactions have the form

$$H_M^{\mathrm{pp}} + H_S^{\mathrm{pp}} = -\frac{1}{2}\sum_{\lambda\mu\sigma\tau} G^{\lambda\mu} P_{\lambda\mu\sigma}^{+}(\tau) P_{\lambda\mu\sigma}(\tau)$$
$$-\frac{1}{2}\sum_{LK\sigma\tau}\sum_{\lambda'=L,L\pm 1} G^{\lambda' LK}(P_{LK\sigma}^{\lambda'}(\tau))^{+} P_{LK\sigma}^{\lambda'}(\tau) \tag{2.49}$$

where

$$P_{\lambda\mu\sigma}(\tau) = \frac{1}{2}\sum_{q_1q_2}{}^{\tau} f^{\lambda\mu}(q_1q_2)\Big\{v_{q_1q_2}^{(+)}[A(q_1q_2;\mu\sigma)$$
$$- A^{+}(q_1q_2;\mu-\sigma)] + v_{q_1q_2}^{(-)}[A(q_1q_2;\mu\sigma)$$
$$+ A^{+}(q_1q_2;\mu-\sigma)] - 4u_{q_2}v_{q_1}B(q_1q_2;\mu-\sigma)\Big\} \tag{2.49'}$$

$$(P_{LK\sigma}^{L\pm 1}(\tau))^{+} = \frac{1}{2}\sum_{q_1q_2}{}^{\tau} f^{L\pm 1LK}(q_1q_2)\Big\{v_{q_1q_2}^{(+)}[\mathcal{A}^{+}(q_1q_2;K\sigma)$$
$$+ \mathcal{A}(q_1q_2;K-\sigma)] + v_{q_1q_1}^{(-)}[\mathcal{A}^{+}(q_1q_2;K\sigma) - \mathcal{A}(q_1q_2;K-\sigma)]$$
$$- 2u_{q_1}v_{q_2}\mathfrak{B}(q_1q_2;K\sigma) \quad + 2v_{q_1}u_{q_2}\mathfrak{B}(q_2q_1;K\sigma)]\Big\} \tag{2.49''}$$

$$G^{\lambda\mu} = G_0^{\lambda\mu} + G_1^{\lambda\mu}$$
$$G^{\lambda' LK} = G_0^{\lambda' LK} + G_1^{\lambda' LK}$$

and $G_0^{\lambda\mu}$, $G_0^{\lambda' LK}$ and $G_1^{\lambda\mu}$, $G_1^{\lambda' LK}$ are the isoscalar and the isovector constants of the multipole and spin-multipole pp interactions.

Let us derive the secular equation for the multipole particle–hole (ph) and particle–particle (pp) interactions. The Hamiltonian can be chosen in the form

$$H_{Mv} = \sum_{q\sigma} \epsilon_q \alpha^+_{q\sigma} \alpha_{q\sigma} - \frac{1}{4} \sum_{\lambda\mu i i'} \sum_{\sigma\tau} \Big\{ \sum_{\rho \pm 1} (\kappa_0^{(\lambda\mu} + \rho\kappa_1^{(\lambda\mu)}) D_\tau^{\lambda\mu i} D_{\rho\tau}^{\lambda\mu i'}$$
$$+ G^{\lambda\mu} [D_{g\tau}^{\lambda\mu i} D_{g\tau}^{\lambda\mu i'} + D_{w\tau}^{\lambda\mu i} D_{w\tau}^{\lambda\mu i'}] \Big\} Q^+_{\lambda\mu i\sigma} Q_{\lambda\mu i'\sigma} \tag{2.50}$$

where

$$D_{g\tau}^{\lambda\mu i} = \sum_{qq'}{}^{\tau} f^{\lambda\mu}(qq') v_{qq'}^{(-)} g_{qq'}^{\lambda\mu i}$$
$$D_{w\tau}^{\lambda\mu i} = \sum_{qq'}{}^{\tau} f^{\lambda\mu}(qq') v_{qq'}^{(+)} w_{qq'}^{\lambda\mu}. \tag{2.51}$$

Let us calculate the mean value of H_{Mv} over state (2.36); making use of the variational principle, we then obtain the equations for states with $K^\pi \neq 0^+$:

$$\epsilon_{qq'} g_{qq'}^{\lambda\mu i} - \omega_{\lambda\mu i} w_{qq'}^{\lambda\mu i} - (\kappa_0^{(\lambda\mu)} + \kappa_1^{(\lambda\mu)}) f^{\lambda\mu}(qq') D_\tau^{\lambda\mu i}$$
$$- (\kappa_0^{\lambda\mu} - \kappa_1^{\lambda\mu}) f^{\lambda\mu}(qq') u_{qq'}^{(+)} D_{-\tau}^{\lambda\mu i} - G^{\lambda\mu}(qq') v_{qq'}^{(-)} D_{g\tau}^{\lambda\mu i} = 0 \tag{2.52}$$

$$\epsilon_{qq'} w_{qq'}^{\lambda\mu i} - \omega_{\lambda\mu i} g_{qq'}^{\lambda\mu i} - G^{\lambda\mu} f^{\lambda\mu}(qq') v_{qq'}^{(+)} D_{w\tau}^{\lambda\mu i} = 0. \tag{2.52'}$$

2.1.3

The electric-type one-phonon states in spherical nuclei are described by multipole $\lambda\mu$ and spin-multipole $\lambda' LK$ interactions with $\lambda' = L$, while magnetic-type states are described by spin-multipole $\lambda' LK$ interactions with $\lambda' = L-1, L+1$ and by tensor interactions. If independent electric- and magnetic-type phonons are introduced in deformed nuclei, the number of states is doubled. For instance, consider a $K^\pi = 2^-$ state shown in figure 2.3. It can be interpreted as a one-phonon octupole state with $\lambda\mu = 32$, with enhanced E3 transition from the $I^\pi K = 3^-2$ level. This state can also be treated as a one-phonon quadrupole magnetic state with $\lambda' LK = 122$; the M2 transition from the $I^\pi K = 2^-2$ level may be enhanced.

In order to avoid this doubling, the phonon operator was introduced [69]; it has a fixed value K^π and consists of an electric and a magnetic

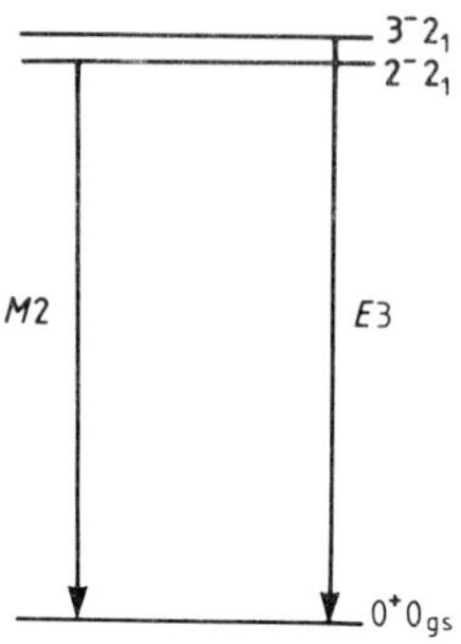

Figure 2.3 The first $K^\pi = 2^-$ state described either as a quadrupole magnetic state with enhancement of M2 transition or as an octupole electric state with enhancement of E3 transition.

component:

$$\begin{aligned}
Q^+_{Ki'_2} = \frac{1}{2\sqrt{2}} \sum_{qq'} \Big\{ & \psi^{Ki_2}_{qq'} \Big[\tilde{A}^+(qq'; K\sigma) + \chi(qq')\bar{A}^+(qq'; K\sigma) \\
& + i\chi(qq')(\tilde{\mathcal{A}}^+(qq'; K\sigma) + \bar{\mathcal{A}}^+(qq'; K\sigma) \Big] \\
& - \varphi^{Ki_2}_{qq'} \Big[\tilde{A}(qq'; K-\sigma) + \chi(qq')\bar{A}(qq'; K-\sigma) \\
& - i\chi(qq')(\tilde{\mathcal{A}}(qq'; K-\sigma) + \bar{\mathcal{A}}(qq'; K-\sigma) \Big] \Big\}. \qquad (2.53)
\end{aligned}$$

$$\begin{aligned}
\chi(qq')\mathcal{A}^+(qq'; K\sigma) &= \mathcal{A}^+(qq'; K\sigma) \\
&= -\chi(q'q)\mathcal{A}^+(qq'; K\sigma) \\
&= \chi(q'q)\mathcal{A}^+(q'q; K\sigma) \\
\chi^2(qq') &= 1.
\end{aligned}$$

Here $\psi^{ki_2}_{qq'} = \psi^{ki_2}_{q'q}$, $\varphi^{ki_2}_{qq'} = \varphi^{ki_2}_{q'q}$ are the functions that are common for the electric and magnetic components and which indicate that there exists a one-phonon state with a fixed state number i_2, where $i_2 = 1, 2, 3, \ldots$. A matrix element of (2.19′) with $\lambda' LK = L-1\, LK$ or $L+1\, LK$ is denoted by $f^{LK}(qq')$. The electric component of the phonon creation operator $Q^+_{Ki_2\sigma}$ is chosen to be real, and the magnetic component, to be imaginary.

The normalization condition of the wave function $Q^+_{Ki_2\sigma}\Psi_0$ has the form

$$\frac{1}{2}\sum_{qq'}[(\psi^{Ki_2}_{qq'})^2 - (\varphi^{Ki_2}_{qq'})^2] = 1. \qquad (2.54)$$

Now we find

$$A^+(q_1q_2;K\sigma) = \frac{1-i\sigma}{\sqrt{2}}\left(\delta_{|K_1-K_2|,K} + \chi(qq')\delta_{|K_1+K_2|,K}\right)$$
$$\times \sum_{i_2}\left[\psi^{Ki_2}_{q_1q_2}Q^+_{Ki_2\sigma} + \varphi^{Ki_2}_{q_1q_2}Q_{Ki_2-\sigma}\right] \quad (2.55)$$
$$\mathcal{A}^+(q_1q_2;K\sigma) = \frac{\sigma-i}{\sqrt{2}}\chi(q_1q_2)\sum_{i_2}\left[\psi^{Ki_2}_{q_1q_2}Q^+_{Ki_2\sigma} + \varphi^{Ki_2}_{q_1q_2}Q_{Ki_2-\sigma}\right].$$

It is not difficult to show that the phonon operators $Q_{Ki_2\sigma}$ and $Q^+_{Ki_2\sigma}$ satisfy all the conditions that are satisfied by the phonons $Q_{\lambda\mu i\sigma}$ and $Q^+_{\lambda\mu i\sigma}$.

We will now express operators (2.24), (2.25′), (2.49′) and (2.49″) in terms of the operators $Q_{Ki_0\sigma}$ and $Q^+_{Ki_0\sigma}$ and construct the Hamiltonian that includes the multipole, spin-multipole and tensor interactions. Making use of the variational principle

$$\delta\left\{\langle Q_{Ki_0\sigma}\left[\sum_{q\sigma}\epsilon_q\alpha^+_{q\sigma}\alpha_{q\sigma} + H^K_v\right]Q^+_{Ki_0\sigma}\rangle - \frac{\omega_{Ki_0}}{2}\left[\sum_{qq'}g^{Ki_0}_{qq'}w^{Ki_0}_{qq'} - 2\right]\right\} = 0 \quad (2.56)$$

we arrive at the following equations:

$$\epsilon_{q_1q_2}g^{Ki_0}_{q_1q_2} - \omega_{Ki_0}w^{Ki_0}_{q_1q_2} - \sum_{\rho=\pm1}(\kappa_0^{\lambda K} + \rho\kappa_1^{\lambda K})u^{(+)}_{q_1q_2}f^{\lambda K}_{\chi}(q_1q_2)$$
$$D^{\lambda Ki_0}_{\rho\tau} + G^{\lambda K}f^{\lambda K}_{\chi}(q_1q_2))$$
$$\times v^{(-)}_{q_1q_2}D^{\lambda Ki_0}_{g\tau} + \sum_{\lambda'=L\pm1}G^{\lambda' LK}f^{\lambda' LK}(q_1q_2)v^{(-)}_{q_1q_2}D^{\lambda' LKi_0}_{g\tau} = 0 \quad (2.57)$$

$$\epsilon_{q_1q_2}w^{Ki_0}_{q_1q_2} - \omega_{Ki_0}g^{Ki_0}_{q_1q_2} - G^{\lambda K}f^{\lambda K}_{\chi}(q_1q_2))v^{(+)}_{q_1q_2}D^{\lambda Ki_0}_{w\tau}$$
$$- \sum_{\lambda'=L\pm1}G^{\lambda' LK}f^{\lambda' LK}(q_1q_2)v^{(+)}_{q_1q_2}D^{\lambda' LKi_0}_{w\tau}$$
$$- \sum_{\rho=\pm1}\left\{\sum_{\lambda'=L\pm1}(\kappa_0^{\lambda LK} + \rho\kappa_1^{\lambda' LK})f^{\lambda' LK}(q_1q_2)\chi(q_1q_2)u^{(-)}_{q_1q_2}D^{\lambda' LKi_0}_{\rho\tau}\right.$$
$$+ (\kappa^{LK}_{T_0} + \rho\kappa^{LK}_{T_1})u^{(-)}_{q_1q_2}\chi(q_1q_2)\left[f^{L-1\,LK}(q_1q_2)D^{L+1\,LKi_0}_{\rho\tau}\right.$$
$$\left.\left. + f^{L+1\,LK}(q_1q_2)D^{L-1\,LKi_0}_{\rho\tau}\right]\right\} = 0. \quad (2.57')$$

Here

$$D^{\lambda Ki_0}_{\tau} = \sum_{q_1q_2}{}^{\tau} f^{\lambda K}_{\chi}(q_1q_2)u^{(+)}_{q_1q_2}g^{Ki_0}_{q_1q_2}$$

$$D_{g\tau}^{\lambda K i_0} = \sum_{q_1 q_2}{}^{\tau} f_\chi^{\lambda K}(q_1 q_2) v_{q_1 q_2}^{(-)} g_{q_1 q_2}^{K i_0}$$

$$D_{w\tau}^{\lambda K i_0} = \sum_{q_1 q_2}{}^{\tau} f_\chi^{\lambda K}(q_1 q_2) v_{q_1 q_2}^{(+)} w_{q_1 q_2}^{K i_0}$$

$$D_{\tau}^{L\pm 1\, L K i_0} = \sum_{q_1 q_2}{}^{\tau} f^{L\pm 1\, LK}(q_1 q_2)\chi(q_1 q_2) u_{q_1 q_2}^{(-)} w_{q_1 q_2}^{K i_0}$$

$$D_{g\tau}^{L\pm 1\, L K i_0} = \sum_{q_1 q_2}{}^{\tau} f^{L\pm 1\, LK}(q_1 q_2)\, v_{q_1 q_2}^{(-)} g_{q_1 q_2}^{K i_0}$$

$$D_{w\tau}^{L\pm 1\, L K i_0} = \sum_{q_1 q_2}{}^{\tau} f^{L\pm 1\, LK}(q_1 q_2) v_{q_1 q_2}^{(+)} w_{q_1 q_2}^{K i_0}$$

$$f_\chi^{\lambda K}(q_1 q_2) = \tilde{f}^{\lambda K}(q_1 q_2) + \chi(q_1 q_2)\bar{f}^{\lambda K}(q_1 q_2). \tag{2.58}$$

The secular equation arises as a determinant of rank $18 n_{\max}$ (provided we choose separable interactions with $n_{\max} > 1$). If we take into account only particle–hole interactions and choose $n_{\max} = 1$, the determinant rank is 6.

Restricting the analysis to particle–hole multipole λK and spin-multipole $L - 1\, LK$ interactions and using the notation $\kappa^{L-1\, LK} \equiv \kappa^{LK}$ and $f^{L-1\, LK}(qq') \equiv f^{LK}(qq')$, we transform the RPA equations to the form

$$\epsilon_{q_1 q_2} g_{q_1 q_2}^{K i_0} - \omega_{K i_0} w_{q_1 q_2}^{K i_0} - \sum_{\rho=\pm 1} (\kappa_0^{\lambda K} + \rho \kappa_1^{\lambda K}) u_{q_1 q_2}^{(+)} \times f_\chi^{\lambda K}(q_1 q_2) D_{\rho\tau}^{L K i_0} = 0 \tag{2.59}$$

$$\epsilon_{q_1 q_2} w_{q_1 q_2}^{K i_0} - \omega_{K i_0} g_{q_1 q_2}^{K i_0} - \sum_{\rho=\pm 1} (\kappa_0^{LK} + \rho \kappa_1^{LK}) \chi(q_1 q_2) \times u_{q_1 q_2}^{(-)} f^{LK}(q_1 q_2) D_{\rho\tau}^{L K i_0} = 0 \tag{2.59'}$$

$$\frac{1}{2}\sum_{q_1 q_2} g_{q_1 q_2}^{K i_0} w_{q_1 q_2}^{K i_0} = 1. \tag{2.59''}$$

The secular RPA equation now changes to

$$\begin{vmatrix} (\kappa_0^{\lambda K} + \kappa_1^{\lambda K}) Z_{\rm p}^{\lambda K i_0} - 1 & (\kappa_0^{\lambda K} - \kappa_1^{\lambda K}) Z_{\rm p}^{\lambda K i_0} & (\kappa_0^{LK} + \kappa_1^{LK}) Z_{\rm p}^{\lambda L K i_0} & (\kappa_0^{LK} - \kappa_1^{LK}) Z_{\rm p}^{\lambda L K i_0} \\ (\kappa_0^{\lambda K} - \kappa_1^{\lambda K}) Z_{\rm n}^{\lambda K i_0} & (\kappa_0^{\lambda K} + \kappa_1^{\lambda K}) Z_{\rm n}^{\lambda K i_0} - 1 & (\kappa_0^{LK} - \kappa_1^{LK}) Z_{\rm n}^{\lambda L K i_0} & (\kappa_0^{LK} + \kappa_1^{LK}) Z_{\rm n}^{\lambda L K i_0} \\ (\kappa_0^{\lambda K} + \kappa_1^{\lambda K}) Z_{\rm p}^{\lambda L K i_0} & (\kappa_0^{\lambda K} - \kappa_1^{\lambda K}) Z_{\rm p}^{\lambda L K i_0} & (\kappa_0^{LK} + \kappa_1^{LK}) Z_{\rm p}^{L K i_0} - 1 & (\kappa_0^{LK} - \kappa_1^{LK}) Z_{\rm p}^{L K i_0} \\ (\kappa_0^{\lambda K} - \kappa_1^{\lambda K}) Z_{\rm n}^{\lambda L K i_0} & (\kappa_0^{\lambda K} + \kappa_1^{\lambda K}) Z_{\rm n}^{\lambda L K i_0} & (\kappa_0^{LK} - \kappa_1^{LK}) Z_{\rm n}^{L K i_0} & (\kappa_0^{LK} + \kappa_1^{LK}) Z_{\rm n}^{L K i_0} - 1 \end{vmatrix} = 0 \tag{2.60}$$

where

$$
\begin{aligned}
Z_\tau^{\lambda K i_0} &= \sum_{qq'}^{\tau} \frac{(f^{\lambda K}(qq')u_{qq'}^{(+)})^2 \epsilon_{qq'}}{(\epsilon_{qq'}^2 - \omega_{K i_0}^2)} \\
Z_\tau^{L K i_0} &= \sum_{qq'}^{\tau} \frac{(f^{L K}(qq')u_{qq'}^{(-)})^2 \epsilon_{qq'}}{(\epsilon_{qq'}^2 - \omega_{K i_0}^2)} \\
Z_\tau^{\lambda L K i_0} &= \sum_{qq'}^{\tau} - \frac{\omega_{K i_0} \chi(q_1 q_2) u_{qq'}^{(+)} f^{LK}(qq') u_{qq'}^{(-)} f_\chi^{\lambda K}(qq')}{(\epsilon_{qq'}^2 - \omega_{K i_0}^2)}.
\end{aligned}
\tag{2.61}
$$

The secular equation (2.60) does not split into two equations for the electric and the magnetic components, since $Z_\tau^{\lambda L K i_2} \neq 0$.

2.1.4

A simpler technique, known as the Tamm–Dankoff method, is used instead of the RPA to study vibrational collective states, especially in light nuclei. Its mathematical foundation was developed by Fock for constructing quantum electrodynamics by the method of functionals. The interactions between quasiparticles are taken into consideration in the Tamm–Dankoff method only in excitation states and are ignored in the ground state; summation is carried out over diagrams shown in figure 2.1 (a). The wave function of the ground state of an even–even nucleus is a quasiparticle vacuum. The non-symmetric treatment of ground and excitation states is a shortcoming eliminated in the RPA.

We will now use the Tamm–Dankoff method to derive the secular equation and the wave functions of one-phonon states. The wave function of the ground state of an even–even nucleus is a quasiparticle vacuum $A(qq';\mu\sigma)\Psi_0^0 = 0$. The phonon operator is

$$
\bar{Q}_{\mu i \sigma} = \frac{1}{2} \sum_{qq'} \psi_{qq'}^{\mu i} A(qq';\mu\sigma). \tag{2.62}
$$

Excited one-phonon states are described here by the wave functions $\bar{Q}_{\mu i \sigma}^{+} \Psi_0^0$. Note that the Tamm–Dancoff formulas can be found from the RPA formulas if we set $\varphi_{qq'}^{\mu i} = 0$.

The secular equation will be derived by the frequently used linearization method. We choose the following form of the Hamiltonian $H^{(1)}$:

$$
H^{(1)} = \sum_q \epsilon_q B(q) - \frac{1}{8}\kappa \Big\{ \sum_{qq'} f(qq') u_{qq'}^{(+)} [A^+(qq';\mu\sigma) + A(qq';\mu\sigma)] \Big\}^2 .
$$

Let $\mathcal{E}_0$ be the ground state energy and $\mathcal{E}_0+\omega_i$ be the energy of a one-phonon state. This means that

$$H^{(1)}\Psi_0^0 = \mathcal{E}_0\Psi_0^0 \quad H^{(1)}\bar{Q}^+_{\mu i\sigma}\Psi_0^0 = (\mathcal{E}_0 + \omega_i)\bar{Q}^+_{\mu i\sigma}\Psi_0^0.$$

Subtracting

$$\bar{Q}^+_{\mu i\sigma}H^{(1)}\Psi_0^0 = \mathcal{E}_0\bar{Q}^+_{\mu i\sigma}\Psi_0^0$$

from this last equality, we find

$$(H^{(1)}\bar{Q}^+_{\mu i\sigma} - \bar{Q}^+_{\mu i\sigma}H^{(1)})\Psi_0^0 = \omega_i\bar{Q}^+_{\mu i\sigma}\Psi_0^0$$

and now can write the operator equation

$$[H^{(1)}, \bar{Q}^+_{\mu i\sigma}] = \omega_i\bar{Q}^+_{\mu i\sigma}.$$

By calculating the commutation relation and comparing the coefficients at $A^+(qq';\mu\sigma)$, we obtain the following equation

$$(\epsilon_{qq'} - \omega_i)\psi^i_{qq'} = \kappa f(qq')u^{(+)}_{qq'}\sum_{q_2q_2'} f(q_2q_2')u^{(+)}_{q_2q_2'}\psi^i_{q_2q_2'}.$$

Defining the constant D^i by

$$D^i = \kappa\sum_{q_2q_2'} f(q_2q_2')u^{(+)}_{q_2q_2'}\psi^i_{q_2q_2'}$$

we obtain

$$\psi^i_{qq'} = D^i\frac{f(qq')u^{(+)}_{qq'}}{\epsilon_{qq'} - \omega^i}.$$

The substitution of $\psi^i_{qq'}$ gives the following secular equation:

$$1 = \kappa\sum_{qq'}\frac{[f(qq')u^{(+)}_{qq'}]^2}{\epsilon_{qq'} - \omega_i}. \tag{2.63}$$

The normalization condition gives

$$\sum_{qq'}\psi^i_{qq'}\psi^{i'}_{qq'} = 2\delta_{ii'}$$

which implies

$$(D^i)^{-2} = \frac{1}{2}\sum_{qq'}\frac{[f(qq')u^{(+)}_{qq'}]^2}{(\epsilon_{qq'} - \omega_i)^2}.$$

The energies of one-phonon states are the roots $\omega_1, \omega_2, \ldots$ of secular equation (2.63). There exists one root smaller than the first pole if

$$\kappa \sum_{qq'} \frac{[f(qq')u_{qq'}^{(+)}]^2}{\epsilon_{qq'}} < 1.$$

The other roots are always sandwiched between the corresponding poles of (2.63).

The wave function of the one-phonon state is

$$\bar{Q}^+_{\mu i\sigma}\Psi_0^0 = \frac{1}{2\sqrt{D^i}} \sum_{qq'} \frac{f(qq')u_{qq'}^{(+)}}{\epsilon_{qq'} - \omega_i} A^+(qq';\mu\sigma)\Psi_0^0. \tag{2.64}$$

It is not difficult to show that if κ is small, so that ω_i is close to the pole, the solutions of equations (2.50) and (2.63) practically coincide (see [70]). When giant resonances are described, the RPA wave functions tend to the Tamm–Dankoff wave functions: $\varphi_{qq'} \ll \psi_{qq'}$, since $\epsilon_{qq'} + \omega_i \gg \epsilon_{qq'} - \omega_i$.

2.1.5

Excited $I^\pi = 0^+$ states occupy a special place in the theory of nuclei. Studying them may produce an impression that the mathematical difficulties are mostly connected with the 0^+ states. The states with $I^\pi = 0^+$ have different structures, others are related to pairing vibrations, some have the properties of the β-vibrations, still others are related to spin-quadrupole forces, two-phonon states, and so on. A considerable number of papers were devoted to methods of analysing the 0^+ states.

A description of the 0^+ states in the one-phonon approximation must exclude the spurious state; hence, we need to take into account those terms of H_0^β which are represented by formula (2.6). Let us introduce phonon operators with $\mu = 0$ and rewrite the phonon creation operators as follows:

$$Q^+_{\lambda 0i} = \frac{1}{2}\sum_{qq'}[\psi_{qq'}^{\lambda 0i} A^+(qq';\mu=0) - \varphi_{qq'}^{\lambda 0i} A(qq';\mu=0)].$$

The operators

$$A^+(qq';\mu=0) = \sum_\sigma \sigma a^+_{q\sigma} a^+_{q'-\sigma} \qquad B(qq';\mu=0) = \sum_\sigma a^+_{q\sigma} a_{q'\sigma} \tag{2.65}$$

are identical to operators (2.20) and (2.21) for $\mu = 0$; note that

$$\begin{aligned}
[A(q_1'q_2';\mu=0), A^+(q_1q_2;\mu=0)] &= 2(\delta_{q_1q_1'}\delta_{q_2q_2'} + \delta_{q_1q_2'}\delta_{q_2q_1'}) \\
&- \delta_{q_1q_1'}B(q_2q_2';\mu=0) - \delta_{q_2q_2'}B(q_1q_1';\mu=0) \\
&- \delta_{q_1q_2'}B(q_2q_1';\mu=0) - \delta_{q_2q_1'}B(q_1q_2';\mu=0).
\end{aligned}$$

Operators $M_{\lambda\sigma\mu}$, $S^L_{L\sigma K}$ [see (2.24), (2.25)] for $\mu = K = 0$ take the form

$$M_{\lambda 0} = \frac{1}{2}\sum_{qq'} \delta_{KK'} f^{\lambda 0}(qq')\Big\{u^{(+)}_{qq'}[A^+(qq';\mu=0)$$
$$+ A(qq';\mu=0)] + 2v^{(-)}_{qq'}B(qq';\mu=0)\Big\}$$
$$S^L_{L0} = \frac{1}{2}\sum_{qq'} \delta_{KK'} f^{LL0}(qq'')\Big\{u^{(-)}_{qq'}[A^+(qq';\mu=0)$$
$$- A(qq';\mu=0)] + 2v^{(-)}_{qq'}B(qq';\mu=0)\Big\}.$$

The condition of normalization of the wave function $Q^+_{\lambda 0i}\Psi_0$ is

$$\sum_{qq'}(\psi^{\lambda' 0i'}_{qq'}\psi^{\lambda 0i}_{qq'} - \varphi^{\lambda' 0i'}_{qq'}\varphi^{\lambda 0i}_{qq'}) = \delta_{\lambda i,\lambda' i'} \tag{2.66}$$

where

$$A(qq';\mu=0) = 2\sum_i(\psi^{\lambda 0i}_{qq'}Q_{\lambda 0i} + \varphi^{\lambda 0i}_{qq'}Q^+_{\lambda 0i}).$$

We now express the operators $A(qq';\mu=0)$ and $A^+(qq';\mu=0)$ in terms of the phonon operators and rewrite the Hamiltonian, taking into account the pairing and quadrupole interactions:

$$H^{(1)}_{M\beta} = \sum_q \epsilon_q B(q) - \frac{1}{2}\sum_{ii'\tau} G_\tau \sum_{qq'}[(u^2_q - v^2_q)(u^2_{q'} - v^2_{q'})g^{2\,0i}_{qq}g^{2\,0i'}_{q'q'}$$
$$+ w^{2\,0i}_{qq}w^{2\,0i'}_{q'q'}]Q^+_{2\,0i}Q_{2\,0i'} - \sum_{ii'\tau\rho=\pm 1}(\kappa^{2\,0}_0 + \rho\kappa^{2\,0}_1)$$
$$\times D^{2\,0i}_\tau D^{2\,0i'}_{\rho\tau}Q^+_{2\,0i}Q_{2\,0i'} \tag{2.67}$$

where $D^{2\,0i}_{\rho\tau}$ is defined by equation (2.41) for $\mu = 0$. We make use now of the variational principle and obtain the secular equation which coincides in form with (2.42):

$$\begin{vmatrix} (\kappa^{2\,0}_0 + \kappa^{2\,0}_1)X^{2\,0i}(n) - 1 & (\kappa^{2\,0}_0 - \kappa^{2\,0}_1)X^{2\,0i}(n) \\ (\kappa^{2\,0}_0 - \kappa^{2\,0}_1)X^{2\,0i}(p) & (\kappa^{2\,0}_0 + \kappa^{2\,0}_1)X^{2\,0i}(p) - 1 \end{vmatrix} = 0 \tag{2.68}$$

where

$$X^{2\,0i}(\tau) = 2\sum_{qq'}{}^{\tau}\frac{f^{2\,0}(qq')\rho^{2\,0i}_\tau(qq')(u^{(+)}_{qq'})^2\epsilon_{qq'}}{\epsilon^2_{qq'} - \omega^2_{2\,0i}} \tag{2.68'}$$
$$\rho^{2\,0i}_\tau(qq') = f^{2\,0}(qq') - \delta_{qq'}\rho^{2\,0i}_\tau(q)/\gamma^{2\,0i}_\tau$$

$$\rho_\tau^{2\,0i}(q) = \sum_{q_2 q_2'}^{\tau} \frac{f^{2\,0}(q_2 q_2')(4C_\tau^2 - \omega_{2\,0i}^2 + 4\xi_{q_2}\xi_{q_2'} - 4\xi_q\xi_{q_2} - 4\xi_q\xi_{q_2'})}{\epsilon_{q_2}(4\epsilon_{q_2}^2 - \omega_{2\,0i}^2)\epsilon_{q_2'}(4\epsilon_{q_2'}^2 - \omega_{2\,0i}^2)}$$

$$\gamma_\tau^{2\,0i} = \sum_{qq'}^{\tau} \frac{4C_\tau^2 - \omega_{2\,0i}^2 + 4\xi_q\xi_{q'}}{\epsilon_q(4\epsilon_q^2 - \omega_{2\,0i}^2)\,\epsilon_{q'}(4\epsilon_{q'}^2 - \omega_{2\,0i}^2)}$$

$$\xi_q = E(q) - \lambda_\tau$$

Making use of the normalization condition for the one-phonon function, we obtain, after cumbersome manipulations [64],

$$\begin{aligned}
\psi_{qq'}^{2\,0i}(\tau) &= \frac{1}{\sqrt{2\mathcal{Y}_\tau^{2\,0i}}}\left[\frac{\rho_\tau^{2\,0i}(qq')u_{qq'}^{(+)}}{\epsilon_{qq'} - \omega_{2\,0i}} - \delta_{qq'}\frac{C_\tau \Xi_\tau^{2\,0i}}{\epsilon_q\omega_{2\,0i}\gamma_\tau^{2\,0i}}\right] \\
\varphi_{qq'}^{2\,0i}(\tau) &= \frac{1}{\sqrt{2\mathcal{Y}_\tau^{2\,0i}}}\left[\frac{\rho_\tau^{2\,0i}(qq')u_{qq'}^{(+)}}{\epsilon_{qq'} + \omega_{2\,0i}} + \delta_{qq'}\frac{C_\tau \Xi_\tau^{2\,0i}}{\epsilon_q\omega_{2\,0i}\gamma_\tau^{2\,0i}}\right]
\end{aligned} \qquad (2.69)$$

or

$$\begin{aligned}
g_{qq'}^{2\,0i}(\tau) &= \frac{\sqrt{2}}{\sqrt{\mathcal{Y}_\tau^{2\,0i}}}\frac{\rho_\tau^{2\,0i}(qq')u_{qq'}^{(+)}\epsilon_{qq'}}{\epsilon_{qq'}^2 - \omega_{2\,0i}^2} \\
w_{qq'}^{2\,0i}(\tau) &= \frac{\sqrt{2}}{\sqrt{\mathcal{Y}_\tau^{2\,0i}}}\left[\frac{\rho_\tau^{2\,0i}(qq')u_{qq'}^{(+)}\omega_{2\,0i}}{\epsilon_{qq'}^2 - \omega_{2\,0i}^2} - \delta_{qq'}\frac{C_\tau \Xi_\tau^{2\,0i}}{\epsilon_q\omega_{2\,0i}\gamma_\tau^{2\,0i}}\right]
\end{aligned} \qquad (2.69')$$

where

$$\begin{aligned}
\mathcal{Y}_\tau^{2\,0i} &= Y_\tau^{2\,0i} + (\mathbf{Y}_{-\tau}^{2\,0i})^2 Y_{-\tau}^{2\,0i} \\
\Xi_\tau^{2\,0i} &= \sum_{qq'}^{\tau} \frac{f_\tau^{2\,0}(qq')(4C_\tau^2 - \omega_{2\,0i}^2 + 4\xi_q\xi_{q'})}{\epsilon_q(4\epsilon_q^2 - \omega_{2\,0i}^2)\epsilon_{q'}(4\epsilon_{q'}^2 - \omega_{2\,0i}^2)} \\
Y_\tau^{2\,0i} &= 2\sum_{qq'}^{\tau} \frac{(\rho_\tau^{2\,0i}(qq')u_{qq'}^{(+)})^2\epsilon_{qq'}\omega_{2\,0i}}{(\epsilon_{qq'}^2 - \omega_{2\,0i}^2)^2}.
\end{aligned} \qquad (2.69'')$$

The quantity $\mathbf{Y}_\tau^{2\,0i}$ was defined in (2.44). If we set $\kappa_1^{2\,0} = 0$, then $\mathcal{Y}_\tau^{2\,0i}$ in (2.69) and (2.69′) must be replaced by $Y_\tau^{2\,0i}$.

Let us analyse secular equation (2.68). To simplify the analysis, we assume that $\kappa_1^{(2\,0)} = 0$ and rewrite equation (2.68):

$$\frac{1}{\kappa_0^{(2\,0)}} = \sum_\tau X^{2\,0i}(\tau) = 2\sum_{qq'} \frac{[f^{2\,0}(qq')u_{qq'}^{(+)}]^2\epsilon_{qq'}}{\epsilon_{qq'}^2 - \omega_{2\,0i}^2}$$
$$-\sum_\tau \frac{2}{\gamma_\tau^{2\,0i}} \sum_q \frac{f_\tau^{2\,0}(qq')\rho_\tau^{2\,0i}(u_{qq'}^{(+)})^2 2\epsilon_q}{4\epsilon_q^2 - \omega_{2\,0i}^2}. \qquad (2.70)$$

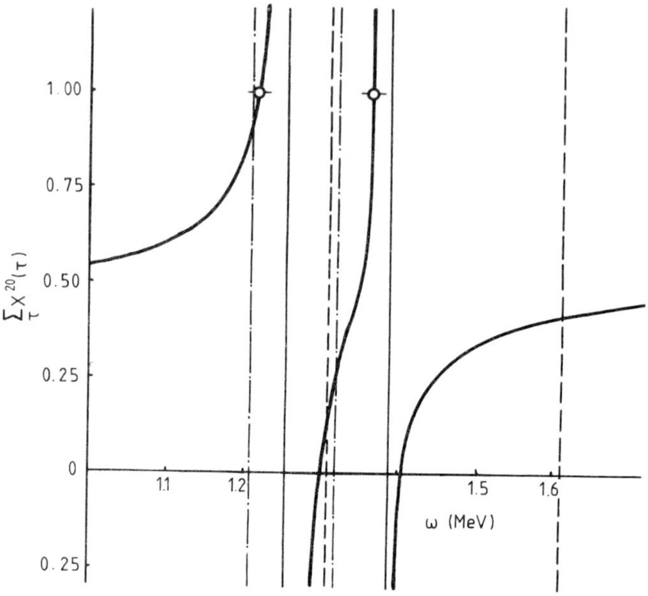

Figure 2.4 The right-hand side of equation (2.70) as a function of energy for 0^+ states in ^{178}Hf: solid vertical lines are the poles given by (2.70′); dashed and dash–dot lines mark the energies of two-quasiparticle neutron and proton states.

The functions $X^{2\,0i}(\tau)$ are regular at the points $\omega_{20} = 2\epsilon_q$ and have first-order poles at

$$\begin{aligned} \omega_{2\,0} &= \epsilon_{qq'} \quad \text{for} \quad q \neq q' \\ \gamma_\tau^{2\,0i} &= 0 \quad\quad \text{for} \quad q = q'. \end{aligned} \tag{2.70′}$$

The values of these poles are independent of the constant $\kappa_0^{(20)}$. The behaviour of the function $\sum_\tau X^{(20)}(\tau)$ is plotted in figure 2.4, which shows that there is one root of secular equation (2.70) between two poles found from (2.70'). The elimination of a spurious state slightly shifts the poles with respect to the energies of the two-quasiparticle states which are the poles of equation (2.50).

Let us look at a model [61] in which all diagonal matrix elements for $\lambda = 2$ are equal, that is, $f^{20}(qq) \equiv f^{20}$. This means that $p_\tau^{2\,0i}(q) = f^{20}\gamma_\tau^{2\,0i}$, $\Xi_\tau^{2\,0i} = f^{20}\gamma_\tau^{2\,0i}$, and secular equation (2.70) becomes

$$\Big\{\sum_{q \neq q'} \frac{[f^{2\,0}(qq')u_{qq'}^{(+)}]^2 \epsilon_{qq'}}{\epsilon_{qq'}^2 - \omega_{2\,0i}^2} - \frac{1}{2\kappa_0^{(2\,0)}}\Big\}\gamma_{\rm p}^{2\,0i}\gamma_{\rm n}^{2\,0i} = 0.$$

The energies of excitation 0^+ states are independent of f^{20}. Two types of excitation 0^+ states are then present. The energies of the first-type states are determined by non-diagonal matrix elements. The states of the second type are pairing vibrational states; their energies are characterized by the H_{pair} interaction. In deformed nuclei, the dependence of the diagonal matrix elements $f^{20}(qq)$ on quantum numbers q of single-particle states results in mixing of quadrupole and pairing vibrational collective motions.

The particle–particle (pp) interactions greatly influence the description of $K^\pi = 0^+$ states. We will follow [70] to derive the relevant equations taking into account the ph and pp interactions. First we supplement Hamiltonian (2.67) by a term

$$-G^{20}\sum_\tau (D^{20i}_{g\tau}D^{20i'}_{g\tau} + D^{20i}_{w\tau}D^{20i'}_{w\tau})Q^+_{20i}Q_{20i'}. \tag{2.71}$$

Using the earlier approach, we obtain the following equations:

$$\begin{aligned}
&\tilde{\epsilon}_{qq'}g^{20i}_{qq'} - \omega_{20i}w^{20i}_{qq'} - G_\tau\delta_{qq'}(u_q^2 - v_q^2)\sum_{q_2}{}^\tau \frac{\xi_{q_2}}{\epsilon_{q_2}} g^{20i}_{q_2q_2} \\
&\quad - 2G^{20}f^{20}(qq')v^{(-)}_{qq'}D^{20i}_{g\tau} - 2(\kappa_0^{20} + \kappa_1^{20})f^{20}(qq')u^{(+)}_{qq'}D^{20i}_\tau \\
&\quad - 2(\kappa_0^{20} - \kappa_1^{20})f^{20}(qq')u^{(+)}_{qq'}D^{20i}_{-\tau} = 0
\end{aligned} \tag{2.72}$$

$$\begin{aligned}
&\tilde{\epsilon}_{qq'}w^{20i}_{qq'} - \omega_{20i}g^{20i}_{qq'} - G_\tau\delta_{qq'}\sum_{q_2}{}^\tau w^{20i}_{q_2q_1} \\
&\quad - 2G^{20}(qq')v^{(+)}_{qq'}D^{20i}_{w\tau} = 0.
\end{aligned} \tag{2.72'}$$

Now we find $g^{20i}_{qq'}$ and $w^{20i}_{qq'}$ from (2.72) and (2.72′), substitute them into D^{20i}_τ, $D^{20i}_{g\tau}$ and $D^{20i}_{w\tau}$ and into other functions, and arrive at the required set of equations.

The condition of exclusion of the spurious state with $\omega_{20i_0} = 0$ produces the following equations for the monopole and the quadrupole pairing:

$$\begin{aligned}
&1 = \frac{G_\tau}{2}\sum_q{}^\tau \frac{C_\tau + f^{20}(qq)C_{2\tau}}{C_\tau\tilde{\epsilon}_q} \\
&1 = 2G^{20}\Big\{\sum_q{}^\tau \frac{f^{20}(q)C_\tau}{2C_{2\tau}\tilde{\epsilon}_q} + \sum_{qq'}{}^\tau \frac{(f^{20}(qq')v^{(+)}_{qq'})^2}{\tilde{\epsilon}_{qq'}}\Big\} \\
&N_\tau = \sum_q{}^\tau [1 - \frac{\xi_q}{\tilde{\epsilon}_q}]
\end{aligned} \tag{2.73}$$

where

$$\begin{aligned}
&C_\tau = G_\tau\sum_q{}^\tau u_q v_q \qquad C_{2\tau} = 2G^{20}\sum_q{}^\tau f^{20}(qq)u_q v_q \\
&\tilde{\epsilon}_q = [\Delta_q^2 + \xi_q^2]^{\frac{1}{2}} \qquad \Delta_q = C_\tau + f^{20}(qq)C_{2\tau}.
\end{aligned} \tag{2.73'}$$

According to [70], the secular equation for the energies $\omega_{2\,0i}$ arises as the equality of a determinant of rank 10 to zero. Note that as the constant G^{20} grows, the poles of this secular equation change and the density of low-lying 0^+ states increases.

2.1.6

The states with $K^\pi = 0^+$ in deformed nuclei are determined by quadrupole forces and the interactions leading to pairing. Multipole interactions with $\lambda\mu = 40$, $\lambda\mu = 60$ etc can undoubtedly be added to them. The unusual feature of deformed nuclei is that their states with a definite value of K^π can be described by multipole interactions with $\lambda_1\mu_1 = KK$, $\lambda_2\mu_2 = (K+2)K$, $\lambda_3\mu_3 = (K+4)K \ldots$ and $(-1)^{\lambda_1} = \pi$. In this way, it is possible to describe, for example, the states with $K^\pi = 2^+$, taking into account the quadrupole and hexadecapole interactions.

Given below are the formulas for describing one-phonon states with $\mu = K$, simultaneously taking into account multipole interactions with $\lambda_1 = K$ and $\lambda_2 = K+2$. If $K^\pi \neq 0^+$, we write the phonon absorption operator in a form which depends on the projection of K, namely,

$$Q_{Ki\sigma} = \frac{1}{2}\sum_{qq'}[\psi^{Ki}_{qq'}A(qq';K\sigma) - \varphi^{Ki}_{qq'}A^+(qq';K-\sigma)]. \tag{2.74}$$

The wave function of a one-phonon state can be taken in the form

$$Q^+_{Ki\sigma}\Psi_0. \tag{2.74$'$}$$

The commutation relations and the normalization conditions for this wave function have a form similar to (2.37), (2.38). The following expression will be chosen for the Hamiltonian:

$$H^{(2)}_M = \sum_q \epsilon_q B(q) - \frac{1}{4}\sum_{\lambda=K,K+2}\sum_{ii'\tau\rho=\pm1}(\kappa_0^{(\lambda K)} + \rho\kappa_1^{(\lambda K)})D^{\lambda Ki}_{\tau}D^{\lambda Ki'}_{\rho\tau}Q^+_{Ki\sigma}Q_{Ki'\sigma} \tag{2.74$''$}$$

where $D^{\lambda Ki}_{\rho\tau} = \sqrt{1+\delta_{K0}}\sum^{\rho\tau}_{qq'} f^{\lambda K}(qq')\,u^{(+)}_{qq'}g^{Ki}_{qq'}$.

Let us find the mean value of (2.74′) over the state $Q^+_{Ki\sigma}\Psi_0$. After transformations of the same type as used to derive equations (2.42) and (2.60), we obtain the following secular equation for finding the energies ω_{Ki}:

$$\begin{vmatrix}
(\kappa_0^{(\lambda_1 K)}+\kappa_1^{(\lambda_1 K)})X^{\lambda_1 Ki}(n)-1 & (\kappa_0^{(\lambda_1 K)}-\kappa_1^{(\lambda_1 K)})X^{\lambda_1 Ki}(n) \\
(\kappa_0^{(\lambda_1 K)}-\kappa_1^{(\lambda_1 K)})X^{\lambda_1 Ki}(p) & (\kappa_0^{(\lambda_1 K)}+\kappa_1^{(\lambda_1 K)})X^{\lambda_1 Ki}(p)-1 \\
(\kappa_0^{(\lambda_1 K)}+\kappa_1^{(\lambda_1 K)})X^{\lambda_1\lambda_2 i}(n) & (\kappa_0^{(\lambda_1 K)}-\kappa_1^{(\lambda_1 K)})X^{\lambda_1\lambda_2 i}(n) \\
(\kappa_0^{(\lambda_1 K)}-\kappa_1^{(\lambda_1 K)})X^{\lambda_1\lambda_2 i}(p) & (\kappa_0^{(\lambda_1 K)}+\kappa_1^{(\lambda_1 K)})X^{\lambda_1\lambda_2 i}(p)
\end{vmatrix}$$

$$\begin{vmatrix} (\kappa_0^{(\lambda_2 K)}+\kappa_1^{(\lambda_2 K)})X^{\lambda_1\lambda_2 i}(n) & (\kappa_0^{(\lambda_2 K)}-\kappa_1^{(\lambda_2 K)})X^{\lambda_1\lambda_2 i}(n) \\ (\kappa_0^{(\lambda_2 K)}-\kappa_1^{(\lambda_2 K)})X^{\lambda_1\lambda_2 i}(p) & (\kappa_0^{(\lambda_2 K)}+\kappa_1^{(\lambda_2 K)})X^{\lambda_1\lambda_2 i}(p) \\ (\kappa_0^{(\lambda_2 K)}-\kappa_1^{(\lambda_2 K)})X^{\lambda_2 i}(n)-1 & (\kappa_0^{(\lambda_2 K)}-\kappa_1^{(\lambda_2 K)})X^{\lambda_2 i}(n) \\ (\kappa_0^{(\lambda_2 K)}-\kappa_1^{(\lambda_2 K)})X^{\lambda_2 i}(p) & (\kappa_0^{(\lambda_2 K)}+\kappa_1^{(\lambda_2 K)})X^{\lambda_2 i}(p)-1 \end{vmatrix} = 0 \tag{2.75}$$

where the function $X^{\lambda K i}(\tau)$ is defined by equation (2.46) in which the replacement $\omega_{\lambda\mu i} \to \omega_{Ki}$ was carried out, and

$$X^{\lambda_1\lambda_2 i}(\tau) = (1+\delta_{K0})\sum_{qq'}^{\tau} \frac{f^{\lambda_1 K}(qq')f^{\lambda_2 K}(qq')(u_q^{(+)})^2\epsilon_{qq'}}{\epsilon_{qq'}^2-\omega_{Ki}^2}.$$

The amplitudes $\psi_{qq'}^{Ki}$ and $\phi_{qq'}^{Ki}$ are found in terms of the algebraic ajunctions of determinant (2.75) in a form similar to (2.59) and (2.59′).

2.1.7

Let us derive formulas for the transition densities and probabilities of electromagnetic transitions for the excitation of one-phonon states in even–even nuclei. The wave function of a deformed nucleus can be written as follows:

$$\Psi_{M\sigma K_0 i}^{I} = \sqrt{\frac{2I+1}{16\pi}}\Big\{D_{M\sigma K_0}^{I}(\theta_e)\Psi_i(K_0^{\pi_0}\sigma) + (-)^{I+K_0}D_{M-\sigma K_0}^{I}(\theta_e)\Psi_i(K_0^{\pi_0}-\sigma)\Big\} \tag{2.76}$$

where $D_{M\sigma K_0}^{I}(\theta_e)$ are generalized spherical functions depending on Euler angles θ_e, and $\Psi_i(K_0^{\pi_0}\sigma)$ are the wave functions of one-phonon states (2.36). The formulas for the operators of the electric $E\lambda$ and magnetic $M\lambda$ transitions are

$$\begin{aligned}\mathfrak{M}(E;\lambda\mu) &= \sum_{qq'\sigma}\langle q|\Gamma(E\lambda\mu)|q'\rangle\{v_{qq'}^{(-)}B(qq';\mu\sigma) + \frac{1}{2}u_{qq'}^{(+)}[A^+(qq';\mu\sigma)+A(qq';\mu-\sigma)]\} \\ &= \sum_{qq'\sigma}\langle q|\Gamma(E\lambda\mu)|q'\rangle\{v_{qq'}^{(-)}B(qq';\mu\sigma) \\ &\quad + (1+\delta_{\mu 0})\sum_i \frac{1}{2}u_{qq'}^{(+)}g_{qq'}^{\lambda\mu i}(Q_{\lambda\mu i\sigma}^{+}+Q_{\lambda\mu i-\sigma})\}\end{aligned} \tag{2.77}$$

$$\mathfrak{M}(M;LK) = \sum_{qq'\sigma}\langle q\Gamma(MLK)|q'\rangle\{v_{qq'}^{(+)}\mathfrak{B}(qq';K\sigma) + \frac{1}{2}u_{qq'}^{(-)}[\mathcal{A}^+(qq';K\sigma)+\mathcal{A}(qq';K-\sigma)]\}$$

$$= \sum_{qq'\sigma} \langle q|\Gamma(MLK)|q'\rangle \{ v^{(+)}_{qq'} \mathfrak{B}(qq';K\sigma)$$

$$+ \sum_{i_2} \frac{1}{2} u^{(-)}_{qq'} g^{Ki_2}_{qq'} (Q^+_{Ki_2\sigma} + Q_{Ki_2-\sigma})\} \tag{2.78}$$

where

$$\Gamma_\tau(E\lambda\mu) = e^{(\lambda)}_{\text{eff}}(\tau) r^\lambda (Y_{\lambda\mu}(\Theta,\varphi) + (-)^\mu Y_{\lambda-\mu})(1+\delta_{\mu 0})^{-1} \tag{2.77$'$}$$

$$e^{(\lambda)}_{\text{eff}}(p) = e(1 + e^{(\lambda)}_{\text{p}}) \qquad e^{(\lambda)}_{\text{eff}}(n) = e e^{(\lambda)}_{\text{n}}. \tag{2.77$''$}$$

For example, for the $E1$ transitions we find

$$1 + e^{(1)}_{\text{p}} = N/A \qquad e^{(1)}_{\text{n}} = -Z/A$$

$$\Gamma(MLK) = \mu_0 \left(g^{\text{eff}}_s(\tau) \boldsymbol{s} + g^{\text{eff}}_l(\tau) \frac{2}{L+1} \boldsymbol{l} \right) \boldsymbol{\nabla} [r^L Y_{LK}(\Theta,\varphi)]$$

$$= \frac{\mu_0}{2} [L(2L+1)]^{\frac{1}{2}} r^{L-1} [g^{\text{eff}}_s(\tau) (\boldsymbol{\sigma} \mathbf{Y}_{L-1})_{LK}$$

$$+ g^{\text{eff}}_l(\tau) \frac{4}{L+1} (\boldsymbol{l} \mathbf{Y}_{L-1})_{LK}] \tag{2.78$'$}$$

$$\langle q|\Gamma(MLK)|q'\rangle = p^{LK}_{s\tau}(qq') \tag{2.78$''$}$$

where μ_0 is the nuclear magneton and g^{eff}_s is the effective g_s factor. Now we will use the notation

$$\langle |q\Gamma(E\lambda\mu)|q'\rangle = e^{(\lambda)}_{\text{eff}}(\tau) p^{\lambda\mu}(qq').$$

Let us calculate the reduced probability of the $E\lambda$ transition from the ground to the one-phonon state. If the multipolarity λ of the γ transition is equal to that of the phonon, we can drop the small term with $B(qq';\mu\sigma)$ in operator (2.77) and rewrite the reduced probability:

$$B(E\lambda; 0^+0 \to I^\pi Ki) = \langle 00\lambda\mu|IK\rangle^2 (\mathcal{M}^E_{\lambda\mu i})^2 (2 - \delta_{\mu 0}) \tag{2.79}$$

$$\mathcal{M}^E_{\lambda\mu i} = \frac{1+\delta_{\mu 0}}{2} \sum_\tau e^{(\lambda)}_{\text{eff}}(\tau) \sum_{qq'}{}^{\tau} p^{\lambda\mu}(qq') g^{\lambda\mu i}_{qq'}(\tau) u^{(+)}_{qq'}. \tag{2.79$'$}$$

If the radial dependence of multipole forces is $R_\lambda(r) = r^{(\lambda)}$, the matrix elements $p^{\lambda\mu}(qq')$ and $f^{\lambda\mu}(qq')$ coincide and equation (2.79) changes to

$$\mathcal{M}^E_{\lambda\mu i} = \frac{1+\delta_{\mu 0}}{\sqrt{2}} \sum_\tau \frac{e^{(\lambda)}_{\text{eff}}(\tau)}{\sqrt{\mathcal{Y}^{\lambda\mu i}_\tau}} \sum_{qq'}{}^{\tau} \frac{[f^{\lambda\mu}(qq') u^{(+)}_{qq'}]^2 \epsilon_{qq'}}{\epsilon^2_{qq'} - \omega^2_{\lambda\mu i}}$$

$$= \frac{1}{\sqrt{2}} \sum_\tau \frac{e^{(\lambda)}_{\text{eff}}(\tau)}{\sqrt{\mathcal{Y}^{\lambda\mu i}_\tau}} X^{\lambda\mu i}(\tau). \tag{2.79$''$}$$

We know (see [6]) that transitions to the first ($i = 1$) quadrupole and octupole one-phonon states have enhanced rates. This enhancement also occurs if $p^{\lambda\mu}(qq') \neq f^{\lambda\mu}(qq')$, since the function $R_\lambda(r) = \partial V/\partial r$ has a maximum on the surface of the nucleus. The calculated reduced probabilities of excitation of the first quadrupole and octupole one-phonon states, $B(E2)$ and $B(E3)$, give a correct description of the enhancement of these transitions.

The expression for the matrix element of the $E\lambda$ transition for one-phonon states in terms of the transition density is

$$\mathcal{M}^E_{\lambda\mu i} = \frac{(1+\delta_{\mu 0})}{2} \int_0^\infty \sum_\tau e^{(\lambda)}_{\text{eff}}(\tau)\rho^{\lambda\mu i}_\tau(r) r^{\lambda+2}\, \mathrm{d}r. \tag{2.80}$$

For one-phonon states we have

$$\rho^{\lambda\mu i}_\tau(r) = \sum_{qq'}^{\tau} \varphi^*_{q'}(r) \sum_k \frac{\delta(r-r_k)}{r_k^2} Y^*_{\lambda\mu}(r_k)\varphi_q(r) g^{\lambda\mu i}_{qq'}(\tau) u^{(+)}_{qq'} \tag{2.80$'$}$$

where summation over k signifies summation over neutrons ($\tau = n$) and over protons ($\tau = p$). Here $\varphi_q(r)$ is a one-particle wave function, for example, for the axially symmetric Saxon–Woods potential (see [71]). Excitation cross sections of one-phonon isoscalar states in the inelastic hadron scattering are given by the sum of the transition proton and neutron densities,

$$\rho^{\lambda\mu}_{\mathrm{p}}(r) + \rho^{\lambda\mu}_{\mathrm{n}}(r)$$

and those of isovector states, by the difference between them,

$$\rho^{\lambda\mu}_{\mathrm{p}}(r) - \rho^{\lambda\mu}_{\mathrm{n}}(r).$$

A multipole $E0$ transition may occur between the states of $I^\pi = 0^+$, owing to the Coulomb interaction between the nucleons of the nucleus and the electrons of the atomic shell. The most probable $E0$ process is the emission of one conversion electron. The probabilities of $E0$ transitions essentially depend on the structure of nuclear states. The matrix element of the $E0$ transition is

$$\mathcal{M}(E0) = \rho(E0) R_0^2. \tag{2.81}$$

For transitions between the ground and one-phonon 0^+ states, we find

$$\rho_i^2(E0) = \frac{1}{2R_0^4} \left| \sum_\tau e^{(20)}_{\text{eff}}(\tau) \sum_{qq'}^{\tau} \langle q|r^2|q'\rangle g^{20i}_{qq'}(\tau) u^{(+)}_{qq'} \right|^2. \tag{2.81$'$}$$

The 0^+ states are characterized by the quantity

$$X = \frac{R_0^4 \rho^2(E0)}{B(E2; 0^+0 \to 2^+0)} \tag{2.81''}$$

which relates the reduced probabilities of the $E0$ and $E2$ transitions from the same level. For a uniformly charged ellipsoid, we have $X = 4\beta_1^2$, and for the independent-particles model with coherent contribution of all protons, $X \approx 9\beta_2^2$.

If we use phonon operator (2.53) consisting of an electric and a magnetic components, the matrix element for the $E\lambda$ transition from the ground to one-phonon state becomes

$$\begin{aligned} \mathcal{M}^E_{\lambda\mu i} = \frac{1 - i\sigma}{2\sqrt{2}} \sum_\tau e^{(\lambda)}_{\text{eff}}(\tau) \sum_{qq'}{}^{\tau} p^{\lambda\mu}(qq')(\delta_{|k-k'|,\mu} \\ + \chi(qq')\delta_{k+k',\mu}) u^{(+)}_{qq'} g^{\mu i}_{qq'} \end{aligned} \tag{2.82}$$

and the matrix element $E\lambda$ for the transition between one-phonon states is

$$\begin{aligned} &\mathcal{M}^E_{\lambda\mu}(K_0^{\pi_0} i_0 \sigma_0 \to K_f^{\pi_f} i_f \sigma_f) \\ &= \sum_\tau e^{(\lambda)}_{\text{eff}} \sum_{q_1 q_2 q_3}{}^{\tau} v^{(-)}_{q_1 q_2} \\ &\times \Big[\tilde{p}^{\lambda\mu}(q_1 q_2)\delta_{\sigma_0 \sigma_f}(\psi^{K_0 i_0}_{q_2 q_3} \psi^{K_f i_f}_{q_3 q_1} - i\sigma_0 \varphi^{K_0 i_0}_{q_3 q_1} \varphi^{K_f i_f}_{q_2 q_3}) \\ &+ \bar{p}^{\lambda\mu}(q_1 q_2)\delta_{\sigma_0, -\sigma_f}(i\sigma_0 \psi^{K_0 i_0}_{q_2 q_3} \psi^{K_f i_f}_{q_3 q_1} + \varphi^{K_0 i_0}_{q_3 q_1} \varphi^{K_f i_f}_{q_2 q_3})\Big] \end{aligned} \tag{2.83}$$

The magnetic components of the operators Q^+_{Ki} constitute an imaginary term in this matrix element.

Consider now the ML transitions. To describe them, we need to operate with a phonon (2.53) contained in (2.78). The reduced probability of the ML transition from the ground to the one-phonon state is

$$B(ML; 0^+0 \to I^\pi K_0 i_0) = \langle 00LK_0 | IK_0 \rangle^2 (2 - \delta_{K_0 0}) |\mathcal{M}^M_{LK_0 i_0}|^2 \tag{2.84}$$

$$\mathcal{M}^M_{LK_0 i_0} = \frac{1 - i\sigma}{2\sqrt{2}} \sum_{qq'}{}^{\tau} p^{LK_0}_{s\tau}(qq')\chi(qq') u^{(-)}_{qq'} w^{K_0 i_0}_{qq'}. \tag{2.85}$$

The function $\omega^{K_0 i_0}_{qq'}$ above belongs to both the electric and the magnetic parts of operator (2.53).

The matrix element of the ML transition between one-phonon states is

$$\mathcal{M}^{M}_{LK}(K_0^{\pi_0} i_0\sigma_0 \to K_f^{\pi_f} i_f\sigma_f) = \sum_{q_1q_2q_3} v^{(+)}_{q_1q_2}\Big\{\bar{p}^{LK}_s(q_1q_2)\delta_{\sigma_0\sigma_f}$$
$$\times\Big[\psi^{K_0i_0}_{q_2q_3}\psi^{K_fi_f}_{q_3q_1}\sigma_0 + i\varphi^{K_0i_0}_{q_3q_1}\varphi^{K_fi_f}_{q_2q_3})\Big]$$
$$+\bar{p}^{LK}_s(qq')\delta_{\sigma_0,-\sigma_f}\Big[i\psi^{K_0i_0}_{q_2q_3}\psi^{K_fi_f}_{q_3q_1}$$
$$+\sigma_0\varphi^{K_0i_0}_{q_3q_1}\varphi^{K_fi_f}_{q_2q_3})\Big]\Big\}. \qquad (2.86)$$

It contains a product of electric and magnetic parts of operators Q^+_{Ki}, that is, it vanishes if magnetic-type interactions only are considered. This means that in contrast to spherical nuclei, magnetic states of deformed nuclei cannot be described in isolation, that is, when electric-type interactions are neglected. A consistent description of electric states and $E\lambda$ transitions can be given ignoring magnetic-type interactions. The mathematical apparatus developed in [69] is valid for describing the $M2$ and $M3$ magnetic strengths distribution and the probabilities of the $M1$, $M2$ and $E2$ transitions between excitation states.

2.1.8

A description of deformed nuclei uses an expression for the Hamiltonian which is derived by taking into account the solutions of secular equations for one-phonon states. This Hamiltonian has no free parameters or constants. First we transform the Hamiltonian terms (2.6), (2.8), (2.28) and (2.28′), taking into consideration the solutions of secular equations (2.42) and (2.68) for multipole phonons. We begin with $H_M^{(1)}$ in the form (2.40′) and $H^{(1)}_{M\beta}$, and obtain a unified description for the states with $\mu \neq 0$ and $\mu = 0$. From (2.41) and (2.43), we find

$$D^{\lambda\mu i}_{\rho\tau} = \sqrt{2}X^{\lambda\mu i}(\rho\tau)/(\mathcal{Y}^{\lambda\mu i}_{\rho\tau})^{\frac{1}{2}}$$

and

$$(1+\delta_{\mu_0})^2 \sum_{\tau\rho=\pm1}(\kappa_0^{(\lambda\mu)} + \rho\kappa_1^{(\lambda\mu)})D^{\lambda\mu i}_{\tau}D^{\lambda\mu i'}_{\rho\tau}$$
$$= \sum_{\tau}\frac{X^{\lambda\mu i}(\tau) + X^{\lambda\mu i'}(\tau)}{(\mathcal{Y}^{\lambda\mu i}_{\tau}\mathcal{Y}^{\lambda\mu i'}_{\tau})^{\frac{1}{2}}}.$$

Now we can rewrite $H_M^{(1)}$ and $H^{(1)}_{M\beta}$ in the form

$$\sum_q \epsilon_q B(q) + H_{Mv}$$

where

$$H_{Mv} = -\frac{1}{4} \sum_{\lambda\mu i i'\tau\sigma} \frac{X^{\lambda\mu i}(\tau) + X^{\lambda\mu i'}(\tau)}{(\mathcal{Y}_\tau^{\lambda\mu i}\mathcal{Y}_\tau^{\lambda\mu i'})^{\frac{1}{2}}} Q^+_{\lambda\mu i\sigma} Q_{\lambda\mu i'\sigma}. \tag{2.87}$$

It follows from (2.38) that

$$0 = \sum_{qq'} (\psi_{qq'}^{\lambda\mu i}\varphi_{qq'}^{\lambda\mu i'} - \psi_{qq'}^{\lambda\mu i'}\varphi_{qq'}^{\lambda\mu i}) = \frac{1}{2}(1+\delta_{\mu 0})(\omega_{\lambda\mu i} - \omega_{\lambda\mu i'})$$
$$\times \sum_{qq'\tau} \frac{1}{(\mathcal{Y}_\tau^{\lambda\mu i}\mathcal{Y}_\tau^{\lambda\mu i'})^{\frac{1}{2}}} \frac{[f^{\lambda\mu}(qq')u_{qq'}^{(+)}]^2 \epsilon_{qq'}}{(\epsilon_{qq'}^2 - \omega_{\lambda\mu i}^2)(\epsilon_{qq'}^2 - \omega_{\lambda\mu i'}^2)}.$$

This equality implies that

$$\langle Q_{\lambda\mu i}\Big\{\sum_q \epsilon_q B(q) + H_{Mv}\Big\} Q^+_{\lambda\mu i'}\rangle$$
$$= \frac{1}{4}(1+\delta_{\mu o})(\omega_{\lambda\mu i} + \omega_{\lambda\mu i'})^2 \sum_\tau \frac{1}{(\mathcal{Y}_\tau^{\lambda\mu i}\mathcal{Y}_\tau^{\lambda\mu i'})^{\frac{1}{2}}}$$
$$\times \sum_{qq'} \frac{[f^{\lambda\mu}(qq')u_{qq'}^{(+)}]^2 \epsilon_{qq'}}{(\epsilon_{qq'}^2 - \omega_{\lambda\mu i}^2)(\epsilon_{qq'}^2 - \omega_{\lambda\mu i'}^2)} = \delta_{ii'}\omega_{\lambda\mu i}. \tag{2.88}$$

We can show in a similar manner (see [6]) that the following conditions hold for the solutions of equations (2.45) and (2.68):

$$\langle H_M^{(1)} Q^+_{\lambda\mu i\sigma} Q^+_{\lambda\mu i'\sigma}\rangle = \langle H_{M\beta}^{(1)} Q^+_{\lambda\mu i\sigma} Q^+_{\lambda\mu i'\sigma}\rangle = 0. \tag{2.88$'$}$$

We thus approximately have

$$\sum_q \epsilon_q B(q) + H_{Mv} = \sum_{\lambda\mu i} \omega_{\lambda\mu i} Q^+_{\lambda\mu i\sigma} Q_{\lambda\mu i\sigma}. \tag{2.89}$$

In view of (2.42) and (2.68), we can transform H_{M2}^{ph} from (2.28′) to

$$H_{M_2}^{\mathrm{ph}} = -\frac{1}{4} \sum_{\lambda\mu i\sigma\tau\rho=\pm 1} (\kappa_0^{(\lambda\mu)} + \rho\kappa_1^{(\lambda\mu)}) D_{\rho\tau}^{\lambda\mu i} \sum_{qq'} f^{\lambda\mu}(qq') v_{qq'}^{(-)}$$
$$\times [(Q^+_{\lambda\mu i\sigma} + Q_{\lambda\mu i-\sigma}) B(qq';\mu-\sigma) + \text{h.c.}]$$

which gives

$$H_{Mvq} = -\sum_{\lambda\mu i\sigma\tau} \sum_{qq'}^{\tau} \Gamma_{qq'}^{\lambda\mu i}(\tau)[(Q^+_{\lambda\mu i\sigma} + Q_{\lambda\mu i-\sigma})$$
$$\times B(qq';\mu-\sigma) + \text{h.c.}] \tag{2.90}$$

where

$$\Gamma^{\lambda\mu i}_{qq'}(\tau) = \frac{1}{2\sqrt{2}} \frac{f^{\lambda\mu}(qq')v^{(-)}_{qq'}}{(\mathcal{Y}^{\lambda\mu i}_{\tau})^{\frac{1}{2}}}.$$

This term describes the interaction between quasiparticles and phonons and plays a decisive role in the description of how the structure gets more complex as the excitation energy increases.

The mathematical apparatus for the description of one-phonon states in deformed even–even nuclei is the basis for the microscopic treatment of non-rotational states at low, intermediate and high excitation energies.

2.2 ONE-PHONON STATES IN SPHERICAL NUCLEI

2.2.1

The properties of spherical nuclei are very different from those of deformed ones. The spherical symmetry makes it possible to widely apply the mathematical apparatus of the quantum theory of angular momentum. The degeneracy of each j-th subshell is $2j+1$ while the mean-field single-particle levels of deformed nuclei are twice degenerate. The quantum numbers that characterize single-particle states of spherical nuclei differ from those for deformed nuclei. These differences result in different formulas for spherical and deformed nuclei. For this reason, we give separate expositions to these two groups of nuclei.

In addition to one-phonon states which are generated by individual components of multipole or spin-multipole particle–hole interactions, more complex situations are analysed, such as simultaneous consideration of multipole and spin-multipole interactions or particle–hole and particle–particle interactions, and so forth. For brevity of the presentation, some complex cases are analysed for deformed nuclei, and others for spherical nuclei.

The description of the energies and wave functions of the first 2^+_1 and 3^-_1 collective states in spherical nuclei has been obtained in the RPA framework sufficiently long ago [6, 62, 63, 72–74]. The RPA is also used to study the giant resonances. Nevertheless, it appears to be necessary to outline the problem of one-phonon states in spherical nuclei, since it constitutes a foundation for describing not only the low-lying states but also the states at intermediate and high excitation energies.

To describe one-phonon states in spherical and deformed nuclei, the Hamiltonian is presented in the form of (2.1). In order to take into account the specifics of spherical nuclei, we choose the following form of the terms of the Hamiltonian in (2.1):

$$H_{\text{av}} + H_{\text{pair}} = \sum_{\tau}\Big\{\sum_{jm}^{\tau}{}'[E(j) - \lambda]a^+_{jm}a_{jm} - \frac{G_\tau}{4}\sum_{jj'mm'}(-)^{j-m}(-)^{j'-m'}a^+_{jm}a_{j-m}a_{j'-m'}a_{j'm'}\Big\} \quad (2.91)$$

$$H^{\text{ph}}_M = -\frac{1}{2}\sum_{\lambda}[\kappa_0^{(\lambda)} + \kappa_1^{(\lambda)}(\boldsymbol{\tau}^{(1)}\boldsymbol{\tau}^{(2)})]\sum_{\mu} M^+_{\lambda\mu}M_{\lambda\mu} \quad (2.92)$$

$$M^+_{\lambda\mu} = \sum_{jj'mm'}\langle jm|\mathrm{i}^\lambda R_\lambda(r)Y_{\lambda\mu}(\Theta\varphi)|j'm'\rangle a^+_{jm}a_{j'm'} \quad (2.92')$$

$$H^{\text{ph}}_S = -\frac{1}{2}\sum_{L\lambda=L,L\pm1}[\kappa_0^{(\lambda L)} + \kappa_1^{(\lambda L)}(\boldsymbol{\tau}^{(1)}\boldsymbol{\tau}^{(2)})]\sum_M (S^\lambda_{LM})^+ S^\lambda_{LM} \quad (2.93)$$

$$(S^\lambda_{LM})^+ = \sum_{jj'mm'}\langle jm|i^\lambda R_\lambda(r)\{\sigma Y_{\lambda\mu}(\theta\varphi)\}_{LM}|j'm'\rangle a^+_{jm}a_{j'm'} \quad (2.93')$$

$$\{\sigma Y_{\lambda\mu}(\theta\varphi)\}_{LM} = \sum_{\mu\mu'}\langle 1\mu'\lambda\mu|LM\rangle\sigma_{\mu'}Y_{\lambda\mu}(\theta,\varphi).$$

If multipole interactions in the particle–particle channel are introduced into the model, the following term H^{pp}_M is added to Hamiltonian (2.1):

$$H^{\text{pp}}_M = -\frac{1}{2}\sum_{\lambda\mu}\sum_{\tau} G^{(\lambda)}_\tau P^+_{\lambda\mu}(\tau)P_{\lambda\mu}(\tau) \quad (2.94)$$

$$P^+_{\lambda\mu}(\tau) = \sum_{jj'mm'}\langle jm|i^\lambda R_\lambda(r)Y_{\lambda\mu}(\theta,\varphi)|j'm'\rangle(-)^{j'-m'}a^+_{jm}a^+_{j'-m'}. \quad (2.94')$$

The Hamiltonian can also be supplemented by an isovector tensor interaction in the particle–hole channel:

$$H^{\text{ph}}_T = \frac{1}{2}(\boldsymbol{\tau}^{(1)}\boldsymbol{\tau}^{(2)})\sum_{LM}\kappa^{(L)}_T[(S^{L-1}_{LM})^+S^{L+1}_{LM} + (S^{L+1}_{LM})^+S^{L-1}_{LM}]. \quad (2.95)$$

The single-particle states are characterized by the quantum numbers $nljm$. If no ambiguity is caused, they are denoted by jm. In the absence of special notation, the summation over j simultaneously signifies the summation over $\tau = p$ and $\tau = n$. In (2.91)–(2.94′), $E(j)$ stand for single-particle energies, $G^{(\lambda)}_\tau$ are the constants of particle–particle interaction ($\lambda \neq 0$) and the remaining symbols have meanings identical to those of section 2.1. The matrix elements are taken over single-particle wave functions of a spherical basis and are hereafter denoted by

$$\begin{aligned} f^{\lambda}(jj') &= \langle j \| \mathrm{i}^{\lambda} R_{\lambda}(r) Y_{\lambda\mu} \| j' \rangle \\ f^{\lambda}(jj') &= (-)^{j'-j-\lambda} f^{\lambda}(j'j) \\ f^{\lambda L}(jj') &= \langle j \| \mathrm{i}^{\lambda} R_{\lambda}(r) \{\sigma Y_{\lambda\mu}\}_{LM} \| j' \rangle \\ f^{\lambda L}(jj') &= (-)^{j'+j-\lambda} f^{\lambda L}(j'j) \end{aligned}$$

Let us transform Hamiltonian (2.1). To do this, we carry out the canonical Bogoliubov transformation

$$a_{jm} = u_j \alpha_{jm} + (-)^{j-m} v_j \alpha^{+}_{j-m}$$

and introduce the operators

$$\begin{aligned} A^{+}(jj';\lambda\mu) &= \sum_{mm'} \langle jmj'm' | \lambda\mu \rangle \alpha^{+}_{jm} \alpha^{+}_{j'm'} \\ A^{+}(j) &= \frac{1}{\sqrt{2(2j+1)}} \sum_{m} (-)^{j-m} \alpha^{+}_{jm} \alpha^{+}_{j-m} \end{aligned} \tag{2.96}$$

$$\begin{aligned} B(jj';\lambda\mu) &= \sum_{mm'} (-)^{j'+m'} \langle jmj'm' | \lambda\mu \rangle \alpha^{+}_{jm} \alpha_{j'-m'} \\ B(j) &= \sum_{m} \alpha^{+}_{jm} \alpha_{jm} \quad B^{+}(jj';\lambda\mu) = (-)^{j'-j+\mu} B(j'j;\lambda-\mu). \end{aligned} \tag{2.97}$$

The operators $A(j_1 j_2;\lambda\mu)$ and $A^{+}(j_1' j_2';\lambda'\mu')$ satisfy the commutation relations

$$\begin{aligned} [A(j_1 j_2;\lambda\mu), A^{+}(j_1' j_2';\lambda'\mu')] = {} & \delta_{\lambda\mu,\lambda'\mu'} (\delta_{j_1 j_1'} \delta_{j_2 j_2'} + (-)^{j_2 - j_1 + \lambda} \\ & \times \delta_{j_1 j_2'} \delta_{j_2 j_1'}) - \sum_{m_1 m_2 m_1' m_2'} \langle j_1 m_1 j_2 m_2 | \lambda\mu \rangle \\ & \times \langle j_1' m_1' j_2' m_2' | \lambda'\mu' \rangle (\delta_{j_1 m_1, j_1' m_1'} \alpha^{+}_{j_2' m_2'} \alpha_{j_2 m_2} \\ & + \delta_{j_2' m_2', j_2 m_2} \alpha^{+}_{j_1' m_1'} \alpha_{j_1 m_1} \\ & - \delta_{j_2 m_2, j_1' m_2'} \alpha^{+}_{j_2' m_2'} \alpha_{f_1 m_1} \\ & - \delta_{j_1 m_1, j_2' m_2'} \alpha^{+}_{j_1' m_1'} \alpha_{j_2 m_2}). \end{aligned} \tag{2.98}$$

The terms H^{ph}_{M} and H^{ph}_{S} of the operator can be changed to (see [74–76])

$$H^{\mathrm{ph}}_{M} = -\frac{1}{2} \sum_{\lambda\mu} \sum_{\tau\rho=\pm 1} (\kappa_0^{(\lambda)} + \rho\kappa_1^{(\lambda)}) M^{+}_{\lambda\mu}(\tau) M_{\lambda\mu}(\rho\tau) \tag{2.99}$$

$$M^+_{\lambda\mu}(\tau) = \frac{(-)^{\lambda-\mu}}{\sqrt{2\lambda+1}}\sum_{ff'}^{\tau} f^{\lambda}(jj')\Big\{\frac{1}{2}u^{(+)}_{ff'}[A^+(jj';\lambda\mu)$$
$$+(-)^{\lambda-\mu}A(jj';\lambda-\mu)] + v^{(-)}_{jj'}B(jj';\lambda\mu)\Big\} \qquad (2.99')$$
$$M^+_{\lambda\mu}(\tau) = (-)^{\lambda-\mu}M_{\lambda\mu}(\tau)$$
$$H_S^{\rm ph} = -\frac{1}{2}\sum_{LM\lambda=L\pm1,L}\sum_{\rho\tau=\pm1}(\kappa_0^{(\lambda L)}+\rho\kappa_1^{(\lambda L)})S^{\lambda}_{LM}(\tau)^+S^{\lambda}_{LM}(\rho\tau)$$
$$(2.100)$$
$$(S^{\lambda}_{LM}(\tau))^+ = \frac{(-)^{L-M-1}}{\sqrt{2L+1}}\sum_{jj'}^{\tau} f^{\lambda L}(jj')\Big\{\frac{1}{2}u^{(-)}_{jj'}[A^+(jj';LM)$$
$$-(-)^{L-M}A(jj';L-M)] + v^{(+)}_{jj'}B(jj';L-M)\Big\}$$
$$(S^{\lambda}_{LM}(\tau))^+ = (-)^{L-M-1}S^{\lambda}_{L-M}(\tau). \qquad (2.100')$$

Now the multipole and spin-multipole effective interactions can be written in the same form as in (2.27)–(2.31) for the case of deformed nuclei.

2.2.2

We will derive RPA equations for which condition (2.33) holds and in which the terms of commutative relation (2.98) that contain $\alpha^+_{jm}\alpha_{j'm'}$ are dropped. The phonon creation operator will be

$$Q^+_{\lambda\mu i} = \frac{1}{2}\sum_{jj'}[\psi^{\lambda i}_{jj'}A^+(jj';\lambda\mu) - (-)^{\lambda-\mu}\varphi^{\lambda i}_{jj'}A(jj';\lambda-\mu)] \qquad (2.101)$$

where the projection μ assumes values from λ to $-\lambda$. The commutation relations between phonon operators in the form (2.37) impose the following conditions on the amplitudes $\psi^{\lambda i}_{jj'}$ and $\varphi^{\lambda i}_{jj'}$:

$$\frac{1}{2}\sum_{jj'}(\psi^{\lambda i}_{ijj'}\psi^{\lambda' i'}_{jj'} - \varphi^{\lambda i}_{jj'}\varphi^{\lambda' i'}_{jj'}) = \delta_{\lambda i,\lambda' i'}$$
$$\sum_{jj'}(\psi^{\lambda i}_{jj'}\varphi^{\lambda' i'}_{jj'} - \psi^{\lambda' i'}_{jj'}\psi^{\lambda i}_{jj'}) = 0. \qquad (2.102)$$

Let us derive equations for the energies and wave functions of one-phonon states, neglecting the contribution of spin-multipole interactions. Take the term $\sum_{jm}\epsilon_j\alpha^+_{jm}\alpha_{jm}$ from (2.91) and those terms of (2.99) that do not contain $B(jj';\lambda\mu)$. We use the phonon operators and rewrite the Hamiltonian,

$$H_M^{(1)} = \sum_{jm} \epsilon_j \alpha_{jm}^+ \alpha_{jm} - \frac{1}{8} \sum_{\lambda\mu ii'} \sum_{\tau\rho=\pm 1} \frac{1}{2\lambda+1}(\kappa_0^{(\lambda)} + \rho\kappa_1^{(\lambda)})$$
$$\times D_\tau^{\lambda i} D_{\rho\tau}^{\lambda i'} [Q_{\lambda-\mu i} + (-)^{\lambda-\mu} Q_{\lambda\mu i}^+][Q_{\lambda-\mu i'}^+ + (-)^{\lambda-\mu} Q_{\lambda\mu i'}] \tag{2.103}$$

where

$$D_{\rho\tau}^{\lambda i} = \sum_{jj'}^{\rho\tau} f^\lambda(jj') u_{jj'}^{(+)} g_{jj'}^{\lambda i}$$
$$g_{jj'}^{\lambda i} = \psi_{jj'}^{\lambda i} + \varphi_{jj'}^{\lambda i} \quad w_{jj'}^{\lambda i} = \psi_{jj'}^{\lambda i} - \varphi_{jj'}^{\lambda i}. \tag{2.104}$$

The mean value of $H_M^{(1)}$ over a one-phonon state is

$$\langle Q_{\lambda\mu i} H_M^{(1)} Q_{\lambda\mu i}^+ \rangle = \frac{1}{4} \sum_{jj'} \epsilon_{jj'} [(g_{jj'}^{\lambda i})^2 + (w_{jj'}^{\lambda i})^2]$$
$$- \frac{1}{4} \frac{1}{2\lambda+1} \sum_{\tau\rho=\pm 1} (\kappa_0^{(\lambda)} + \rho\kappa_1^{(\lambda)}) D_\tau^{\lambda i} D_{\rho\tau}^{\lambda i}.$$

Making use of the variational principle

$$\delta\Big\{\langle Q_{\lambda\mu i} H_M^{(1)} Q_{\lambda\mu i}^+ \rangle - \frac{\omega_{\lambda i}}{2} \Big[\sum_{jj'} g_{jj'}^{\lambda i} w_{jj'}^{\lambda i} - 2\Big]\Big\}$$

we obtain the secular equation,

$$(\kappa_0^{(\lambda)} + \kappa_1^{(\lambda)})[X^{\lambda i}(n) + X^{\lambda i}(p)] - 4\kappa_0^{(\lambda)} \kappa_1^{(\lambda)} X^{\lambda i}(n) X^{\lambda i}(p) = 1 \tag{2.105}$$

where

$$X^{\lambda i}(\tau) = \frac{1}{2\lambda+1} \sum_{jj'}^{\tau} \frac{[f^\lambda(jj') u_{jj'}^{(+)}]^2 \epsilon_{jj'}}{\epsilon_{jj'}^2 - \omega_{\lambda i}^2}. \tag{2.106}$$

We also obtain the phonon amplitudes,

$$\psi_{jj'}^{\lambda i} = \frac{1}{\sqrt{2\mathcal{Y}_\tau^{\lambda i}}} \frac{f^\lambda(jj') u_{jj'}^{(+)}}{\epsilon_{jj'} - \omega_{\lambda i}} \qquad \varphi_{jj'}^{\lambda i} = \frac{1}{\sqrt{2\mathcal{Y}_\tau^{\lambda i}}} \frac{f^\lambda(jj') u_{jj'}^{(+)}}{\epsilon_{jj'} + \omega_{\lambda i}} \tag{2.107}$$

$$g_{jj'}^{\lambda i} = \sqrt{\frac{2}{\mathcal{Y}_\tau^{\lambda i}}} \frac{f^\lambda(jj') u_{jj'}^{(+)} \epsilon_{jj'}}{\epsilon_{jj'}^2 - \omega_{\lambda i}^2} \qquad w_{jj'}^{\lambda i} = \sqrt{\frac{2}{\mathcal{Y}_\tau^{\lambda i}}} \frac{f^\lambda(jj') u_{jj'}^{(+)} \omega_{\lambda i}}{\epsilon_{jj'}^2 - \omega_{\lambda i}^2} \tag{2.107$'$}$$

$$
\begin{aligned}
Y_{\tau}^{\lambda i} &= \frac{1}{2}\frac{\partial}{\partial\omega}X^{\lambda i}(\tau)\Big|_{\omega=\omega_{\lambda i}} \\
&= \frac{1}{2\lambda+1}\sum_{jj'}^{\tau}\frac{[f^{\lambda}(jj')u_{jj'}^{(+)}]^2\epsilon_{jj'}\omega_{\lambda i}}{(\epsilon_{jj'}^2-\omega_{\lambda i}^2)^2} \\
\mathcal{Y}_{\tau}^{\lambda i} &= Y_{\tau}^{\lambda i}+(\mathbf{Y}_{-\tau}^{\lambda i})^2 Y_{-\tau}^{\lambda i} \\
\mathbf{Y}_{\mathrm{p}}^{\lambda i} &= \frac{1}{\mathbf{Y}_{\mathrm{n}}^{\lambda i}} = \frac{(\kappa_0^{(\lambda)}-\kappa_1^{(\lambda)})X^{\lambda i}(n)}{1-(\kappa_0^{(\lambda)}+\kappa_1^{(\lambda)})X^{\lambda i}(p)} \\
D_{\tau}^{\lambda i} &= (2\lambda+1)\sqrt{2}(\mathcal{Y}_{\tau}^{\lambda i})^{-\frac{1}{2}}X^{\lambda i}(\tau).
\end{aligned}
\tag{2.108}
$$

If we take into account only the isoscalar forces, secular equation (2.105) is considerably simplified,

$$
\frac{\kappa_0^{(\lambda)}}{2\lambda+1}\sum_{jj'}\frac{[f^{\lambda}(jj')u_{jj'}^{(+)}]^2\epsilon_{jj'}}{\epsilon_{jj'}^2-\omega_{\lambda i}^2}=1 \tag{2.105$'$}
$$

The specific features of its solution are almost identical to those of the solution for deformed nuclei.

The formulas given above describe a simple separable interaction. They are readily generalized to the case of a separable interaction of rank $n_{\mathrm{max}} > 1$. Let us rewrite the radial function in the form (2.13′), replace $f^{\lambda}(jj')$ with matrix elements $f_{\mathrm{n}}^{\lambda}(jj') = \langle j \parallel i^{\lambda}R_{\mathrm{n}}^{\lambda}(r)Y_{\lambda\mu} \parallel j'\rangle$, and rewrite Hamiltonian (2.103) in the form

$$
\begin{aligned}
H'^{(1)}_{M} = {} & \sum_{jm}\epsilon_j\alpha_{jm}^{+}\alpha_{jm} - \frac{1}{8}\sum_{\mu ii'}\sum_{n=1}^{n_{\max}}\sum_{\tau\rho=\pm1}\frac{1}{2\lambda+1}(\kappa_0^{(\lambda)}+\rho\kappa_1^{(\lambda)}) \\
& \times D_{n\tau}^{\lambda i}D_{n\rho\tau}^{\lambda i'}[Q_{\lambda-\mu i}+(-)^{\lambda-\mu}Q_{\lambda\mu i}^{+}][Q_{\lambda-\mu i}^{+}+(-)^{\lambda-\mu}Q_{\lambda\mu i'}]
\end{aligned}
\tag{2.103$'$}
$$

where

$$
D_{n\tau}^{\lambda i} = \sum_{jj'}^{\tau} f_{\mathrm{n}}^{\lambda}(jj')u_{jj'}^{(+)}g_{jj'}^{\lambda i}. \tag{2.104$'$}
$$

Applying the variational principle, we arrive at the following equations (see [77]):

$$
\epsilon_{jj'}g_{jj'}^{\lambda i} - \omega_{\lambda i}w_{jj'}^{\lambda i} - \frac{u_{jj'}^{(+)}}{2\lambda+1}\sum_{\rho=\pm1}\sum_{n=1}^{n_{\max}}(\kappa_0^{(\lambda)}+\rho\kappa_1^{(\lambda)})f_{\mathrm{n}}^{\lambda}(jj')D_{n\rho\tau}^{\lambda i} = 0 \tag{2.109}
$$

$$
\epsilon_{jj'}w_{jj'}^{\lambda i} - \omega_{\lambda i}g_{jj'}^{\lambda i} = 0. \tag{2.109$'$}
$$

The corresponding secular equation for $\omega_{\lambda i}$ is a rank-$2n_{\max}$ determinant equal to zero. This means that the use of a separable interaction of rank $n_{\max} > 1$ increases the rank of the determinant by a factor of $n_{\max}$ as compared with simple separable interactions. The general form of the interactions with finite-rank separable particle–hole and particle–particle multipole and spin-multipole interactions is given in [77].

2.2.3

Let us consider electric-type one-phonon states when multipole and spin-multipole interactions are also taken into account. From (2.100), we add to Hamiltonian (2.103) the terms for $L = \lambda$ of the same type as in (2.99). Then the Hamiltonian becomes

$$H_{MS}^{(1)} = \sum_{jm} \epsilon_j \alpha_{jm}^+ \alpha_{jm} - \frac{1}{8} \sum_{\lambda\mu i i'} \sum_{\rho\tau=\pm 1} \frac{1}{2\lambda+1} [(\kappa_0^{(\lambda)} + \rho\kappa_1^{(\lambda)}) \times D_\tau^{\lambda i} D_{\rho\tau}^{\lambda i'} + (\kappa_0^{(\lambda\lambda)} + \rho\kappa_1^{(\lambda\lambda)}) D_\tau^{\lambda\lambda i} D_{\rho\tau}^{\lambda\lambda i'}] \times [Q_{\lambda-\mu i'} + (-)^{\lambda-\mu} Q_{\lambda\mu i}^+][Q_{\lambda-\mu i'}^+ + (-)^{\lambda-\mu} Q_{\lambda\mu i'}] \quad (2.110)$$

$$D_{\rho\tau}^{\lambda L i} = \sum_{jj'}^{\rho\tau} f^{\lambda L}(jj') u_{jj'}^{(-)} w_{jj'}^{Li}. \quad (2.111)$$

The mean value over a one-phonon state is

$$\langle Q_{\lambda\mu i} H_{MS}^{(1)} Q_{\lambda\mu i}^+ \rangle = \frac{1}{4} \sum_{jj'} \epsilon_{jj'} [(g_{jj'}^{\lambda i})^2 + (w_{jj'}^{\lambda i})^2] - \frac{1}{4} \frac{1}{2\lambda+1} \sum_{\tau\rho=\pm 1} [(\kappa_0^{(\lambda)} + \rho\tau_1^{(\lambda)}) D_\tau^{\lambda i} D_{\rho\tau}^{\lambda i} + (\kappa_0^{(\lambda\lambda)} + \rho\tau_1^{(\lambda\lambda)}) D_\tau^{\lambda\lambda i} D_{\rho\tau}^{\lambda\lambda i}].$$

The application of the variational principle generates the secular equation [74]:

$$\begin{vmatrix} (\kappa_0^{(\lambda)} + \kappa_1^{(\lambda)}) X^{\lambda i}(n) - 1 & (\kappa_0^{(\lambda)} - \kappa_1^{(\lambda)}) X^{\lambda i}(n) & (\kappa_0^{(\lambda\lambda)} + \kappa_1^{(\lambda\lambda)}) X_{MS}^{\lambda i}(n) & (\kappa_0^{(\lambda\lambda)} - \kappa_1^{(\lambda\lambda)}) X_{MS}^{\lambda i}(n) \\ (\kappa_0^{(\lambda)} - \kappa_1^{(\lambda)}) X^{\lambda i}(p) & (\kappa_0^{(\lambda)} + \kappa_1^{(\lambda)}) X^{\lambda i}(p) - 1 & (\kappa_0^{(\lambda\lambda)} - \kappa_1^{(\lambda\lambda)}) X_{MS}^{\lambda i}(p) & (\kappa_0^{(\lambda\lambda)} + \kappa_1^{(\lambda\lambda)}) X_{MS}^{\lambda i}(p) \\ (\kappa_0^{(\lambda)} + \kappa_1^{(\lambda)}) X_{MS}^{\lambda i}(n) & (\kappa_0^{(\lambda)} - \kappa_1^{(\lambda)}) X_{MS}^{\lambda i}(n) & (\kappa_0^{(\lambda\lambda)} + \kappa_1^{(\lambda\lambda)}) X_\lambda^{\lambda i}(n) - 1 & (\kappa_0^{(\lambda\lambda)} - \kappa_1^{(\lambda\lambda)}) X_\lambda^{\lambda i}(n) \\ (\kappa_0^{(\lambda)} - \kappa_1^{(\lambda)}) X_{MS}^{\lambda i}(p) & (\kappa_0^{(\lambda)} + \kappa_1^{(\lambda)}) X_{MS}^{\lambda i}(p) & (\kappa_0^{(\lambda\lambda)} - \kappa_1^{(\lambda\lambda)}) X_\lambda^{\lambda i}(p) & (\kappa_0^{(\lambda\lambda)} + \kappa_1^{(\lambda\lambda)}) X_\lambda^{\lambda i}(p) - 1 \end{vmatrix} = 0 \quad (2.112)$$

$$
\begin{aligned}
X_L^{\lambda i}(\tau) &= \frac{1}{2\lambda+1}\sum_{jj'}^{\tau}\frac{(f^{\lambda L}(jj')u_{jj'}^{(-)})^2\epsilon_{jj'}}{\epsilon_{jj'}^2-\omega_{\lambda i}^2} \\
X_{MS}^{\lambda i}(\tau) &= \frac{1}{2\lambda+1}\sum_{jj'}^{\tau}\frac{f^{\lambda}(jj')u_{jj'}^{(+)}f^{\lambda\lambda}(jj')u_{jj'}^{(-)}\omega_{\lambda i}}{\epsilon_{jj'}^2-\omega_{\lambda i}^2}.
\end{aligned}
\tag{2.113}
$$

The phonon amplitudes, for instance, the neutron amplitudes, are

$$
\begin{aligned}
\psi_{j_n j'_n}^{\lambda i} &= \frac{\mathfrak{N}}{\epsilon_{j_n j'_n}-\omega_{\lambda i}}\Big\{f^{\lambda}(j_n j'_n)u_{j_n j'_n}^{(+)}[(\kappa_0^{(\lambda)}+\kappa_1^{(\lambda)})A_{41} \\
&+(\kappa_0^{(\lambda)}-\kappa_1^{(\lambda)})A_{42}]+f^{\lambda\lambda}(j_n j'_n)u_{j_n j'_n}^{(-)}[(\kappa_0^{(\lambda\lambda)}+\kappa_1^{(\lambda\lambda)})A_{43} \\
&+(\kappa_0^{(\lambda\lambda)}-\kappa_1^{(\lambda\lambda)})A_{44}]\Big\} \\
\varphi_{j_n j'_n}^{\lambda i} &= \frac{\epsilon_{j_n j'_n}-\omega_{\lambda i}}{\epsilon_{j_n j'_n}+\omega_{\lambda i}}\psi_{j_n j'_n}^{\lambda i}
\end{aligned}
\tag{2.114}
$$

where $\mathfrak{N}$ is a normalization coefficient found from condition (2.102) and $A_{41},\ldots,A_{44}$ are the algebraic ajunctions of determinant (2.112).

2.2.4

Let us derive RPA equations for magnetic-type one-phonon states. The Hamiltonian can be written in the form

$$
\begin{aligned}
H_S^{(1)} &= \sum_{jm}\epsilon_j a_{jm}^{+}a_{jm}-\frac{1}{2}\sum_{LM\lambda=L\pm1}\sum_{\tau\rho=\pm1}(\kappa_0^{(\lambda L)} \\
&+\rho\kappa_1^{(\lambda\lambda)})(S_{LM}^{\lambda}(\tau))^{+}S_{LM}^{\lambda}(\rho\tau).
\end{aligned}
\tag{2.115}
$$

We drop the terms with $B(jj';\lambda\mu)$ and introduce the operators of spin-multipole phonons Q_{LMi}, Q_{LMi}^{+}, obtaining the Hamiltonian

$$
\begin{aligned}
H_S^{(1)} &= \sum_{jm}\epsilon_j a_{jm}^{+}a_{jm}-\frac{1}{4}\sum_{LM\lambda=L\pm1}\frac{1}{2L+1} \\
&\times\sum_{jj'}(\kappa_0^{(\lambda L)}+\rho\kappa_1^{(\lambda L)})D_{\tau}^{\lambda Li}D_{\rho\tau}^{\lambda Li'}Q_{LMi}^{+}Q_{KMi'}
\end{aligned}
\tag{2.116}
$$

where the function $D_{\rho\tau}^{\lambda Li}$ is given by (2.111). After the mean value (2.116) over a one-phonon state, $Q_{LMi}^{+}\Psi_0$, has been calculated and variation with respect to $g_{jj'}^{Li}$ and $w_{jj'}^{Li}$ has been carried out, we obtain the secular equation [74]

$$\begin{vmatrix} (\kappa_0^{(L-1\,L)}+\kappa_1^{(L-1\,L)})X_L^{L-1\,i}(n)-1 & (\kappa_0^{(L-1\,L)}-\kappa_1^{(L-1\,L)})X_L^{L-1\,i}(n) & (\kappa_0^{(L+1\,L)}+\kappa_1^{(L+1\,L)})X_L^{L-1\,L+1\,i}(n) & (\kappa_0^{(L+1\,L)}-\kappa_1^{(L+1\,L)})X_L^{L-1\,L+1\,i}(n) \\ (\kappa_0^{(L-1\,L)}-\kappa_1^{(L-1\,L)})X_L^{L-1\,i}(p) & (\kappa_0^{(L-1\,L)}+\kappa_1^{(L-1\,L)})X_L^{L-1\,i}(p)-1 & (\kappa_0^{(L+1\,L)}-\kappa_1^{(L+1\,L)})X_L^{L-1\,L+1\,i}(p) & (\kappa_0^{(L+1\,L)}+\kappa_1^{(L+1\,L)})X_L^{L-1\,L+1\,i}(p) \\ (\kappa_0^{(L-1\,L)}+\kappa_1^{(L-1\,L)})X_L^{L-1\,L+1\,i}(n) & (\kappa_0^{(L-1\,L)}-\kappa_1^{(L-1\,L)})X_L^{L-1\,L+1\,i}(n) & (\kappa_0^{(L+1\,L)}+\kappa_1^{(L+1\,L)})X_L^{L+1\,i}(n)-1 & (\kappa_0^{(L+1\,L)}-\kappa_1^{(L+1\,L)})X_L^{L+1\,i}(n) \\ (\kappa_0^{(L-1\,L)}-\kappa_1^{(L-1\,L)})X_L^{L-1\,L+1\,i}(p) & (\kappa_0^{(L-1\,L)}+\kappa_1^{(L-1\,L)})X_L^{L-1\,L+1\,i}(p) & (\kappa_0^{(L+1\,L)}-\kappa_1^{(L+1\,L)})X_L^{L+1\,i}(p) & (\kappa_0^{(L+1\,L)}+\kappa_1^{(L+1\,L)})X_L^{L+1\,i}(p)-1 \end{vmatrix} = 0 \tag{2.117}$$

in which the function $X_L^{\lambda i}(\tau)$ is given by (2.113)

$$X_L^{L-1\,L+1\,i}(\tau) = \frac{1}{2L+1}\sum_{jj'}^{\tau} \frac{f^{L-1\,L}(jj')f^{L+1\,L}(jj')(u_{jj'}^{(-)})^2\epsilon_{jj'}}{\epsilon_{jj'}^2-\omega_{Li}^2}. \tag{2.118}$$

The phonon (neutron) amplitudes are

$$\begin{aligned} \psi_{j_n j_n'}^{Li} &= \frac{\mathfrak{N}u_{j_n j_n'}^{(-)}}{\epsilon_{j_n j_n'}-\omega_{Li}}\{f^{L-1\,L}(j_n j_n')[(\kappa_0^{(L-1\,L)} \\ &\quad +\kappa_1^{(L-1\,L)})A_{41}+(\kappa_0^{(L-1\,L)}-\kappa_1^{(L-1\,L)})A_{42}] \\ &\quad + f^{L+1\,L}(j_n j_n')[(\kappa_0^{(L+1\,L)}+\kappa_1^{(L+1\,L)})A_{43} \\ &\quad +(\kappa_0^{(L+1\,L)}-\kappa_1^{(L+1\,L)})A_{44}]\} \\ \varphi_{j_n j_n'}^{Li} &= -\frac{\epsilon_{j_n j_n'}-\omega_{Li}}{\epsilon_{j_n j_n'}+\omega_{Li}}\psi_{j_n j_n'}^{Li} \end{aligned} \tag{2.119}$$

where $A_{41},\ldots,A_{44}$ are the algebraic ajunctions of determinant (2.117).

When magnetic-type one-phonon states are studied, the terms of (2.115) containing $(S_{LM}^{L+1})^+S_{LM}^{L+1}$ are often neglected. The secular equation and the amplitudes $\psi_{jj'}^{Li}$, $\varphi_{jj'}^{Li}$ are then simplified:

$$\begin{aligned} &(\kappa_0^{(L-1\,L)}+\kappa_1^{(L-1\,L)})[X_L^{L-1\,i}(n)+X_L^{L-1\,i}(p)] \\ &\quad -4\kappa_0^{(L-1\,L)}\kappa_1^{(L-1\,L)}X_L^{L-1\,i}(n)X_L^{L-1\,i}(p)=1 \end{aligned} \tag{2.120}$$

$$\psi_{jj'}^{Li} = \frac{1}{\sqrt{2\mathcal{Y}_\tau^{Li}}} \frac{f^{L-1\,L}(jj')u_{jj'}^{(-)}}{\epsilon_{jj'} - \omega_{Li}}$$

$$\varphi_{jj'}^{Li} = -\frac{1}{\sqrt{2\mathcal{Y}_\tau^{Li}}} \frac{f^{L-1\,L}(jj')u_{jj'}^{(-)}}{\epsilon_{jj'} + \omega_{Li}} \tag{2.121}$$

$$\mathcal{Y}_\tau^{Li} = Y_\tau^{Li} + (\mathbf{Y}_{-\tau}^{Li})Y_{-\tau}^{Li}$$

$$Y_\tau^{Li} = \frac{1}{2}\frac{\partial}{\partial\omega} X_L^{L-1\,i}(\tau)\Big|_{\omega=\omega_{Li}} = -\frac{1}{2L+1} \sum_{jj'}^{\tau} \frac{[f^{L-1\,L}(jj')u_{jj'}^{(-)}]^2 \epsilon_{jj'}\omega_{Li}}{(\epsilon_{jj'}^2 - \omega_{Li}^2)^2}$$

$$\mathbf{Y}_\tau^{Li} = \frac{1 - (\kappa_0^{(L-1\,L)} + \kappa_1^{(L-1\,L)})X_L^{L-1\,i}(-\tau)}{(\kappa_0^{(L-1\,L)} - \kappa_1^{(L-1\,L)})X_L^{L-1\,i}(\tau)}.$$

It is not difficult to derive the secular equations and the wave functions for the Tamm–Dankoff method, just as we did in section 2.1.

2.2.5

Let us consider the probabilities of the γ transitions to one-phonon states and the transition densities. The operators of the electric $E\lambda$ and the magnetic $M\lambda$ transitions are given by the following familiar formulas (see, e.g., [6, 12]):

$$\begin{aligned}\mathfrak{M}(E;\lambda\mu) = {} & \frac{1}{\sqrt{2\lambda+1}} \sum_{jj'} \langle j'|\Gamma(E\lambda)|j\rangle(-)^{j+j'-\lambda+1} \\ & \times \{v_{jj'}^{(-)} B(jj';\lambda\mu) + \frac{1}{2}u_{jj'}^{(+)}[A^+(jj';\lambda\mu) \\ & + (-)^{\lambda-\mu}A(jj';\lambda-\mu)]\}\end{aligned} \tag{2.122}$$

$$\begin{aligned}\mathfrak{M}(M;\lambda\mu) = {} & \frac{1}{\sqrt{2\lambda+1}} \sum_{jj'} \langle j'|\Gamma(M\lambda)|j\rangle(-)^{j+j'-\lambda} \\ & \times \{v_{jj'}^{(+)} B(jj';\lambda\mu) + \frac{1}{2}u_{jj'}^{(-)}[A^+(jj';\lambda\mu) - \\ & (-)^{\lambda-\mu}A(jj';\lambda-\mu)]\}\end{aligned} \tag{2.123}$$

where $\Gamma(E\lambda)$ and $\Gamma(M\lambda)$ are defined by (2.77′)–(2.78″).

In terms of the phonon operators, (2.122) and (2.123) become

$$\mathfrak{M}(E;\lambda\mu) = \frac{1}{\sqrt{2\lambda+1}} \sum_{jj'} \langle j'|\Gamma(E\lambda)|j\rangle(-)^{j+j'-\lambda+1}$$

$$\times \{v^{(-)}_{jj'} B(jj';\lambda\mu) + \frac{1}{2} u^{(+)}_{jj'} \times \sum_i g^{\lambda i}_{jj'} [Q^+_{\lambda\mu i} + (-)^{\lambda-\mu} Q_{\lambda-\mu i}]\} \tag{2.122'}$$

$$\mathfrak{M}(M\lambda) = \frac{1}{\sqrt{2\lambda+1}} \sum_{jj'} \langle j' | \Gamma(M\lambda) | j \rangle (-)^{j+j'-\lambda} \times \{v^{(+)}_{jj'} B(jj';\lambda\mu) + \frac{1}{2} u^{(-)}_{jj'} \times \sum_i w^{\lambda i}_{jj'} [Q^+_{\lambda\mu i} - (-)^{\lambda-\mu} Q_{\lambda-\mu i}]\}. \tag{2.123'}$$

The difference ($w^{\lambda i}_{jj'}$ replacing $g^{\lambda i}_{jj'}$) is illusory: it is caused by a different definition of the phase in the amplitude $\varphi^{\lambda i}_{jj'}$.

If we neglect $B(jj';\lambda\mu)$-containing terms in (2.122′) and (2.123′), the reduced probabilities of γ transitions from the ground to the one-phonon states of even–even nuclei become

$$B(E\lambda; 0^+_{\rm gs} \to \lambda^\pi_i) = |\sum_\tau e^{(\lambda)}_{\rm eff}(\tau) \sum_{jj'}^{\tau} p^\lambda(jj') g^{\lambda i}_{jj'} u^{(+)}_{jj'}|^2 \tag{2.124}$$

$$B(M\lambda; 0^+_{\rm gs} \to \lambda^\pi i) = |\sum_{jj'} (-)^{j+j'-\lambda} \langle j' | \Gamma(M\lambda) | j \rangle w^{\lambda i}_{jj'} u^{(-)}_{jj'}|^2$$

where $\langle j' | \Gamma(E\lambda) | j \rangle = e^{(\lambda)}_{\rm eff}(\tau) p^\lambda(jj')$. If we set $R_\lambda(r) = r^\lambda$ in (2.92), then $p^\lambda(jj') = f^\lambda(jj')$, so that we can rewrite (2.124):

$$B(E\lambda; 0^+_{\rm gs} \to \lambda^\pi i) = 2 \left| \sum_\tau e^{(\lambda)}_{\rm eff}(\tau) \frac{X^{\lambda i}(\tau)}{\sqrt{\mathcal{Y}^{\lambda i}_\tau}} \right|^2. \tag{2.124'}$$

The electric-type one-phonon states have been well studied both theoretically and experimentally [4, 6, 74]. As an example, we will consider the octupole one-phonon excitations in ^{140}Ce. The calculations in [74] were carried out for the forces $R_3(r) = r^3$; the constants $\kappa^{(3)}_0$ and $\kappa^{(3)}_1$ were found using the experimentally determined energy of the 3^-_1 level and the ratio

$$\kappa^{(\lambda)}_1 / \kappa^{(\lambda)}_0 = -0.2\,(2\lambda + 3).$$

The single-particle basis includes all bound states and all relatively narrow quasi-bound states. The total number of two-quasiparticle states (and hence, one-phonon states as well) with $J^\pi = 3^-$ and energy up to 35 MeV is 130. Two peaks, corresponding to transitions across one and three shells are well pronounced on the density histogram of two-quasiparticle states (figure 2.5 (a)). The action of multipole forces results in the concentration

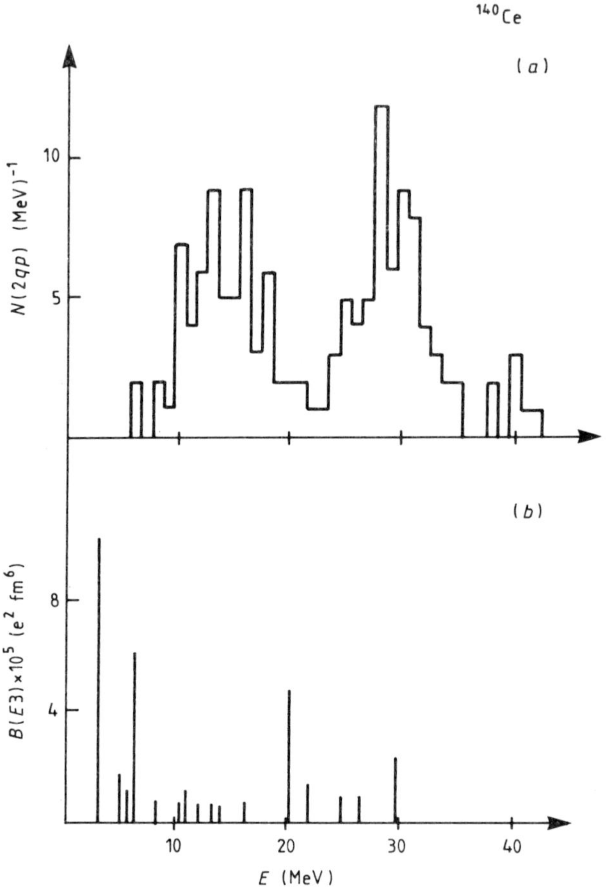

Figure 2.5 (*a*) Histogram of the number of two-quasiparticle poles with $J^\pi = 3^-$ at a 1 MeV step, and (*b*) probability distribution of $B(E3,\ 0^+_{gs} \to 3^-_i)$ transitions to one-phonon states with $\lambda^\pi_i = 3^-_i$ in ^{140}Ce.

of the E3-transition strengths on a moderate number of collective states. Figure 2.5 (*b*) clearly shows the first 3^-_1 state, the low- and high-lying isoscalar $E3$ resonances, and a high-lying isovector $E3$ resonance.

Let us consider the isotopic structure of one-phonon $E3$ states. Even though the multipole interaction in isotopically invariant, this is not so for the model Hamiltonian as a whole. In addition to the Coulomb term in the expression for the mean potential, various parameters of the Saxon–Woods potential and isotopically non-invariant multipole pairing are used for neutrons and protons. Nevertheless, the structure of one-phonon states preserves to an appreciable extent the isotopic symmetry. This is illustrated by the data in figure 2.6(a–c). The maxima on the distributions of isoscalar $E3$ transitions are precisely at the locations of the 3_1^- level and of the low- and high-lying isoscalar $E3$ resonances. It is for these states that the strength of isovector transitions is practically zero. Hence, both protons and neutrons produce coherent contributions to their structure. The situation is reversed for the isovector $E3$ resonance. At the same time, the probabilities of isoscalar and isovector transitions for collective states in the energy range $10 < E < 15$ MeV are of the same order of magnitude.

The transition densities are important characteristics of excitation states; they determine the excitation probabilities in various nuclear reactions. The isotopic structure of one-phonon states is clearly pronounced in their transition densities. The charge density $\rho(\boldsymbol{r})$, the current convective density $j^{\mathrm{c}}(\boldsymbol{r})$, and the magnetic, $j^{\mathrm{m}}(\boldsymbol{r})$, transition densities for transitions between the initial Ψ_{i} and the final Ψ_{f} states of a nucleus are given by the following formulas [74]:

$$\begin{aligned}\rho(\boldsymbol{r}) &= e\sum_k \delta(\boldsymbol{r}-\boldsymbol{r}_k)g_l^k\{\Psi_{\mathrm{f}}^*\Psi_{\mathrm{i}}\}\\ j^{\mathrm{c}}(\boldsymbol{r}) &= \mu_0\sum_k \delta(\boldsymbol{r}-\boldsymbol{r}_k)g_l^k(\Psi_{\mathrm{f}}^*\boldsymbol{\nabla}_k\Psi_{\mathrm{i}}-\Psi_{\mathrm{i}}\boldsymbol{\nabla}_k\Psi_{\mathrm{f}}^*)\\ j^{\mathrm{m}}(\boldsymbol{r}) &= \mu_0\sum_k \delta(\boldsymbol{r}-\boldsymbol{r}_k)g_s^k\boldsymbol{\nabla}_k\{\Psi_j^*\boldsymbol{\sigma}_k\Psi_i\}.\end{aligned} \tag{2.125}$$

The braces in (2.125) denote integration over the coordinates of all nucleons except the kth. The transition densities will be written as expansions in partial densities $\rho_\lambda(r)$ and $\rho_{\lambda L}(r)$:

$$\begin{aligned}\rho(\boldsymbol{r}) &= e\sum_{\lambda\mu}(-)^\lambda\langle j_{\mathrm{i}}M_{\mathrm{i}}\lambda\mu|J_{\mathrm{f}}M_{\mathrm{f}}\rangle\rho_\lambda(r)Y_{\lambda\mu}^*(\theta,\varphi)\\ \boldsymbol{j}^{\mathrm{c,m}}(\boldsymbol{r}) &= \mu_0\sum_{L\lambda\mu}(-)^L\langle j_{\mathrm{i}}M_{\mathrm{i}}\lambda\mu|j_{\mathrm{f}}M_{\mathrm{f}}\rangle\rho_{\lambda L}^{\mathrm{c,m}}(r)\\ &\quad\times\sum_{M\rho=0,\pm1}\langle LM1\rho|\lambda\mu\rangle Y_{LM}(\theta,\varphi)\boldsymbol{n}_\rho\end{aligned} \tag{2.126}$$

where $\boldsymbol{n}_\rho$ are unit vectors.

The transition charge density of the one-phonon state λ_i is

$$\begin{aligned}\rho_{\lambda i}(r) &= \sum_{j\geqslant j'}\frac{1}{1+\delta_{jj'}}\rho_{jj'}^{(\lambda)}(r)\,g_{jj'}^{\lambda i}u_{jj'}^{(+)}\\ \rho_{jj'}^{(\lambda)}(r) &= -[1+(-)^{l+l'+\lambda}](-)^{j+\lambda+\frac{1}{2}}\frac{\sqrt{(2j+1)(2j'+1)}}{\sqrt{4\pi(2\lambda+1)}}\\ &\quad\times\langle j\tfrac{1}{2}j'-\tfrac{1}{2}|\lambda 0\rangle u_j^*(r)u_{j'}(r)i^{l'-l+\lambda}\end{aligned} \tag{2.127}$$

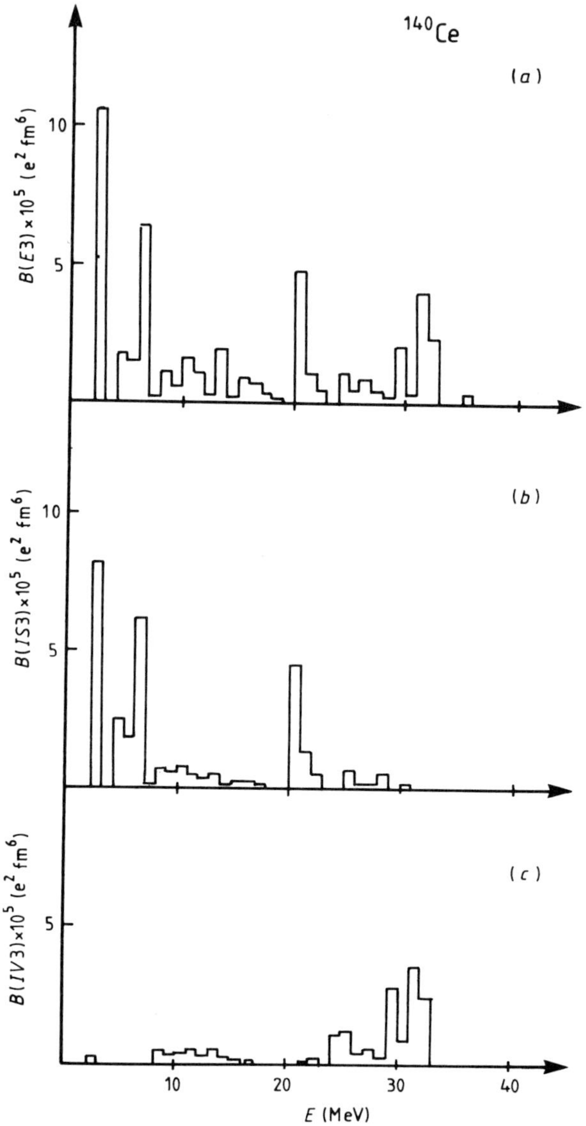

Figure 2.6 Histograms of reduced probability distributions of (a) electromagnetic, (b) isoscalar and (c) isovector $E3$ transitions in ^{140}Ce.

where $u_j(r)$ is the radial part of the single-particle wave function of the nlj state.

All charge transition densities of some collective 3^- states, with the exception of the transition density of the state with $E = 12.4\,\text{MeV}$, have a well-pronounced surface peak (figure 2.7). For the isoscalar low-lying (LEOR) and high-lying (HEOR) states, the proton and neutron charge transition densities as functions of r change phases, at least on the surface of the nucleus. The proton and neutron transition densities of the $E = 29.8\,\mathrm{MeV}$ state (isovector $E3$ resonance) are in anti-phase. States with coherent and states with incoherent changes in $\rho_{\mathrm{n}}(r)$ and $\rho_{\mathrm{p}}(r)$ are encountered among those in the range $10 < E < 15\,\mathrm{MeV}$.

What will be the change in the properties of electric-type one-phonon states if we take into account the contribution of the spin-multipole interaction to their structure? Formulas (2.114) for $\psi_{jj'}^{\lambda i}$ and $\phi_{jj'}^{\lambda i}$ show that they are renormalized, owing to the terms which contain $f^{\lambda\lambda}(jj')$. A detailed analysis, carried out for several spherical nuclei with the forces $R_\lambda(r) = \partial V(r)/\partial r$, taking into consideration only the isovector component of the spin-multipole interaction, has demonstrated that the properties of low-lying states and isoscalar resonance $E\lambda$ states do not change appreciably. Changes in the regions of isovector resonances, especially the $E1$ resonance, are more prominent. The spin-multipole forces also cause changes in the behaviour of transition densities. Note that the charge transition densities barely change while the changes in the current densities are quite considerable.

Let us look at the reduced probabilities of magnetic transitions to one-phonon states. Using the interaction containing $(S_{LM}^{L-1})^+ S_{LM}^{L-1}$ automatically produces coherent enhancement of the ML transitions from a collective magnetic state of multipolarity L; indeed, the operator S_{LM}^{L-1} coincides with the spin component of the magnetic transition operator $\mathfrak{M}(ML)$. Typical probability distributions of the $M1$ and $M2$ transitions obtained in calculations with simple spin and spin-dipole forces are shown in figure 2.8 [74]. The results were obtained for forces with the radial dependence $R_L(r) = r^{L-1}$, the effective gyromagnetic factors $g^{\mathrm{eff}} = 0.8 g_s^{\mathrm{free}}$, $g_l^{\mathrm{eff}} = g_l^{\mathrm{free}}$ and the constants $\kappa_0^{(\lambda L)} = 0$ and $\kappa_0^{(\lambda L)} = \kappa_1^{(\lambda L)}$, $\kappa_1^{(\lambda L)} = -4\pi(28/A)\langle r^2\rangle\,\mathrm{MeV\,fm}^{-2\lambda}$. Since the isoscalar and the isovector components of the spin-multipole interaction have identical signs, the isoscalar component does not appreciably affect the probability distribution $B(ML)$. Its role reduces to changing the distribution probabilities of ML transitions between resonance one-phonon states.

The structure of the 1^+ phonons was well clarified by numerous calculations. The one-phonon states that form the $M1$ resonance give a typical example of weakly collective states. Two or three two-quasiparticle components, formed by quasiparticles at the level of the spin-orbit doublet, contribute to the wave functions of these states. Large values of $B(M1)$ are explained by large single-particle matrix elements of the magnetic dipole operator $\mathfrak{M}(M1)$ of the transition between the states of the doublet. The transition current densities of one-phonon resonance 1^+ states have a max-

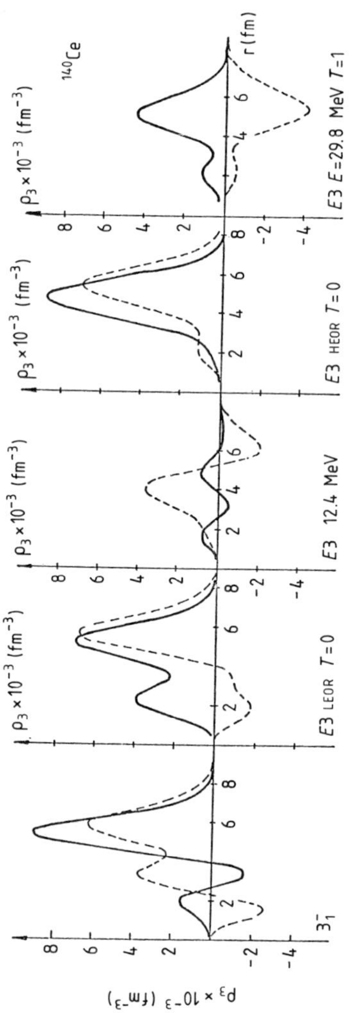

Figure 2.7 Proton (full curves) and neutron (broken curves) transition densities of some one-phonon states with $J^\pi = 3^-$ in ^{140}Ce.

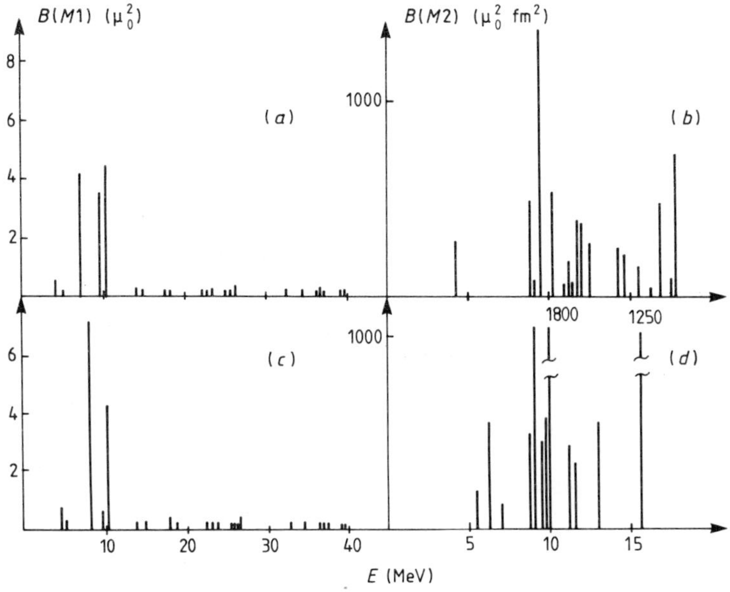

Figure 2.8 Distribution of reduced probabilities of (a, c) $M1$ transitions in ^{124}Te and (b, d) $M2$ transitions in ^{90}Zr. (a, b) Calculated for $\kappa_0^{(\lambda L)} = 0$ and (c, d) for $\kappa_0^{(\lambda L)} = \kappa_1^{(\lambda L)}$.

imum near the surface of the nucleus [74]. The structure and properties of 1^+ states with higher excitation energies are greatly affected by spin-quadrupole forces.

The $B(M2)$ probability distribution in the spectra of spherical nuclei differs significantly from the probability distribution $B(M1)$ discussed above. States with high $B(M2)$ occupy a broad range of excitation energies, which expands with increasing mass number from 4 MeV (for ^{58}Ni) to 10 MeV (for ^{208}Pb). The structure of the resonance 2^- states is that of a collective state, and the behaviour of their current transition densities is very different from that of 1^+ states. The current transition densities of 2^- states are concentrated inside the nucleus and fall off towards its surface. The structure of magnetic-type one-phonon states and the probability distribution $B(ML)$ in the spectra of spherical nuclei have quite different characteristics in comparison with the structure of electric-type states and the probability distribution $B(E\lambda)$. The probability distributions $B(ML)$ and the probabilities of the ee' excitation of 1^+ states are also considerably different.

2.2.6

Let us consider the effect of particle–particle interactions on one-phonon multipole states. It will be convenient to choose operator (2.94′) in the form

$$P^{+}_{\lambda\mu}(\tau) = \frac{1}{\sqrt{2\lambda+1}} \sum_{jj'}^{\tau} f^{\lambda}(jj')[u_j u_{j'} A^{+}(jj';\lambda\mu) - (-)^{\lambda-\mu} v_j v_{j'} A(jj';\lambda-\mu) - 2u_j v_j B(jj';\lambda\mu)] \quad (2.128)$$

then change from operators A^{+} and A to the phonon operators, drop the terms proportional to B,

$$\frac{1}{2\sqrt{2\lambda+1}} \sum_{i} [(D^{\lambda i(-)}_{\tau} + D^{\lambda i(+)}_{i}) Q^{+}_{\lambda\mu i} + (-)^{\lambda-\mu} \times (D^{\lambda i(-)}_{\tau} - D^{\lambda i(+)}_{\tau}) Q_{\lambda-\mu i}$$

and substitute the result into Hamiltonian (2.94) where

$$D^{\lambda i(+)}_{\tau} = \sum_{jj'}^{\tau} f^{\lambda}(jj') v^{(+)}_{jj'} w^{\lambda i}_{jj'} \qquad D^{\lambda i(-)}_{\tau} = \sum_{jj'}^{\tau} f^{\lambda}(jj') v^{(-)}_{jj'} g^{\lambda i}_{jj'}. \quad (2.129)$$

Adding to Hamiltonian (2.103) the corresponding interactions in the particle–particle channel and retaining only terms proportional to $Q^{+}Q$, we finally obtain

$$\begin{aligned} H^{(1)}_{M}(\mathrm{ph}+\mathrm{pp}) = {} & \sum_{jm} \epsilon_j a^{+}_{jm} a_{jm} - \frac{1}{4} \sum_{\lambda\mu i i'} \frac{1}{2\lambda+1} \sum_{\tau} \{ \sum_{\rho=\pm 1} (\kappa^{(\lambda)}_{0} \\ & + \rho\kappa^{(\lambda)}_{1})(D^{\lambda i}_{\tau} D^{\lambda i'}_{\rho\tau} + D^{\lambda i'}_{\tau} D^{\lambda i}_{\rho\tau}) \\ & + G^{(\lambda)}_{\tau}[(D^{\lambda i(-)}_{\tau} + D^{\lambda i(+)}_{\tau})(D^{\lambda i'(-)}_{\tau} + D^{\lambda i'(+)}_{\tau}) \\ & + (D^{\lambda i'(-)}_{\tau} - D^{\lambda i'(+)}_{\tau})(D^{\lambda i(-)}_{\tau} - D^{\lambda i(+)}_{\tau})]\} Q^{+}_{\lambda\mu i} Q_{\lambda\mu}. \end{aligned} \quad (2.130)$$

The mean value $H^{(1)}_{M}(\mathrm{ph}+\mathrm{pp})$ over a one-phonon state now is

$$\begin{aligned} \langle Q_{\lambda\mu i} H^{(1)}_{M}(\mathrm{ph}+\mathrm{pp}) Q^{+}_{\lambda\mu i} \rangle = {} & \frac{1}{4} \sum_{jj'} \epsilon_{jj'} [(g^{\lambda i}_{jj'})^2 + (w^{\lambda i}_{jj'})^2] \\ & - \frac{1}{4} \frac{1}{2\lambda+1} \sum_{\tau} [\sum_{\rho=\pm 1} (\kappa^{(\lambda)}_{0} + \rho\kappa^{(\lambda)}_{1}) D^{\lambda i}_{\tau} D^{\lambda i}_{\rho\tau} \\ & + G^{(\lambda)}_{\tau} (D^{\lambda i(-)}_{\tau} D^{\lambda i(-)}_{\tau} + D^{\lambda i(+)}_{\tau} D^{\lambda i(+)}_{\tau})]. \end{aligned} \quad (2.131)$$

The secular equation for the calculation of the one-phonon state energies $\omega_{\lambda i}$ becomes [74]

$$
\left|\begin{array}{cccccc}
(\kappa_0^{(\lambda)}+\kappa_1^{(\lambda)})X^{\lambda i}(n)-1 & (\kappa_0^{(\lambda)}+\kappa_1^{(\lambda)})X^{\lambda i}(n) & G_{\rm n}^{(\lambda)}X_u^{(+-)i}(n) & 0 & G_{\rm n}^{(\lambda)}X^{(+)i}(n) & 0 \\
(\kappa_0^{(\lambda)}-\kappa_1^{(\lambda)})X^{\lambda i}(p) & (\kappa_0^{(\lambda)}+\kappa_1^{(\lambda)})X^{\lambda i}(p)-1 & 0 & G_{\rm p}^{(\lambda)}X_u^{(+-)i}(p) & 0 & G_{\rm p}^{(\lambda)}X^{(+)i}(p) \\
(\kappa_0^{(\lambda)}+\kappa_1^{(\lambda)})X_u^{(+-)i}(n) & (\kappa_0^{(\lambda)}-\kappa_1^{(\lambda)})X_u^{(+-)i}(n) & G_{\rm n}^{(\lambda)}X_v^{(-)i}(n)-1 & 0 & G_{\rm n}^{(\lambda)}X_v^{(+-)i}(n) & 0 \\
(\kappa_0^{(\lambda)}-\kappa_1^{(\lambda)})X_u^{(+-)i}(p) & (\kappa_0^{(\lambda)}+\kappa_1^{(\lambda)})X_u^{(+-)i}(p) & 0 & G_{\rm p}^{(\lambda)}X_v^{(-)i}(p)-1 & 0 & G_{\rm p}^{(\lambda)}X_v^{(+-)i}(p) \\
(\kappa_0^{(\lambda)}+\kappa_1^{(\lambda)})X^{(+)i}(n) & (\kappa_0^{(\lambda)}-\kappa_1^{(\lambda)})X^{(+)i}(n) & G_{\rm n}^{(\lambda)}X_v^{(+-)i}(n) & 0 & G_{\rm n}^{(\lambda)}X_v^{(+)i}(n)-1 & 0 \\
(\kappa_0^{(\lambda)}-\kappa_1^{(\lambda)})X^{(+)i}(p) & (\kappa_0^{(\lambda)}+\kappa_1^{(\lambda)})X^{(+)i}(p) & 0 & G_{\rm p}^{(\lambda)}X_v^{(+-)i}(p) & 0 & G_{\rm p}^{(\lambda)}X_v^{(+)i}(p)-1
\end{array}\right| = 0. \tag{2.132}
$$

The notation that we have used above is

$$
\begin{aligned}
X^{(+)i}(\tau) &= \frac{1}{\sqrt{2\lambda+1}}\sum_{jj'}^{\tau}\frac{[f^{\lambda}(jj')]^2 u_{jj'}^{(+)} v_{jj'}^{(+)}\omega_{\lambda i}}{\epsilon_{jj'}^2-\omega_{\lambda i}^2} \\
X_v^{(+-)i}(\tau) &= \frac{1}{\sqrt{2\lambda+1}}\sum_{jj'}^{\tau}\frac{[f^{\lambda}(jj')]^2 u_{jj'}^{(+)} v_{jj'}^{(-)}\epsilon_{jj'}}{\epsilon_{jj'}^2-\omega_{\lambda i}^2} \\
X_v^{(\pm)i}(\tau) &= \frac{1}{\sqrt{2\lambda+1}}\sum_{jj'}^{\tau}\frac{[f^{\lambda}(jj') v_{jj'}^{(\pm)}]^2 \epsilon_{jj'}}{\epsilon_{jj'}^2-\omega_{\lambda i}^2} \\
X_v^{(+-)i}(\tau) &= \frac{1}{\sqrt{2\lambda+1}}\sum_{jj'}^{\tau}\frac{[f^{\lambda}(jj')]^2 v_{jj'}^{(+)} v_{jj'}^{(-)}\omega_{\lambda i}}{\epsilon_{jj'}^2-\omega_{\lambda i}^2}
\end{aligned} \tag{2.133}
$$

and $X^{\lambda i}(\tau)$ are defined by (2.106). The phonon amplitudes are

$$
\begin{aligned}
\psi_{j_{\rm n}j'_{\rm n}}^{\lambda i} &= \frac{\mathfrak{N}f^{\lambda}(j_{\rm n}j'_{\rm n})}{\epsilon_{j_{\rm n}j'_{\rm n}}-\omega_{\lambda i}}\{u_{j_{\rm n}j'_{\rm n}}^{(+)}[(\kappa_0^{(\lambda)}+\kappa_1^{(\lambda)})A_{11}+(\kappa_0^{(\lambda)}-\kappa_1^{(\lambda)})A_{12}] \\
&\quad + v_{j_{\rm n}j'_{\rm n}}^{(-)}G_{\rm n}^{(\lambda)}A_{13}+v_{j_{\rm n}j'_{\rm n}}^{(+)}G_{\rm n}^{(\lambda)}A_{14}\} \\
\varphi_{j_{\rm n}j'_{\rm n}}^{\lambda i} &= \frac{\epsilon_{j_{\rm n}j'_{\rm n}}-\omega_{\lambda i}}{\epsilon_{j_{\rm n}j'_{\rm n}}+\omega_{\lambda i}}\psi_{j_{\rm n}j'_{\rm n}}^{\lambda i}
\end{aligned} \tag{2.134}
$$

where $A_{11}\dots, A_{14}$ are the algebraic ajunctions of determinant (2.132).

Equation (2.132) shows how the problem of calculating the structure and energy of one-phonon states is getting progressively more complex as more and more components of the residual interaction are taken into consideration. For example, if we also include the spin-multipole component , we arrive at a secular equation which sets to zero the order-eight determinant.

The problem of particle–particle (pp) interactions has not been exhaustively researched, even though quite a few studies have been carried out and the quadrupole pairing has become an accepted tool in the theory of nuclear fields [78]. Most papers are devoted to studying the effects of the particle–particle interaction on the low-lying nuclear excitations. Calculations demonstrated that the interaction in the particle–particle channel decreases the probability of transition from the ground state to the low-lying vibrational levels of spherical nuclei [79]. For example, if the quadrupole interaction in the particle–particle channel is increased and at the same time we change the constant $\kappa_0^{(2)}$ in such a way that the energy of the 2_1^+ level remains equal to the experimental value at all times, the probability $B(E2, 0_{gs}^+ \to 2_1^+)$ decreases. If we recall that when only the particle–hole interactions are taken into account, so that the resulting values of $B(E2, 0_{gs}^+ \to 2_1^-)$ are often too large, then the introduction of the particle–particle interaction allows one to describe simultaneously both the energies and the excitation probabilities of the 2_1^+ levels. One should not forget, however, that the probability of transition from the ground state to the level 2_1^+ diminishes appreciably if anharmonic corrections are taken into account. Furthermore, no choice of $G^{(2)}$ results in a strongly suppressed $B(E2)$.

2.2.7

Let us introduce separable interactions of rank $n_{\max} > 1$. For instance, consider a central spin-independent interaction

$$V(|\boldsymbol{r}_1 - \boldsymbol{r}_2|) + \boldsymbol{\tau}^{(1)}\boldsymbol{\tau}^{(2)} V_\tau(|\boldsymbol{r}_1 - \boldsymbol{r}_2|).$$

Expanding it in multipoles according to (2.10), we recast it to the secondary-quantization form. If the separable rank-$n_{\max}$ interaction for the particle–hole interaction is given by

$$R^\lambda(r_1, r_2) = \kappa^\lambda \sum_{n=1}^{n_{\max}} R_n^\lambda(r_1) R_n^\lambda(r_2) \tag{2.135}$$

and for the particle–particle interaction, by

$$R^\lambda(r_1, r_2) = G^\lambda \sum_{n=1}^{n_{\max}} \tilde{R}_n^\lambda(r_1) \tilde{R}_n^\lambda(r_2) \tag{2.135'}$$

the expansion in multipoles becomes

$$\sum_{\lambda\mu}\sum_{n=1}^{n_{\max}}\Big\{\sum_{\tau\rho=\pm1}(\kappa_0^\lambda+\rho\kappa_1^\lambda)M_{\lambda\mu n}^+(\tau)M_{\lambda\mu n}(\rho\tau)+\sum_\tau(G_0^\lambda+G_1^\lambda)$$
$$\times P_{\lambda\mu n}^+(\tau)P_{\lambda\mu n}(\tau)+\kappa_1^\lambda(M_{\lambda\mu n}^{CH})^+M_{\lambda\mu n}^{CH}+G_1^\lambda(P_{\lambda\mu n}^{CH})^+P_{\lambda\mu n}^{CH}$$
$$+\sum_\tau G_1^\lambda P_{\lambda\mu n}^+(\tau)P_{\lambda\mu n}(\tau)\Big\}. \tag{2.136}$$

The first term takes into account the particle–hole (ph) interaction, the second, the particle–particle (pp) interaction, the third and the fourth, the charge-exchange (ph) and (pp) interactions, and the last term accounts for the two-nucleon exchange.

In calculations with separable rank-$n_{\max}>1$ interactions, the matrix elements

$$\begin{aligned} f^\lambda(jj') &= \langle j\|i^\lambda R_\lambda(r)Y_{\lambda\mu}\|j'\rangle \\ f^{\lambda L}(jj') &= \langle j\|i^\lambda R_\lambda(r)\big\{\sigma Y_{\lambda\mu}\big\}_{LM}\|j'\rangle \end{aligned} \tag{2.137}$$

are replaced with the following matrix elements:

$$\begin{aligned} f_n^\lambda(jj') &= \langle j\|i^\lambda R_n^\lambda(r)Y_{\lambda\mu}\|j'\rangle \\ f_n^{\lambda L}(jj') &= \langle j\|i^\lambda \tilde{R}_n^\lambda(r)\big\{\sigma Y_{\lambda\mu}\big\}_{LM}\|j'\rangle. \end{aligned} \tag{2.138}$$

The operators $M_{\lambda\mu}^+(\tau)$ (2.99′), $(S_{LM}^\lambda(\tau))^+$ (2.100′), $P_{\lambda\mu}^+(\tau)$ (2.128) and others are replaced with $M_{\lambda\mu n}^+(\tau)$, $(S_{LMn}^\lambda(\tau))^+$, $P_{\lambda\mu n}^+(\tau)$ and so on, in which matrix elements (2.137) are replaced with matrix elements (2.138). For example, $M_{\lambda\mu n}^+(\tau)$ has the form

$$M_{\lambda\mu n}^+(\tau)=\frac{(-)^{\lambda-\mu}}{\sqrt{2\lambda+1}}\sum_{jj'}^{\tau} f_n^\lambda(jj')\Big\{\frac{1}{2}u_{jj'}^{(+)}\Big[A^+(jj';\lambda\mu)$$
$$+(-)^{\lambda-\mu}A(jj';\lambda-\mu)\Big]+v_{jj'}^{(-)}B(jj';\lambda\mu)\Big\}. \tag{2.139}$$

The introduction of a separable interaction of finite rank $n_{\max}$ results in an additional summation over n, as compared with the case of $n_{\max}=1$. Its introduction is justified if $n_{\max}$ is much less than the rank of the determinant of the secular equation for the non-separable interaction in the RPA. Note that a simple separable interaction

$$R^\lambda(r_1,r_2)=\kappa^\lambda R^\lambda(r_1)R^\lambda(r_2)$$

for $R^\lambda(r) = \sum_n R_n^\lambda(r)$ offers even greater freedom for a description of nuclear structure.

2.2.8

Let us look first at the accuracy of the RPA description. As the constant κ_0 increases, the energy of the first root of the secular equation decreases and the degree of collective behaviour increases. As we move away from closed-shell nuclei and approach transition nuclei, the energies of the first 2_1^+ states decrease and the probabilities $B(E2)$ increase. Condition (2.33) holds in the RPA; for spherical nuclei, we rewrite it in the form

$$n_j^\lambda = (2j+1)^{-1}\Big\langle \sum_m \alpha_{jm}^+ \alpha_{jm} \Big\rangle = 0.$$

We assume that the number of quasiparticles in the ground state is negligible, even though the wave function of the ground state of an even–even nucleus is a phonon, not a quasiparticle vacuum.

We will now derive an explicit expression for the number of quasiparticles n_j^λ in the ground state, and analyse how it varies as we go from spherical to deformed nuclei. The ground state wave function of an even–even spherical nucleus, treated as a multipolarity λ phonon vacuum, has the form

$$\Psi_0^\lambda = \frac{1}{\mathfrak{N}} \exp\Big\{-\frac{1}{4}\sum_\mu \sum_{j_1 j_2 j_3 j_4} (\psi_{j_3 j_4}^{\lambda i})^{-1} \times \varphi_{j_1 j_2}^{\lambda i} (-)^{\lambda-\mu} A^+(j_1 j_2; \lambda\mu) A^+(j_3 j_4; \lambda-\mu)\Big\}\Psi_0^0 \qquad (2.140)$$

where $\mathfrak{N}$ is a normalization factor. Using the formula $e^A B e^{-A} = B - [A, B] + \ldots$, the commutation relation

$$[B(j), A^+(j_1 j_2; \lambda\mu)] = (2j+1)^{\frac{1}{2}}(\delta_{jj_1} + \delta_{jj_2}) A^+(j_1 j_2; \lambda\mu)$$

and the fact that, by definition,

$$B(jj'; \lambda\mu)\Psi_0^0 = 0,$$

we obtain [80]

$$n_j^\lambda = (2j+1)^{-1}\frac{1}{4}\sum_{\mu i}\sum_{j_1 j_2 j_3 j_4} (-)^{\lambda-\mu}\varphi_{j_1 j_2}^{\lambda i} \times (\psi_{j_3 j_4}^{\lambda i})^{-1} \langle [A^+(j_1 j_2; \lambda\mu) A^+(j_3 j_4; \lambda-\mu), B(j)] \rangle$$

$$
\begin{aligned}
&= (2j+1)^{-1} \sum_{\mu i} \sum_{j_2 j_3 j_4} (-)^{\lambda-\mu} \varphi^{\lambda i}_{jj_2} (\psi^{\lambda i}_{j_3 j_4})^{-1} \\
&\quad \times \langle A^+(jj_2;\lambda\mu) A^+(j_3 j_4; \lambda-\mu)\rangle \\
&= (2j+1)^{-1} \sum_{\mu i i' i''} \varphi^{\lambda i}_{jj_2} \varphi^{\lambda i'}_{jj_2} (\psi^{\lambda i}_{j_3 j_4})^{-1} \psi^{\lambda i''}_{j_3 j_4} \\
&\quad \times \langle Q_{\lambda-\mu i'} Q^+_{\lambda-\mu i''}\rangle = \frac{2\lambda+1}{2j+1} \sum_{ij_2} (\varphi^{\lambda i}_{jj_2})^2 .
\end{aligned}
\tag{2.141}
$$

This derivation made use of formula (2.101). We see from (2.141) that n_j^λ depend on $(\varphi^{\lambda i}_{jj'})^2$. For low-lying states with $i = 1, 2, \ldots$, the amplitude $\varphi^{\lambda i}_{jj'}$ is not small; as the excitation energy increases, it falls off since the denominator in (2.107) increases. As a result, the contribution of high-lying states to n_j^λ, including the contribution of giant resonances, is small.

Since the wave function of the ground state is the vacuum for all multipole and spin-multipole phonons, that is,

$$
\Psi_0 = \prod_\lambda \Psi_0^\lambda
$$

the number of quasiparticles in the ground state with quantum numbers j (rather, nlj) is

$$
n_j = \sum_\lambda n_j^\lambda = (2j+1)^{-1} \sum_{\lambda i} (2\lambda+1) \sum_{j_2} (\varphi^{\lambda i}_{jj_2})^2 .
$$

In analysing the applicability of the RPA, the values of greatest interest are the maximum values of n_j^λ and n_j with respect to j, $(n_j^\lambda)_{\max}$ and $(n_j)_{\max}$.

The calculation in [80] of the number n_j of quasiparticles in ground states of even–even nuclei has shown that n_j in strongly deformed nuclei, in spherical nuclei with one closed shell, and in close neighbours of these nuclei is not large, so that the RPA can be used to describe one-phonon states. As we approach transition nuclei, $(n_j)_{\max}$ increases and reaches 0.3 in transition nuclei. Correlations in ground states were also studied in [81–83]. If interactions in the particle-particle channel are taken into consideration, the number of quasiparticles in the ground states is reduced.

To clarify the role of the ground state correlations, it is not sufficient to find n_j. One needs to obtain for $n_j^\lambda \neq 0$ a set of equations and a formula for the pairing function, and to solve them. This problem was discussed, for example, in [84], where the effect of correlations on the first collective 2_1^+ and 3_1^- states in ground states and their relation to phonon amplitudes were analysed. Calculations for samarium isotopes and other nuclei demonstrated that both for the spherical and deformed nuclei, corrections to the energies and the values of $B(E\lambda)$ were small. Corrections proved to be high for transition nuclei, such as ^{150}Sm.

The Hamiltonian $H_M^{\rm ph}$ given by (2.99) and (2.99′) contains terms with products of operators, $B(jj';\lambda\mu)B(j_2j_2';\lambda-\mu)$, that are dropped in the RPA. An expression for the mean value of this Hamiltonian over a one-phonon state was obtained in [85] and corrections to RPA energies were calculated. In transition nuclei, they rose to 0.2 MeV. We will show later in the book that other large corrections exist in the RPA for transition nuclei. As a result, the RPA cannot serve as a basis for describing collective states in transition nuclei. The RPA is the more accurate, the smaller the difference between the root of the secular equation and the corresponding two-quasiparticle pole, as compared with the excitation energy. For states which form various giant resonances, this difference is much smaller than for the first 2_1^+ and 3_1^- states. Hence, the RPA is successfully applied to describe the centroid energies of the giant resonances.

2.3 NEUTRON–PROTON ONE-PHONON STATES

2.3.1

There exist collective charge-exchange statesdue to the neutron–proton interaction. They are observed in (p,n) and (n,p) reactions and in the β decay. The Gamow–Teller and electric spin–dipole charge-exchange resonances in spherical and deformed nuclei have been identified and studied [86, 87].

The central part of the interaction between quasiparticles (see (2.9)) contains isovector terms which are proportional to

$$(\boldsymbol{\tau}^{(1)}\boldsymbol{\tau}^{(2)}) = \tau_z^{(1)}\tau_z^{(2)} + \frac{1}{2}(\tau_+^{(1)}\tau_-^{(2)} + \tau_-^{(1)}\tau_+^{(2)})$$

where $\tau_\pm = \tau_x \pm i\tau_y$, and the terms containing $\tau_z^{(1)}\tau_z^{(2)}$ describe the interactions that we discussed in sections 2.1 and 2.2. The neutron–proton multipole and spin-multipole interactions are described by the following terms:

$$\begin{aligned}&\kappa_1^{(\lambda)}(\tau_+^{(1)}\tau_-^{(2)} + \tau_-^{(1)}\tau_+^{(2)})R_\lambda(r_1)R_\lambda(r_2)Y_{\lambda\mu}(\theta_1,\varphi_1)Y_{\lambda-\mu}(\theta_2,\varphi_2)\\ &-\kappa_1^{(\lambda L)}(\tau_+^{(1)}\tau_-^{(2)} + \tau_-^{(1)}\tau_+^{(2)})R_\lambda(r_1)R_\lambda(r_2)\\ &\times\{\boldsymbol{\sigma}^{(1)}\mathbf{Y}_\lambda(\theta_1,\varphi_1)\}_{LM}\{\boldsymbol{\sigma}^{(2)}\mathbf{Y}_\lambda(\theta_2,\varphi_2)\}_{L-M}. \end{aligned} \tag{2.142}$$

Terms of the same type are present in the tensor interaction. Hence, the interactions treated in sections 2.1 and 2.2 already contain terms that generate charge-exchange states, so that we need not introduce any new constants. In order to describe charge-exchange states in the RPA, we introduce neutron–proton phonons and arrive at secular equations and at the probabilities of transitions in which these one-phonon states are excited.

Let us consider the case of deformed nuclei. The charge-exchange components of multipole (2.14), spin-multipole (2.15) and tensor interactions can be written in the form

$$H^{\rm ph}_{CM} = -\frac{1}{2}\sum_{\lambda\mu\sigma} \kappa_1^{(\lambda\mu)} (M^{CM}_{\lambda\sigma\mu})^+ M^{CH}_{\lambda\sigma\mu} \tag{2.143}$$

$$H^{\rm ph}_{CS} = -\frac{1}{2}\sum_{\substack{LK\sigma \\ \lambda=L,L\pm1}} \kappa_1^{(\lambda LK)} (S^{\lambda CM}_{L\sigma K})^+ S^{\lambda CH}_{L\sigma K} \tag{2.144}$$

$$H^{\rm ph}_{CT} = -\frac{1}{2}\sum_{LK\sigma} \kappa_T^{(L)} [(S^{L-1\,CH}_{L\sigma K})^+ S^{L+1\,CH}_{L\sigma K} + (S^{L+1\,CH}_{L\sigma K})^+ S^{L-1\,CH}_{L\sigma K}] \tag{2.145}$$

where

$$M^{CH}_{\lambda\sigma\mu} = \sum_{rs\sigma_1\sigma_2} \langle r\sigma_1 | f^{\lambda\sigma\mu}\tau_- | s\sigma_2\rangle a^+_{r\sigma_1} a_{s\sigma_2}$$

$$S^{\lambda CH}_{L\sigma K} = \sum_{rs\sigma_1\sigma_2} \langle r\sigma_1 | f^{\lambda L\sigma K}\tau_- | s\sigma_2\rangle a^+_{r\sigma_1} a_{s\sigma_2}.$$

The quantum numbers of single-particle states of the neutron system are denoted by $s\sigma$ and those of the proton system, by $r\sigma$; they include the projections $K_{\rm n}$ and $K_{\rm p}$ of angular momenta on the symmetry axis of the nucleus. Now we introduce, as in section 2.1, the operators A, $\mathfrak{A}$, B and $\mathfrak{B}$, which differ from (2.20)–(2.21′) in that $q_1 = r$, $q_2 = s$. For example,

$$A(rs;\mu\sigma) = \begin{cases} \tilde{A}(rs;\mu\sigma) = \sum_{\sigma'} \delta_{\sigma'(K_{\rm p}-K_{\rm n}),\sigma\mu}\sigma'\alpha_{s'-\sigma'}\alpha_{r\sigma'} \\ \qquad \text{for} \quad \mu = |K_{\rm p} - K_{\rm n}| \\ \bar{A}(rs;\mu\sigma) = \sum_{\sigma'} \delta_{\sigma'(K_{\rm p}+K_{\rm n}),\sigma\mu}\alpha_{r\sigma'}\alpha_{s\sigma'} \\ \qquad \text{for} \quad \mu = |K_{\rm p} + K_{\rm n}| \end{cases} \tag{2.146}$$

The matrix elements are written as in (2.19) and (2.19′) and are denoted by $f^{\lambda\mu}(rs)$ and $f^{\lambda LK}(rs)$. The difference lies in that the matrix element $\langle r\sigma_1 | f^{\lambda\sigma\mu}\tau_- | s\sigma_2\rangle$ describes the neutron-to-proton transition while $\langle s\sigma_2 | f^{\lambda\sigma\mu}\tau_+ | r\sigma_1\rangle$ describes the proton-to-neutron transition.

Neutron-to-proton operators of type (2.146) satisfy the following commutation relations:

$$[\tilde{A}(rs;\mu\sigma), \tilde{A}^+(r_2s_2;\mu_2\sigma_2)]$$

$$= \delta_{ss_2}\delta_{rr_2}\delta_{\mu\mu_2}\delta_{\sigma\sigma_2}\delta_{|K_P-K_n|,\mu}(1+\delta_{\mu_0})$$
$$-\sum_{\sigma'}\delta_{\sigma'(K_p-K_n),\sigma\mu}\delta_{\sigma'(K_{2p}-K_{2n}),\sigma_2\mu_2}$$
$$\times(\delta_{ss_2}\alpha^+_{r_2\sigma'}\alpha_{r\sigma'}+\delta_{rr_2}\alpha^+_{s_2-\sigma'}\alpha_{s-\sigma'}) \qquad (2.147)$$

$$[\bar{A}(rs;\mu\sigma),\bar{A}^+(r_2s_2;\mu_2\sigma_2)]$$
$$= \delta_{ss_2}\delta_{rr_2}\delta_{\mu\mu_2}\delta_{\sigma\sigma_2}\delta_{K_P+K_n,\mu}$$
$$-\sum_{\sigma'}\delta_{\sigma'(K_p+K_n),\sigma\mu}\delta_{\sigma'(K_{2p}+K_{2n}),\sigma_2\mu_2}(\delta_{ss_2}\alpha^+_{r_2\sigma'}\alpha_{r\sigma'}$$
$$+\delta_{rr_2}\alpha^+_{s_2-\sigma'}\alpha_{s-\sigma'}). \qquad (2.147')$$

The commutation relations $[\widetilde{\mathcal{A}}(rs;\mu\sigma),\ \widetilde{\mathcal{A}}^+(r_2s_2;\mu_2\sigma_2)]$ coincide with (2.147), and $[\bar{\mathcal{A}}(rs;\mu\sigma),\bar{\mathcal{A}}^+(r_2s_2;\mu_2\sigma_2)]$ coincide with (2.147′).

Taking into consideration the formulas given above, we rewrite the operators $M^{CH}_{\lambda\sigma\mu}$ and $S^{CH}_{L\sigma K}$ as follows:

$$M^{CH}_{\lambda\sigma\mu} = \sum_{rs}\{f^{\lambda\mu}(rs)[u_rv_sA^+(rs;\mu\sigma)+v_ru_sA^+(rs;\mu-\sigma)$$
$$-u_ru_sB(rs;\mu\sigma)]-f^{\lambda\mu}(sr)v_rv_sB(sr;\mu\sigma)\} \qquad (2.148)$$
$$S^{L\pm1\,CH}_{L\sigma K} = \sum_{rs}f^{L\pm1\,LK}(rs)\{u_rv_s\mathcal{A}^+(rs;K\sigma)-v_ru_s\mathcal{A}^+(rs;K-\sigma)$$
$$+u_ru_s\mathfrak{B}(rs;K\sigma)]+v_rv_s\mathfrak{B}(sr;K\sigma)\} \qquad (2.148')$$

where $S^{LCH}_{L\sigma K}$ is obtained from (2.148) by replacing $f^{\lambda\mu}(rs)$ with the function $f^{LLK}(rs)$ and by reversing the sign in front of the last term.

Now we transform the Hamiltonians $H^{\rm ph}_{CM}$ and $H^{\rm ph}_{CS}$ to the form (2.27)–(2.31′), where

$$H^{\rm ph}_{CM\,1} = -\frac{1}{2}\sum_{\lambda\mu\sigma}\kappa_1^{(\lambda\mu)}\sum_{rsr's'}f^{\lambda\mu}(rs)f^{\lambda\mu}(r's')[v_su_rA^+(sr;\mu\sigma)$$
$$+u_sv_rA(sr;\mu-\sigma)][u_{r'}v_{s'}A^+(r's';\mu-\sigma)+v_{r'}u_{s'}A(r's';\mu\sigma)]$$
$$(2.149)$$

$$H^{\rm ph}_{CM\,2} = -\frac{1}{2}\sum_{\lambda\mu\sigma}\kappa_1^{(\lambda\mu)}\sum_{rsr's'}\{f^{\lambda\mu}(rs)[v_su_rA^+(sr;\mu\sigma)$$
$$+u_sv_rA(sr;\mu-\sigma)][f^{\lambda\mu}(r's')u_{r'}u_{s'}B(r's';\mu-\sigma)]$$
$$-f^{\lambda\mu}(s'r')v_{r'}v_{s'}B(s'r';\mu-\sigma)]+\text{h.c.}\} \qquad (2.149')$$

and where $H^{\rm ph}_{CSE1}$ differs from (2.149) in the values of the matrix elements:

$$H^{\rm ph}_{CSE\,2} = -\frac{1}{2}\sum_{LK\sigma}\kappa_1^{(LLK)}\sum_{rsr's'}\{f^{LLK}(sr)[v_su_rA^+(sr;K\sigma)$$

$$+ u_s u_r A(sr; K-\sigma)][f^{LLK}(r's')u_{r} u_{s} B(r's'; K-\sigma)$$
$$+ f^{LLK}(s'r')v_{r'} v_{s'} B(s'r'; K-\sigma)] + \text{h.c.}\} \qquad (2.150)$$

$$H^{\text{ph}}_{CSM\,1} = -\frac{1}{2}\sum_{LK\lambda=L\pm1}\kappa_1^{(\lambda LK)}\sum_{rsr's'\sigma} f^{\lambda LK}(sr)$$
$$\times f^{\lambda LK}(r's')[-v_s u_r \mathcal{A}^+(sr; K\sigma) + u_s v_r \mathcal{A}(sr; K-\sigma)$$
$$\times [u_{r'} v_{s'} \mathcal{A}^+(r's'; K-\sigma) - v_{r'} u_{s'} \mathcal{A}(r's'; K\sigma)] \qquad (2.151)$$

$$H^{\text{ph}}_{CSM\,2} = -\frac{1}{2}\sum_{LK\lambda=L\pm1}\kappa_1^{(\lambda LK)}\sum_{rsr's'\sigma} f^{\lambda LK}(sr)$$
$$\times f^{\lambda LK}(r's')\{[-v_s u_r \mathcal{A}^+(sr; K\sigma) + u_s v_r \mathcal{A}(sr; K-\sigma)]$$
$$\times [u_{r'} u_{s'} \mathfrak{B}(r's'; K-\sigma) + v_{r'} v_{s'} \mathfrak{B}(r's'; K-\sigma)] + \text{h.c.}\}. \qquad (2.151')$$

We will not need the remaining terms of the Hamiltonian.

2.3.2

Let us derive RPA equations for multipole and spin-multipole interactions with fixed values of $\lambda\mu$ and λLK. We assume that the operators $A(rs;\mu\sigma)$ and $\mathcal{A}(rs;K\sigma)$ are boson operators; furthermore, we drop the terms with the operators $\alpha^+_{r\sigma}\alpha_{r'-\sigma}$ in commutation relations (2.147) and (2.147′). Let us introduce the absorption operator for the neutron–proton multipole phonon:

$$\Omega_{\mu i\sigma} = \sum_{rs}[\psi^{\mu i}_{rs} A(rs;\mu\sigma) - \varphi^{\mu i}_{rs} A^+(rs;\mu-\sigma)]. \qquad (2.152)$$

Phonon operators satisfy commutation relations (2.37). We can write the wave functions of one-phonon states in the form

$$\Omega^+_{\mu i\sigma}\Psi_0. \qquad (2.153)$$

The condition of orthonormalization of the wave functions of the ground and the one-phonon excitation states imply the equality

$$\sum_{rs}(\psi^{\mu i}_{rs}\psi^{\mu' i'}_{rs} - \varphi^{\mu i}_{rs}\varphi^{\mu' i'}_{rs}) = \delta_{\mu\mu'}\delta_{ii'} \qquad (2.154)$$

and a condition of the type of (2.38′). The neutron–proton phonon operator for spin-multipole interactions with $\lambda = L\pm1$ is expressed in a similar way in terms of the operators $\mathcal{A}$ and $\mathcal{A}^+$. Using relations of the type (2.39), we can express the operators A, A^+, $\mathcal{A}$ and $\mathcal{A}^+$ in terms of the phonon operators Ω and Ω^+.

The definition of the spin-multipole phonon is

$$\Omega_{LKi\sigma} = \sum_{rs}[\psi^{LKi}_{rs}\mathcal{A}(rs;K\sigma) \mp \varphi^{LKi}_{rs}\mathcal{A}^+(rs;K-\sigma)]$$

where the minus sign is chosen for $\lambda = L$ and the plus sign for $\lambda = L \pm 1$. Likewise,

$$\mathcal{A}(rs; K\sigma) = \sum_i (\psi_{rs}^{LKi}\Omega_{LKi\sigma} \pm \varphi_{rs}^{LKi}\Omega_{LKi-\sigma}^{+}).$$

Let us consider states with a certain value of K^π, limiting the analysis to the contribution of a single type of multipole or spin-multipole interaction. The Hamiltonian will be

$$H_c^{(1)} = \sum_s \epsilon_s B(s) + \sum_r \epsilon_r B(r) - \frac{1}{2}\sum_{ii'\sigma} \kappa_1 (D_1^i D_1^{i'} + D_2^i D_2^{i'})\Omega_{i\sigma}^{+}\Omega_{i'\sigma}. \tag{2.155}$$

In the case of multipole forces with $\lambda\mu$ or spin-multipole forces with $\lambda = L$, we have

$$D_1^{\lambda\mu i} = \sum_{rs} f^{\lambda\mu}(rs)(\psi_{rs}^{\lambda\mu i} u_r v_s + \varphi_{rs}^{\lambda\mu i} v_r u_s)$$
$$D_2^{\lambda\mu i} = \sum_{rs} f^{\lambda\mu}(rs)(\psi_{rs}^{\lambda\mu i} v_r u_s + \varphi_{rs}^{\lambda\mu i} u_r v_s).$$

For spin-multipole forces with $\lambda = L = \pm 1$, we have

$$D_1^{L\pm1\,LKi} = \sum_{rs} f^{L\pm1\,LK}(rs)(\psi_{rs}^{LKi} u_r v_s - \varphi_{rs}^{LKi} v_r u_s)$$
$$D_2^{L\pm1\,LKi} = \sum_{rs} f^{L\pm1\,LK}(rs)(-\psi_{rs}^{LKi} v_r u_s + \varphi_{rs}^{LKi} u_r v_s).$$

As in section 2.1, we find the mean value $H_C^{(1)}$ over state (2.153),

$$\langle \Omega_{i\sigma} H_C^{(1)} \Omega_{i\sigma}^{+} \rangle = \frac{1}{2}\sum_{sr} \epsilon_{rs}[(\psi_{rs}^i)^2 + (\varphi_{rs}^i)^2] - \frac{1}{2}\kappa_1[(D_1^i) + (D_2^i)^2]$$

make use of the variational principle, and after transformations arrive at a secular equation from which to determine the energies Ω_i of one-phonon states:

$$\mathcal{F}(\Omega_i) = (1 - \kappa_1 X_1^i)(1 - \kappa_1 X_2^i) - (\kappa_1 X_{12}^i)^2 = 0 \tag{2.156}$$

$$X_1^i = \frac{1 + \delta_{\mu 0}}{2}\sum_{rs} f^2(rs)\left(\frac{u_r^2 v_s^2}{\epsilon_{rs} - \Omega_i} + \frac{v_r^2 u_s^2}{\epsilon_{rs} + \Omega_i}\right)$$
$$X_2^i = \frac{1 + \delta_{\mu 0}}{2}\sum_{rs} f^2(rs)\left(\frac{v_r^2 u_s^2}{\epsilon_{rs} - \Omega_i} + \frac{u_r^2 v_s^2}{\epsilon_{rs} + \Omega_i}\right) \tag{2.157}$$
$$X_{12}^i = \frac{1 + \delta_{\mu 0}}{2}\sum_{rs} f^2(rs) u_r v_r u_s v_s \left(\frac{1}{\epsilon_{rs} - \Omega_i} + \frac{1}{\epsilon_{rs} + \Omega_i}\right).$$

The secular equation has identical form for all multipole and spin-multipole interactions. The difference between (2.156) and secular equations (2.45) and (2.60) arises because the summation in (2.156) is carried simultaneously over the levels of the neutron and proton systems, and these systems are different. In this case, we do not use the relations between matrix elements of the types (2.19) and (2.19′), which resulted in the difference between equations (2.45) and (2.60).

In order to find the amplitudes ψ^i_{rs} and φ^i_{rs}, we use condition (2.154) and obtain, after the appropriate calculations (see [88]),

$$\begin{aligned}\psi_{rs} &= \sqrt{\frac{2\kappa_1}{\mathcal{Y}^i}}\frac{f(rs)}{\epsilon_{rs}-\Omega_i}(u_r v_s + \mathbf{Y}^i v_r u_s)\\ \varphi_{rs} &= \sqrt{\frac{2\kappa_1}{\mathcal{Y}^i}}\frac{f(rs)}{\epsilon_{rs}+\Omega_i}(v_r u_s + \mathbf{Y}^i u_r v_s)\\ (\mathcal{Y}^i)^{-1} &= \frac{1-\kappa_1 X^i_2}{-\partial\mathcal{F}(\Omega)/\partial\Omega|_{\Omega=\Omega_i}} \qquad \mathbf{Y}^i = \frac{\kappa_1 X^i_{12}}{1-\kappa_1 X^i_2}.\end{aligned} \tag{2.158}$$

Let us discuss the specific features of secular equation (2.156). If we assume that Ψ_0 is the wave function of the ground state of an even–even nucleus $\{Z_0, N_0\}$, then the wave function $\Omega^+_i\Psi_0$ describes one-phonon states of odd–odd nuclei $\{Z_0+1, N_0-1\}$ and $\{Z_0-1, N_0+1\}$. The wave functions $\Omega^+_i\Psi_0$ are superpositions of excitations in the nuclei $\{Z_0+1, N_0-1\}$ and $\{Z_0-1, N_0+1\}$ but there is a dominant branch for each state. Some solutions of secular equation (2.156) describe the nucleus $\{Z_0+1, N_0-1\}$, others describe the nucleus $\{Z_0-1, N_0+1\}$. However, wave functions belonging to the nucleus $\{Z_0+1, N_0-1\}$ are mixed with components belonging to the nucleus$\{Z_0-1, N_0+1\}$, and vice versa. If we exclude pairing and change to the Tamm–Dancoff approximation, that is, if we set $\varphi^i_{rs}=0$ in (2.152), then equation (2.156) splits in two. One refers to the nucleus $\{Z_0+1, N_0-1\}$, the other to the nucleus $\{Z_0-1, N_0+1\}$.

We will now derive the equations for the case of simultaneously taking into account the particle–hole and particle–particle multipole interactions. Instead of (2.155), we rewrite the Hamiltonian in the form

$$\begin{aligned}H'^{(1)}_C &= \sum_s \epsilon_s B(s) + \sum_r \epsilon_r B(r) - \frac{1}{2}\sum_{ii'\sigma}\{\kappa_1(D^i_1 D^{i'}_1 + D^i_2 D^{i'}_2)\\ &\quad + G_1(D^i_3 D^{i'}_3 + D^i_4 D^{i'}_4)\}\Omega^+_{i\sigma}\Omega_{i'\sigma}\end{aligned} \tag{2.159}$$

where

$$\begin{aligned}D^{\lambda\mu i}_3 &= \sum_{rs} f^{\lambda\mu}(rs)(\psi^{\lambda\mu i}_{rs} u_r u_s - \varphi^{\lambda\mu i}_{rs} v_r v_s)\\ D^{\lambda\mu i}_4 &= \sum_{rs} f^{\lambda\mu}(rs)(-\psi^{\lambda\mu i}_{rs} v_r v_s + \varphi^{\lambda\mu i}_{rs} u_r u_s).\end{aligned} \tag{2.159′}$$

Finding the mean value (2.159) over state (2.153) and using the variational principle, we obtain the following equations:

$$\begin{aligned}(\epsilon_r + \epsilon_s - \Omega_i)\psi^i_{rs} - f(rs)[\kappa_1(D^i_1 u_r v_s + D^i_2 v_r u_s) \\ + G_1(D^i_3 u_r u_s - D^i_4 v_r v_s)] = 0 \qquad (2.160)\\ (\epsilon_r + \epsilon_s + \Omega_i)\varphi^i_{rs} - f(rs)[\kappa_1(D^i_1 v_r u_s + D^i_2 u_r v_s) \\ - G_1(D^i_3 v_r v_s - D^i_4 u_r u_s)] = 0. \qquad (2.160')\end{aligned}$$

Let us transform the Hamiltonians H^{ph}_{CM} given by (2.143) and H^{ph}_{CS} given by (2.144) in view of the secular equations of type (2.156). Using the earlier notation and formulas (2.158), we can change the operators $M^{CH}_{\lambda\sigma\mu}$ and $S^{L\pm 1\,CH}_{L\sigma K}$ to

$$\begin{aligned}M^{CH}_{\lambda\sigma\mu} &= \sum_i (D^{\lambda\mu i}_1 \Omega^+_{\lambda\mu i\sigma} + D^{\lambda\mu i}_2 \Omega_{\lambda\mu i-\sigma}) + \dots \\ &= \frac{\sqrt{2}}{1+\delta_{\mu 0}} \sum_i (\kappa^{(\lambda)}_1 \mathcal{Y}_{\lambda\mu i})^{-\frac{1}{2}} (\Omega^+_{\lambda\mu i\sigma} + \mathbf{Y}^{\lambda\mu i}\Omega_{\lambda\mu i-\sigma}) \\ &\quad + \sum_{rs} [f^{\lambda\mu}(rs) u_r u_s B(rs;\mu\sigma) - f^{\lambda\mu}(sr) v_r v_s B(sr;\mu\sigma)] \qquad (2.161)\\ S^{L\pm 1\,CH}_{L\sigma K} &= \frac{\sqrt{2}}{1+\delta_{\mu 0}} \sum_i (\kappa^{(L\pm 1\,L)}_1 \mathcal{Y}_{LKi})^{-\frac{1}{2}} (\Omega^+_{LMi\sigma} + \mathbf{Y}^{LKi}\Omega_{LKi-\sigma}) \\ &\quad + \sum_{rs} f^{L\pm 1\,LK}(rs)[u_r u_s \mathfrak{B}(rs;K\sigma) + v_r v_s \mathfrak{B}(sr;K\sigma)].\end{aligned}$$

A more convenient form of expression (2.149) is

$$H_{CMv} = -\sum_{\lambda\mu i i'\sigma} \frac{1}{(1+\delta_{\mu 0})^2} \frac{1 + \mathbf{Y}^{\lambda\mu i}\mathbf{Y}^{\lambda\mu i'}}{(\mathcal{Y}^{\lambda\mu i}\mathcal{Y}^{\lambda\mu i'})^{\frac{1}{2}}} \Omega^+_{\lambda\mu i\sigma}\Omega_{\lambda\mu i'\sigma}. \qquad (2.162)$$

By replacing $\lambda\mu \to LK$ and also using the appropriate constants and matrix elements, we obtain an expression for H_{CSv}, similar to (2.162). The interactions of quasiparticles with phonons, (2.149′) and (2.151′), take the form

$$\begin{aligned}H_{CMvq} = -\sum_{\lambda\mu i\sigma} \frac{1}{\sqrt{2}(1+\delta_{\mu 0})} \sqrt{\kappa^{(\lambda\mu)}_1} (\mathcal{Y}^{\lambda\mu i})^{\frac{1}{2}} \\ \times \{(\Omega^+_{\lambda\mu i\sigma} + \mathbf{Y}^{\lambda\mu i}\Omega_{\lambda\mu i-\sigma}) \sum_{rs} [f^{\lambda\mu}(rs) u_r u_s B(rs;\mu-\sigma) \\ - f^{\lambda\mu}(sr) v_r v_s B(sr;\lambda-\sigma)] + \text{h.c.}\} \qquad (2.163)\end{aligned}$$

and for $\lambda = L \pm 1$,

$$H_{CSvq} = -\sum_{LKi\sigma} \frac{1}{\sqrt{2}(1+\delta_{\mu 0})} \left(\frac{\kappa_1^{(L\pm 1\, LK)}}{\mathcal{Y}^{LKi}}\right)^{\frac{1}{2}}$$
$$\times \{(\Omega^+_{LKi\sigma} + \mathbf{Y}^{LKi}\Omega_{LKi-\sigma}) \sum_{rs} [f^{L\pm 1\, LK}(rs)[u_r u_s \mathfrak{B}(rs; K-\sigma)$$
$$+ v_r v_s \mathfrak{B}(sr; K-\sigma)] + \text{h.c.}\}. \qquad (2.163')$$

2.3.3

Neutron–proton phonons are used to describe charge-exchange states which manifest themselves in β^+ and β^- decays and in nuclear (p,n) and (n,p) reactions. It is in these reactions that the corresponding collective states are revealed.

The operators describing the excitations of collective states from the ground states of even–even nuclei are: $D_1^{\lambda\mu i}\Omega^+_{\lambda\mu i\sigma}$ in (p,n) transitions to multipole states, $D_1^{L-1\,Ki}\Omega^+_{LKi\sigma}$ in (p,n) transitions to spin-multipole states, and $D_2^{\lambda\mu i}\Omega^+_{\lambda\mu i\sigma}$ and $D_2^{L-1\,Ki}\Omega^+_{LKi\sigma}$, respectively, in (n,p) transitions. It is with the transition operator that the component corresponding to the (p,n) and (n,p) transition is singled out of the wave function of the one-phonon state $\Omega_i^+\Psi_0$. The corresponding matrix elements equal $\langle\Omega_{i\sigma} D^i_{1(2)}\Omega^+_{i\sigma}\rangle$. Let us make use of equation (2.156) and carry out the necessary transformations to express the strength distribution of the (p,n) and (n,p) transitions in the following form:

$$B(p,n;\Omega_i) = (2-\delta_{\mu 0})|\sum_{rs} f(rs)(\psi^i_{rs} u_r v_s + \varphi^i_{rs} v_r u_s)|^2 \qquad (2.164)$$

$$B(n,p;\Omega_i) = (2-\delta_{\mu 0})|\sum_{rs} f(rs)(\psi^i_{rs} v_r u_s + \varphi^i_{rs} u_r v_s)|^2. \qquad (2.165)$$

The equations have an identical form for the excitation of the multipole and spin-multipole states.

Let us discuss the specific features of the expressions for $B(p,n;\Omega_i)$ and $B(n,p;\Omega_i)$. As we see from (2.164), there is an admixture of the (n,p) transition [the term $f(rs)v_r u_s \varphi^i_{rs}$] in the (p,n) transition [the term $f(rs)u_r v_s \psi^i_{rs}$]. A similar mixed term exists in the (n,p) transition. We will evaluate these admixtures as follows: set $\kappa_1 = 0$, which means that no terms are mixed in and that $B(p,n;\Omega_i)$ and $B(n,p;\Omega_i)$ equal the sum of squared matrix elements with the corresponding Bogoliubov transformation coefficients. We introduce total transition strengths

$$S(p,n) = \sum_i B(p,n;\Omega_i) \qquad S(n,p) = \sum_i B(n,p;\Omega_i) \qquad (2.165')$$

and compare their values in calculations for $\kappa_1 \neq 0$ and $\kappa_1 = 0$. Such estimates were made in [88]. In deformed nuclei of rare-earth and actinides regions, $S(p,n)$ is greater than $S(n,p)$ by a factor of 50 to 200. If we assume $\kappa_1 = 0$, the difference $S(p,n) - S(n,p)$ changes at most by 0.5%, and the values of $S(p,n)$ by not more than 2–4%. Since $S(n,p)$ is considerably smaller than $S(p,n)$, it changes by 30–50%. As a result, the formulas given above are valid for (p,n)-transition and β^--decay calculations, since the mixing is very low. In the (n,p)-transition and β^+-decay calculations, the formulas have to be modified by imposing an additional condition of the isospin projection conservation.

Particle–particle interactions change the functions $S(p,n)$ only slightly but affect $S(n,p)$ quite strongly, reducing them by approximately a half.

Note that no new constants are introduced for the calculation of the energies and wave functions of charge-exchange states, since the equations include the constants $\kappa_1^{(\lambda\mu)}$ and $\kappa_1^{(\lambda LK)}$ which enter the interaction between quasiparticles in the general form (2.10), (2.11) and (2.13). For example, the constant κ_1^{01K} for the calculation of the characteristics of the Gamow–Teller resonance can be taken from the $M1$ resonance data.

The results of calculations of $B(p,n;\Omega_i)$ and $B(n,p;\Omega_i)$ for the target nucleus ^{162}Dy are given in figure 2.9. The low-energy part, the region of the maximum, and the high-energy part are singled out in the transition strength distribution for the (p,n) transitions in which the Gamow–Teller resonances are excited. The resonance strength is distributed in the energy range 5–40 MeV, 1% of the strength being at the energy 50 MeV. As a result of the interaction between quasiparticles, the Gamow–Teller strength shifts towards higher energies and a maximum is formed in the region of resonance. This shift is responsible for the reduction in the probabilities of the allowed unhindered *au* β transitions in deformed nuclei, as compared with the independent quasiparticles model.

The (n,p) transition strength is distributed fairly uniformly in the region of 2–35 MeV. The distribution shows peaks at 5–8, 10–15 and 25–30 MeV. Only the lower-energy peak is well-pronounced in rare-earth nuclei. As A increases, the maximum of the strength distribution shifts towards higher energies and the total strength $S(n,p)$ decreases. The value of $S(n,p)$ for actinides is roughly 20% lower than for rare-earth nuclei.

The regions of charge-exchange $E1$ and spin-dipole resonances overlap in deformed nuclei. Hence, calculations were carried out in [88] taking simultaneously into account the dipole and spin-dipole electric forces; the secular equation for them is defined by (2.160). The shape of the resonance curve is practically unaltered when spin-dipole forces are added. Mostly, the heights of some peaks in the low-energy part are changed, and the energy centroid increases by 0.1 MeV. We can assume, therefore, that the addition of spin-dipole forces does not affect the shape and position of the charge-exchange $E1$ resonance. Taking into consideration the dipole

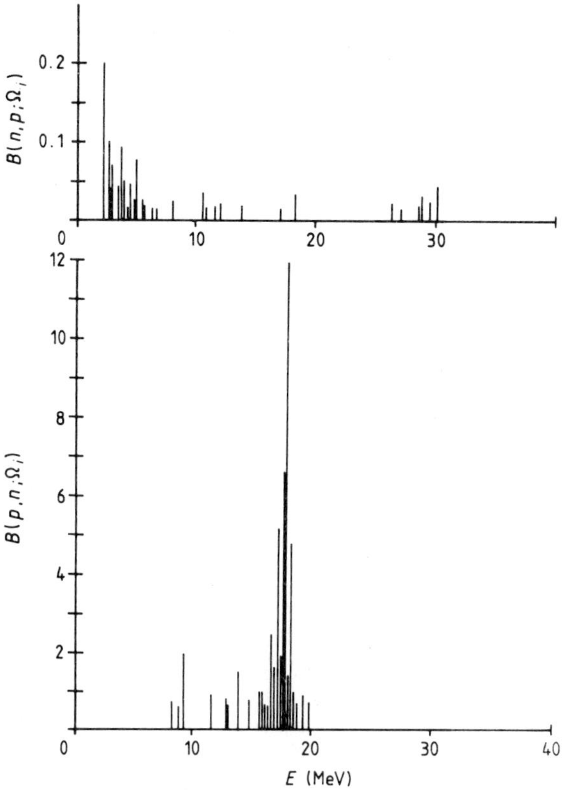

Figure 2.9 Strength distributions of (p,n) and (n,p) transitions with the excitation of the 1^+ states on ^{162}Dy. States with transition probabilities above 0.07 of maximum are shown. The energies are measured off the ground state of ^{162}Dy.

electric forces, when calculating the strength functions for the transitions to the spin-dipole electric states, produces a similarly small result.

2.3.4

Let us consider spherical nuclei. The charge-exchange terms of the multipole, spin-multipole and tensor interactions ((2.92), (2.93) and (2.95) respectively) can be written in the form

$$H^{\text{ph}}_{CM} = -\frac{1}{2}\sum_{\lambda\mu}\kappa_1^{(\lambda)}(M^{CH}_{\lambda\mu})^+ M^{CH}_{\lambda\mu} \tag{2.166}$$

$$H^{\text{ph}}_{CS} = -\frac{1}{2}\sum_{LM\lambda=L,L\pm1}\kappa_1^{(\lambda L)}(S^{\lambda CH}_{LM})^+ S^{\lambda CH}_{LM} \tag{2.167}$$

$$H_T^{\rm ph} = -\frac{1}{2}\sum_{LM} \kappa_T^{(L)}(S_{LM}^{L-1})^+ S_{LM}^{L+1} + (S_{LM}^{L+1})^+ S_{LM}^{L-1}] \tag{2.168}$$

where

$$M_{\lambda\mu}^{CH} = \sum_{jmj'm'} \langle jm|i^\lambda R_\lambda(r) Y_{\lambda\mu}\tau_-|j'm'\rangle a_{jm}^+ a_{j'm'} \tag{2.169}$$

$$S_{LM}^{\lambda CH} = \sum_{jmj'm'} \langle jm|i^\lambda R_\lambda(r)\{\sigma Y_{\lambda\mu}\}_{LM}\tau_-|j'm'\rangle a_{jm}^+ a_{j'm'}. \tag{2.169'}$$

Here the matrix elements with the operator τ_- describe the neutron-to-proton transition, those with τ_+ describe the proton-to-neutron transition; they are denoted by $f^\lambda(j_{\rm p}j_{\rm n})$ and $f^{\lambda L}(j_{\rm p}j_{\rm n})$; $j_{\rm p}m_{\rm p}$ and $j_{\rm n}m_{\rm n}$ are the quantum numbers of single-particle proton and neutron states.

The operators $A^+(j_{\rm p}j_{\rm n};\lambda\mu)$ and $B(j_{\rm p}j_{\rm n};\lambda\mu)$ are similar to (2.96) and (2.97). The operators $A(j_{\rm p}'j_{\rm n}';\lambda'\mu')$ and $A^+(j_{\rm p}j_{\rm n};\lambda\mu)$ satisfy the commutation relation

$$\begin{aligned}
&[A(j_{\rm p}'j_{\rm n}';\lambda'\mu'), A^+(j_{\rm p}j_{\rm n};\lambda\mu)] \\
&\quad = \delta_{\lambda\mu,\lambda'\mu'}\delta_{j_{\rm p}j_{\rm p}'}\delta_{j_{\rm n}j_{\rm n}'} - \delta_{j_{\rm p}j_{\rm p}'} \\
&\quad \times \sum_{m_{\rm p}m_{\rm n}m_{\rm n}'} \langle j_{\rm p}m_{\rm p}j_{\rm n}'m_{\rm n}'|\lambda'\mu'\rangle\langle j_{\rm p}m_{\rm p}j_{\rm n}m_{\rm n}|\lambda\mu\rangle \\
&\quad \alpha_{j_{\rm n}m_{\rm n}}^+\alpha_{j_{\rm n}'m_{\rm n}'} - \delta_{j_{\rm n}j_{\rm n}'} \sum_{m_{\rm p}m_{\rm p}'m_{\rm n}} \langle j_{\rm p}'m_{\rm p}'j_{\rm n}m_{\rm n}|\lambda'\mu'\rangle \\
&\quad \times \langle j_{\rm p}m_{\rm p}j_{\rm n}m_{\rm n}|\lambda\mu\rangle\alpha_{j_{\rm p}m_{\rm p}}^+\alpha_{j_{\rm p}'m_{\rm p}'}.
\end{aligned} \tag{2.170}$$

The operator $B(j_{\rm p}j_{\rm n};\lambda\mu)$ satisfies the relation

$$B^+(j_{\rm p}j_{\rm n};\lambda\mu) = (-)^{j_{\rm n}-j_{\rm p}+\mu} B(j_{\rm n}j_{\rm p};\lambda-\mu). \tag{2.171}$$

Let us introduce the operator of absorption of a neutron–proton phonon [89]

$$\Omega_{\lambda\mu i} = \sum_{j_{\rm p}j_{\rm n}} [\psi_{j_{\rm p}j_{\rm n}}^{\lambda i} A(j_{\rm p}j_{\rm n};\lambda\mu) - (-)^{\lambda-\mu}\varphi_{j_{\rm p}j_{\rm n}}^{\lambda i} A^+(j_{\rm p}j_{\rm n};\lambda-\mu)]. \tag{2.172}$$

and rewrite the operator $M_{\lambda\mu}^{CH}$ in the form

$$M_{\lambda\mu}^{CH} = \sum_{j_{\rm p}j_{\rm n}} \frac{1}{\sqrt{2\lambda+1}} f^\lambda(j_{\rm p}j_{\rm n})\Big\{\frac{1}{2}u_{j_{\rm p}j_{\rm n}}^{(+)}[A^+(j_{\rm p}j_{\rm n};\lambda\mu)$$

$$+(-)^{\lambda+\mu}A(j_pj_n;\lambda-\mu)]+\frac{1}{2}u^{(-)}_{j_pj_n}[A^+(j_pj_n;\lambda\mu)$$
$$-(-)^{\lambda+\mu}A^+(j_pj_n;\lambda-\mu)]+\ldots\Big\}$$
$$=\sum_{j_pj_n}\frac{1}{\sqrt{2\lambda+1}}f^\lambda(j_pj_n)$$
$$\times\Big\{\frac{1}{2}\sum_i[(u^{(+)}_{j_pj_n}g^{\lambda i}_{j_pj_n}+u^{(-)}_{j_pj_n}w^{\lambda i}_{j_pj_n})\Omega^+_{\lambda\mu i}+(-)^{\lambda+\mu}(u^{(+)}_{j_pj_n}g^{\lambda i}_{j_pj_n}$$
$$-u^{(-)}_{j_pj_n}w^{\lambda i}_{j_pj_n})\Omega_{\lambda-\mu i}]+u_{j_p}u_{j_n}B(j_pj_n;\lambda\mu)$$
$$+v_{j_p}v_{j_n}(-)^{j_p+j_n-\lambda}B(j_nj_p;\lambda\mu)\Big\}. \qquad (2.169'')$$

The expression for $S^{\lambda CH}_{LM}$ is obtained from (2.169) by the replacement $f^\lambda(j_pj_n)\rightarrow f^{\lambda L}(j_pj_n)$. The neutron–proton phonon operators satisfy the commutation relations

$$[\Omega_{\lambda'\mu'i'},\Omega^+_{\lambda\mu i}]=\delta_{\lambda\mu,\lambda'\mu'}\sum_{j_pj_n}(\psi^{\lambda'i'}_{j_pj_n}\psi^{\lambda i}_{j_pj_n}-\varphi^{\lambda'i'}_{j_pj_n}\varphi^{\lambda i}_{j_pj_n})$$
$$-\sum_{j_mm_nj'_nm'_n}l^{j_nj'_n}_{m_nm'_n}(\lambda'i',\lambda i)\alpha^+_{j_nm_n}\alpha_{j'_nm'_n}$$
$$-\sum_{j_pm_pj'_pm'_p}l^{j_pj'_p}_{m_pm'_p}(\lambda'i',\lambda i)\alpha^+_{j_pm_p}\alpha_{j'_pm'_p} \qquad (2.173)$$

where

$$l^{j_nj'_n}_{m_nm'_n}(\lambda'i',\lambda i)=\sum_{j_pm_p}\Big\{\psi^{\lambda'i'}_{j_pj'_n}\psi^{\lambda i}_{j_pj_n}\langle j_pm_pj'_nm'_n|\lambda'\mu'\rangle$$
$$\times\langle j_pm_pj_nm_n|\lambda\mu\rangle-(-)^{\lambda+\lambda'-\mu-\mu'}\varphi^{\lambda'i'}_{j_pj'_n}\varphi^{\lambda i}_{j_pj_n}$$
$$\times\langle j_pm_pj'_nm'_n|\lambda'-\mu'\rangle\langle j_pm_pj_nm_n|\lambda-\mu\rangle\Big\}.$$

Let us derive RPA expressions for the energies and wave functions of one-phonon neutron–proton states. We write the wave functions of one-phonon states in the form

$$\Omega^+_{\lambda\mu i}\Psi_0 \qquad (2.174)$$

with the following normalization condition:

$$\sum_{j_pj_n}(\psi^{\lambda i}_{j_pj_n}\psi^{\lambda'i'}_{j_pj_n}-\varphi^{\lambda i}_{j_pj_n}\varphi^{\lambda'i'}_{j_pj_n})=\delta_{\lambda i,\lambda'i'}. \qquad (2.175)$$

In the right-hand side of commutation relations (2.170), we drop the terms containing $\alpha^+\alpha$. We also omit in (2.169″) the terms containing the operators B and rewrite the corresponding part of the Hamiltonian:

$$H_C^{(1)} = \sum_{j_p} \epsilon_{j_p} B(j_p) + \sum_{j_n} \epsilon_{j_n} B(j_n)$$
$$- \sum_{ii'} \frac{1}{2\lambda+1} \kappa_1 (D^i_{(+)} D^{i'}_{(+)} + D^i_{(-)} D^{i'}_{(-)}) \Omega_i^+ \Omega_{i'} \qquad (2.176)$$
$$D^i_{(+)} = \frac{1}{2} \sum_{j_p j_n} f(j_p j_n) u^{(+)}_{j_p j_n} g^i_{j_p j_n}$$
$$D^i_{(-)} = \frac{1}{2} \sum_{j_p j_n} f(j_p j_n) u^{(-)}_{j_p j_n} w^i_{j_p j_n}. \qquad (2.176')$$

In (2.176) and (2.177), the expressions for multipole forces include the constant $\kappa_1^{(\lambda)}$ and the matrix elements $f^\lambda(j_p j_n)$, and those for the spin-multipole forces include $\kappa_1^{(\lambda L)}$ and $f^{\lambda L}(j_p j_n)$.

As before, we find the mean value $H_C^{(1)}$ over one-phonon state (2.174) and generate the secular equation using the variational principle (see [89]):

$$\mathcal{F}(\Omega_i) = (1 - \kappa_1 X^i_{(+)})(1 - \kappa_1 X^i_{(-)}) - (\kappa_1 X^i_{(+-)})^2 = 0 \qquad (2.177)$$

where

$$X^i_{(\pm)}(\Omega_i) = \frac{1}{(2\lambda+1)} \sum_{j_p j_n} \frac{[f(j_p j_n) u^{(\pm)}_{j_p j_n}]^2 \epsilon_{j_p j_n}}{\epsilon^2_{j_p j_n} - \Omega_i^2}$$
$$X^i_{(+-)}(\Omega_i) = \frac{1}{(2\lambda+1)} \sum_{j_p j_n} \frac{[f(j_p j_n)]^2 u^{(+)}_{j_p j_n} u^{(-)}_{j_p j_n} \Omega_i}{\epsilon^2_{j_p j_n} - \Omega_i^2}. \qquad (2.177')$$

This equation has the same form for multipole interactions with $\kappa_1^{(\lambda)}$ and $f^\lambda(j_p j_n)$, and for spin-multipole interactions with $\kappa_1^{(\lambda L)}$ and $f^{\lambda L}(j_p j_n)$; it is also characterized by the same specific features as equation (2.156). Using (2.175) and performing the required manipulation, we obtain

$$\psi^{\lambda i}_{j_p j_n} = \frac{1}{\sqrt{2\lambda+1}} \sqrt{\frac{\kappa_1^{(\lambda)}}{\mathcal{Y}^{\lambda i}}} \frac{f^\lambda(j'_p j_n)}{\epsilon_{j_p j_n} - \Omega_{\lambda i}} (u^{(+)}_{j_p j_n} - Y^{\lambda i} u^{(-)}_{j_p j_n}) \qquad (2.178)$$
$$\varphi^{\lambda i}_{j_p j_n} = \frac{1}{\sqrt{2\lambda+1}} \sqrt{\frac{\kappa_1^{(\lambda)}}{\mathcal{Y}^{\lambda i}}} \frac{f^\lambda(j_p j_n)}{\epsilon_{j_p j_n} + \Omega_{\lambda i}} (u^{(+)}_{j_p j_n} + Y^{\lambda i} u^{(-)}_{j_p j_n}) \qquad (2.178')$$
$$\mathcal{Y}^{\lambda i} = \left[\frac{1 - \kappa_1^{(\lambda)} X^{\lambda i}_{(-)}}{-\partial \mathcal{F}(\Omega)/\partial\Omega|_{\Omega=\Omega_{\lambda i}}}\right]^{-1} = \frac{4\kappa_1^{(\lambda)}}{2\lambda+1} \sum_{j_p j_n} \frac{(f^\lambda(j_p j_n))^2}{(\epsilon^2_{j_p j_n} - \Omega^2_{\lambda i})^2}$$
$$\times \{\epsilon_{j_p j_n} \Omega_{\lambda i} [(u^{(+)}_{j_p j_n})^2 + \mathbf{Y}^{\lambda i} (u^{(-)}_{j_p j_n})^2]$$

$$- (\epsilon^2_{j_p j_n} + \Omega^2_{\lambda i}) \mathbf{Y}^{\lambda i} u^{(+)}_{j_p j_n} u^{(-)}_{j_p j_n} \} \tag{2.179}$$

$$\mathbf{Y}^{\lambda i} = \frac{\kappa_1^{(\lambda)} X^{\lambda i}_{(+-)}}{1 - \kappa_1^{(\lambda)} X^{\lambda i}_{(-)}} = \frac{1 - \kappa_1^{(\lambda)} X^{\lambda i}_{(+)}}{\kappa_1^{(\lambda)} X^{\lambda i}_{(+-)}}. \tag{2.179'}$$

Let us derive the corresponding RPA equations taking the particle–hole and particle–particle interactions into consideration. Correspondingly, we supplement Hamiltonian (2.76) with the term

$$- \sum_{ii'} G_1 \frac{1}{2\lambda + 1} [D^{\lambda i}_g D^{\lambda i'}_g + D^{\lambda i}_w D^{\lambda i'}_w] \Omega^+_i \Omega_i \tag{2.180}$$

where

$$D^{\lambda i}_g = \frac{1}{2} \sum_{j_p j_n} f^\lambda (j_p j_n) v^{(-)}_{j_p j_n} g^{\lambda i}_{j_p j_n}$$

$$D^{\lambda i}_w = \frac{1}{2} \sum_{j_p j_n} f^\lambda (j_p j_n) v^{(+)}_{j_p j_n} w^{\lambda i}_{j_p j_n}. \tag{2.180'}$$

As before, we arrive at the equations

$$\epsilon_{j_p j_n} g^{\lambda i}_{j_p j_n} - \Omega_{\lambda i} w^{\lambda i}_{j_p j_n} - 2\kappa_1^\lambda f^\lambda (j_p j_n) u^{(+)}_{j_p j_n} D^{\lambda i}_{(+)}$$
$$- 2G_1 f^\lambda (j_p j_n) v^{(-)}_{j_p j_n} D^{\lambda i}_g = 0 \tag{2.181}$$

$$\epsilon_{j_p j_n} w^{\lambda i}_{j_p j_n} - \Omega_{\lambda i} g^{\lambda i}_{j_p j_n} - 2\kappa_1^\lambda f^\lambda (j_p j_n) u^{(-)}_{j_p j_n} D^{\lambda i}_{(-)}$$
$$- 2G_1 f^\lambda (j_p j_n) v^{(+)}_{j_p j_n} D^{\lambda i}_w = 0. \tag{2.181'}$$

As a result, the secular equation for $\Omega_{\lambda i}$ requires that the fourth-order determinant equal zero.

We seek the RPA solutions for the Hamiltonian in the form

$$H^{(1)}_{CMS} = \sum_{j_p} \epsilon_{j_p} B(j_p) + \sum_{j_n} \epsilon_{j_n} B(j_n) - \sum_{ii'} \Big\{ \sum_{\lambda\mu} \frac{1}{2\lambda + 1} \kappa_1^{(\lambda)}$$
$$(D^{\lambda i}_{(+)} D^{\lambda i'}_{(+)} + D^{\lambda i}_{(-)} D^{\lambda i'}_{(-)}) + \sum_{\lambda LM} \frac{1}{2L + 1} \kappa_1^{(\lambda L)}$$
$$[D^{\lambda L i}_{(+)} D^{\lambda L i'}_{(+)} + D^{\lambda L i}_{(-)} D^{\lambda L i'}_{(-)}] \Big\} \Omega^+_i \Omega_{i'}. \tag{2.182}$$

Two types of solutions are possible: the electric type if multipole forces with λ and spin-multipole forces with $\lambda = L$ are taken into account simultaneously, and the magnetic type if the spin-multipole forces with $\lambda = L - 1$ and $\lambda = L + 1$ are considered. We are to derive an equation for both types, so that the superscript a will denote multipole forces with λ and spin-multipole forces with $\lambda = L - 1$, and b will denote spin-multipole forces with $\lambda = L$ and $\lambda = L + 1$. The mean value of (2.182) over one-phonon state $\Omega^{+}_{\lambda\mu i}\Psi_0$ (or $\Omega^{+}_{LMi}\Psi_0$) is then given by

$$\langle \Omega_{\lambda\mu i} H^{(1)}_{CMS} \Omega^{+}_{\lambda\mu i} \rangle = \frac{1}{4} \sum_{j_p j_n} \epsilon_{j_p j_n} [(g^{i}_{j_p j_n})^2 + (w^{i}_{j_p j_n})^2] - \frac{1}{2\lambda + 1} \{\kappa^{a}_{1}[(D^{ai}_{(+)})^2 + (D^{ai}_{(-)})^2] + \kappa^{b}_{1}[(D^{bi}_{(+)})^2 + (D^{bi}_{(-)})^2]\}.$$

Using the variational principle, we finally obtain the following secular equation:

$$\begin{vmatrix} X^{ai}_{(+)} - \frac{1}{\kappa^{a}_{1}} & X^{ai}_{(+-)} & X^{abi}_{(+)} & X^{abi}_{(+-)} \\ X^{ai}_{(+-)} & X^{ai}_{(-)} - \frac{1}{\kappa^{a}_{1}} & X^{abi}_{(+-)} & X^{abi}_{(-)} \\ X^{abi}_{(+)} & X^{abi}_{(+-)} & X^{bi}_{(+)} - \frac{1}{\kappa^{b}_{1}} & X^{bi}_{(+-)} \\ X^{abi}_{(+-)} & X^{abi}_{(-)} & X^{bi}_{(+-)} & X^{bi}_{(-)} - (\kappa^{b}_{1})^{-1} \end{vmatrix} = 0 \tag{2.183}$$

where X^{ai} and X^{bi} are defined by formulas (2.179) and

$$\begin{aligned} X^{abi}_{(\pm)} &= \frac{1}{2(2\lambda + 1)} \sum_{j_p j_n} \frac{f^{a}(j_p j_n) f^{b}(j_p j_n) (u^{(\pm)}_{j_p j_n})^2 \epsilon_{j_p j_n}}{\epsilon^{2}_{j_p j_n} - \Omega^{2}_{i}} \\ X^{abi}_{(+-)} &= -\frac{1}{2(2\lambda + 1)} \sum_{j_p j_n} \frac{f^{a}(j_p j_n) f^{b}(j_p j_n) u^{(+)}_{j_p j_n} u^{(-)}_{j_p j_n} \Omega_i}{\epsilon^{2}_{j_p j_n} - \Omega^{2}_{i}}. \end{aligned} \tag{2.183'}$$

If we set $X^{abi}_{(\pm)} = 0$ and $X^{abi}_{(+-)} = 0$, equation (2.183) splits into two equations of type (2.178).

2.3.5

In a number of cases, tensor forces may appreciably affect the spin-multipole-type charge-exchange states, that is, states with $L^{\pi} = 1^{+}, 2^{-}, 3^{+}, \ldots$. Let us derive equations for simultaneously taking into account the spin-multipole and tensor forces. Using formulas (2.167), (2.168), (2.169″) and (2.177), we rewrite the Hamiltonian in the form

$$\begin{aligned} H^{(1)}_{CST} = &\sum_{j_p} \epsilon_{j_p} B(j_p) + \sum_{j_n} \epsilon_{j_n} B(j_n) - \sum_{LMii'} \frac{1}{2L + 1} [\kappa^{(L-1\,L)}_{1} \\ &\times (D^{L-1\,Li}_{(+)} D^{L-1\,Li'}_{(+)} + D^{L-1\,Li}_{(-)} D^{L-1\,Li'}_{(-)}) + \kappa^{(L+1\,L)}_{1} \\ &\times (D^{L+1\,Li}_{(+)} D^{L+1\,Li'}_{(+)} + D^{L+1\,Li}_{(-)} D^{L+1\,Li'}_{(-)}) + \kappa^{(L)}_{T} (D^{L-1\,Li}_{(+)} D^{L+1\,Li'}_{(+)} \\ &+ D^{L+1\,Li}_{(+)} D^{L-1\,Li'}_{(+)} + D^{L-1\,Li}_{(-)} D^{L+1\,Li'}_{(-)} \\ &+ D^{L+1\,Li}_{(-)} D^{L-1\,Li'}_{(-)})] \Omega^{+}_{LMi} \Omega_{LMi'} \end{aligned} \tag{2.184}$$

where the functions D are defined by (2.177). The mean value $H_{CST}^{(1)}$ over the one-phonon state $\Omega_{LMi}^{+}\Psi_0$ is

$$
\begin{aligned}
\langle \Omega_{LMi} H_{CMS}^{(1)} \Omega_{LMi}^{+} \rangle = \frac{1}{4} \sum_{j_p j_n} \epsilon_{j_p j_n} [(g_{j_p j_n}^{Li})^2 + (w_{j_p j_n}^{Li})^2] \\
- \frac{1}{2L+1} \{ \kappa_1^{(L-1\,L)} [(D_{(+)}^{L-1\,Li})^2 + (D_{(-)}^{L-1\,Li})^2] \\
+ \kappa_1^{(L+1\,L)} [(D_{(+)}^{L+1\,Li})^2 + (D_{(-)}^{L+1\,Li})^2] + 2\kappa_T^{(L)} \\
[D_{(+)}^{L-1\,Li} D_{(+)}^{L+1\,Li} + D_{(-)}^{L-1\,Li} D_{(-)}^{L+1\,Li}] \}.
\end{aligned}
$$

Using the variational principle again and transforming, we obtain a secular equation for calculating Ω_{Li}:

$$
\begin{vmatrix}
\kappa_1^{(L-1\,L)} X_{(+)}^{L-1\,Li} + \kappa_T^{(L)} X_{(+)}^{L+1\,L-1\,i} - 1 & \kappa_1^{(L+1\,L)} X_{(+)}^{L+1\,L-1\,i} + \kappa_T^{(L)} X_{(+)}^{L-1\,Li} & \kappa_1^{(L-1\,L)} X_{(+-)}^{L-1\,Li} + \kappa_T^{(L)} X_{(+-)}^{L+1\,L-1\,i} & \kappa_1^{(L+1\,L)} X_{(+-)}^{L+1\,L-1\,i} + \kappa_T^{(L)} X_{(+-)}^{L-1\,Li} \\
\kappa_1^{(L-1\,L)} X_{(+)}^{L+1\,L-1\,i} + \kappa_T^{(L)} X_{(+)}^{L+1\,Li} & \kappa_1^{(L+1\,L)} X_{(+)}^{L+1\,Li} + \kappa_T^{(L)} X_{(+)}^{L-1\,Li} - 1 & \kappa_1^{(L-1\,L)} X_{(+-)}^{L+1\,L-1\,i} + \kappa_T^{(L)} X_{(+-)}^{L+1\,Li} & \kappa_1^{(L+1\,L)} X_{(+-)}^{L+1\,Li} + \kappa_T^{(L)} X_{(+-)}^{L+1\,L-1\,i} \\
\kappa_1^{(L-1\,L)} X_{(+-)}^{L-1\,Li} + \kappa_T^{(L)} X_{(+-)}^{L+1\,L-1\,i} & \kappa_1^{(L+1\,L)} X_{(+-)}^{L+1\,L-1\,i} + \kappa_T^{(L)} X_{(+-)}^{L-1\,Li} & \kappa_1^{(L-1\,L)} X_{(-)}^{L-1\,Li} + \kappa_T^{(L)} X_{(-)}^{L+1\,L-1\,i} - 1 & \kappa_1^{(L+1\,L)} X_{(-)}^{L+1\,L-1\,i} + \kappa_T^{(L)} X_{(-)}^{L-1\,Li} \\
\kappa_1^{(L-1\,L)} X_{(+-)}^{L+1\,L-1\,i} + \kappa_T^{(L)} X_{(+-)}^{L+1\,Li} & \kappa_1^{(L+1\,L)} X_{(+-)}^{L+1\,Li} + \kappa_T^{(L)} X_{(+-)}^{L+1\,L-1\,i} & \kappa_1^{(L-1\,L)} X_{(-)}^{L+1\,L-1\,i} + \kappa_T^{(L)} X_{(-)}^{L+1\,Li} & \kappa_1^{(L+1\,L)} X_{(-)}^{L+1\,Li} + \kappa_T^{(L)} X_{(-)}^{L+1\,L-1\,i} - 1
\end{vmatrix} = 0
\tag{2.185}
$$

where the functions X are given by (2.179) and (2.183′) with $a = L - 1\,L$ and $b = L + 1\,L$ and the matrix elements $f^{\lambda L}(j_p j_n)$. If we set $\kappa_T^{(L)} = 0$, we obtain equation (2.183) for spin-multipole interactions with $a = L - 1\,L$ and $b = L + 1\,L$.

2.3.6

Let us transform the Hamiltonian taking into account the solutions of the secular equation and analyse the strength distribution for charge-exchange states. We begin with Hamiltonians (2.166) and (2.167) and secular equations (2.178). Making use of formulas (2.180) and (2.180′), we rewrite (2.169) as

$$
\begin{aligned}
M_{\lambda\mu}^{CH} = \frac{1}{2} \sum_i (\kappa_1^{(\lambda)} y^{\lambda i})^{-\frac{1}{2}} [(1 - \mathbf{Y}^{\lambda i}) \Omega_{\lambda \mu i}^{+} \\
+ (-)^{\lambda-\mu} (1 + \mathbf{Y}^{\lambda i}) \Omega_{\lambda -\mu i}] + \sum_{j_p j_n} \frac{1}{\sqrt{2\lambda+1}} f^{\lambda}(j_p j_n) \\
\times [u_{j_p} u_{j_n} B(j_p j_n; \lambda\mu) + (-)^{j_p + j_n - \lambda} v_{j_p} v_{j_n} B(j_n j_p; \lambda\mu)]
\end{aligned}
$$

and change the component $H^{\rm ph}_{CM}$ in (2.176) as follows:

$$H_{CMv} = -\sum_{\lambda\mu i i'} \frac{1 + \mathbf{Y}^{\lambda i}\mathbf{Y}^{\lambda i'}}{(\mathcal{Y}^{\lambda i}\mathcal{Y}^{\lambda i'})^{\frac{1}{2}}} \Omega^{+}_{\lambda\mu i}\Omega_{\lambda\mu i'}. \tag{2.186}$$

It is not difficult to show that the part corresponding to the quasiparticle–phonon interaction now becomes

$$\begin{aligned} H_{CMvq} = &-\frac{1}{2}\sum_{\lambda\mu i}\sqrt{\frac{\kappa_1^{(\lambda)}}{\mathcal{Y}^{\lambda i}}}\sum_{j_{\rm p}j_{\rm n}}\frac{f^{\lambda}(j_{\rm p}j_{\rm n})}{\sqrt{2\lambda+1}}\{[(-)^{\lambda-\mu}(1+\mathbf{Y}^{\lambda i}) \\ &\times \Omega^{+}_{\lambda\mu i} + (1-\mathbf{Y}^{\lambda i})\Omega_{\lambda-\mu i}][u_{j_{\rm p}}u_{j_{\rm n}}B(j_{\rm p}j_{\rm n};\lambda-\mu) \\ &+ (-)^{j_{\rm p}+j_{\rm n}-\lambda}v_{j_{\rm p}}v_{j_{\rm n}}B(j_{\rm n}j_{\rm p};\lambda-\mu)] + \text{h.c.}\}. \end{aligned} \tag{2.187}$$

These formulas differ by a factor of 1/2 from those in [75, 89], owing to different definitions of the matrix element $f(j_{\rm p}j_{\rm n})$: here we include $\tau_{\pm}$ while in [75, 89] $\tau_{\pm}/2$ was used. The expressions for H_{CSv} and that for H_{CSvq} for spin-multipole forces with $\lambda = L-1$ differ from (2.186) and (2.187) in the constants $\kappa_1^{(\lambda L)}$ which correspond to the matrix elements $f^{\lambda L}(j_{\rm p}j_{\rm n})$.

2.3.7

Now we need to derive formulas for the strength distributions of the (p,n) and (n,p) transitions. The operator of the (p,n) transition from the even–even target nucleus $\{Z_0, N_0\}$ to the i states of the odd–odd nucleus $\{Z_0+1, N_0-1\}$, with the isospin projection reduced by unity, is

$$\mathfrak{M}_i^{(-)} = \frac{1}{2}\sum_{j_{\rm p}j_{\rm n}}\frac{f(j_{\rm p}j_{\rm n})}{\sqrt{2\lambda+1}}(\psi^i_{j_{\rm p}j_{\rm n}}u_{j_{\rm p}}v_{j_{\rm n}} + \varphi^i_{j_{\rm p}j_{\rm n}}v_{j_{\rm p}}u_{j_{\rm n}})\Omega_i^{+}. \tag{2.188}$$

The operator of the (n,p) transition to the state i of the nucleus $\{Z_0-1, N_0+1\}$ can be transformed to

$$\mathfrak{M}_i^{(+)} = \frac{1}{2}\sum_{j_{\rm p}j_{\rm n}}\frac{f(j_{\rm p}j_{\rm n})}{\sqrt{2\lambda+1}}(-)^{\lambda-\mu}(\psi^i_{j_{\rm p}j_{\rm n}}v_{j_{\rm p}}u_{j_{\rm n}} + \varphi^i_{j_{\rm p}j_{\rm n}}u_{j_{\rm p}}v_{j_{\rm n}})\Omega_i^{+}. \tag{2.188$'$}$$

The amplitudes of the (p,n) and (n,p) transitions to the states $\Omega_i^{+}\Psi_0$ are

$$\Phi_i^{(\mp)} = \sqrt{2\lambda+1}\,\frac{(1\pm\mathbf{Y}^{\lambda i})}{(\kappa_1\mathcal{Y}^{i})^{\frac{1}{2}}}. \tag{2.189}$$

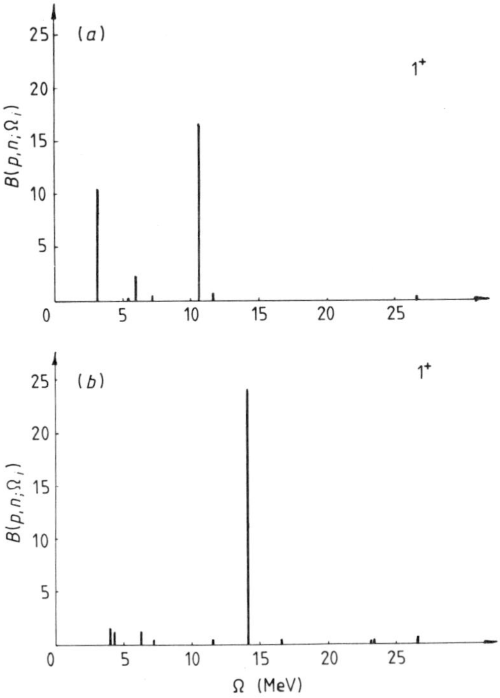

Figure 2.10 Distribution of the Gamow–Teller strength of the (p,n) transition (or the β^- decay) in ^{90}Zr; (a) neglecting the interaction , $\kappa_1^{(01)}=0$; (b) taking the interaction , $\kappa_1^{(01)}=23/A$ MeV into account. The energy is measured with respect to the ground state of ^{90}Zr.

Amplitudes (2.189) enter the corresponding probabilities of the β^- and β^+ decays.

The strength distribution for the (p,n) and (n,p) transitions can be given in the form

$$B(p,n;\Omega_i)=\Big|\sum_{j_p j_n} f(j_p j_n)(u_{j_p}v_{j_n}\psi^i_{j_p j_n}+v_{j_p}u_{j_n}\varphi^i_{j_p j_n})\Big|^2 \qquad (2.190)$$

$$B(p,n;\Omega_i)=\Big|\sum_{j_p j_n} f(j_p j_n)(v_{j_p}u_{j_n}\psi^i_{j_p j_n}+u_{j_p}v_{j_n}\varphi^i_{j_p j_n})\Big|^2. \qquad (2.191)$$

The interaction between quasiparticles as described in the RPA results in strength concentration on one or several one-phonon states that form giant Gamow–Teller and spin-dipole charge-exchange resonances. These resonances are especially well-pronounced in (p,n) reactions with small momentum transfer. For example, a large part of the Gamow–Teller strength generated by spin-multipole interactions with $L = 1$ and $\lambda = 0$ is concentrated on one one-phonon state. As a result of this concentration, the Gamow–Teller strength in the region of low-lying states is reduced in comparison with the results of calculation s in the model of independent quasiparticles (see figure 2.10). If the constant $\kappa_1^{(01)}$ is set equal to zero, we obtain a description in terms of the independent quasiparticles model. In this case, a considerable part of the strength is in the range of low-lying 1^+ states of ^{90}Nb, so that the calculated probabilities of the β^- transitions are found to be much greater than the experimentally observed values. The figure demonstrates the significance of the decrease in the Gamow–Teller strength at low excitation energies if the interaction between particles is taken into consideration in the RPA. Nevertheless, the decrease is not sufficient to fit the theory to the experimental data. The same type of strength concentration occurs for spin-dipole charge-exchange resonances [89].

The total strength generated in (n,p) reactions on even–even spherical nuclei is lower than that in (p,n) reactions by a factor of 30–50. When calculating the (n,p) strength distribution in RPA, we find the corrections due to the coupling with the (p,n) channel to be quite high, so that the formulas have to be elaborated. The concentration of the strength excited in (n,p) reactions is not as high as in (p,n) reactions. This strength is distributed in the range from 5 to 20 MeV. The interaction between quasiparticles and phonons reduces the Gamow–Teller and spin-dipole strengths in the region of low-lying states. The reduction is not as significant, in fact, as for (p,n) transition strengths.

The concept of one-phonon neutron–proton states is fundamental for the description of charge-exchange resonances, as well as for the calculation of the probabilities of the β^- and β^+ decays and of the double β decay [90].

3

Interacting bosons approximation

3.1 BOSON EXPANSIONS

3.1.1

The success of the Bohr–Mottelson collective model [4] (see also [7, 91]) in the interpretation of low-lying states was a stimulus to developing a microscopic description of the collective motion. A model of the quadrupole harmonic vibrator was produced: the energy of the one-phonon 2_1^+ state equals ω_1, the two-phonon 0_2^+, 2_2^+, 4_1^+ triplet is characterized by twice this energy, and so on. Phonons describing collective vibrations are expressed in terms of two-quasiparticle operators:

in the Tamm–Dancoff method

$$\bar{Q}^+_{\lambda\mu i} = \frac{1}{2}\sum_{jj'} \psi^{\lambda i}_{jj'} A^+(jj';\lambda\mu) \tag{3.1}$$

and in the RPA

$$Q^+_{\lambda\mu i} = \frac{1}{2}\sum_{jj'} [\psi^{\lambda i}_{jj'} A^+(jj';\lambda\mu) - (-)^{\lambda\mu}\varphi^{\lambda i}_{jj'} A(jj';\lambda - \mu)]. \tag{3.2}$$

The spectra of low-lying states of spherical nuclei differ appreciably from those of an harmonic quadrupole vibrator. First, anharmonic effects are connected with expansion terms of an order higher than two, and second, they are due to the Pauli exclusion principle, since the phonons described by operators (3.1) and (3.2) are not ideal bosons. Higher-order expansion terms are caused by the interaction of quadrupole phonons with themselves and with other degrees of freedom. Anharmonic effects are high for transition nuclei, in which the vibrational spectrum changes to a rotational one. For such nuclei, the RPA has to be rejected. The methods specially adjusted for transition nuclei are the boson expansions.

In the boson expansion methods, the bilinear combinations of fermion operators are expressed in terms of ideal boson operators as finite or infinite series. This is the problem of mapping the fermion collective pairs into particle–particle bosons. Boson expansion methods would not be advantageous if all boson operators introduced were important for the description of low-lying states. The main idea of all boson expansions is to limit the analysis to several collective bosons which form a collective subspace of the total space. It is in this subspace that the Hamiltonian is diagonalized. If it is possible to neglect the coupling of the bosons selected to other degrees of freedom, a description of the required states encounters no problems. The Hamiltonian in quadratic form is then quite simple and the higher-order terms of the boson representation of the Hamiltonian decrease fairly rapidly. Obviously, this assumption will be met for only several states and not necessarily for all nuclei. In many cases, it is necessary to take into consideration the coupling to collective states.

3.1.2

In microscopic analysis of collective subspace, which plays a most important role in the boson expansion methods, it is necessary to establish a one-to-one correspondence of the pair of fermion operators $a_a^+ a_c^+$ to the ideal boson operators b_{ac}^+ ($b_{ac}^+ = -b_{ca}^+$). Consider even–even nuclei. Let the states $|n\rangle$ form an orthonormalized basis in the fermion state space. The total number of boson states that can be constructed using the operators b_{ac}^+ is much higher than the number of fermion states. Among these, we can identify a subspace formed by orthonormalized states $|n)$ whose number coincides with the number of fermion states $|n\rangle$ and which possess the same symmetry properties with respect to the permutation of the indices a_1c_1, a_2c_2 ... as the fermion states do. This subspace of the boson space is known as the physical space. The remaining part of the boson space is called the non-physical one, since the $|u)$ states have nothing in common with the fermion system. The mapping of the fermion state $|n\rangle$ on the corresponding physical boson state $|n)$ is an exact one-to-one correspondence. This correspondence of the fermion and boson spaces is illustrated in figure 3.1. The completeness condition for the boson space is given by the equation

$$|0)\,(0| + \sum_{n\neq 0} |n)\,(n| + \sum_{u} |u)\,(u| = 1. \tag{3.3}$$

The next relation to be established is the one between the fermion states $|n\rangle$ and the operators F, and also between the corresponding physical boson states $|n)$ and the operators $\mathcal{F}$, that is

$$|n\rangle \leftrightarrow |n) \qquad F \leftrightarrow \mathcal{F}. \tag{3.4}$$

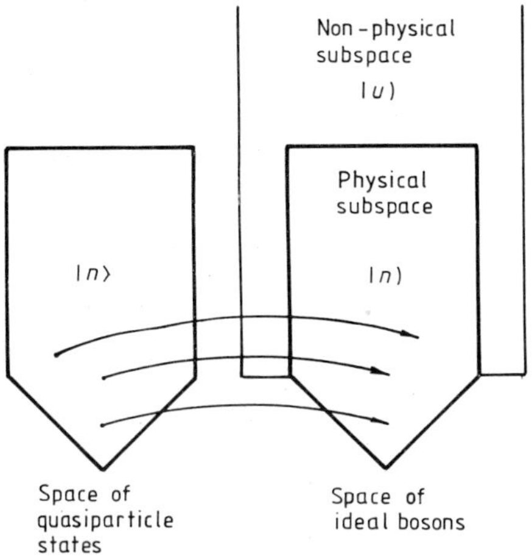

Figure 3.1 Mapping of the quasiparticle state space onto the ideal boson space.

The boson mapping requires that the following relations between matrix elements be satisfied:

$$\langle n|F|n'\rangle = (n|\mathcal{F}|n') \quad (u|\mathcal{F}|n) = (n|\mathcal{F}|u) = 0 \tag{3.5}$$

that is, the operator $\mathcal{F}$ leaves the physical boson space invariant. Two basic methods are available to perform this programme: the Belyaev–Zelevinsky [92] and the Marumori [93] methods.

In the Belyaev–Zelevinsky method, the basis fermion operators

$$A^+(jj';JM) \qquad A(jj';JM) \qquad B(jj';JM)$$

are expressed in terms of the creation operators b_a^+ and absorption operators b_a in such a way that all commutation relations are preserved. The commutation relations between the basis fermion operators form a closed algebra, isomorphous to the Lie group algebra SO(2Ω) with $\Omega = \sum_a (j_a + \frac{1}{2})$. The basis operators are given in terms of bosons, so as to preserve the SO(2Ω)-group algebra. Any mapped operator is expressed as a Taylor series expansion

$$\mathcal{F}^{|BZ|} = \mathcal{F}^{(0)} + \mathcal{F}^{(1)} + \ldots + \mathcal{F}^{(n)} + \ldots \tag{3.6}$$

and any commutation relation $[\mathcal{F}_1, \mathcal{F}_2] = \mathcal{F}_3$ must hold in each order of expansion in the boson mapping

$$\begin{aligned} &[\mathcal{F}_1^{(1)}, \mathcal{F}_2^{(1)}] = \mathcal{F}_3^{(0)} \\ &[\mathcal{F}_1^{(1)}, \mathcal{F}_2^{(2)}] + [\mathcal{F}_1^{(2)}, \mathcal{F}_2^{(1)}] = \mathcal{F}_3^{(1)}. \\ &\ldots\ldots\ldots \end{aligned} \tag{3.7}$$

Such expansions are to a certain degree arbitrary since the commutation relations define operators up to a unitary transformation.

In the Belyaev–Zelevinsky method, the fermion operators are expanded into infinite series in the boson operators:

$$\begin{aligned} A^+(jj';JM) \to A_a^+ &= x^{10}(a)b_a^+ \\ &+ \sum_{a_1a_2a_3} x_a^{21}(a_1a_2a_3)b_{a_1}^+b_{a_2}^+b_{a_3} + \dots \\ B(jj';JM) \to B_a &= y^{00}(a) + \sum_{a_1a_2} y_a''(a_1a_2)b_{a_1}^+b_{a_2} \\ &+ \sum_{a_1a_2a_3a_4} y_a^{22}(a_1a_2a_3a_4)b_{a_1}^+b_{a_2}^+b_{a_3}b_{a_4} + \dots . \end{aligned} \tag{3.8}$$

The coefficients of these expansions are found iteratively by demanding that commutation relations be satisfied in each approximation. The resulting coefficients are not found unambiguously. For example, if we assume that the ground state of the system is at the same time a boson and a quasiparticle vacuum, then $y^{00}(a) = 0$. The small expansion parameter is $[\Sigma_a(2j_a + 1)]^{-1/2}$. The convergence of the expansion depends also on the ratio $\epsilon_{a_1a_2}/\omega_1$, that is, the convergence rate deteriorates for strongly collective states. The method works better if instead of the operators A_a^+ and A_a we expand the operators $\bar{Q}_{\lambda\mu i}^+$ and $\bar{Q}_{\lambda\mu i}$ or $Q_{\lambda\mu i}^+$ and $Q_{\lambda\mu i}$.

3.1.3

We will present below the Marumori boson expansion method Let us define a direct product of the fermion and boson spaces. In this case, the vacuum is the product of a fermion and a boson vacua, $|0\rangle$ and $|0)$. Let us also define a unitary mapping operator

$$U_M = |0\rangle \sum_n |n)\langle n|(0| \tag{3.9}$$

which converts a fermion state $|n\rangle$ into a boson state $|n)$, that is,

$$U_M|n\rangle|0) = |0\rangle|n). \tag{3.10}$$

The operator

$$U_M^+ = |0) \sum_n |n\rangle(n|\langle 0|$$

converts a boson state $|n)$ into a fermion state $n\rangle$, that is,

$$U_M^+|n)|0\rangle = |0)\cdot|n\rangle$$

in which

$$U_M^+|u)|0\rangle = 0.$$

The operators U_M and U_M^+ satisfy the relations

$$\begin{aligned} U_M^+ U_M &= \Gamma_B \qquad & U_M U_M^+ &= \Gamma_B P \\ \Gamma_B &\equiv |0)(0| \qquad & \Gamma_F &\equiv |0\rangle\langle 0| \qquad & P = \sum_n |n)(n|. \end{aligned} \tag{3.11}$$

The operator of projection to the physical boson subspace can be written as follows:

$$P = U_M \sum_n |n\rangle\langle n| U_M^+ = U_M U_M^+. \tag{3.12}$$

Since the fermion space is complete

$$\sum_n |n\rangle\langle n| = 1.$$

Equation (3.12) implies that

$$U_M^+ P = \Gamma_B U_M^+ = U_M^+ \qquad P U_M = U_M \Gamma_B = U_M. \tag{3.12$'$}$$

The normalization condition

$$\langle 0|(n|n')|0\rangle = \delta_{mn'} = (0|\langle n|U_M^+ U_M|n'\rangle|0)$$

implies that U^+U is an identity operator in the fermion space. The following transformation converts any fermion operator F by the operators U_M^+ and U_M into the boson operator $\mathcal{F}$:

$$\begin{aligned} (0|\langle n|F|n'\rangle|0) &= \langle 0|(n|U_M F U_M^+|n')|0\rangle \\ &= \langle 0|(n|\mathcal{F}|n')|0\rangle \end{aligned} \tag{3.13}$$

where $\mathcal{F} = U_M F U_M^+$. The operator $\mathcal{F}$ has non-zero matrix elements only between the states of the physical boson subspace. Indeed, formulas (3.12$'$) give us

$$\mathcal{F} = U_M F U_M^+ = P U_M F U^+ P = P\mathcal{F}P. \tag{3.14}$$

Following [94], we find the explicit form of the operator U_{LM}. The following boson operators b_{ac}^+ and b_{ac} will satisfy the commutation relations:

$$\begin{aligned} [b_{ac}^+, b_{a'c'}] &= \delta_{aa'}\delta_{cc'} - \delta_{ac'}\delta_{a'c} \\ [b_{ac}, b_{a'c'}] &= [b_{a'c'}^+, b_{ac}^+] = 0. \end{aligned} \tag{3.15}$$

The wave function of the boson physical state will be chosen in the form

$$|n) = \frac{1}{\sqrt{(2n-1)!!}} \sum_p (-)^p P b^+_{a_1c_1} b^+_{a_2c_2} \dots b^+_{a_nc_n} |0) \tag{3.16}$$

where $\sum_p (-)^p P$ guarantees that the wave function of $|n)$ is antisymmeric under the permutation of any indices from the set $a_1c_1, \dots, a_nc_n$. The wave function of fermion states is

$$|n\rangle = a^+_{a_1} a^+_{c_1} \dots a^+_{a_n} a^+_{c_n} |0\rangle. \tag{3.17}$$

The operator U_M is then chosen in the form

$$U_M = |0\rangle(0| \sum_{n=0}^{\infty} \frac{1}{(2n)!} \frac{1}{\sqrt{(2n-1)!!}} (\sum_{ac} b^+_{ac} a_c a_a)^n |0)\langle 0|. \tag{3.18}$$

Applying this operator to the fermion wave function, we obtain

$$\begin{aligned} U_M |n\rangle &= U_M a^+_{a_1} a^+_{c_1} \dots a^+_{a_n} a^+_{c_n} |0\rangle |0) \\ &= |0\rangle(0| \frac{1}{(2n)! \sqrt{(2n-1)!!}} \\ &\quad \times (\sum_{ac} b^+_{ac} a_c a_a)^n \, a^+_{a_1} a^+_{c_1} \dots a^+_{a_n} a^+_{c_n} |0\rangle |0)(0|0) \\ &= \frac{1}{\sqrt{(2n-1)!!}} \sum_p (-)^p P b^+_{a_1c_1} b^+_{a_2c_2} \dots b^+_{a_nc_n} |0)|0\rangle \\ &= |n)|0\rangle. \end{aligned} \tag{3.19}$$

We see that U_M transforms a fermion wave function $|n\rangle$ into a boson wave function $|n)$.

It will be useful to give here the explicit form of the boson operator $\mathcal{F}$; to obtain it, we express the operator in terms of the matrix elements of the fermion operator$F$, namely:

$$\begin{aligned} \mathcal{F} &= \sum_{nn'} |n)\langle n|F|n'\rangle(n'| = \langle 0|F|0\rangle |0)(0| + \sum_{n=1} \frac{1}{(2n)!} \\ &\quad \times \sum_{ac} \langle 0| a_{c_n} a_{a_n} \dots a_{c_1} a_{a_1} F |0\rangle \frac{1}{\sqrt{(2n-1)!!}} \\ &\quad \times \sum_p (-)^p \, P b^+_{a_1c_1} \dots b^+_{a_nc_n} |0)(0| \\ &\quad + \sum_{n=1} \frac{1}{(2n)!} \sum_{ac} \langle 0| F a^+_{a_1} a^+_{c_1} \dots a^+_{a_n} a^+_{c_n} |0\rangle \end{aligned}$$

$$\times \frac{1}{\sqrt{(2n-1)!!}} |0)(0| \sum_p (-)^p P b_{a_n c_n} \ldots b_{a_1 c_1}$$
$$+ \sum_{nn'} \frac{1}{(2n)!(2n')!} \sum_{(ac)(a'c')} \langle 0 | a_{c'_{n'}} a_{a'_{n'}}$$
$$\ldots a_{c'_1} a_{a'_1} F a^+_{a_1} a^+_{c_1} \ldots a^+_{a_n} a^+_{c_n} |0\rangle$$
$$\times \frac{1}{\sqrt{(2n-1)!!(2n'-1)!!}} \sum_p (-)^p P b^+_{a_1 c_1}$$
$$\ldots b^+_{a_n c_n} |0)(0| \sum_{\mathrm{P}} (-)^p P b_{a'_n c'_n} \ldots b_{a'_1 c'_1}. \tag{3.20}$$

In the particular case of $F = a^+_a a^+_c$, formula (3.20) changes to

$$U_M a^+_a a^+_c U^+_M = b^+_{ac} - \frac{1}{2}\left(1 - \frac{1}{\sqrt{3}}\right) b^+_{ac}$$
$$\times \sum_{a_2 c_2} b^+_{a_2 c_2} b_{a_2 c_2} - \frac{1}{\sqrt{3}} \sum_{a_2 c_2} b^+_{a a_2} b^+_{c c_2} b_{a_2 c_2} + \ldots .$$

As shown in [3], two-fermion operators can be given in the following compact form:

$$U_M a^+_a a^+_c U^+_M = P \sum_{a_2} b^+_{a a_2} \sqrt{\delta_{c a_2} - \sum_{c_2} b^+_{c c_2} b_{a_2 c_2}}$$
$$U_M a^+_a a_c U^+_M = \sum_{a_2} b^+_{a a_2} b_{c a_2} P \tag{3.21}$$

where the operator in the radicand is interpreted as an expansion in Taylor series.

The Belyaev–Zelevinsky (BZ) and Marumori (M) boson expansions for the operators are related by the formula [94, 95]

$$\mathcal{F}^{(\mathrm{M})} = P \mathcal{F}^{(\mathrm{BZ})} P. \tag{3.22}$$

The Dyson boson expansion [96] has been widely used in recent years (see [11]). Its advantage lies in the presentation of the fermion operators by finite boson expansions. The Dyson transformation is not unitary, however, so that the transformed Hamiltonian is non-Hermitian.

3.1.4

The boson mapping of the fermion pairs operators $a^+_a a^+_c$, $a_a a_c$, $a^+_a a_c$ is not widely used because of a very slow convergence of the corresponding boson expansion. It is advisable, therefore, to choose the correlated fermion pairs

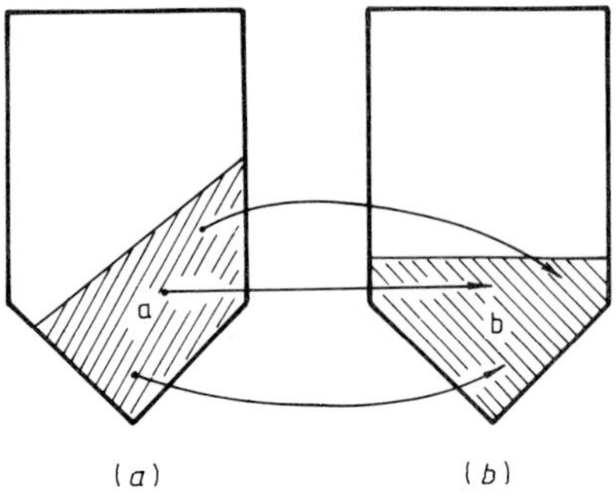

Figure 3.2 The boson mapping of the collective fermion subspace: (*a*) collective subspace of the space of quasiparticle states; (*b*) collective subspace of the boson space.

operators, for instance, phonon operators in the Tamm–Dancoff method or in the RPA, so as to guarantee a rapid convergence of the series. The phenomenological Bohr–Mottelson model is actually based on the following proposition: only some special subspace of the fermion states need be presented as a collective bosons subspace (figure 3.2).

Let us introduce operators of collective fermion pairs used in the Tamm–Dancoff method:

$$\bar{Q}_i^+ = \frac{1}{2} \sum_{jj'} \psi_{jj'}^i A^+(jj'; \lambda\mu).$$

The functions $\psi_{jj'}^i$ satisfy the conditions

$$\sum_{j>j'} \psi_{jj'}^i \psi_{jj'}^{i'} = \delta_{ii'}$$

$$\sum_i \psi_{jj'}^i \psi_{j_2 j_2'}^i = \delta_{jj_2}\delta_{j'j_2'} - \delta_{jj_2'}\delta_{j'j_2}.$$

Let us define an operator $B_{ii'}$:

$$B_{ii'} = \sum_{jj'j_2} \psi_{j'j_2}^i \psi_{jj_2}^{i'} \sum_{mm'm_2} \langle j'm'j_2m_2|\lambda\mu\rangle \times \langle jmj_2m_2|\lambda'\mu'\rangle a_{jm}^+ a_{j'm'}.$$

The operators Q_i^+, $\bar{Q}_i$ and $B_{ii'}$ satisfy the following commutation relations:

$$[\bar{Q}_i, \bar{Q}_{i'}] = [\bar{Q}_{i'}^+, \bar{Q}_i^+] = 0 \qquad [\bar{Q}_i, \bar{Q}_{i'}^+] = \delta_{ii'} - B_{ii'}$$
$$[\bar{Q}_i, B_{i'i_2}] = \sum_{i_2'} \Gamma_{i_2 i_2'}^{ii'} \bar{Q}_{i_2'} \qquad [\bar{Q}_i^+, B_{i'i_2}] = -\sum_{i_2'} \Gamma_{i_2 i_2'}^{ii'} \bar{Q}_{i_2'}^+ \tag{3.23}$$

where

$$\Gamma^{ii'}_{i_2 i'_2} = \sum_{jj'j_2j'_2} \psi^{i}_{jj_2}\psi'^{i}_{j_2j'_2}\psi^{i_2}_{j'_2j'}\psi^{i'_2}_{j'j}.$$

An additional assumption is often used [97]: a double commutator of the operators $\bar{Q}_i$ and $\bar{Q}^+_{i'}$ does not scatter out of the collective subspace.

Let us construct a normalized wave function of $2n$-quasiparticle states:

$$|n\rangle = N_F(n)|i_1, \ldots, i_n\rangle = N_F(n)\bar{Q}^+_{i_1}\bar{Q}^+_{i_2} \ldots \bar{Q}^+_{i_n}|0\rangle \tag{3.24}$$

where the normalization constant is

$$N_F(n) = \langle i_1 i_2 \ldots i_n | i_1 i_2 \ldots i_n\rangle^{-\frac{1}{2}}.$$

The states (3.24) with $n \neq 1$ are non-orthogonal. Indeed,

$$\langle i_1 i_2 | i'_1 i'_2\rangle = \delta_{i_1 i'_1}\delta_{i_2 i'_2} + \delta_{i_1 i'_2}\delta_{i_2 i'_1} - \Gamma^{i_1 i'_1}_{i_2 i'_2}. \tag{3.25}$$

The diagonal matrix elements determine the normalization of states (3.24). Thus we have for the state $|i_1 i_2\rangle$

$$N_F(2) = (1 + \delta_{i_1 i_2} - \Gamma^{i_1 i_2}_{i_1 i_2})^{-\frac{1}{2}}. \tag{3.25'}$$

Marumori has introduced the boson subspace of normalized orthogonal states

$$|n) = N^{(n)}_B |i_1 \ldots i_n) = N_B(n)(b^+_{i_1})^{n_1}(b^+_{i_2})^{n_2} \ldots |0) \tag{3.26}$$

where $N_B(n) = (n_1!\, n_2! \ldots)^{-1/2}$; the operators b^+_i and b_i satisfy the boson commutation relations. Using the unitary operator U_M in the form (3.9), we transform the collective component of the fermion Hamiltonian

$$H^{\text{coll}}_F = \sum_{nn'} |n\rangle\langle n|H_F|n'\rangle\langle n'| \tag{3.27}$$

into a Hamiltonian in terms of the boson operators,

$$H^{\text{coll}}_F = U_M H^{\text{coll}}_F U^+_M = \sum_{nn'} |n)\langle n|H_F|n'\rangle(n'|. \tag{3.28}$$

Since

$$\begin{aligned} |0)(0| &=: \exp\{-\sum_i b^+_i b_i\} : \\ &= 1 - \sum_i b^+_i b_i + \frac{1}{2!}\sum_{i_1 i_2} b^+_{i_1} b^+_{i_2} b_{i_1} b_{i_2} + \ldots \end{aligned}$$

the operator $|n)(n'|$ can be rewritten as follows:

$$|n)(n'| = N_B(n)N_B(n')(b_{i_1}^+)^{n_1}(b_{i_2}^+)^{n_2} \ldots\{1-\sum_i b_i^+ b_i + \ldots\}(b_{i_1'})^{n_1'}(b_{i_2'})^{n_2'} \ldots . \tag{3.29}$$

The collective Hamiltonian H_B^{coll} will be found in terms of the fermion matrix elements $\langle n|H_F|n'\rangle$ and the expansion $|n)(n'|$, in the form (3.29). An approximation is considered to be a good one if the sum over n and n' can be limited to small values of n and n', that is, if $n + n' \leqslant n_{\text{max}}$, which is not large. In this case,

$$(n_0|H_B^{\text{coll}}|n_0') \propto \sum_{n,n'=0}^{n+n'\leqslant n_{\max}} N_B(n)N_B(n') \times \langle n|H_F|n'\rangle(n_0|(b_{i_1}^+)^{n_1}(b_{i_2}^+)^{n_2} \ldots \times \{1-\sum_i b_i^+ b_i + \ldots\}(b_{i_1'})^{n_1'}(b_{i_2'})^{n_2'} \ldots |n_0') \tag{3.30}$$

and the matrix elements $(n_0|H_B^{\text{coll}}|n_0')$ of a high order in n_0, n_0' are given approximately by a sum containing fermion matrix elements $\langle n|H_F|n'\rangle$ of a low order in n and n'. The matrix elements $\langle n|H_F|n'\rangle$ are given by the expressions

$$\langle i_1 \ldots i_n|i_1 \ldots i_n\rangle \qquad \langle i_1 \ldots i_{n+1}|i_1 \ldots i_{n+1}\rangle$$
$$\langle i_1 \ldots i_{n+2}|i_1 \ldots i_{n+2}\rangle.$$

A recurrent relation between such expressions was obtained in [97, 98]; the normalization of multi-phonon fermion states is proved to be reducible to the normalization of the two-phonon state in the form of (3.25) and (3.25′). It is assumed that the matrix elements $\langle n|H_F|n'\rangle$ do not take the system from the subspace of collective states.

The operators $\bar{Q}_i^+$ are not ideal bosons; hence, a rigorous application of the Pauli exclusion principle shows that $\langle i_1 \ldots i_n|i_1 \ldots i_n\rangle$ decrease rapidly as n increases. As a result, the Marumori expansion for collective states converges rapidly as n increases, so that the terms with high values of n can be dropped. This expansion is found to be efficient if the collective subspace defined by the states $|i_1 \ldots i_n\rangle$ is close to the space defined by the approximate eigen functions of the multi-particle Hamiltonian, that is, if H produces a weak coupling to the remaining part of the state space.

If all the $\bar{Q}_i^+$, including the non-collective ones, are taken into account, the Marumori method becomes identical, as a first approximation, to the Tamm–Dancoff method. A boson expansion method based on the Tamm–Dancoff approximation was also developed in [99].

Certain difficulties are encountered when boson expansion methods are employed. For example, it was shown in [100] that if we limit the analysis to the lowest orders of the expansion of the collective fermion pairs in particle-particle bosons, then no correct quantitative description of the quadrupole states of nuclei can be obtained. This is an indication that higher-order terms must be introduced into the model.

3.2 MICROSCOPIC DESCRIPTION OF QUADRUPOLE EXCITATIONS

3.2.1

Quadrupole degrees of freedom, introduced by Bohr and Mottelson, clearly manifest themselves in nuclear spectra [4]. The collective states are coupled by large matrix elements of the quadrupole moment operator. We will use the RPA as a basis for the boson representation of quadrupole excitations, as in [99, 100]. Following [101], we introduce the operators

$$\begin{aligned} q_i^{\lambda\mu} &= \frac{1}{2}\sum_{jj'} q_{jj'}^{\lambda i}[A^+(jj';\lambda\mu) + (-)^{\lambda-\mu}A(jj';\lambda-\mu)] \\ p_i^{\lambda\mu} &= -\frac{i}{2}\sum_{jj'} p_{jj'}^{\lambda i}[A(jj';\lambda\mu) - (-)^{\lambda-\mu}A^+(jj';\lambda-\mu)] \end{aligned} \tag{3.31}$$

where $i = 1, 2, \ldots$ is the ordinal number of a root of secular equation (2.105′). The amplitudes $p_{jj'}^{\lambda i}$ and $q_{jj'}^{\lambda i}$ are related by the conditions

$$\begin{aligned} &\sum_{jj'} q_{jj'}^{\lambda i} p_{jj'}^{\lambda i'} = \delta_{ii'} \\ &\sum_{i} q_{jj'}^{\lambda i} p_{j_2 j_2'}^{\lambda i} = \frac{1}{2}\left[\delta_{jj_2}\delta_{j'j_2'} - (-)^{j+j'-\lambda}\delta_{jj_2'}\delta_{j'j_2}\right]. \end{aligned} \tag{3.32}$$

The operators $q_i^{\lambda\mu}$ and $p_i^{\lambda\mu}$ satisfy the following commutation relation:

$$\begin{aligned} [q_i^{\lambda\mu}, p_{i'}^{\lambda'\mu'}] &= i\delta_{\lambda\mu i,\lambda'\mu' i'} - i\sum_{jj'j_2JM} q_{jj_2}^{\lambda i} p_{j'j_2}^{\lambda' i'} B(jj';JM) \\ &\times \begin{Bmatrix} j & j' & J \\ \lambda' & \lambda & j_2 \end{Bmatrix} \langle \lambda\mu\lambda' - \mu'|JM\rangle(-)^{\lambda'-\mu'}[1 + (-)^{j+j'+J}]. \end{aligned} \tag{3.33}$$

If we neglect the second term in (3.33), as we did in the RPA, we obtain a commutation relation between a coordinate and a momentum; hence, they are treated as the generalized coordinates and momenta.

In order to have a complete set of operators (a closed algebra), we need to add to the operators $q_i^{\lambda\mu}$ and $p_i^{\lambda\mu}$ the linearly independent commutators $[p_i^{\lambda\mu}, p_{i'}^{\lambda'\mu'}]$, $[q_i^{\lambda\mu}, q_{i'}^{\lambda'\mu'}]$ and $[q_i^{\lambda\mu}, p_{i'}^{\lambda'\mu'}]$. It was shown in [101] that this set of operators is equivalent to a set of two-quasiparticle operators $\alpha_{jm}^+ \alpha_{j'm}^+$, $\alpha_{j'm'} \alpha_{jm}$ and $\alpha_{jm}^+ \alpha_{j'm'}$.

As an illustration, consider the β oscillations of nuclei. In this case, the complete set consists of the operators q_{coll}^{20}, p_{coll}^{20}, and $[q_{\text{coll}}^{20}, p_{\text{coll}}^{20}]$, that is, of operators with $i = 1$. The expression for the double commutators is

$$[[q_{\text{coll}}^{20}, p_{\text{coll}}^{20}], q_{\text{coll}}^{20}] = -L_1 p_{\text{coll}}^{20} - \sum_{i \neq 1} L_i p_i^{2,0}$$

$$[[q_{\text{coll}}^{20}, p_{\text{coll}}^{20}], p_{\text{coll}}^{20}] = \tilde{L}_1 q_{\text{coll}}^{20} + \sum_{i \neq 1} \tilde{L}_i q_i^{2,0} \qquad (3.34)$$

where

$$L_1 = 4 \sum_{jj'j_2j_2'} q_{jj'}^{20} p_{j_2j'}^{20} g_{jj_2'}^{20} q_{j_2j_2'}^{20}$$

$$\tilde{L}_1 = 4 \sum_{jj'j_2j_2'} q_{jj'}^{20} p_{j_2j'}^{20} p_{jj_2'}^{20} p_{j_2j_2'}^{20}.$$

They show that the algebra of the operators q_{coll}^{20}, p_{coll}^{20} and $[q_{\text{coll}}^{20}, p_{\text{coll}}^{20}]$ is not closed. If we consider only the collective states and neglect the operators p_i^{20} and q_i^{20} with $i \neq 1$, then these operators form a closed algebra which coincides with the SU(2)-group algebra.

The following commutators and operators are introduced when the β and γ oscillations are taken into account:

$$q_{\text{coll}}^{20}, \; q_{\text{coll}}^{22}, \; p_{\text{coll}}^{20}, \; p_{\text{coll}}^{22}, \; [q_{\text{coll}}^{20}, q_{\text{coll}}^{22}], \; [p_{\text{coll}}^{20}, p_{\text{coll}}^{22}]$$

$$[q_{\text{coll}}^{22}, p_{\text{coll}}^{22}], \; [q_{\text{coll}}^{20}, p_{\text{coll}}^{20}], \; [q_{\text{coll}}^{20}, p_{\text{coll}}^{22}], \; [q_{\text{coll}}^{22}, p_{\text{coll}}^{20}].$$

Among these operators, there are eight linearly independent ones, which form the SU(3) algebra. If all five quadrupole degrees of freedom are considered, that is, if we add rotation, then the operators $q_{\text{coll}}^{2\mu}$, $p_{\text{coll}}^{2\mu}$ and their commutators (as given in [101]) form a closed algebra identical to the SU(6) algebra.

3.2.2

Let us construct a collective Hamiltonian which we want to be rotationally invariant and also invariant with respect to the time reversal operation. Since the fermion Hamiltonian is expressed in terms of the pairs of operators $A(jj'; \lambda\mu)$, $A^+(jj'; \lambda\mu)$ and $B(jj'; \lambda\mu)$, the collective Hamiltonian must be quadratic with respect to the operators that form the SU(6) algebra In

[101], the fermion Hamiltonian is written, using formulas (3.31) and (3.33), in terms of the operators $q^{2\mu}_{\rm coll}$ and $p^{2\mu}_{\rm coll}$ and their linearly independent commutators that close the SU(6) algebra. For the operators $q^{2\mu}_{\rm coll}$ and $p^{2\mu}_{\rm coll}$ and their commutators, there is an exact boson realization using the quadrupole boson operators d^+_μ, d_μ ($\mu = 0, \pm 1, \pm 2$):

$$\begin{aligned}
q^{2\mu}_{\rm coll} &= \sqrt{N - \hat{N}_d}(-)^\mu d^+_{-\mu} + d^+_\mu \sqrt{N - \hat{N}_d} \\
p^{2\mu}_{\rm coll} &= {\rm i}[(-)^\mu d^+_{-\mu}\sqrt{N - \hat{N}_d} - \sqrt{N - \hat{N}_d}\, d_\mu] \\
{\rm i}[\, q^{2\mu}_{\rm coll}, p^{2\mu'}_{\rm coll}\,] &= d^+_\mu d_{\mu'} + (-)^{\mu+\mu'} d^+_{-\mu'} \cdot d_{-\mu} - \delta_{\mu\mu'}(N - \hat{N}_d) \\
[\, q^{2\mu}_{\rm coll}, q^{2\mu'}_{\rm coll}\,] &= -(-)^{\mu'} d^+_\mu d_{-\mu'} - (-)^\mu d^+_\mu \cdot d_{-\mu} \\
\hat{N}_d &= \sum_\mu d^+_\mu d_\mu
\end{aligned} \qquad (3.34')$$

where $[d_\mu, d^+_\mu] = \delta_{\mu\mu'}$, $[d_\mu, d_{\mu'}] = [d^+_{\mu'}, d^+_\mu] = 0$.

Here N is a positive integer equal to the maximum number of bosons in the wave function. The fermion operators are represented by infinite Taylor series in powers of the operator $\hat{N}_d$. This boson representation of the fermion operators is of the same type as the representation for the SU(2) algebra in [102]. Boson realization (3.34′) corresponds to the completely symmetric irreducible representation of the SU(6) group, the integer N being the eigenvalue of the linear Casimir operator for this group.

We will use formulas (3.34′) to construct the collective Hamiltonian in the form

$$\begin{aligned}
H^{\rm coll} = h_0 &+ h_1 \hat{N}_d + h_2[(d^+ \cdot \tilde{d}^+)\sqrt{(N - \hat{N}_d)(N - 1 - \hat{N}_d)} \\
&+ {\rm h.c.}] + h_3\{d^+ \cdot \widetilde{[d^+ \times d\,]}^{(2)}\sqrt{N - \hat{N}_d} + {\rm h.c.}\} \\
&+ \sum_{L=0,2,4} \{h_{4L}\widetilde{[d^+ \times d^+]}^{(L)} \cdot [d \times d]^{(L)}\}
\end{aligned}$$

where

$$\tilde{d}_\mu = (-)^\mu d_{-\mu} \qquad \widetilde{[d^+ \times d\,]}^{(2)} = (-)^\mu [d^+ \times d]^{2-\mu}. \qquad (3.35)$$

Here $\cdot$ stands for scalar product, and $\times$, for the tensor product. The information on the mean field and on the residual interactions is included in the parameters h_0, h_1, h_2, h_3 and h_{4L}. The quadrupole moment operator is

$$Q_{2\mu} = \tilde{e}(d^+_\mu \sqrt{N - \hat{N}_d} + \sqrt{N - \hat{N}_d}\, \tilde{d}_\mu) + \chi [d^+ \times d]^{(2\mu)} \qquad (3.36)$$

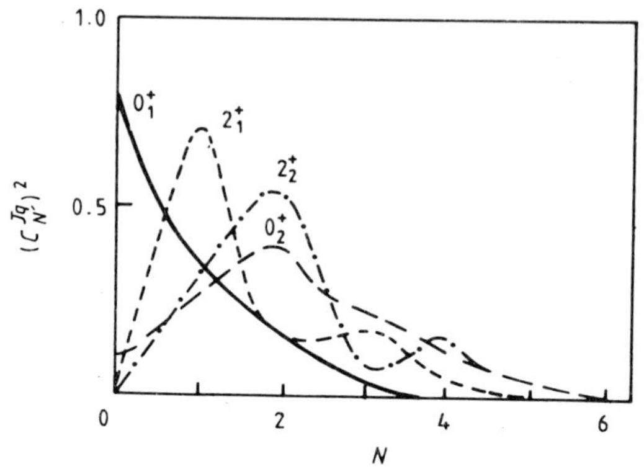

Figure 3.3 The contribution of bosonless components and of the components with one, two and a larger number of bosons to the normalization of wave functions of the ground states 0_1^+ and the excitation 2_1^+, 2_2^+ and 0_2^+ states in ^{152}Gd; the results were calculated in [101] for $N = 7$.

where $\tilde{e}$ and χ are parameters. Typically, N is assumed to equal one half of the number of nucleons in an open shell. Hamiltonian (3.35) is diagonalized in the state space which realizes the basis of the completely symmetric irreducible representation SU(6) with the quantum number N, that is, the wave function of the collective state has the form

$$|NJMq) = \sum_{N'=0}^{N} C^{Jq}_{N'} (d^+_\mu)^{N'}_{JMq} |0) \tag{3.37}$$

where $d_\mu|0) = 0$; q denotes other quantum numbers. Orthonormalization conditions are superposed on the coefficients C^{Jq}_N.

Wave function (3.37) contains components with different numbers of bosons. Figure 3.3 shows that in addition to the contribution of the dominating one-boson component, appreciable contributions are produced in the wave function of the first 2_1^+ state by the components with $N' = 2$ and $N' = 3$. In the 0_2^+ and 2_2^+ states, a large contribution is produced, in addition to the two-boson component, by the components with $N' = 1$, 3 and 4.

Hamiltonian (3.35) is a combination of terms which are linear and quadratic in powers of the operators

$$d^+_\mu \sqrt{N - \widehat{N}_d} \quad \sqrt{N - N_d} d_\mu, d^+_\mu d_{\mu'}. \tag{3.38}$$

This presentation is done in the SU(6) approximation. It is based (*a*) on the assumption that double commutators similar to (3.34′) do not lead beyond the collective subspace, that is, that we assume only collective quadrupole degrees of freedom with $i = 1$, and (*b*) on neglecting the dependence of the double commutator on J. Operators (3.38) and their commutators form 35 linearly independent basis operators of the SU(6) group. This means that we have obtained a boson realization of the SU(6) algebra for the completely symmetric representation characterized by the quantum number N.

Arima and Iachello [103] gave an elegant interpretation of the quadratic root in (3.35) using the transformation

$$d^+_\mu \sqrt{N - \widehat{N}_d} \to d^+_\mu s \qquad \sqrt{N - \widehat{N}_d}\, d_\mu \to s^+ d_\mu$$
$$N - \widehat{N}_d \to s^+ s \equiv N_s \tag{3.39}$$

where s^+ is the creation operator for an s boson with angular momentum $J = 0$. (By virtue of the conservation of the total number of bosons, the scalar s boson can be eliminated.) No new degrees of freedom arise if the s boson is introduced in this way. The collective Hamiltonian H^{coll} in the SU(6) approximation can be written in the form

$$H^{\text{coll}} \to U_M H_F U_M^+ = \Gamma_0 P \tilde{H} \tag{3.40}$$

where $\widetilde{H}$ is expressed in terms of the s and d boson operators. Arima and Iachello [103] have postulated a phenomenological Hamiltonian of the interacting bosons model (IBM). Arima and Iachello assume the SU(6) group to be a dynamic symmetry group.

The IBM Hamiltonian cannot be obtained directly from the fermion Hamiltonian via the Marumori mapping. According to [104], the mapping of the fermion state space to a boson state space goes in two stages. At the first stage, the space of fermions coupled to $J = 0$ is mapped to an s boson space, and at the second stage, the state space with non-zero seniority is mapped to a d boson space. As a result of the two-stage Marumori mapping, we obtain a Hamiltonian as an infinite series in d bosons. The SU(6) approximation is used to convert to the IBM Hamiltonian.

Note that the pairing subspace contains both the ground and the excitation states of zero seniority. One of the important modes of excitation of the pair subspace is the pairing-vibration mode [105] whose manifestation was first observed in [106] as an enhancement of favourable α decays. It has been shown [6, 61] that the pairing-vibration mode in deformed nuclei is mixed with the quadrupole mode, and that they both form β vibration states.

The equivalence of the model developed in [101] and outlined in this subsection to the IBM model was proved in [107] where the Hamiltonians and

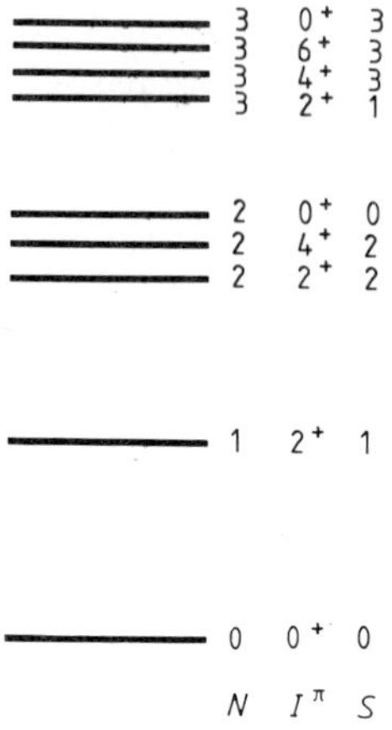

Figure 3.4 Spectrum of collective states of a weakly anharmonic oscillator.

the corresponding operators were shown to be identical. This was done by explicitly constructing a Schwinger realization of the generators $q^{2\mu}_{\rm coll}$, $p^{2\mu}_{\rm coll}$, $[q^{2\mu}_{\rm coll}, p^{2\mu'}_{\rm coll}]$, $[q^{2\mu}_{\rm coll}, q^{2\mu'}_{\rm coll}]$, $[p^{2\mu}, p^{2\mu''}]$ of the SU(6) collective quadrupole algebra in terms of the boson operators d^+_μ, d_μ, s^+, s. A relation was established between the phenomenological parameters of the two Hamiltonians.

3.2.3

The method with Hamiltonian (3.35), developed in [101] and outlined above, was meant to describe low-lying states of transition nuclei for which the RPA is not valid. This method can also be used to analyse the transition from spherical to deformed nuclei and to describe a number of spherical nuclei.

Let us consider a particular solution of the Schrödinger equation with Hamiltonian (3.35), for $h_2 = h_3 = 0$. In this case, the Hamiltonian commutates with the operator $\widehat{N}_d$ and the eigenfunctions are characterized by the number of quadrupole quanta N. The sytem's energy is

$$E(I, N, \mathfrak{S}) = a_1 N + a_2 N^2 + a_3 \mathfrak{S}(\mathfrak{S} + 3) + a_4 I(I + 1)$$

where I is the angular momentum and $\mathfrak{S}$ is the seniority; the coefficients a_1, a_2, a_3, a_4 are expressed in terms of h_1 and h_{4L}. If a_2, a_3, a_4 are small in comparison with a_1, we arrive at the spectrum of a weakly anharmonic oscillator whose schematic diagram is shown in figure 3.4. If we set $\chi = 0$ in operator (3.36), then the same relations as in the model of harmonic quadrupole vibrations are obtained for the probabilities of the $E2$ transitions.

In a number of spherical nuclei, the weakly anharmonic vibrator model fails to explain the ratios of the probabilities $B(E2)$ of transitions from a two-phonon triplet to the ground state and to the one-phonon state, and the ratios for the transitions between the triplet states. For example, it proved to be insufficient for the interpretation of the ratios in ^{108}Pd to take into account the mixing of one-, two- and three-phonon states. An analysis performed in [101] has demonstrated that an admixture of states with a large number of bosons plays a very important role. Thus the potential energy of quadrupole vibrations in ^{108}Pd has a maximum, not a minimum, at $\beta_2 = 0$. The minimum lies at $\beta_2 \neq 0$ but its depth is considerably smaller than the energy of the zero vibrations. As a result, the properties of low-lying states of ^{108}Pd differ dramatically from those of spherical and deformed nuclei. It is in such complex cases that a description based on the boson expansion method proves its effectiveness.

The properties of many transition nuclei are not explained by interpolating between the vibrational and the rotational limiting cases. The ratio of energies can be quite close to the value predicted by the rotational model while the ratio of $B(E2)$ may be better predicted by the vibrational model, and vice versa. Thus calculations were carried out in [101] for the low-lying collective states of ^{150,152}Sm and ^{152}Gd using Hamiltonian (3.35) with $h_{4L} = 0$. The Hamiltonian's parameters are fixed on the basis of experimental data, namely, $N = 7$ for ^{150}Sm and ^{152}Gd and $N = 9$ for Sm^{152}. The calculations resulted in a satisfactory description of the energies of the states 0_1^+, 2_1^+, 4_1^+, 6_1^+, 8_1^+ and also of 0_2^+, 2_2^+, 4_2^+ and 2_3^+, 3_1^+, 4_3^+. Successful description was also achieved for the ratios of $B(E2)$ in good agreement with the rotational model, while the energy ratios are found to be close to those predicted by the vibrational model. The reason for this is that the wave functions of these states have a complex structure. The wave functions of excitation states contain terms with different numbers of bosons; this was seen in figure 3.3, where no component (except the 2_1^+ state) exceeds 50%. As a rule, only two or three components are responsible for the main contribution. If the ground states of ^{150}Sm and ^{152}Gd are nearly spherical, then the wave functions of the excitation states with various components possess rotational and vibrational properties.

The rotational motion in transition nuclei does not decouple from the internal motion. However, the experimental data indicate that the levels can be grouped into quasi-rotational bands [108] within which the values of $B(E2)$ are considerably greater than between the levels of different quasi-rotational bands. The explanation is that the levels of one quasi-rotational band have similar boson structures. Their dependence on the number of bosons can be approximately described by the same curve which shifts by unity along the boson number axis as we go from one level of a quasi-rotational band to another. The similarity leads to high values of $B(E2)$ within quasi-rotational bands. The boson structures of the levels in different quasi-rotational bands differ greatly, which explains the low values of $B(E2)$ for the transitions between the levels of different bands. Various modifications of the boson models are widely used to describe the collective quadrupole motion (see, e.g., [109]).

3.3 INTERACTING BOSONS MODEL

3.3.1

We have mentioned already that the phenomenological IBM model was developed by Arima and Iachello [103, 110, 111] who used the group theory methods. They introduced two types of bosons: s bosons with $J = 0$ and quadrupole d bosons (d_μ, $\mu = 0, \pm 1, \pm 2$), and assumed that the 0^+ and 2^+ collective nucleon pairs play the dominant role in the description of quadrupole collective states. They had used the Schwinger realization of the 35 generators of the SU(6) group and postulated that the collective Hamiltonian is the most general rotational invariant that can be constructed out of the generators of the SU(6) group. The bosons are expressed in terms of correlated nucleon pairs, so that the IBM model is related to the conventional shell model. In fact, the IBM is based on a greatly truncated fermion space of the shell model. The IBM model was developed for the description of low-lying states of open-shell nuclei. We will restrict the analysis to even–even nuclei.

In the initial (and the more widespread) version of the model (referred to as IBM-1), the difference between the proton and neutron bosons is neglected. The Hamiltonian of the model conserves the total number of bosons:

$$N = N_s + N_d = s^+ s + \sum_\mu d_\mu^+ d_\mu. \tag{3.41}$$

The Hamiltonian is limited to fourth-order terms and contains the following products:

$$s^+ s, \; d^+ d, \; d^+ d^+ dd, \; s^+ s^+ ss, \; d^+ d^+ ss, \; d^+ d^+ ds + \text{h.c.}$$

The operators s and d satisfy the boson commutation relations.

The part $H_{\text{IBM}-1}$, which contributes to the excitation energy, is often presented in the form

$$\begin{aligned} H_{\text{IBM}-1} = {} & E_d \widehat{N}_d + \kappa''(P^+ \cdot P) - \kappa'(L \cdot L) \\ & - \kappa(Q \cdot Q) + C_3(T_3 \cdot T_3) + C_4(T_4 \cdot T_4) \end{aligned} \tag{3.42}$$

where

$$\begin{aligned} & \widehat{N}_d = (d^+ \cdot \tilde{d}) \quad P = (\tfrac{1}{2})[(\tilde{d} \cdot \tilde{d}) - (s \cdot s)] \\ & L = \sqrt{10}\,[d^+ \times d]^{(1)} \end{aligned} \tag{3.43}$$

$$Q = [d^+ \times s + s^+ \times \tilde{d}]^{(2)} - \sqrt{7/2}\,[d^+ \times \tilde{d}]^{(2)} \tag{3.44}$$

$$T_3 = [d^+ \times \tilde{d}]^{(3)} \quad T_4 = [d^+ \times \tilde{d}]^{(4)} \tag{3.45}$$

and E_d, κ, κ', κ'', C_3 and C_4 are phenomenological parameters. It is convenient to present the Hamiltonian in the form (3.42) since we can retain, for example, only three terms.

If the interaction between bosons is neglected, the wave function of the non-perturbed ground state of the nucleus is the product of N s-bosons:

$$|g\rangle = (1/\sqrt{N!})(s^+)^N|0\rangle \tag{3.46}$$

that is, it is a boson condensate corresponding to the situation when only the lowest-energy states are occupied. The boson vacuum is identified with the double-magic core. In excitation states, some s bosons are replaced with d bosons. The wave function of the quadrupole excitation is

$$|\mu\rangle = [1/\sqrt{N-1)!}]\, d_\mu^+(s^+)^{N-1}|0\rangle. \tag{3.47}$$

The non-perturbed spectrum is the spectrum of the harmonic oscillator.

The boson–boson interactions lead to more complex wave functions of the ground and excitation states. The wave function of the collective state with boson number N, angular momentum J, its projection M and additional quantum number q is

$$|NJMq\rangle = \sum_{N_d=0}^{N} C_{N_d}^{Jq}(d^+)_{JMq}^{N_d}(s^+)^{N-N_d}|0\rangle. \tag{3.48}$$

Orthonormalization conditions are imposed on the coefficients C_{Nd}^{Jq}. The total number of bosons is

$$N = \begin{cases} n/2 & n \leqslant \Omega \\ \Omega - n/2 & n > \Omega \end{cases}$$

where $\Omega = \Sigma_i(j_i + \frac{1}{2})$ is the number of the mean field single-particle levels taken into account; in many cases, Ω is chosen equal to the number of levels in the shell being filled. In the formula above, n is the sum of the numbers of neutrons and protons in excess of the closed shell. Cases are possible in which N is equal to the sum of one half the number of neutrons in excess of the closed shell and one half of the number of protons needed to close the shell (the number of proton holes). When Hamiltonian (3.42) is diagonalized, one calculates the energies E_N of the collective states and the coefficients $C_{N_d}^{Jq}$ of wave function (3.48).

The operators corresponding to the electromagnetic transitions with $\lambda \leqslant 4$ and to the static moments is expressed in terms of boson operators:

$$\mathfrak{M}(\lambda\mu) = \tilde{e}\delta_{\lambda,2}[d^+ \times s + s^+ \times d]^{(2\mu)} + \chi_\lambda[d^+ \times d]^{\lambda\mu} + \chi_0\delta_{\lambda\mu,00}[s^+ \times s]^{(00)} \tag{3.49}$$

with the operator of the $E2$ transition being

$$\mathfrak{M}(E2\mu) = \tilde{e}\{[d^+ \times s + s^+ \times \tilde{d}]^{(2\mu)} + \chi[d^+ \times \tilde{d}]^{2\mu}\} \tag{3.50}$$

and with the selection rules $\Delta N_d = 0, \pm 1$. In this formulation of the model, all $M1$ transitions are forbidden and equation (3.49) contains only the diagonal component of the $M1$ operator responsible for the magnetic moment. It is also possible to perform IBM calculations of the intensity of two-nucleon transfer reactions.

The wave eigenfunctions of Hamiltonian (3.42) are the basis functions of irreducible representations of the SU(6) group. In order to find the quantum numbers and construct the complete basis of multiboson states, it is necessary to determine the invariant subgroups of SU(6). The following chains of subgroups are used to classify the groups of SU(6):

$$\mathrm{SU}(6) \supset \mathrm{SU}(5) \supset \mathrm{O}(5) \supset \mathrm{O}(3)$$
$$\mathrm{SU}(6) \supset \mathrm{SU}(3) \supset \mathrm{O}(3)$$
$$\mathrm{SU}(6) \supset \mathrm{O}(6) \supset \mathrm{O}(5) \supset \mathrm{O}(3).$$

The IBM has several limits corresponding to various subgroups of SU(6), depending on the values of the parameters.

Consider the vibration limit with the SU(5) symmetry. In this case, the s bosons are of no significance, only quadrupole bosons are to be considered, and the Hamiltonian of the model becomes

$$H'_{\mathrm{IBM-1}} = E_d\widehat{N}_d + \sum_{L=0,2,4} C_L[(d^+ \times d^+]^{(L)}[\tilde{d} \times \tilde{d}]^{[L]}). \tag{3.51}$$

The energy eigenvalues and transition probabilities are found in analytic form.

Let us consider a case in which we assume in (3.42) that $\kappa'' = C_3 = C_4 = 0$; Hamiltonian (3.42) then changes to

$$H''_{\mathrm{IBM-1}} = E_d(d^+\tilde{d}) - \kappa'(L \cdot L) - \kappa(Q \cdot Q). \tag{3.52}$$

It is a function of the parameters E_d, κ' and κ. The rotational limit based on the SU(3) group is reached for $E_d = 0$. The spectrum of an axially symmetric rigid rotator with degenerate β and γ states is then produced.

Finally, the O(6) subgroup plays an important role in state classification. Eigenvalues can also be obtained in an analytic form. In this limit, the IBM can describe nuclei that are soft with respect to the γ deformation.

Nuclear spectra are not satisfactorily reproduced in the framework of the indicated limits of the model. The spectra of some nuclei, for example, samarium isotopes, are described using Hamiltonian (3.52) which is intermediate between the SU(5) and SU(3) limits. The states 2_1^+, 0_2^+, 2_2^+ and 4_1^+ of osmium isotopes are calculated using the Hamiltonian which is intermediate between the SU(3) and O(6) limits. In xenon isotopes, such states are described using a Hamiltonian which is intermediate between the O(6) and SU(5) limits. For some nuclei, the calculation of spectra requires the diagonalization of the total Hamiltonian (3.42). If the IBM-1 is used, its large number of free parameters leads to a satisfactory description of low-lying levels and of the probabilities of $E2$ transitions between them in spherical, transition and deformed nuclei (see [110–112]).

3.3.2

The s and d bosons introduced are similar to the correlated pairs of nucleons whose angular momenta are $J = 0$ and $J = 2$. According to the shell model, neutrons and protons of complex nuclei occupied different shells. The structure of low-lying states of even–even nuclei is found in terms of the correlated pairs of neutrons and protons in open shells. A more complex version of the model, known as IBM-2, was formulated in order to establish a relation between the boson and the nucleon degrees of freedom [110, 111, 113, 114]. Two types of bosons were introduced: the proton s_{p}, d_{p} bosons and the neutron s_{n}, d_{n} bosons. The group structure of the model becomes $\mathrm{SU}^{\mathrm{p}}(6) \otimes \mathrm{SU}^{\mathrm{n}}(6)$. The transition to the IBM-2 was also stimulated by the spectra of nuclei with a single closed shell. If the spectrum is only slightly affected by the addition of valence nucleon pairs, a pair added in excess of a closed shell sharply reduces the energy of the 2_1^+ state. This is an indication of an important role played by the neutron–proton interaction, and of the desirability of its explicit introduction into the collective Hamiltonian.

The IBM-2 Hamiltonian has the form

$$H_{\mathrm{IBM-2}} = E_{\mathrm{p}} N_{d\mathrm{p}} + E_{\mathrm{n}} N_{d\mathrm{n}} + \kappa (Q_{\mathrm{p}} \cdot Q_{\mathrm{n}}) + V_{\mathrm{pp}} + V_{\mathrm{nn}} + M_{\mathrm{pn}} \tag{3.53}$$

where

$$\begin{aligned} N_{d\tau} &\doteq (d^+ \cdot \tilde{d})_\tau \qquad \tau = n, p \\ Q_\tau &= [d^+ \times s + s^+ \times \tilde{d}]_\tau^{(2)} + \chi_\tau [d^+ \times \tilde{d}]_\tau^{(2)} \end{aligned} \tag{3.54}$$

$$V_{\tau\tau} = \sum_{L=0,2,4} C_{L\tau}\frac{1}{2}(2L+1)^{\frac{1}{2}}[[d^+ \times d^+]^{(L)}_\tau [\tilde{d} \times \tilde{d}]^{(L)}_\tau]^{(0)}$$

$$M_{\rm pn} = \xi_2[[s^+_{\rm n} \times d^+_{\rm p} - d^+_{\rm n} \times s^+_{\rm p}]^{(2)} \times [s_{\rm n} \times \tilde{d}_{\rm p} - \tilde{d}_{\rm n} \times s_{\rm p}]^{(2)}]^{(0)} + \sum_{k=1,3} \xi_k[[d^+_{\rm n} \times d^+_{\rm p}]^{(k)}[\tilde{d}_{\rm n} \times d_{\rm p}]^{(k)}]^{(0)}. \quad (3.55)$$

Here $E_{\rm p}$ and $E_{\rm n}$ are the energies of the proton and neutron bosons, $\kappa(Q_{\rm p} \cdot Q_{\rm n})$ is the quadrupole-quadrupole interaction between the proton and neutron bosons, and the terms $V_{\rm nn}$ and $V_{\rm pp}$ describe the interactions between similar bosons. The Majorana term $M_{\rm pn}$ was introduced to shift the states that are not completely symmetric with respect to the proton and neutron degrees of freedom, towards higher energies. The total number of bosons is conserved:

$$N = N_{\rm p} + N_{\rm n}$$
$$N_{\rm p} = N_{s\rm p} + N_{d\rm p} \quad N_{\rm n} = N_{s\rm n} + N_{d\rm n}. \quad (3.56)$$

Typically, $2N_{\rm p}$ and $N_{\rm n}$ are assumed to equal the numbers of protons and neutrons, respectively, outside the filled shells. The wave function of a collective state having the number N of bosons with JM and additional quantum numbers $q_{\rm n}$ and $q_{\rm p}$ takes the form

$$|NN_{\rm p}N_{\rm n}; JMq_{\rm p}q_{\rm n}) = \sum_{N_{d\rm p}=0}^{N_{\rm p}} \sum_{N_{d\rm n}=0}^{N_{\rm n}} C^{Jq_{\rm p}q_{\rm n}}_{N_{d\rm p}N_{d\rm n}} \times \left[(d^+_{\rm p})^{N_{d\rm p}}_{J_{\rm p}M_{\rm p}q_{\rm p}}(d^+)^{N_{d\rm n}}_{J_{\rm n}M_{\rm n}q_{\rm n}}\right]^{(JM)} \times (s^+_{\rm p})^{N_{\rm p}-N_{d\rm p}}(s^+_{\rm n})^{N_{\rm n}-N_{d\rm n}}|0). \quad (3.57)$$

Orthonormalization conditions are imposed on the coefficients $C^{Jq_{\rm p}q_{\rm n}}_{N_{d\rm p}N_{d\rm n}}$. It is on such a basis that Hamiltonian (3.53) is diagonalized.

A mixture of two configurations that differ in the maximum number of proton bosons (the difference was equal to two proton bosons) was introduced in a number of papers (see, e.g., [115, 116]). If a pair of protons are excited to the corresponding closed shell above, two types of bosons seem to be present: particle and hole bosons. This distinction is not implemented, however, and normally one introduces an additional term,

$$H_{\rm mix} = \zeta_1(s^+_{\rm p} \cdot s^+_{\rm p} + s_{\rm p} \cdot s_{\rm p}) + \zeta_2[d^+_{\rm p} \times d^+_{\rm p} + \tilde{d}_{\rm p} \times \tilde{d}_{\rm p}]^{(0)} \quad (3.58)$$

which is responsible for the mixture of the $N_{\rm p}$ and $N_{\rm p}+2$ configurations. Furthermore, a parameter $\Delta E_{\rm p}$ is introduced, which is the energy required to excite this configuration.

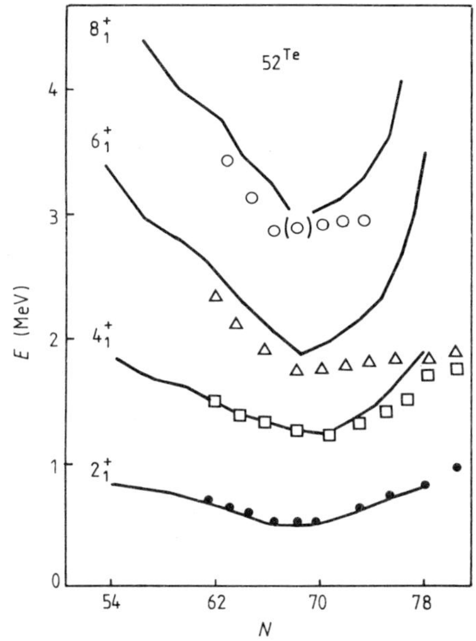

Figure 3.5 Experimental and IBM-2 energies of the 2_1^+, 4_1^+, 6_1^+ and 8_1^+ states of tellurium isotopes [116].

Hamiltonian (3.53) contains a large number of parameters. In order to reduce their number, one assumes $E_{\rm p} = E_{\rm n} = E$ and omits the terms $M_{\rm pn}$ and $H_{\rm mix}$. Even after this, ten parameters survive. Their number is usually reduced to six by assuming that $C_{4\tau} = 0$ and taking into consideration either $V_{\rm pp}$ or $V_{\rm nn}$. The remaining parameters are E, κ, $\chi_{\rm p}$, $\chi_{\rm n}$, $C_{0\tau}$, $C_{2\tau}$ for $\tau = {\rm p}$ or $\tau = {\rm n}$. As a rule, these parameters are found in the process of describing the spectrum of one nucleus (see, e.g., figure 3.5). It cannot be said that the energies of the 6_1^+ and 8_1^+ states for nuclei with $N > 70$ were reproduced well. The behaviour of the parameters ϵ, κ and $\chi_{\rm n}$ in tellurium isotopes (figure 3.6) is quite similar to that of the parameters of the cadmium isotopes. Figure 3.6 shows that as the number N of neutrons changes, some parameters change slightly while others are very appreciably affected. Obviously, the IBM-2 description seems more reliable, the smaller the changes in the parameters and the more monotone these changes are in the transition from one nucleus to another. It should be kept in mind that normally one has experimental data on the energies of 10–12 levels of a fixed nucleus, that is, about two measured results per one parameter. Hence, it is typical to determine several sets of parameters which give equally good descriptions of experimental data.

The $E2$ transition operator is chosen in the form

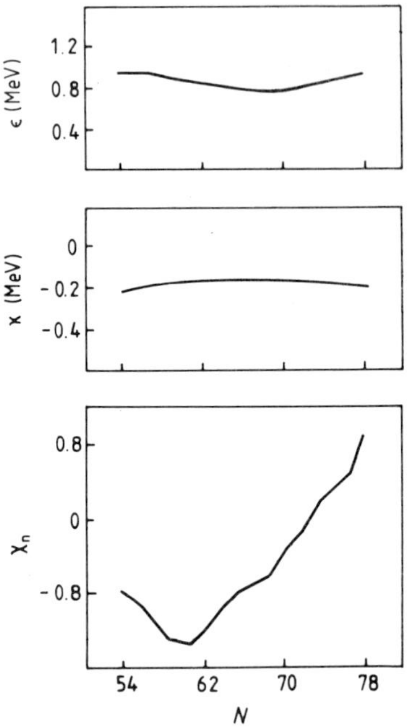

Figure 3.6 The behaviour of the IBM-2 parameters ϵ, κ and $\chi_{\rm n}$ in tellurium isotopes.

$$\mathfrak{M}(E2) = e_{\rm p}^{(2)} Q_{\rm p} + e_{\rm n}^{(2)} Q_{\rm n} \tag{3.59}$$

where $e_{\rm p}^{(2)}$ and $e_{\rm n}^{(2)}$ are effective quadrupole charges of proton and neutron bosons, and Q_τ is given by (3.54). Some authors assume that $e_{\rm p}^{(2)}$ and $e_{\rm n}^{(2)}$ depend only on $N_{\rm p}$ and $N_{\rm n}$, respectively. In other publications, authors assume that $e_{\rm p}^{(2)} = e_{\rm n}^{(2)} = \tilde{e}$ and that they are constant for all isotopes of the element. It is also assumed that $e_{\rm n}^{(2)}$ is proportional to κ and that $e_{\rm p}^{(2)}$ is a function only of $N_{\rm p}$ and is thus constant for all isotopes. For example, quite a good description was obtained in [116] for the probabilities of the $E2$ transitions $2_1^+ \to 0_1^+$ and $4_1^+ \to 2_1^+$, and a somewhat poorer description for the transitions $2_2^+ \to 2_1^+$ and $0_2^+ \to 2_1^+$ in cadmium isotopes with $\tilde{e} = 0.095 \times 10^{-24} e\,{\rm cm}^2$ and in tellurium isotopes with $\tilde{e} = 0.118 \times 10^{-24} e\,{\rm cm}^2$. If a mixture of the $N_{\rm p} = 1$ and $N_{\rm p} = 3$ configurations is taken into account by using (3.58) for $H_{\rm mix}$, the $E2$ transition operator is taken in the form

$$\mathfrak{M}(E2) = \tilde{e}_1(Q_{1\rm p} + Q_{1\rm n}) + \tilde{e}_3(Q_{3\rm p} + Q_{3\rm n}) \tag{3.60}$$

where $\tilde{e}_1$ and $\tilde{e}_3$ are the effective charges of the proton and neutron bosons in the $N = 1$ and $N = 3$ configurations. It was only by introducing these two configurations that it became possible to describe in [116] the energies and the $B(E2)$ probabilities for 2_1^+, 2_2^+, 2_3^+, 0_2^+, 0_3^+ and 4_1^+ states in ^{112,114}Cd; it was assumed that $\tilde{e}_1 = 0.095 \times 10^{-24} e\,\mathrm{cm}^2$ and $\tilde{e}_3 = 0.152 \times 10^{-24} e\,\mathrm{cm}^2$.

The IBM-2 parameters that have not been determined quite unambiguously or with sufficient accuracy were compared with the values that were calculated microscopically in the framework of the shell model. Such microscopic calculations are performed in three stages. The first is the truncation of the space of valence nucleons to the subspace of pairs with $J = 0$ and $J = 2$ and positive parity. These operators are related to the lowest 0^+ and 2^+ collective states. The second stage is the transformation of this fermion subspace to a boson subspace, mostly by using the generalized seniority concept [117]. Although the number of bosons equals one half of the number of fermions, the boson operators are not just fermion pairs with $J = 0$ and $J = 2$; they carry a high degree of the collective interaction of the fermions. The third stage is the evaluation of renormalization effects due to the states (especially those with $J = 4$) which do not belong to the truncated fermion subspace but fall under the larger space of the shell model. In a number of calculations, the renormalization effects were found to be considerable. Microscopic calculation of IBM parameters has been carried out by a number of groups [111, 117, 118] but a great deal of work has still to be done to cover the renormalization effect.

3.3.3

The IBM-2 model reproduces all the results of the IBM-1 but also contains the so-called mixed symmetry states. Such states are not completely symmetric in the *sd* space; they are allowed in IBM-2 owing to the additional neutron-proton degrees of freedom. We can expect their energies to lie above those of completely symmetric states. In deformed nuclei, the mixed symmetry states include bands with $K^\pi = 1^+$, absent from IBM-1. Experimental analysis of inelastic scattering of electrons [119] identified in the ^{156}Gd 3.075 MeV state with $K^\pi = 1^+$ and a large value of $B(M1)$ for the transition to the ground state. It is interpreted as a mixed symmetry state. An analysis of the γ decay of 2_3^+ states [120] in ^{140}Ba, ^{142}Ce, ^{144}Nd, and ^{148,150}Sm indicates that they can be regarded as mixed-symmetry IBM-2 states.

The introduction of the F spin [113] with SU(2) symmetry proved to be useful for the description of mixed-symmetry states. It is assumed that the F spin plays the same role in the neutron–proton boson system as the isospin plays in the proton–neutron fermion system. The IBM-1 operates with a single species of bosons, and the F spin is considered as a good quantum number. The SU(2) generators of the F spin are

$$F_+ = \sum_\mu d^+_{p\mu} d_{n\mu} + s^+_p s_n \quad F_- = F^+_+$$
$$F_0 = \frac{1}{2} \sum_\mu (d^+_{p\mu} d_{p\mu} - d^+_{n\mu} d_{n\mu}) + s^+_p s_p - s^+_n s_n. \tag{3.61}$$

The zero component of the F spin operator is

$$F_0 = (1/2)(N_p - N_n). \tag{3.62}$$

N_p and N_n being fixed for each nucleus, we have

$$[F_0, H_{IBM-2}] = 0.$$

If the Hamiltonian is invariant with respect to the F spin, then

$$[H_{IBM-2}, F_\pm] = 0 \qquad [H_{IBM-2}, F_0] = 0. \tag{3.63}$$

The F-spin invariance imposes severe constraints on the Hamiltonian. For instance, equation (3.53) must satisfy the conditions $E_p = E_n$, $\chi_p = \chi_n$, $V_{pp} = V_{nn}$, $M_{pn} = 0$, $\kappa = 0$. In this case, there should exist F-spin multiplets of identical energies. The F spin multiplets in a sequence of nuclei consist of a group of levels that have the same value of F and different values of F_0. F-spin multiplets have a constant number of bosons, $N = N_p + N_n$, and it is required that there are N_n neutron bosons of a certain kind (particle or hole bosons) and N_p proton bosons, also of a certain kind. Owing to a change in the boson kind from the particle to the hole in the middle of the shell being filled, an F-spin multiplet with a fixed value of F contains less than $2F + 1$ nuclei. This is what distinguishes F-spin multiplets from isospin ones.

As the F-spin invariance imposes severe constraints, some F-non-invariant terms still survive in the Hamiltonian. In spite of this, F remains a sufficiently good quantum number; furthermore, the excitation state energies of each multiplet depend on F_0 in complete analogy with the isospin multiplets. This situation is realized in the IBM-2, in which the Majorana terms shift the excitation states having other values of F spins towards higher energies. If an isotopic multiplet consists of isobaric nuclei, two types of F-spin multiplets are present. The first type consists of isobaric nuclei $\{A, Z\}$, $\{A, Z+2\}$, $\{A, Z+4\}, \ldots$, when neutrons and protons are at the same time either particles at the beginning of shell filling, or holes at the end of the process. The second type includes the sequence $\{A, Z\}$, $\{A+4, Z+2\}$, $\{A+8, Z+4\}, \ldots$, when some bosons are particle bosons and others are hole bosons. It was shown [121] that there is a similarity in the behaviour of the levels of quasi-rotational bands constructed using the ground and the quasi-γ-vibrational states in nuclei that belong to the same

F-spin multiplet and are distributed in a wide region of mass numbers A. Magic numbers are excluded for obvious reasons. The IBM-2 parameters are practically constant for nuclei that belong to one multiplet.

It can be stated that the introduction of the F spin as a quantum number proved to be useful for characterizing low-lying states of nuclei.

3.3.4

The interacting bosons model is applied to describe low-lying states of deformed nuclei [110, 111, 122, 123]. Typically, deformed nuclei are treated in terms of IBM-1; if s and d bosons are introduced, the Hamiltonian is written in the form

$$H^{\text{def}}_{\text{IBM}-1} = -\kappa(Q \cdot Q) - \kappa'(L \cdot L) + \kappa''(P^+ \cdot P) \tag{3.64}$$

where

$$Q = [d^+ \times s + s^+ \times \tilde{d}]^{(2)} + \chi[d^+ \times \tilde{d}]^{(2)} \tag{3.65}$$

and L and P are defined by (3.43). Setting $\kappa'' = 0$ gives us the SU(3) limit. In this case, the energy is given by

$$E = (0.75\kappa - \kappa')I(I+1) - \kappa C(2N, N')$$

where $C(2N, N')$ is the Casimir operator for the SU(3) group. The first term in (3.64) corresponds to the quadrupole interaction which produces, for $\kappa > 0$ and $\chi \neq 0$, the axial-symmetric rotator spectrum. The second term depends on the total angular momentum and is responsible for the predominant contribution to the rotational energy; it does not affect the internal structure of the wave function. There exist rotational bands (g, β, γ, $\gamma\gamma$, $\beta\beta$ etc) which correspond to various representations of the SU(3) group. The last term in (3.64) describes the pair interaction between bosons; it eliminates the degeneracy of various bands that is found in the SU(3) limit. As κ'' increases, there is a transition from the axial rotator spectrum to the limit in which the potential energy is independent of γ (the SO(5) symmetry). The results of calculations are influenced by the number of bosons N, which is taken equal to one half of the sum of the numbers of protons and neutrons beyond the closed shells; N may equal the sum of one half of the number of protons and neutrons required to fill the shells, or it may equal one half of the sum of the number of particles in the neutron system and the number of holes in the proton system.

The internal wave function of the ground state in the harmonic approximation for an axial-symmetric deformed nucleus is

$$\begin{aligned} |g) &= (N!)^{-1/2}(b^+)^N|0) \\ b^+ &= (1+\beta^2)^{-1/2}(s^+ + \beta d_0^+). \end{aligned} \tag{3.66}$$

This wave function is the head of the rotational band with $I = 0, 2, 4, \ldots, 2N$, which belongs to the $(2N, 0)$ representation of SU(3). The parameter β is found from the condition of minimum of the mean value of Hamiltonian (3.64) over state (3.66); it is a function of the parameters of (3.64). The term proportional to $(Q \cdot Q)$ produces a minimum at $\beta = \sqrt{2}$; the term proportional to $(P^+ \cdot P)$ gives a minimum at $\beta = 1$. If we follow [124] and set $\kappa = 8\,\text{keV}$, $\kappa' = -20\,\text{keV}$, $\kappa'' = 30\,\text{keV}$, $\chi = -\sqrt{7}/2$, we obtain $\beta = 1.24$.

In the harmonic approximation, we find excitation states of two types that possess the symmetry of the β and γ vibrations and are characterized by the quantum numbers n_β and n_γ. The state with $n_\beta = 1$ has $K^\pi = 0^+$ and is described by the wave function

$$|n_\beta = 1) = N^{-1/2} b_\beta^+ b|g) = [(N-1)!]^{-1/2} b_\beta^+ (b^+)^{N-1}|0)$$
$$b_\beta^+ = (1+\beta^2)^{-1/2}(-\beta s^+ + d_0^+). \tag{3.67}$$

The state with $n_\gamma = 1$ has $K^\pi = 2^+$ and is described by the wave function

$$|n_\gamma = 1) = N^{-1/2} d_2^+ b|g) = [(N-1)!]^{-\frac{1}{2}} d_2^+ (b^+)^{N-1}|0). \tag{3.68}$$

The energies of the states with $n_\beta = 1$ and $n_\gamma = 1$ in the leading terms in N are

$$\begin{aligned} \omega_\beta &= N(1+\beta^2)^{-2}[2\kappa(-1 - 2^{3/2}\beta + 9\beta^2 \\ &\quad + 2^{5/2}\beta^3 - \beta^4/2) + \kappa''(-1/2 + 5\beta^2 - \beta^4/2)] \\ \omega_\gamma &= N(1+\beta^2)^{-2}[2\kappa(-1 - 2^{3/2}\beta + 9\beta^2 \\ &\quad + 2^{5/2}\beta^3 - \beta^4/2) - (1/2)\kappa''(1-\beta^2)^2] \end{aligned} \tag{3.69}$$

If $\kappa'' = 0$, the energies of β and γ vibrations coincide. The state with $K^\pi = 1^+$, in whose wave function the operator d_1^+ replaces the operator $(1+\beta^2)^{-1/2}(s^+ + \beta d_0^+)$, describes rotation; d_1^+ is unrelated to the excitation of the internal degrees of freedom.

The classification of collective excitations by the quantum numbers n_β and n_γ is a standard one in the IBM. In view of the presence of dominating components, the wave functions of the states with $K^\pi = 4_1^+$ are characterized by $n_\gamma = 2$, those with $K^\pi = 2_2^+$ by $\{n_\gamma = 1,\, n_\beta = 1\}$, those with $K^\pi = 0_3^+,\, 0_4^+$ by the superposition of $n_\beta = 2$ and $n_\gamma = 2$, and so on. The energies and wave functions of collective states are written as follows:

$$E_c = E_g + n_\beta\omega_\beta + n_\gamma\omega_\gamma + E_{\text{anh}} \tag{3.70}$$

$$|Kq) = \sum_{N_d N_\beta} C_{N_d N_\beta}^{Kq} \left\{(d_2^+)_K^{N_d}(b_\beta^+)^{N_\beta}\right\}_q (b^+)^{N-N_d-N_\beta}|0) \tag{3.71}$$

where $N_d + N_\beta \leqslant N$. The parameters κ and κ'' are found from the energies of the low-lying states with $n_\gamma = 1$ and $n_\beta = 1$. As a rule, the calculated anharmonic corrections in strongly deformed nuclei are not large. Each representation of SU(3) corresponds to rotational bands which terminate at $I_{\max} = 2N$, with angular momenta of all d bosons aligned. Owing to the interaction between bosons, the strength of states with $n_\beta = 1$ and $n_\gamma = 1$ is fragmented among several levels. For example, the strength of the $n_\beta = 1$ state in ^{168}Er gives the following contributions to the state normalization: $0_g^+ = 0_1^+$—6.7%, 0_2^+—69%, 0_3^+—13% and 0_4^+—4%.

The $E2$-transition operator is

$$\mathfrak{M}(E2,\mu) = \tilde{e}\{[d^+ \times s + s^+ \times \tilde{d}]^{2\mu} + \chi'[d^+ \times \tilde{d}]^{(2\mu)}\} \tag{3.72}$$

where the parameter χ' is typically chosen to be different from χ in (3.65). If we set $\chi' = \chi$, the ratio of matrix elements corresponding to the excitation of vibration states to those for rotational states is much lower than the value obtained from experimental data. This follows from the selection rules in the SU(3) limit. The matrix elements of $E2$ transitions in the leading terms in N are

$$\begin{aligned}
(g|\mathfrak{M}(E2,0)|g) &= \tilde{e}N(1+\beta^2)^{-1}(2\beta - \sqrt{2/7}\,\chi'\beta^2) \\
(n_\beta = 1|\mathfrak{M}(E2,0)|g) &= \tilde{e}N^{1/2}(1+\beta^2)^{-1}(1 - \sqrt{2/7}\,\chi'\beta' - \beta^2) \\
(n_\gamma = 1|\mathfrak{M}(E2,2)|g) &= \tilde{e}N^{1/2}(1+\beta^2)^{-1}(1 + \sqrt{2/7}\,\chi'\beta).
\end{aligned}$$

The transitions within each rotational band (within one representation) depend weakly on χ', while the transitions between different rotational bands (between different representations) are to a large degree determined by the parameters χ'. In contrast to the Bohr–Mottelson model, the probabilities of the transitions $n_\beta = 2 \to n_\gamma = 1$ and $n_\gamma = 2 \to n_\beta = 1$ are not zero in the IBM, and the transition $n_\beta = 2 \to n_\gamma = 1$ is stronger than $n_\gamma = 2 \to n_\beta = 1$ [122]. Furthermore, the transition $n_\beta = 1 \to n_\gamma = 1$ may prove stronger that the transition $n_\beta = 1 \to |g)$.

Adding a term proportional to P^+P to the Hamiltonian increases the energy of β states relative to the γ states, reduces the probability of the transition $n_\beta = 1 \to |g)$ and violates the relations implied by the SU(3) limit. As shown in [124], IBM does not provide correct values that depend on the total moment I of the matrix elements of the $E2$ transitions. This shortcoming is due to the term $\kappa'(L \cdot L)$ added to the Hamiltonian to represent most of the rotational energy; this term does not include, however, the dynamics of the interacting s and d bosons. In order to describe correctly I-dependent matrix elements, one has to introduce additional phenomenological terms.

An analysis of the spectra of deformed nuclei makes it necessary to introduce the g boson with $\lambda = 4$ [126, 127]. The g boson is taken into account in two ways: by using a renormalized boson–boson interaction while not introducing the g-boson degrees of freedom explicitly and by introducing the g boson explicitly. With this method of defining the g boson, states with $K^\pi = 1^+$, 3^+ and 4^+ arise in deformed nuclei; these states contain two-quasiparticle components, and the subspace of two-quasiparticle states with $K^\pi = 0^+$ and 2^+ is expanded. The symmetry group has to be expanded from SU(6) to U(15) if we want to include the degrees of freedom connected with the g boson.

For states whose band heads have energies below 2 MeV, the weight of the g boson in the wave functions of the states with $I \leqslant 6$ does not exceed 30%. This is true for states with $K^\pi = 4_1^+$ and 4_2^+, that is, the sd dominance still holds in the U(15) scheme. The introduction of the g boson increases I_{max} by a factor of 2, and the contribution of the g boson becomes predominant for states with $I > 40$.

The IBM-2 is frequently employed to analyse low-lying states of deformed nuclei [126, 128]. It is a basis for calculating the probabilities of $M1$ transitions, the g factors of 2^+ states and the $E2/M1$ ratios. IBM-2 calculations give correct values of energy and the probabilities $B(M1)$ for the collective 1^+ state in ^{156}Gd (this state was discovered in [119]). The distribution of the $M1$ transition strength is connected with the difference in the collective motion of protons and neutrons. The $M1$ transition operator is written in the form

$$\mathfrak{M}(M1) = \sqrt{3/4\pi}\,(g_p L_p + g_n L_n) \qquad (3.73)$$

where L is given by (3.43); g_p and g_n are the boson g factors. Several sets of the boson g factors are used in the calculations, such as: $g_p = \mu_0$, $g_n = 0$, $g_p = 0.8\mu_0$, $g_n = 0.1\mu_0$, $g_p = 0.1\mu_0$, $g_n = 0.4\mu_0$ and some others, where μ_0 is the nuclear magneton. Note that the concept of the F spin is applied to deformed nuclei as well. At the same time, the formation of a number of additional rotational bands is an obstacle to the use of the IBM-2 for the description of deformed nuclei.

Let us discuss the relationship between the phenomenological IBM and the Bohr–Mottelson geometric collective model. We can do this since the IBM is also connected with the quantization of the shape variables. Nucleons in nuclei tend to form pairs and to generate a deformed field. The IBM assigns priority to pair formation while the Bohr–Mottelson model favours the deformed field generation. Additional degrees of freedom in the IBM due to the s boson are illusory since they are virtually cancelled out by the restriction on the number of bosons N. As shown in [129], the IBM differs from the Bohr–Mottleson model in the finite number N of bosons; it becomes identical to it as $N \to \infty$. In both models, analytic solutions are found in three limiting cases: the five-dimensional harmonic quadrupole vibrator, the rotor, and the case of γ instability. These models gives different results if the number of bosons is small. To a certain degree, the finiteness of N, that is, the adjustment of the model to the number of nucleons taken into account, makes the IBM more general than the Bohr–Mottelson model. At the same time, it is not possible in the IBM with Hamiltonian (3.64) to describe a three-axial shape of a nucleus. A certain equivalence of the IBM and the Bohr–Mottelson model guarantees the IBM's success.

Attempts to give microscopic justification to these models lead to the following question: can the wave functions of low-lying states of the fermion Hamiltonian be represented in the subspace in which all nucleons are coupled into pairs with $J = 0$ and $J = 2$, and each angular momentum is composed of such pairs. This question was analysed in [124]. In strongly deformed nuclei, described by a Hamiltonian containing the Nilsson potential and pairing, the contribution of pairs with $J = 0$ and $J = 2$ to the normalization of the wave function is greater than 90%; even in such cases, there is a considerable difference between the quadrupole moment and the pairing function which have been calculated for the cases of $J > 2$ pairs included or ignored. The introduction of pairs with $J > 4$ proved to be quite important [127] for the description of several physical quantities and for the elimination of contradictions described in [124].

3.3.5

The IBM progresses in several directions. One of them is the introduction of terms of higher order in the number of boson operators (the Hamiltonian is supplemented with terms like $d^+d^+d^+ddd$). Using them, it is possible to describe, for example, three-axial deformed nuclei. Another approach is to introduce other degrees of freedom: hexadecapole g bosons, dipole p and octupole f bosons, and so on. To improve the description of nuclear spectra, especially those in which two protons or two neutrons are in the higher shell, additional s'_{p}, d'_{p}, s'_{n}, d'_{n} boson operators are introduced. Of course, the introduction of higher-order operators and other degrees of freedom is accompanied with the appearance of a large number of new parameters.

In order to describe nuclei with an odd number of nucleons, a model of interacting bosons and fermions was developed [111], which contains fermions in addition to the s and d bosons. The model Hamiltonian consists of three parts: the boson component, the fermion component, and the boson–fermion interaction. The boson–fermion interaction consists of a large number of terms and thus has numerous parameters. The spectra calculated in this model were found to be very complex; a phenomenological analysis of the spectra observed is quite difficult. In some papers (see, e.g., [111]), the boson–fermion interaction is restricted to three terms: the effective monopole, the quadrupole, and the exchange terms. A relatively simple picture arises in the limiting cases for the boson part of the Hamiltonian. In the SU(5) limit, the resulting picture is typical of the particle-vibration model. For instance, the ^{101}Rh spectrum was described in the framework of the SU(5) $\otimes$ 9/2 scheme, that is, with a proton in the $1/g_{9/2}$ subshell. In the SU(3) limit, the main features of the Nilsson scheme are reproduced. The model of interacting bosons and fermions is successfully applied to describe the spectra of transition nuclei, such as the odd-mass isotopes of samarium, europium, platinum and gold. The spectra of odd-mass nuclei can be described by taking into account the coupling of the odd particle to the core described by the SU(6) quadrupole model [101] or to the core described by the IBM. We can point out recent attempts to describe the spectra of odd–odd nuclei in terms of an approach based on dynamic supersymmetries [130].

The IBM was widely used to analyse experimental data on the energies and probabilities of $E2$ transitions in a large number of spherical, transition and deformed nuclei. The IBM successfully described specific features of spectra of transition nuclei which other models failed to reproduce. It must be borne in mind that a good description of the energies and probabilities of the $E2$ transitions does not mean by itself that the model gives a correct reproduction of the structure of these collective states. Vibrational states also have, in addition to integral characteristics, differential characteristics which manifest themselves in reactions with one-nucleon transfer and in the β and γ transitions to these states.

4

Quasiparticle–phonon nuclear model

4.1 THE FUNDAMENTALS OF THE MODEL

4.1.1

Wave functions of low-lying states are dominated by one component: the single-quasiparticle component in odd nuclei, and one-phonon state or two-quasiparticle component in even nuclei. The simplicity of the structure of low-lying states facilitates their detailed experimental and theoretical study. As the excitation energy increases, the state density in atomic nuclei increases and their structures become more complex. At medium and high excitation energies, there is a transition from simple low-lying states to more complex ones. The fragmentation of single-particle states, that is, the distribution of the single-particle strength over many nuclear levels, is an important factor for the analysis of the structure of states with intermediate and high excitation energies. The single-particle strength is concentrated on a single level in the models of independent particles and quasiparticles. In the extreme statistical model, it is randomly distributed over all nuclear levels. A broad poorly studied region of states with medium and high excitation energies lies between low-lying states, for which the characteristics of each individual level are studied, and the states of the extreme statistical model which erases the individuality of a nucleus (e.g., the shell effect vanishes).

An experimental investigation of the structure of states with medium and high excitation energies meets with considerable difficulties. It is quite clear that we cannot measure the characteristics of each level among many thousands, especially because, as the excitation energy increases, the state structures get more complex and the number of components of the wave functions that need be measured increases. The growing complexity of structures is felt already at fairly low excitation energies. In deformed odd nuclei, quasiparticle⊗phonon components are added to single-quasiparticle components at excitation energies above 0.5 MeV. In odd spherical nuclei with open shells, this admixture is already observed in the ground states.

The existing theories and the available computers do not make it possible to obtain the correct description of the structure of each level at excitation energies above 3 MeV (except for the levels of light and double-magic nuclei, and their nearest neighbours). One reason for this is the need to diagonalize matrices of orders $10^{14}-10^{20}$; another reason is that we use a crude description of nuclear forces and the approximate solution of the nuclear many-body problem. Furthermore, calculating each of the many millions of wave function components for each state would be pointless since quantitative data on the nuclear structure are available solely for wave function configurations with only a few quasiparticles. The most accurate experimental data are connected with the fragmentation of single-quasiparticle, one-phonon states and quasiparticle $\otimes$ phonon states. High-spin states are an exception. The fragmentation of single-quasiparticle states at medium energies results in the formation of local maximums or of a substructure in the cross sections of one-nucleon transfer reactions. The fragmentation of the $s_{1/2}$, $p_{1/2}$ and $p_{3/2}$ subshells determines the s- and p-wave neutron strength functions. Giant resonances are determined by the position of collective one-phonon states, while the widths of the giant resonances are related to their fragmentation. Few-quasiparticles components reflect the shell-structure effects. The problem of the nuclear theory is not so much to obtain the most rigorous general solution of the many-body problem, as to describe with maximum accuracy those nuclear characteristics which are experimentally measurable and could be measured within several years. The progress in the description of fragmentation was stimulated by improved understanding of the key role played by taking into consideration the coupling of the single-particle and the collective vibrational motions, that is, the quasiparticle–phonon interaction. The role of the quasiparticle–phonon interaction was demonstrated during 1965–1975 [131–135]. It is on the basis of the results of these papers that the quasiparticle–phonon nuclear model (QPNM) has been formulated.

The QPNM was formulated for the description of the few-quasiparticle components of the wave functions at low, medium and high excitation energies [65, 66, 74, 75, 133–147]. The model operates with the strength function method, calculating the fragmentation of single-quasiparticle, one-phonon states and the quasiparticle $\otimes$ phonon states over a large number of nuclear levels. Those characteristics of complex nuclei are calculated which are determined by these components.

Let us give the general scheme of finding the solution of the nuclear many-body problem (figure 4.1); it precedes the formulation of the QPNM. The nuclear Hamiltonian in its general form is written in terms of the neutron and proton creation and absorption operators, a_f^+ and a_f; a system of coupled equations (1.9′), (1.10′), (1.11) is derived. The approximation is chosen in the form (1.12); this gives a closed set of equations (1.13), (1.14). A large number of equations are dropped in this approximation. It

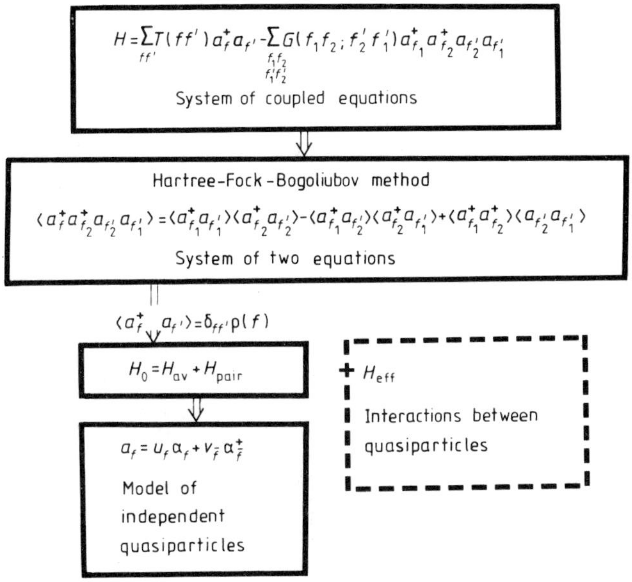

Figure 4.1 Flow chart to assist in solving the nuclear many-body problem.

is assumed that the effect of omitted equations is small and that they can be partially compensated for by introducing effective forces with constants fitted to experimental data. The Hartree–Fock–Bogoliubov (HFB) approximation and the condition of diagonality of the density matrix makes it possible to single out explicitly the mean field of the nucleus and the interactions leading to superfluid pairing correlations. One then uses the Bogoliubov canonical transformation and arrives at the independent quasiparticle model.

The QPNM Hamiltonian is constructed on the basis of the approximate solution of the nuclear many-body problem; it is schematically shown by the diagram of figure 4.1. It consists of the mean field of the neutron and proton system, in the form of the Saxon–Woods potentials and the interactions leading to the superfluid pairing correlations. The QPNM Hamiltonian contains multipole and spin-multipole isoscalar and isovector interactions, including the charge-exchange interactions, interactions in the particle–hole and particle–particle channels, and also the tensor interaction.

The Saxon–Woods potential parameters are fixed so as to take into account the quasiparticle–phonon interaction and to obtain a correct description of the low-lying states of odd nuclei. There is no doubt that one can choose a different form of the mean field potential or make use of the Hartree–Fock method to calculate the energies and wave functions of single-particle states for subsequent QPNM calculations. These differences are not essential. The application of the Hartree–Fock method signifies an earlier stage of parametrization, that is, the parametrization of the effective interaction (chosen, e.g., in the form of the Skyrme forces). Instead of functions, constants G_N and G_Z were taken in the interactions which result in pairing; their values were found from the nuclear mass difference [12]. This approximation does not limit the accuracy of QPNM calculations.

The effective interactions between quasiparticles are represented by expansions (2.10) and (2.11) in multipoles and spin-multipoles. Effective interactions play an important role in compensating for the equations omitted in the HFB approximation. They are also connected with the nucleon–nucleon interactions in the nuclear matter; individual terms can be put in correspondence with the exchange of one or two mesons. It is essential for QPNM calculations that interactions between quasiparticles are given in a separable (factorizable) form. Instead of separable interactions (2.13), it is possible to use separable interactions of rank $n_{\max} > 1$, for example, in the form (2.13′). Separable representations were constructed for a number of nucleon–nucleon potentials. For instance, separable representations of rank $n_{\max} \leqslant 5$ give a satisfactory approximation for the Paris and Bonn potentials. It is well known [42, 148] that the separable potentials are widely used to describe the nucleon–nucleon interactions and to analyse three-body nuclear systems and light nuclei; that is, they are helpful when the results of calculations are more sensitive to the form of the radial dependence of forces than the QPNM-calculated characteristics of complex nuclei. It must be remembered that the calculations operate with the matrix elements of effective interactions. The single-particle wave functions ‘cut away’ a small fraction of interactions. It is possible to construct separable interactions whose matrix elements are close to the matrix elements of more complex interactions [149]. We assume that the accuracy of the calculations is not restricted by a suitable choice of separable interactions between quasiparticles.

There is a certain arbitrariness in the choice of the radial dependence of separable interactions. The existence of collective vibrational quadrupole and octupole states indicates that there must be a maximum in the radial dependence of multipole forces in the region of the surface of the nucleus. Hence, $R_\lambda(r)$ is taken in the form $R_\lambda(r) = r^\lambda$ or $R_\lambda(r) = \partial V(r)/\partial r$, where $V(r)$ is the central part of the Saxon–Woods potential. This form is quite suitable for multipole forces. The same radial dependence is also used for spin-multipole forces. Owing to the absence of clearly pronounced magnetic-type collective states, the uncertainty in the radial dependence of the separable spin-multipole interaction is quite high. Owing to an important role of the one-pion (and ρ-meson) exchange in the nucleon–nucleon interaction at relatively large distances, the QPNM Hamiltonian is supplemented by an isovector tensor interaction. A large number of papers have been devoted to searching for effective interactions between quasiparticles (see, e.g., [55, 150, 151]).

We can assume that in the QPNM with constants fitted to experimental data or phenomenological estimates, the effective finite-rank separable interactions between quasiparticles describe the nuclear characteristics as well as much more complex effective interactions used in many papers. They have unquestionable advantages in comparison with the Landau–Migdal zero-radius forces used to calculate the structure of nuclei with closed shells. Various terms were added to the Landau–Migdal forces. For example, the spin–isospin forces determined by the parameter g' are supplemented by one-pion exchange terms and sometimes by terms for the exchange by other mesons.

It appears that one should not be too worried about the self-consistency of the mean field and of the effective interactions, since a large number of equations have been dropped in the HFB approximation. The advantage of self-consistent calculations lies in the qualitatively correct description of nuclear characteristics rather than in the quantitative description that would fit the experimental data (it is never obtained). Self-consistent calculations have demonstrated that for the solution of the nuclear many-body problem, the HFB approximation can serve as a good basis for constructing various nuclear models.

4.1.2

Let us describe the system of QPNM calculations (figure 4.2). The explicit form of the model Hamiltonian will be given later. When transforming the QPNM Hamiltonian using the canonical Bogoliubov transformation, one changes from nucleon operators to quasiparticle operators α^+_{jm} and α_{jm}. The pairs of operators $\alpha^+_{jm}\alpha^+_{j'm'}$ and $\alpha_{j'm'}\alpha^+_{jm}$ are expressed in terms of the phonon operators, while the quasiparticle operators remain solely in the form $\alpha^+_{jm}\alpha_{j'm'}$. When the phonon operators are introduced in this manner, the difficulties with counting twice the number of diagrams are avoided (they do occur, e.g., in the nuclear field theory) [152]. Then one has to solve the RPA equation for the energies and the wave functions of the one-phonon state. All parameters of the model are fixed at this stage. To a certain extent, the effect of mapping in the HFB hierarchy approximation is effectively taken into account by making use of the experimental data in order to fix the constants of the pairing, multipole and spin-multipole isoscalar and isovector interactions.

One peculiarity and also an advantage of the QPNM is the fact that one-phonon states, not single-particle states, are used as the basis. This is possible since the RPA gives a unified description of collective, weakly collective, and two-quasiparticle states. This is the *first feature* of the QPNM.

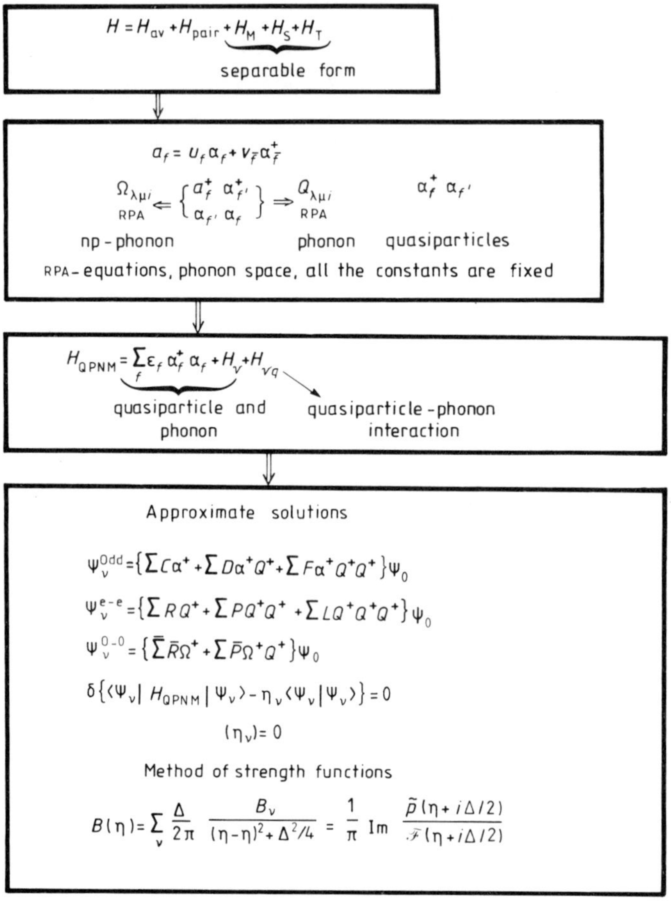

Figure 4.2 QPNM calculation scheme.

The phonon space corresponds to the complete space of particle–hole-type two-quasiparticle states and to a part of particle–particle-type states. When the particle–particle interactions are taken into account, the entire space of two-quasiparticle states is covered. When the phonon basis is constructed in deformed nuclei for $K^{\pi} = 0^{\pm}, 1^{\pm}, 2^{\pm}, \ldots, 7^{\pm}$, multipole forces are used. In the case of spherical nuclei, multipole forces are chosen for constructing one-phonon states with $J^{\pi} = 1^{-}, 2^{+}, 3^{-}, \ldots, 7^{-}$, and spin-multipole forces are chosen for constructing states with $J^{\pi} = 1^{+}, 2^{-}, 3^{+}, \ldots, 7^{+}$. For each value of K^{π} or J^{π}, one calculates several hundred roots of the seqular equation and the corresponding wave functions. The results of calculations of nuclear state density [153] indicate that the phonon space is complete. All QPNM constants were found to be fixed as a result of calculating the phonon space.

The second feature of the model lies in the quasiparticle–phonon interaction being responsible for the fragmentation of the quasiparticle and collective motions and, therefore, for the complications in the structure of nuclear states as excitation energies increase.

To solve the problem, it is necessary to expand the wave functions of the excitation states in series for the number of phonon operators; for odd nuclei, each term is additionally multiplied by a quasiparticle operator. The approximation consists in cutting off this series, which constitutes the *third feature of the model*. Truncating the series in phonon number creates an approximation of the same type as truncating the HFB hierarchy of equations. For the time being, the expansion is restricted to two phonons (see figure 4.2). It is as difficult to determine the effect of multi-phonon terms of the wave functions on the quantities to be calculated, as it is to evaluate the role of the equations omitted in the HFB hierarchy approximation of the nuclear many-body problem. In both cases, it is customary to assume that approximate equations of the hierarchy correctly describe the properties of nuclear excitations and that the neglected terms are partly taken into account by using constants fitted to the experimental data. In calculations, the Pauli principle is included by using exact commutation relations between the phonon and quasiparticle operators.

The *fourth feature of the model* is the use of the strength function method. Its modified version [66, 137] directly gives the reduced transition probabilities, spectroscopic factors, transition densities, cross sections and other nuclear characteristics without solving the relevant secular equations. Owing to the strength function method, the required computer time is reduced by a factor of 10^3; it becomes possible to calculate the fragmentation of single-quasiparticle, one-phonon states and the quasiparticle $\otimes$ phonon states for a large number of nuclei. The characteristics of excitation states are displayed for spherical nuclei both with open and with closed shells, and for deformed nuclei as well.

Unequal difficulties are encountered in describing low- and high-lying states of complex nuclei. The results of calculations for low-lying non-rotational states strongly depend on the behaviour of single-particle levels close to the Fermi surface. Maximum difficulties are connected with a microscopic description of the first, the most collective, vibrational states. In transition nuclei, the RPA 'works' at the limit of its capabilities. Corrections connected with taking strict account of the Pauli exclusion principle and with the ground-state correlations are large. It is desirable, when describing low-lying states, to project onto particle number and angular momentum, which considerably complicates the calculations. No such difficulties are encountered for the giant-resonance-type collective states. Difficulties for describing high-lying states are connected with the need to take into consideration a large phonon space, a continuous spectrum, and wave function components with a large number of phonons.

4.1.3

Let us look at the general scheme of QPNM calculations as applied to spherical nuclei. The starting QPNM Hamiltonian contains the mean field of the proton and neutron systems in the form of the Saxon–Woods potential, the superconducting pairing interactions, and also multipole, spin-multipole and tensor isoscalar and isovector, particle–hole and particle–particle interactions between quasiparticles, that is,

$$H = H_{\text{av}} + H_{\text{pair}} + H_M + H_S + H_T. \tag{4.1}$$

QPNM calculations are carried out in four stages. At the *first stage*, single-particle energy eigenvalues and the wave functions of the Saxon–Woods potential are found. The Saxon–Woods potential parameters are so fixed that we obtain a correct description of low-lying states of odd-A nuclei, taking into account the quasiparticle–phonon interactions. At the *second stage*, the canonical Bogoliubov transformation is performed, converting from the particle operators a^+_{jm}, a_{jm} to the quasiparticle operators α^+_{jm}, α_{jm}, and calculations are carried out in the independent quasiparticles model [6]. The pairing constants G_τ are fixed using experimental data on pairing energies.

At the *third stage*, the phonon basis is constructed. The phonon basis is calculated in the RPA. Both multipole and spin-multipole phonons are used in the construction of the phonon basis for spherical nuclei. To obtain the one-phonon states with angular momenta and parities $\lambda^\pi = 1^-, 2^+, 3^-, 4^+, \ldots$, multipole forces are used. To obtain one-phonon states with $L^\pi = 1^+, 2^-, 3^+, 4^-, \ldots$, spin-multipole forces are chosen. Each phonon type is generated by a single type of force, corresponding to an electromagnetic transition operator. For example, quadrupole forces are used to calculate phonons with $I^\pi = 2^+$ while spin-quadrupole forces are ignored.

The main advantage of introducing the phonon basis is the unified description of collective, weakly collective, and two-quasiparticle states; this gives one the possibility of constructing models for analysing the behaviour of few-quasiparticle components of the wave functions at low, intermediate and high excitation energies. It is essential that the rank of the appropriate determinants is greatly reduced and computer simulations become feasible. One has to keep in mind that phonons are not 'ideal' bosons. They are built of fermions, so that the problem gets considerably more complex since the Pauli exclusion principle has to be taken into account in a correct manner. A disadvantage of converting to the phonon basis lies in the overdefinition of the multi-phonon states as compared with correctly symmetrized multi-quasiparticle states of similar complexity. For instance, the total number of two-phonon states with a given spin is greater than the total number of four-quasiparticle states with the same spin, owing to the recoupling of quasiparticles in the two-phonon states. The problem also becomes more difficult when spurious states are eliminated.

The construction of the phonon basis fixes the QPNM constants. The constant of the particle-hole multipole interaction κ_0^λ, κ_1^λ in the radial dependence $R_\lambda(r) = \partial V(r)/\partial r$, where $V(r)$ is the central part of the Saxon–Woods potential, is weakly dependent on λ, while the matrix elements $\langle j|\partial V(r)/\partial r|j'\rangle$ are independent of the upper integration limit. For electric-type dipole forces, the constant κ_0^1 is chosen to cause vanishing of the energy of the lowest 1^- state in the RPA. In other words, it is such that, following [154], the 1^- spurious state admixture due to the violation of the translation invariance is eliminated. The constant $\kappa_1^{(1)}$ is chosen in accord with the position of the energy centroid of the giant dipole isovector electric resonance. The constants κ_0^2, κ_0^3 and κ_0^4 are chosen so as to obtain a correct QPNM description of the experimental values of the energies and of the $B(E\lambda)$ values of the first 2_1^+, 3_1^- and 4_1^+ states. The constants κ_0^λ slightly grow with increasing λ and somewhat decrease as A increases.

After the RPA energies $\omega_{\lambda i}$ and wave functions $Q_{\lambda i}\Psi_0$ have been found and, therefore, the phonon basis has been constructed, Hamiltonian (4.1) has to be transformed. As a result, the QPNM Hamiltonian changes to

$$H_{\text{QPNM}} = \sum_{jm} \epsilon_j \alpha_{jm}^+ \alpha_{jm} + H_v + H_{vq}. \tag{4.2}$$

It describes free quasiparticles and phonons and the quasiparticle–phonon interaction. When constructing (4.2), we take into account that the following conditions are met if the RPA equations are to have solutions:

$$\langle Q_{\lambda i'} H^{(1)} Q_{\lambda i}^+\rangle = \omega_{\lambda i}\delta_{ii'}\delta_{\lambda\lambda'}, \langle \Omega_{\lambda i'} H^{(1)} \Omega_{\lambda i}^+\rangle = \Omega_{\lambda i}\delta_{ii'}, \tag{4.3}$$
$$\langle H^{(1)} Q_{\lambda i}^+ Q_{\lambda i'}^+\rangle = \langle Q_{\lambda i'} Q_{\lambda i} H^{(1)}\rangle = 0$$
$$\langle H^{(1)} \Omega_i^+ \Omega_{i'}^+\rangle = \langle \Omega_{i'} \Omega_i H^{(1)}\rangle = 0 \tag{4.3$'$}$$

where $H^{(1)} = \sum_{jm} \varepsilon_j \alpha_{jm}^+ \alpha_{jm} + H_v$ and averaging is carried out over the phonon vacuum. Hence, the terms that contain two phonon creation or absorption operators are dropped from the QPNM Hamiltonian. The terms in H_M and H_S that contain the products $B(jj';\lambda\mu)B(j_2j_2';\lambda\mu)$ are also dropped. We have already mentioned in Chapter 2 that RPA corrections due to terms of this type are not large and their contribution can be neglected.

At the *fourth stage* of the calculations, the quasiparticle–phonon interaction is taken into account. The wave functions of the excitation states of odd, even–even and odd–odd spherical nuclei are written in the following form:

$$\Psi_\nu(JM) = C_{J\nu}\Big\{\alpha^+_{JM} + \sum_{\lambda ij} D^{\lambda i}_j(J\nu)[\alpha^+_{jm}Q^+_{\lambda\mu i}]_{JM}$$
$$+ \sum_{\lambda_1 i_1 \lambda_2 i_2 jI} F^{\lambda_1 i_1 \lambda_2 i_2}_{jI}(J\nu)[\alpha^+_{jm}$$
$$\times [Q^+_{\lambda_1\mu_1 i_1}Q^+_{\lambda_2\mu_2 i_2}]_{IM'}]_{JM}\Big\}\Psi_0 \tag{4.4}$$

$$\Psi_\nu(JM) = \Big\{\sum_i R_i(J\nu)Q^+_{\lambda\mu i} + \sum_{\lambda_1 i_1 \lambda_2 i_2} P^{\lambda_1 i_1}_{\lambda_2 i_2}(J\nu)$$
$$\times [Q_{\lambda_1\mu_1 i_1}Q_{\lambda_2\mu_2 i_2}]_{JM}\Big\}\Psi_0 \tag{4.5}$$

$$\Psi_\nu(JM) = \Big\{\sum_i \bar{R}_i(J\nu)\Omega^+_{\lambda\mu i} + \sum_{\lambda_1 i_1 \lambda_2 i_2} \bar{P}^{\lambda_1 i_1}_{\lambda_2 i_2}(J\nu)$$
$$\times [\Omega^+_{\lambda_1\mu_1 i_1}Q^+_{\lambda_2\mu_2 i_2}]_{JM}\Big\}\Psi_0 \tag{4.6}$$

where Ψ_0 is the wave function of the ground state of the even–even nucleus (the phonon vacuum), and α^+_{jm}, Q^+_{jmi} and Ω^+_{jmi} are the quasiparticle and phonon creation operators. Then we find the mean value using (4.4) or (4.5) or (4.6), employ the variational principle, and deduce the secular equation for the energies η_ν of the excitation states,

$$\mathcal{F}(\eta_\nu) = 0. \tag{4.7}$$

We also obtain the set of equations for the coefficients, including the normalizations of the wave functions, wave functions (4.4), or (4.5), or (4.6).

4.1.4

Let us construct a QPNM Hamiltonian for spherical nuclei. Using rank-$n_{\max} > 1$ separable interactions between quasiparticles, we arrive at the summation over n from $n = 1$ to $n = n_{\max}$. Following [77], we rewrite the QPNM Hamiltonian in the following, most general form:

$$H_{\mathrm{QPNM}} = \sum_{jm} \epsilon_j \alpha^+_{jm}\alpha_{jm} + H_v + H_{vq} \tag{4.2$'$}$$

$$H_v = H_{Ev} + H_{Sv} \tag{4.8}$$

$$H_{Ev} = \sum_{\lambda\mu ii'} W^\lambda_{Eii'} Q^+_{\lambda\mu i}Q_{\lambda\mu i'} \tag{4.9}$$

$$W^\lambda_{Eii'} = -\frac{1}{4}\frac{1}{2\lambda+1}\sum_{n\tau}\Big\{\sum_{\rho=\pm1}[(\kappa^\lambda_0 + \rho\kappa^\lambda_1)D^{\lambda i}_{n\tau}D^{\lambda i'}_{n\rho\tau}$$

$$+ (\kappa_0^{\lambda\lambda} + \rho\kappa_1^{\lambda\lambda}) D_{n\tau}^{\lambda\lambda i} D_{n\rho\tau}^{\lambda\lambda i'}) + G^{\lambda}[D_{nL\tau}^{\lambda i-} D_{n\tau}^{\lambda i'-}$$
$$+ D_{n\tau}^{\lambda i+} D_{n\tau}^{\lambda i'+}] + G_{\tau}^{\lambda\lambda}[D_{n\tau}^{\lambda\lambda i-} D_{n\tau}^{\lambda\lambda i'-} + D_{n\tau}^{\lambda\lambda i+} D_{n\tau}^{\lambda\lambda i'+}]\Big\} \qquad (4.9')$$

$$H_{Sv} = \sum_{LMii'} W_{Sii'}^{L} Q_{LMi}^{+} Q_{LMi'} \qquad (4.10)$$

$$W_{Sii'}^{L} = -\frac{1}{4}\frac{1}{2L+1}\sum_{n\tau}\Big\{\sum_{\lambda=L\pm1,\rho=\pm1}[(\kappa_0^{\lambda L} + \rho\kappa_1^{\lambda L}) D_{n\tau}^{\lambda Li} D_{n\rho\tau}^{\lambda Li'}]$$
$$- \kappa_T^{L}\sum_{\rho=\pm1}(D_{n\tau}^{L-1\,Li} D_{n\rho\tau}^{L+1\,Li'} + D_{n\rho\tau}^{L+1\,Li} D_{n\tau}^{L-1\,Li'})$$
$$+ \sum_{\lambda=L\pm1} G^{\lambda L}(D_{n\tau}^{\lambda Li-} D_{n\tau}^{\lambda Li'-} + D_{n\tau}^{\lambda Li+} D_{n\tau}^{\lambda Li'+})$$
$$- G_T^{L}(D_{n\tau}^{L-1\,Li-} D_{n\tau}^{L+1\,Li'-} + D_{n\tau}^{L+1\,Li-} D_{n\tau}^{L-1\,Li'-}$$
$$+ D_{n\tau}^{L-1\,Li+} D_{n\tau}^{L+1\,Li'+} + D_{n\tau}^{L+1\,Li+} D_{n\tau}^{L-1\,Li'+})\Big\} \qquad (4.10')$$

$$H_{vq} = H_{Evq} + H_{Svq} \qquad (4.11)$$

$$H_{Evq} = \sum_{\lambda\mu i\tau}\sum_{jj'}^{\tau}\Big\{V_{1\,E\tau}^{\lambda\mu i}(jj')[(Q_{\lambda\mu i}^{+} + (-1)^{\lambda-\mu} Q_{\lambda\mu i})$$
$$\times B(jj';\lambda-\mu) + \text{h.c.}] + V_{2\,E\tau}^{\lambda\mu i}(jj')[(Q_{\lambda\mu i}^{+} - (-1)^{\lambda-\mu} Q_{\lambda-\mu i})$$
$$\times B(jj';\lambda-\mu) + \text{h.c.}]\Big\} \qquad (4.12)$$

$$V_{1E\tau}^{\lambda\mu i}(jj') = -\frac{1}{4}\frac{1}{2\lambda+1}\sum_{n}\Big\{\sum_{\rho=\pm1}(\kappa_0^{\lambda} + \rho\kappa_1^{\lambda}) f_n^{\lambda}(jj') v_{jj'}^{(-)} D_{n\rho\tau}^{\lambda i}$$
$$- G^{\lambda} f_n^{\lambda}(jj')(-1)^{\lambda-\mu}(u_{jj'}^{(+)} - u_{jj'}^{(-)}) D_{n\tau}^{\lambda i-}$$
$$- G^{\lambda} f_n^{\lambda\lambda}(jj')(-1)^{\lambda-\mu}(u_{jj'}^{(+)} - u_{jj'}^{(-)}) D_{n\tau}^{\lambda\lambda i-}\Big\} \qquad (4.12')$$

$$V_{2E\tau}^{\lambda\mu i}(jj') = -\frac{1}{4}\frac{1}{2\lambda+1}\sum_{n}\Big\{\sum_{\rho=\pm1}(\kappa_0^{\lambda\lambda} + \rho\kappa_1^{\lambda\lambda}) f_n^{\lambda\lambda}(jj') v_{jj'}^{(+)} D_{n\rho\tau}^{\lambda\lambda i}$$
$$- G^{\lambda} f_n^{\lambda}(jj')(-1)^{\lambda-\mu}(u_{jj'}^{(+)} - u_{jj'}^{(-)}) D_{n\tau}^{\lambda i+}$$
$$- G^{\lambda} f_n^{\lambda\lambda}(jj')(-1)^{\lambda-\mu}(u_{jj'}^{(+)} - u_{jj'}^{(-)}) D_{n\tau}^{\lambda\lambda i+}\Big\} \qquad (4.12'')$$

$$H_{Svq} = \sum_{LMi\tau}\sum_{jj'}^{\tau}\Big\{V_{1\,S\tau}^{Mi}(jj')[(Q_{LMi}^{+} - (-1)^{L-M} Q_{L-Mi}) B(jj';L-M)$$
$$+ \text{h.c.}] + V_{2\,S\tau}^{LMi}(jj')[(Q_{LMi}^{+} + (-1)^{L-M} Q_{L-Mi})$$
$$\times B(jj';L-M) + \text{h.c.}]\Big\} \qquad (4.13)$$

$$V_{1\,S\tau}^{LMi}(jj') = -\frac{1}{4}\frac{1}{2L+1}\sum_{n}\Big\{\sum_{\lambda=L\pm1,\rho=\pm1}(\kappa_0^{\lambda L} + \rho\kappa_1^{\lambda L})$$

$$\times f_n^{\lambda L}(jj')v_{jj'}^{(+)}D_{n\rho\tau}^{\lambda Li} - \kappa_T^L \sum_{\rho=\pm 1} \rho v_{jj'}^{(+)}[D_{n\rho\tau}^{L-1\,Li} f_n^{L+1\,L}(jj')$$
$$+ D_{n\rho\tau}^{L+1\,Li} f_n^{L-1\,L}(jj')]$$
$$- \sum_{\lambda=L\pm 1} G^{\lambda L} f_n^{\lambda L}(jj')(-1)^{L-M}(u_{jj'}^{(+)} - u_{jj'}^{(-)})D_{n\tau}^{\lambda Li+}$$
$$- G_T^L(-1)^{L-M}(u_{jj'}^{(+)} - u_{jj'}^{(-)})[f_n^{L+1\,L}(jj')$$
$$\times D_{n\tau}^{L-1\,Li+} + f_n^{L-1\,L}(jj')D_{n\tau}^{L+1\,Li+}]\Big\} \qquad (4.13')$$

$$V_{2\,S\tau}^{LMi}(jj') = -\frac{1}{4}\frac{1}{2L+1}\sum_n\Big\{\sum_{\lambda=L\pm 1} G^{\lambda L}$$
$$f_n^{\lambda L}(jj')(-1)^{L-M}(u_{jj'}^{(+)} - u_{jj'}^{(-)})D_{n\tau}^{\lambda Li-} + G_T^L(-1)^{L-M}$$
$$\times (u_{jj'}^{(+)} - u_{jj'}^{(-)})[f_n^{L+1\,L}(jj')D_{n\tau}^{L-1\,Li-}$$
$$+ f_n^{L-1\,L}(jj')D_{n\tau}^{L+1\,Li-}]\Big\}. \qquad (4.13'')$$

Here κ_0^λ, $\kappa_0^{\lambda L}$ and κ_1^λ, $\kappa_1^{\lambda L}$ are the isoscalar and isovector constants of the multipole and spin-multipole ph interactions, $G^\lambda = G_0^\lambda + G_1^\lambda$ and $G^{\lambda L} = G_0^{\lambda L} + G_1^{\lambda L}$ are the constants of the multipole and spin-multipole pp interactions, and κ_T^L and G_T^L are the constants of the isovector ph and pp tensor interactions. The matrix elements $f_n^\lambda(jj')$, $f_n^{\lambda L}(jj')$ are defined by formulas (2.138)

$$D_{n\tau}^{\lambda i} = \sum_{jj'}^{\tau} f_n^\lambda(jj')u_{jj'}^{(+)}g_{jj'}^{\lambda i} \qquad D_{n\tau}^{\lambda Li} = \sum_{jj'}^{\tau} f_n^{\lambda L}(jj')u_{jj'}^{(-)}w_{jj'}^{Li}$$
$$D_{n\tau}^{\lambda i-} = \sum_{jj'}^{\tau} f_n^\lambda(jj')v_{jj'}^{(-)}g_{jj'}^{\lambda i} \qquad D_{n\tau}^{\lambda i+} = \sum_{jj'}^{\tau} f_n^\lambda(jj')v_{jj'}^{(+)}w_{jj'}^{\lambda i}$$
$$D_{n\tau}^{\lambda Li-} = \sum_{jj'}^{\tau} f_n^{\lambda L}(jj')v_{jj'}^{(-)}g_{jj'}^{Li} \qquad D_{n\tau}^{\lambda Li+} = \sum_{jj'}^{\tau} f_n^{\lambda L}(jj')v_{jj'}^{(+)}w_{jj'}^{Li}. \qquad (4.14)$$

The Hamiltonian above must also be supplemented with charge-exchange interactions.

Calculations operate with a Hamiltonian with simple separable interactions $n_{\max} = 1$. We take the following final form of the QPNM Hamiltonian after it has been transformed for the RPA secular equations:

$$H_{\text{QPNM}} = \sum_{jm} \epsilon_j a_{jm}^+ a_{jm} + H_{Mv} + H_{Mvq} + H_{Sv} + H_{Svq}$$
$$+ H_{CMv} + H_{CMvq} + H_{CSv} + H_{CSvq} + H_{Tv} + H_{Tvq}$$
$$+ H_{CTv} + H_{CTvq} \qquad (4.15)$$

$$H_{Mv} = -\frac{1}{4} \sum_{\lambda\mu ii'\tau} \frac{X^{\lambda i}(\tau) + X^{\lambda i'}(\tau)}{\sqrt{\mathcal{Y}_\tau^{\lambda i} \mathcal{Y}_\tau^{\lambda i'}}} Q^+_{\lambda\mu i} Q_{\lambda\mu i'} \tag{4.16}$$

$$H_{Mvq} = -\sum_{\lambda\mu i\tau} \sum_{jj'}^{\tau} \Gamma^{\lambda i}_{jj'}(\tau) \{[(-)^{\lambda-\mu} Q^+_{\lambda\mu i} + Q_{\lambda-\mu i}] B(jj';\lambda-\mu) + \text{h.c.}\} \tag{4.16'}$$

$$H_{Sv} = -\frac{1}{4} \sum_{LMii'\tau} \frac{X_L^{L-1\,i}(\tau) + X_L^{L-1\,i'}(\tau)}{\sqrt{\mathcal{Y}_\tau^{\lambda i} \mathcal{Y}_\tau^{\lambda i'}}} Q^+_{LMi} Q_{LMi'} \tag{4.17}$$

$$H_{Svq} = -\sum_{LMi\tau} \sum_{jj'}^{\tau} \Gamma^{L-1\,Li}_{jj'}(\tau) \{[(-)^{L-M-1} Q^+_{LMi} + Q_{L-Mi}] B(jj';L-M) + \text{h.c.}\} \tag{4.17'}$$

$$\Gamma^{\lambda i}_{jj'}(\tau) = \frac{1}{2\sqrt{2}} \frac{f^{\lambda}(jj') v^{(-)}_{jj'}}{(\mathcal{Y}_\tau^{\lambda i})^{1/2}} \qquad \Gamma^{L-1\,Li}_{jj'}(\tau) = \frac{1}{2\sqrt{2}} \frac{f^{L-1\,L}(jj') v^{(+)}_{jj'}}{(\mathcal{Y}_\tau^{\lambda i})^{1/2}}$$

$$H_{CMv} = -\sum_{\lambda\mu ii'} \frac{1 + \mathbf{Y}^{\lambda i} \mathbf{Y}^{\lambda i'}}{\sqrt{\mathcal{Y}_\tau^{\lambda i} \mathcal{Y}_\tau^{\lambda i'}}} \Omega^+_{\lambda\mu i} \Omega_{\lambda\mu i'} \tag{4.18}$$

$$\begin{aligned} H_{CMvq} = -\frac{1}{2} \sum_{\lambda\mu i} \left(\frac{\kappa_1^{(\lambda)}}{\mathcal{Y}^{\lambda i}} \right)^{1/2} & \sum_{j_p j_n} (2\lambda+1)^{-1} \\ & \times f^{\lambda}(j_p j_n) \{[(-)^{\lambda-\mu} (1 + \mathbf{Y}^{\lambda i}) \Omega^+_{\lambda\mu i} \\ & + (1 - \mathbf{Y}^{\lambda i}) \Omega_{\lambda-\mu i}][u_{j_p} u_{j_n} B(j_p j_n; \lambda-\mu) \\ & + (-)^{j_p + j_n - \lambda} v_{j_p} v_{j_n} B(j_n j_p; \lambda-\mu)] + \text{h.c.}\}. \end{aligned} \tag{4.18'}$$

The expressions H_{CSv} and H_{CSvq} for spin-multipole forces with $\lambda = L-1$ are obtained from (4.18) and (4.18′) by replacing the constants and matrix elements. The notation used in formulas (4.16)–(4.18′) is given in Chapter 2. Hamiltonian (4.15) has no free parameters.

In some cases, more complex expressions are used for one-phonon states. For instance, spin-multipole forces with $\lambda = L$ can be taken into account in addition to multipole forces. For magnetic-type states, spin-multipole forces with $\lambda = L-1$ and $\lambda = L+1$ can be used simultaneously. For magnetic-type charge-exchange states, the tensor forces and also the interactions in the particle-particle channel can be taken into account in addition to the spin-multipole forces with $\lambda = L-1$ and $\lambda = L+1$; we can also solve (2.132)-type equations. For example, the following expressions are taken instead of (4.17) and (4.17′) for magnetic-type states, simultaneously taking into account the spin-multipole forces with $\lambda = L-1$ and $\lambda = L+1$:

$$H_{Sv} = -\frac{1}{4} \sum_{LM\lambda=L\pm1} (2L+1)^{-1} \sum_{ii'\tau\rho=\pm1} (\kappa_0^{(\lambda L)} + \rho\kappa_1^{(\lambda L)}) \times D_\tau^{\lambda Li} D_{\rho\tau}^{\lambda Li'} Q_{LMi}^+ Q_{LMi'}$$

$$H_{Svq} = -\frac{1}{4} \sum_{LM\lambda=L\pm1} (2L+1)^{-1} \sum_{i\tau\rho=\pm1} (\kappa_0^{\lambda L} + \rho\kappa_1^{(\lambda L)}) \times \{D_\tau^{\lambda Li} \sum_{jj'}^{\rho\tau} f^{\lambda L}(jj') v_{jj'}^{(+)} [Q_{LMi}^+ - (-)^{L-M} Q_{L-Mi}] B(jj'; L-M) + \text{h.c.}\}$$

where $D_{\rho\tau}^{\lambda Li}$ are defined by (4.14) and the amplitudes $\psi_{jj'}^{Li}$ and $\varphi_{jj'}^{Li}$ are given by (2.119). If the spin-multipole and isovector tensor forces are simultaneously taken into account, the tensor interactions are taken in the form

$$H_{Tv} = -\frac{1}{4} \sum_{LM} \frac{\kappa_T^{(L)}}{2L+1} \sum_{ii'\tau\rho=\pm1} \left[D_\tau^{L-1\,Li} D_{\rho\tau}^{L+1\,Li'} + D_\tau^{L+1\,Li} D_{\rho\tau}^{L-1\,Li'}\right] Q_{LMi}^+ Q_{LMi'} \tag{4.19}$$

$$H_{Tvq} = -\frac{1}{4} \sum_{LM} \frac{\kappa_T^{(L)}}{2L+1} \sum_{i\tau\rho=\pm1} \sum_{jj'}^{\rho\tau} v_{jj'}^{(+)} \times \{[D_\tau^{L-1\,Li} f^{L+1\,L}(jj') + D_\tau^{L+1\,Li} f^{L-1\,L}(jj')] \times [Q_{LMi}^+ - (-)^{L-M} Q_{L-Mi}] B(jj'; L-M) + \text{h.c.}\}. \tag{4.19'}$$

Spin-multipole and tensor forces can be considered simultaneously in the analysis of charge-exchange states. The RPA equations have the form (2.185). The charge-exchange tensor interactions are

$$H_{CTv} = -\sum_{LM} \frac{\kappa_T^{(L)}}{2L+1} \sum_{ii'} (D_{(+)}^{L-1\,Li} D_{(+)}^{L+1\,Li'} + D_{(+)}^{L+1\,Li} D_{(+)}^{L-1\,Li'} + D_{(-)}^{L-1\,Li} D_{(-)}^{L+1\,Li'} + D_{(-)}^{L+1\,Li} D_{(-)}^{L-1\,Li'}) \Omega_{LMi}^+ \Omega_{LMi'}$$

$$
\begin{aligned}
H_{CTvq} = &-\sum_{LM} \frac{\kappa_T^{(L)}}{2L+1} \sum_{ij_pj_n} \Big\{ \Big([f^{L+1\,L}(j_pj_n) \\
&\times (D_{(+)}^{L-1\,Li} + D_{(-)}^{L-1\,Li}) + f^{L-1\,L}(j_pj_n)(D_{(+)}^{L+1\,Li} \\
&+ D_{(-)}^{L+1\,Li})]\Omega^+_{LMi} + (-)^{L+M}[f^{L+1\,L}(j_pj_n)(D_{(+)}^{L-1\,Li} \\
&- D_{(-)}^{L-1\,Li}) + f^{L-1\,L}(D_{(+)}^{L+1\,Li} - D_{(-)}^{L+1\,Li})]\Omega_{L-Mi} \Big) \\
&\times [u_{j_p}u_{j_n}B(j_pj_n; L-M) + (-)^{j_p+j_n-\lambda}v_{j_p}v_{j_n} \\
&\times B(j_nj_p; L-M) + \text{h.c.} \Big\}
\end{aligned}
$$

where $D_{(\pm)}^{\lambda Li}$ are defined by (2.176′).

The expression $\sum_{jm} \epsilon_j \alpha^+_{jm}\alpha_{jm}$ contributes to all multipole and spin-multipole terms of the Hamiltonian of the model. Indeed, we can write

$$
\alpha^+_{jm}\alpha_{jm} = \sum_n \alpha^+_{jm}|n\rangle\langle n|\alpha_{jm}
$$

where $|n\rangle$ is a complete set of functions consisting of one-, three- and five-particle configurations and configurations with larger numbers of particles. If we limit the analysis to one-quasiparticle configurations, we obtain

$$
\begin{aligned}
\sum_{jm\tau} \epsilon_j a^+_{jm}a_{jm} = &-\frac{1}{2}\sum_{\lambda\mu\tau}\sum_{jj'}{}^{\tau} \epsilon_{jj'}A^+(jj';\lambda\mu)A(jj';\lambda\mu) \\
&+ \frac{1}{2}\sum_{\lambda\mu}\sum_{j_pj_n} \epsilon_{j_pj_n}A^+(j_pj_n;\lambda\mu)A(j_pj_n;\lambda\mu)
\end{aligned}
\tag{4.20}
$$

where the summation over $\lambda\mu$ signifies summation over multipole and spin-multipole indices. If we switch to phonon operators, then the formula

$$
H^{(1)} = \sum_{jm} \epsilon_j a^+_{jm}a_{jm} + H_v
$$

can be rewritten as

$$
H^{(1)} = \sum_{\lambda\mu i} \omega_{\lambda i} Q^+_{\lambda\mu i}Q_{\lambda\mu i} + \sum_{\lambda\mu i} \Omega_{\lambda i}\Omega^+_{\lambda\mu i}\Omega_{\lambda\mu i}.
$$

This expression can be used when the Pauli exclusion principle in the multiphonon parts of wave functions is ignored.

4.1.5

Wave functions of the type (4.4)–(4.6) do not contain multiphonon components and are therefore quite far from the exact wave functions of states with intermediate and high excitation energies. When the QPNM was formulated, the objective was not to find correct wave functions of high-excitation states. The formulation is such that the aim is to describe in the most accurate manner the distribution of the strength of a few-quasiparticle states over nuclear levels. Hence, when the results of calculations are compared with experimental data, averaging over an appropriate energy range must be carried out.

A secular equation like (4.7) has from several tens to several hundreds of roots at intermediate and high excitation energies in a range of about 100 keV. The number of components of the wave function of each state runs into many thousands. In order to find the values of one or two components, one has to calculate all the components of the wave function for each state. This requires diagonalization of high-order matrices. In fact, a very small part of the information obtained is utilized. Finding eigenvalues and wave functions is superfluous for a large number of problems in physics. It will be useful now to skip the intermediate stage of calculating the eigenvalues and wave functions of each state and go directly to the determination of the physical quantities averaged over an energy interval; they could be compared with experimental data. Furthermore, it is very difficult to clearly visualize the calculated results for the characteristics of each of the many thousands of states. Consequently, the results are presented as histograms in which the calculated values are summed up over the roots of the corresponding equations within each interval.

In connection with these difficulties, the need arose for a mathematical apparatus capable of calculating, for the selected interval of excitation energies, the distribution of the quantities required, averaged over a prescribed energy interval. The strength function method proved to be suitable as this apparatus: it gives directly averaged characteristics without calculating the eigenvalues and wave functions of each state. A modification of the strength function method was developed for the QPNM in [65, 137, 157].

Let us formulate the strength function method in the most general form. We consider a transition from the initial state of the system, η_{i_0}, to an arbitrary state with the eigenvalue η_i; in terms of the bilinear products in φ_i. The transition is given by the expression

$$B(\eta_{i_0} \to \eta_i) = \sum_{p_0 k_0 \in pk} \varphi_{k_0}(i)\varphi_{p_0}(i) M_{k_0 p_0}(i_0) \tag{4.21}$$

where $M_{k_0 p_0}(i_0)$ is independent of the final state i. Here $\varphi(i)$ are eigenvectors in the $(n-1)$-dimensional space $(k = 1, \ldots, n-1)$ of a set of n equations,

$$(\hat{\mathbf{A}} - \eta \mathbf{I})\varphi = 0$$

where $\mathbf{I} = \|\delta_{ii'}\|$ is the identity matrix; the condition of existence of a non-trivial solution is

$$\det \|\mathbf{A} - \eta \mathbf{I}\| \equiv \|\mathbf{A} - \eta \mathbf{I}\| = 0. \tag{4.22}$$

The eigenvectors satisfy the condition

$$\varphi(1) : \varphi(2) : \ldots : \varphi(n) = |A_{i1} - \eta I| \\ : |A_{i2} - \eta I| : \ldots : |A_{in} - \eta I|$$

where $\mathbf{A}_{ii'}$ is the algebraic adjunction of the matrix $\mathbf{A}$ and at least one of $\mathbf{A}_{ii'}$ is non-zero. The following relations hold for symmetric matrices:

$$\frac{\varphi(i_1)}{\varphi(i_2)} = \frac{|\mathbf{A}_{i_2 i_1} - \eta \mathbf{I}|}{|\mathbf{A}_{i_2 i_2} - \eta \mathbf{I}|} = \frac{|\mathbf{A}_{i_3 i_1} - \eta \mathbf{I}|}{|\mathbf{A}_{i_3 i_2} - \eta \mathbf{I}|} = \frac{|\mathbf{A}_{i_1 i_1} - \eta \mathbf{I}|}{|\mathbf{A}_{i_2 i_1} - \eta \mathbf{I}|}.$$

The vectors $\varphi(i)$ are defined in a unique manner if the normalization condition is satisfied:

$$I = \sum_i [\varphi(i)]^2 = \varphi^2(i_0) \sum_i \left[\frac{\varphi(i)}{\varphi(i_o)}\right]^2 \qquad (4.22')$$

In this case,

$$\begin{aligned} \varphi^2(i_0) &= \left\{\sum_i \left[\frac{\varphi(i)}{\varphi(i_0)}\right]^2\right\}^{-1} \\ &= \left\{\sum_i \frac{|\mathbf{A}_{ii_0} - \eta \mathbf{I}|}{|\mathbf{A}_{i_0 i_0} - \eta \mathbf{I}|} \frac{|\mathbf{A}_{ii} - \eta \mathbf{I}|}{|\mathbf{A}_{ii_0} - \eta \mathbf{I}|}\right\}^{-1} \\ &= \left\{\sum_i \frac{|\mathbf{A}_{ii} - \eta \mathbf{I}|}{|\mathbf{A}_{i_0 i_0} - \eta \mathbf{I}|}\right\}^{-1} = \frac{|\mathbf{A}_{i_0 i_0} - \eta \mathbf{I}|}{\sum_i |\mathbf{A}_{ii} - \eta \mathbf{I}|}. \end{aligned}$$

We make use of the fact that [158]

$$\frac{\mathrm{d}}{\mathrm{d}\eta} |\mathbf{A} - \eta \mathbf{I}| = -\sum_i |\mathbf{A}_{ii} - \eta \mathbf{I}|$$

and obtain, for arbitrary i_0,

$$\begin{aligned} \varphi^2(i_0) &= -|\mathbf{A}_{i_0 i_0} - \eta_{i_0} \mathbf{I}| \left[\frac{\mathrm{d}}{\mathrm{d}\eta} |\mathbf{A}_{ii} - \eta \mathbf{I}| \Big|_{\eta = \eta_{i_0}}\right]^{-1} \\ &= -\left[\frac{\mathrm{d}}{\mathrm{d}\eta} \frac{|\mathbf{A} - \eta \mathbf{I}|}{|\mathbf{A}_{i_0 i_0} - \eta \mathbf{I}|}\right]^{-1}. \end{aligned} \qquad (4.22'')$$

This formula was derived in a number of ways in [64–66, 157].

If set (4.22) and (4.22′) has been solved, the calculation of $B(\eta_{i_0} - \eta_i)$ presents no major difficulties and is simple to perform, provided n is not large. If n is large and we need not calculate $B(\eta_{i_0} - \eta_i)$ for each value of i, it is possible to formulate the problem of finding the strength function which depends on a continuous parameter η. We can give the following definition to the strength function:

$$b(\eta) = \sum_i B(\eta_{i_0} \to \eta_i)\rho(\eta - \eta_i) \tag{4.23}$$

where, as in [4], we use the Lorentz function

$$\rho(\eta) = \frac{1}{2\pi} \frac{\Delta}{\eta^2 + \Delta^2/4} \tag{4.24}$$

and the normalization condition

$$\int_{\infty}^{\infty} \rho(\eta)\,\mathrm{d}\eta = 1.$$

The choice of the energy averaging interval Δ determines the manner of representing the results of calculations. Let us make use of formulas (4.21) and (4.22″) and substitute them into (4.23); we obtain

$$\begin{aligned} b(\eta) &= -\sum_i \sum_{p_0 k_0} \frac{|\mathbf{A}_{p_0 k_0} - \eta_i \mathbf{I}|}{\frac{\mathrm{d}}{\mathrm{d}\eta}|\mathbf{A} - \eta \mathbf{I}|\Big|_{\eta = \eta_{i_0}}} M_{k_0 p_0}(i_0)\rho(\eta - \eta_i) \\ &= \sum_i \frac{P(\eta_i)}{\frac{\mathrm{d}}{\mathrm{d}\eta}|\mathbf{A} - \eta \mathbf{I}|} \rho(\eta - \eta_i) \end{aligned}$$

where

$$P(\eta_i) = \sum_{p_0 k_0} |\mathbf{A}_{p_0 k_0} - \eta \mathbf{I}| M_{k_0 p_0}(i_0).$$

Let us resort to the theorem of residues and rewrite $b(\eta)$ as a contour integral

$$b(\eta) = -\frac{1}{2\pi \mathrm{i}} \oint_{l_\eta} \frac{P(z)\rho(\eta - z)}{|\mathbf{A} - z\mathbf{I}|}\,\mathrm{d}z. \tag{4.25}$$

Inside the I_η contour, there are poles $z = \eta_i$ which correspond to the roots of equation (4.22). The function $\rho(\eta - z)$ has two poles, $z = \eta \pm i\Delta/2$; the two contours corresponding to them are denoted in figure 4.3 by l_1 and l_2. The poles of the function $P(z)$ are denoted by $\bar{\eta}$, $\bar{\eta}_2, \ldots$; $l_{\bar{\eta}}$ is the corresponding contour and l_∞ is the contour at infinity. Using the

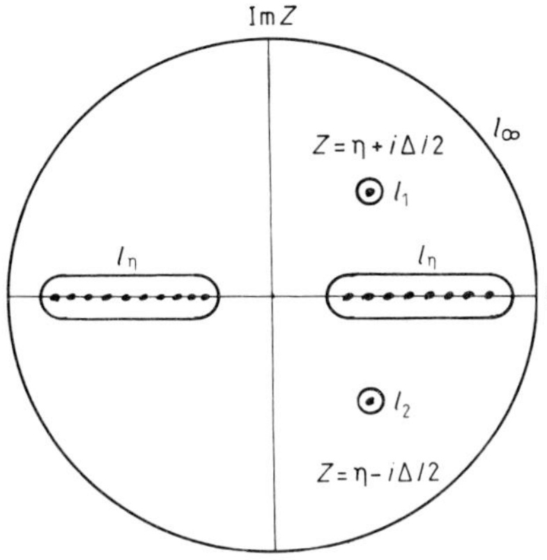

Figure 4.3 Integration contours in the complex plane.

analytical properties of the integrands in (4.25), we transform to other contour integrals via the Cauchy theorem (see figure 4.3):

$$b(\eta) = -\frac{1}{2\pi i}\oint_{l_\eta} = \frac{1}{2\pi i}\oint_{l_1 l_2} + \frac{1}{2\pi i}\oint_{l_{\bar\eta}} + \frac{1}{2\pi i}\oint_{l_\infty} .$$

In a particular case of considerable practical importance, when $P(z)$ has no singularities and $\oint_{l_\infty} = 0$, we obtain

$$\begin{aligned}
b(\eta) &= \frac{1}{2\pi i}\oint_{l_1 l_2} \frac{P(z)}{|\mathbf{A} - z\mathbf{I}|}\,\rho(\eta - z)\,\mathrm{d}z \\
&= \frac{\Delta}{2\pi}\sum_{1,2} \mathrm{res}\left[\frac{P(z)}{|\mathbf{A} - z\mathbf{I}|}\,\frac{1}{(\eta - z)^2 + \Delta^2/4}\right]_{z_{1,2} = \eta \pm i\Delta/2} \\
&= \frac{i}{2\pi}\left\{\left.\frac{P(z)}{|\mathbf{A} - z\mathbf{I}|}\right|_{z = \eta - i\Delta/2} - \left.\frac{P(z)}{|\mathbf{A} - z\mathbf{I}|}\right|_{z = \eta + i\Delta/2}\right\} \\
&= -\frac{1}{\pi}\,\mathrm{Im}\left.\frac{P(z)}{|\mathbf{A} - z\mathbf{I}|}\right|_{z = \eta + i\Delta/2} .
\end{aligned} \tag{4.25'}$$

In the general case, one looks additionally for residues at $z = \bar\eta_i$. Note that finding residues is typically an easier task than solving high-order secular equation (4.22). For a factorized function $M_{k_0 p_0} = M_{k_0} M_{p_0}$, the numerator in (4.25) is transformed into a bordered determinant, which gives

$$b(\eta) = \frac{1}{\pi} \operatorname{Im} \left. \frac{\begin{vmatrix} 0 & -M_{p_0} \\ M_{k_0} & \|\mathbf{A} - z\mathbf{I}\| \end{vmatrix}}{\|\mathbf{A} - z\mathbf{I}\|} \right|_{z=\eta+i\Delta/2} . \qquad (4.25'')$$

As a result, the energy eigenvalues and wave functions of (4.22) and (4.22′) need not be calculated to find the strength function (4.23). The calculation of high-order determinants in (4.25′) or (4.25″) is considerably less difficult, especially because in QPNM, determinants (4.25′) and (4.25″) can often be transformed to low-order determinants.

Let us compare strength functions obtained by using different averaging functions. QPNM calculations are carried out using Lorentz function (4.24). This choice was made because $\lim_{\Delta \to 0} \rho(\eta) = \delta(\eta)$, that is, because the limiting value is the delta function. Hence,

$$b(\eta)|_{\Delta \to 0} = \sum_i B(\eta_i)\delta(\eta - \eta_i)$$

and we obtain the values of physical quantities for the solutions of equation (4.22). If Δ is small, we obtain the function $b(\eta)$ as an envelope of the values $B(\eta_i)$ for $i = 1, 2, \ldots$. In most cases, it is easier to calculate this envelope than to find $B(\eta_i)$. The Lorentz function has two poles and the strength function acquires a simple form. A disadvantage of the Lorentz function $\rho(\eta)$ is that it falls off very steeply as η increases, so that $\int_{\eta+\Delta/2}^{\eta-\Delta/2} \rho(x)\, dx$ is found to be much less than unity and the accuracy of calculations is sometimes diminished. To obviate this, calculations are sometimes carried out, as in [159], with the function

$$\rho_2(\eta - \eta_i) = \frac{\Delta^3}{4\pi} \frac{1}{[(\eta - \eta_i)^2 + \Delta^2/4]^2} .$$

The distributions of the physical quantities calculated using the Lorentz function and the following three functions,

$$\rho_R(x) = \begin{cases} 0 & |x| > \Delta \\ 1/2\Delta & |x| \leqslant \Delta \end{cases} \qquad (4.26)$$

$$\rho_G(x) = \frac{2}{\sqrt{2\pi}\,\Delta} \exp\left\{ -\frac{x^2}{2(\Delta/2)^2} \right\} \qquad (4.26')$$

$$\rho_s(x) = \frac{1}{\pi} \frac{1}{x} \sin \frac{x}{\Delta/2} \qquad (4.26'')$$

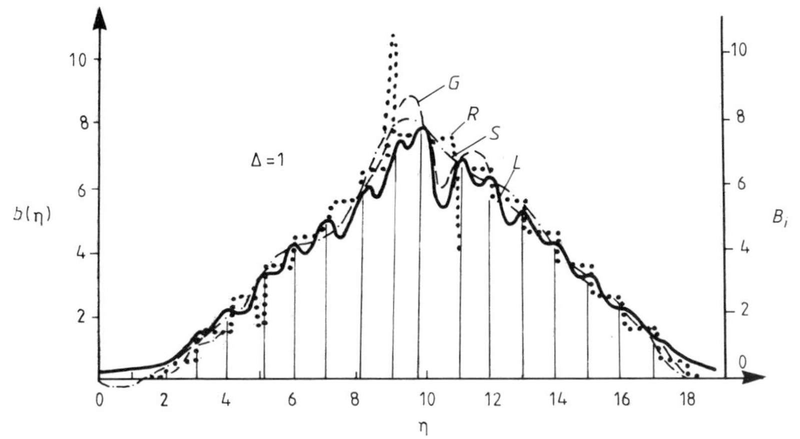

Figure 4.4 The values of B_i and the strength functions, after averaging by (4.24), (4.26), (4.26′), (4.26″), for $\Delta = 1$: B_i—vertical lines, L—Lorentz function (4.24), R—rectangular function (4.26); G—Gauss function (4.26′), S—sine function (4.26″).

were compared in [157]. The results of calculation of the strength functions $b(\eta)$ for a practically equidistant distribution of B_i values for $i = 1, 2, \ldots, 17$ are shown in figure 4.4. The strength functions are compared with the exact values of B_i. Obviously, all strength functions give a qualitatively correct fit to the B_i distribution, although the Lorentz and Gauss functions are preferable. The differences are the greatest at the tails of the distribution.

The choice of the averaging interval Δ is very important; it was discussed in [64–66, 137, 157]. There can be no doubt that the averaging interval must be less than both the region of localization of a physical quantity and the interval on which we observe the effects of the fine structure of its distribution. At the same time, the averaging interval must be greater than the uncertainty in the calculation of a given quantity within a specific model. When the calculated values are compared with the experimental data, the averaging interval can be chosen equal to the energy resolution. In a number of papers, effects of ignored interactions, or higher-order configurations which 'smear' B_i values, are simulated by choosing Δ for averaging.

4.2 EQUATIONS FOR ODD-MASS SPHERICAL NUCLEI

4.2.1

In the QPNM, the wave function of an excitation state is represented by an expansion in the phonon numbers. In odd nuclei, the expansion begins with single-quasiparticle terms, followed by quasiparticle ⊗ phonon terms, quasiparticle⊗two-phonons terms, and so forth. The quasiparticle–phonon interaction, H_{vq}, mixes the wave function components that differ by one phonon. For instance, the fragmentation of a single-quasiparticle state is mostly caused by coupling to quasiparticle ⊗ phonon states. The effect of wave function components of quasiparticle ⊗ three phonons state on the fragmentation of a single-particle state is seen in a changed fragmentation of the quasiparticle⊗two-phonons components; these changes, in their turn, affect the fragmentation of the quasiparticle ⊗ phonon components. Hence, a change in the fragmentation of a single-quasiparticle state is realized by modifying the the fragmentation of the quasiparticle ⊗ phonon components. If a problem is solved in this way, the effect of multiphonon components on the fragmentation of a single-quasiparticle state and of quasiparticle ⊗ phonon states should be insignificant. If one is to analyse the fragmentation of the quasiparticle ⊗ n-phonons states, it is necessary to include all components of the wave function, up to those components that consist of a quasiparticle and $n+1$ phonons.

Since the phonon operators are not ideal boson operators but consist of pairs of fermion quasiparticles, we have to bear in mind the following exact commutation relations:

$$[Q_{\lambda\mu i}, Q^{+}_{\lambda'\mu' i'}] = \delta_{\lambda\mu,\lambda'\mu'}\frac{1}{2}\sum_{jj'}(\psi^{\lambda i}_{jj'}\psi^{\lambda i'}_{jj'} - \varphi^{\lambda i}_{jj'}\varphi^{\lambda i'}_{jj'})$$
$$- \sum_{jj'j_2}\sum_{mm'm_2}[\psi^{\lambda i}_{j'j_2}\psi^{\lambda' i'}_{jj_2}\langle j'm'j_2m_2|\lambda\mu\rangle$$
$$\times \langle jmj_2m_2|\lambda'\mu'\rangle - (-)^{\lambda+\lambda'-\mu-\mu'}\varphi^{\lambda i}_{jj_2}\varphi^{\lambda' i'}_{j'j_2}$$
$$\times \langle jmj_2m_2|\lambda-\mu\rangle\langle j'm'j_2m_2|\lambda'-\mu'\rangle]\alpha^{+}_{jm}\alpha_{j'm'} \tag{4.27}$$

$$[\alpha_{jm}, Q^{+}_{\lambda\mu i}] = -\sum_{j'm'}\langle j'm'jm|\lambda\mu\rangle\psi^{\lambda i}_{j'j}\alpha^{+}_{j'm'} \tag{4.28}$$

$$\sum_{m_1m_2\mu_1\mu_2}\langle j_1m_1\lambda_1\mu_1|JM\rangle\langle j_2m_2\lambda_2\mu_2|JM\rangle\langle[Q_{\lambda_1\mu_1i_1}, \alpha^{+}_{j_2m_2}][\alpha_{j_1m_1}, Q^{+}_{\lambda_2\mu_2i_2}]\rangle$$
$$= \sum_{j_3}(2\lambda_1+1)^{1/2}(2\lambda_2+1)^{1/2}\begin{Bmatrix} j_3 & j_2 & \lambda_1 \\ j_1 & J & \lambda_2 \end{Bmatrix}$$

$$\times \psi_{j_3 j_2}^{\lambda_1 i_1} \psi_{j_3 j_1}^{\lambda_2 i_2} \equiv -\mathcal{L}^J(j_1 \lambda_1 i_1 | j_2 \lambda_2 i_2) \tag{4.29}$$

$$\begin{aligned}
&[[Q_{\lambda' \mu' i'}, Q^+_{\lambda \mu i}], Q^+_{\lambda_2 \mu_2 i_2}] \\
&\qquad = \sum_{\lambda_2' \mu_2' i_2'} \{\mathcal{K}(\lambda_2' \mu_2' i_2', \lambda' \mu' i' | \lambda \mu i, \lambda_2 \mu_2 i_2) Q^+_{\lambda_2' \mu_2' i_2'} \\
&\qquad + \tilde{\mathcal{K}}(\lambda_2' \mu_2' i_2', \lambda' \mu' i' | \lambda \mu i, \lambda_2 \mu_2 i_2) Q_{\lambda_2' \mu_2' i_2'}\}
\end{aligned} \tag{4.30}$$

where

$$\begin{aligned}
&\mathcal{K}(\lambda_2' \mu_2' i_2', \lambda' \mu' i' | \lambda \mu i, \lambda_2 \mu_2 i_2) \\
&\quad = - \sum_{\substack{j_1 j_2 j_3 j_4 \\ m_1 m_2 m_3 m_4}} [\psi_{j_3 j_4}^{\lambda' i'} \psi_{j_1 j_4}^{\lambda i} \langle j_3 m_3 j_4 m_4 | \lambda' \mu' \rangle \langle j_1 m_1 j_4 m_4 | \lambda \mu \rangle \\
&\quad - (-)^{\lambda + \lambda' - \mu - \mu'} \varphi_{j_1 j_4}^{\lambda' i'} \varphi_{j_3 j_4}^{\lambda i} \langle j_1 m_1 j_4 m_4 | \lambda' - \mu' \rangle \\
&\quad \times \langle j_3 m_3 j_4 m_4 | \lambda - \mu \rangle][\psi_{j_3 j_2}^{\lambda_2 i_2} \psi_{j_1 j_2}^{\lambda_2' i_2'} \langle j_3 m_3 j_2 m_2 | \lambda_2 \mu_2 \rangle \\
&\quad \times \langle j_1 m_1 j_2 m_2 | \lambda_2' \mu_2' \rangle + (-)^{\lambda_2 + \lambda_2' - \mu_2 - \mu_2'} \\
&\quad \times \varphi_{j_1 j_2}^{\lambda_2 i_2} \varphi_{j_3 j_2}^{\lambda_2' i_2'} \langle j_1 m_1 j_2 m_2 | \lambda_2 - \mu_2 \rangle \langle j_3 m_3 j_2 m_2 | \lambda_2' - \mu_2' \rangle].
\end{aligned} \tag{4.31}$$

The following expression will be of use later:

$$\begin{aligned}
&\mathcal{K}^J(\lambda_2', i_2', \lambda' i' | \lambda i, \lambda_2, i_2) \\
&\qquad = \sum_{\mu \mu_2 \mu' \mu_2'} \langle \lambda' \mu' \lambda_2' \mu_2' | JM \rangle \langle \lambda \mu \lambda_2 \mu_2 | JM \rangle \\
&\qquad \times \mathcal{K}(\lambda_2' \mu_2' i_2', \lambda' \mu' i' | \lambda \mu i, \lambda_2 \mu_2 i_2).
\end{aligned} \tag{4.31$'$}$$

The Pauli exclusion principle is taken into account rigorously by introducing the functions $\mathcal{L}^J$ and $\mathcal{K}^J$. The functions $\mathcal{L}^J(j\lambda i | j_2 \lambda_2 i_2)$ and $\mathcal{K}^J(\lambda_2' i_2', \lambda' i' | \lambda i, \lambda_2 i_2)$ are sign-alternating; their absolute diagonal values are maximal and much greater than their non-diagonal values. It is natural that the diagonal functions of $\mathcal{L}^J$ and $\mathcal{K}^J$ are maximal since the probabilitity of violation of the Pauli exclusion principle is the highest for configurations constructed of identical quasiparticles and phonons. The values of several $\mathcal{L}^J(j_1 \lambda_1 i_1 | j_2 \lambda_2 i_2)$ functions are listed in table 4.1. The diagonal values of $\mathcal{L}^J(j\lambda i | j \lambda i)$ are denoted by $\mathcal{L}(Jj\lambda i)$. The expressions for them are simpler:

$$\mathcal{L}(Jj\lambda i) = -\sum_{j'} (2\lambda + 1) \begin{Bmatrix} j & j' & \lambda \\ j & J & \lambda \end{Bmatrix} \left(\psi_{j'j}^{\lambda i}\right)^2. \tag{4.32}$$

Table 4.1 The coefficients $\mathcal{L}^J(\lambda_1 i_1 j_1|j_2\lambda_2 i_2)$ for the states with $J^\pi = 1/2^-$ and $3/2^-$ of ^{58}Ni

J_f^π	$n_1 l_1 j_1$	$\lambda_{1i_1}^\pi$	$n_2 l_2 j_2$	$\lambda_{2i_2}^\pi$	$\mathcal{L}^J(\lambda_1 i_1 j_1\|j_2\lambda_2 i_2)$
$1/2^-$	$1f5/2$	2_5^+	$1f5/2$	2_5^+	−0.95
	$2p3/2$	2_1^+	$2p3/2$	2_1^+	−0.67
	$2p3/2$	2_1^+	$1p3/2$	2_1^+	−0.004
	$2p3/2$	2_1^+	$2p3/2$	2_3^+	−0.04
	$2p3/2$	2_1^+	$1f5/2$	2_1^+	0.17
	$2p3/2$	2_{11}^+	$1d5/2$	3_1^-	−0.05
$3/2^-$	$1f7/2$	2_{10}^+	$1f7/2$	2_{10}^+	−0.36
	$1f7/2$	2_1^+	$2p3/2$	2_2^+	−0.13
	$1f7/2$	2_2^+	$2p1/2$	2_2^+	−0.1
	$1f7/2$	2_8^+	$1d5/2$	3_4^-	0.01

The quasi-diagonal function is given by

$$\mathcal{K}^J(\lambda_2 i_2, \lambda i'|\lambda i, \lambda_2 i_2)$$
$$= -\left(1 - \frac{\delta_{ii'}\delta_{\lambda i,\lambda_2 i_2}}{1+\delta_{\lambda i,\lambda_2 i_2}}\right)$$
$$\times \sum_{j_1 j_2 j_3 j_4} (-)^{j_2+j_4+J} \begin{Bmatrix} j_1 & j_2 & \lambda_2 \\ j_4 & j_3 & \lambda \\ \lambda & \lambda_2 & J \end{Bmatrix} (2\lambda+1)(2\lambda_2+1)$$
$$\times \left[\psi_{j_3 j_4}^{\lambda i}\psi_{j_1 j_4}^{\lambda i'}\psi_{j_3 j_2}^{\lambda_2 i_2}\psi_{j_1 j_2}^{\lambda_2 i_2} - \varphi_{j_3 j_4}^{\lambda i'}\varphi_{j_1 j_4}^{\lambda i}\varphi_{j_3 j_2}^{\lambda_2 i_2}\varphi_{j_1 j_2}^{\lambda_2 i_2}\right].$$

The diagonal function $\mathcal{K}^J$ (for $i = i'$) will be denoted by $\mathcal{K}^J(\lambda i, \lambda_1 i_2)$, so that

$$\mathcal{K}^J(\lambda i, \lambda_2 i_2) = \mathcal{K}^J(\lambda_2 i_2, \lambda i).$$

The values of $\mathcal{K}^J(\lambda i, \lambda_2 i_2)$ for several nuclei are listed in table 4.2.

In addition, we define a function

$$\mathcal{L}(Jj\lambda|\lambda_1 i_1, \lambda_2 i_2) = -(2\lambda+1)\sum_{j'}\left[(2\lambda_1+1)\begin{Bmatrix} J & \lambda & j \\ \lambda & \lambda_2 & \lambda_1 \\ j & \lambda_1 & j' \end{Bmatrix}\right.$$
$$\left.\times \left(\psi_{jj'}^{\lambda_1 i_1}\right)^2 + (2\lambda_2+1)\begin{Bmatrix} J & \lambda & j \\ \lambda & \lambda_1 & \lambda_2 \\ j & \lambda_2 & j' \end{Bmatrix}\left(\psi_{jj'}^{\lambda_2 i_2}\right)^2\right]. \quad (4.33)$$

whose values [145] are listed in table 4.3.

Table 4.2 The functions $\mathcal{K}^J(\lambda_1 i_1, \lambda_2 i_2)$.

Nucleus	$\lambda_1 i_1$	$\lambda_2 i_2$	J^π	$\mathcal{K}^J(\lambda_1 i_1, \lambda_2 i_2)$
^{114}Sn	21	21	2^+	−0.06
	21	22	2^+	−0.17
	21	23	2^+	−0.17
	21	21	4^+	−0.11
	21	22	4^+	−0.13
	21	23	4^+	−0.16
	21	31	3^-	−0.05
	21	32	3^-	−0.005
^{142}Sm	21	21	2^+	−0.07
	22	22	2^+	−0.007
	21	21	4^+	−0.07
	22	22	4^+	−0.85
	21	31	3^-	−0.10
	21	32	3^-	−0.01
^{146}Sm	21	21	2^+	−0.15
	21	22	2^+	−0.26
	21	21	4^+	−0.18
	22	22	4^+	−0.22
	21	31	3^-	−0.14
	21	32	3^-	−0.51
^{208}Pb	31	31	2^+	−0.023
	31	32	2^+	−0.034
	31	35	2^+	−0.018
	21	31	3^-	−0.026
	21	32	3^-	−0.033
	22	31	3^-	−0.089
	23	31	3^-	−0.015

Table 4.3 The functions $\mathcal{L}(Jj\lambda|\lambda_1 i_1, \lambda_2 i_2)$.

Nucleus	J^π	nlj	λ	$\lambda_1 i_1$	$\lambda_2 i_2$	$\mathcal{L}(Jj\lambda|\lambda_1 i_1, \lambda_2 i_2)$
^{61}Ni	$1/2^-$	$2p_{3/2}$	2	21	22	−0.67
				21	23	0.12
				22	22	0.12
				22	24	0.18
^{209}Pb	$9/2^-$	$1i_{13/2}$	2	21	21	−0.018
				21	22	0.006
				22	22	0.003

4.2.2

We choose (4.4) to represent the wave function of an odd spherical nucleus. In spherical nuclei, the single-particle states with fixed values of j^π are located on the next-nearest shell and separated by a large energy gap; hence, only one single-quasiparticle state is typically taken in (4.4). In a diagonal approximation of the functions $\mathcal{K}^J$ and $\mathcal{L}$, normalization condition (4.4) is

$$1 = C_{J\nu}^2\Big\{1 + \sum_{\lambda ij}[D_j^{\lambda i}(J\nu)]^2[1 + \mathcal{L}(Jj\lambda i)] + 2\sum_{\lambda i\lambda_2 i_2 j\lambda'}[F_{j\lambda'}^{\lambda i\lambda_2 i_2}(J\nu)]^2[1 + \mathcal{K}^{\lambda'}(\lambda i, \lambda_2 i_2)] \times [1 + \mathcal{L}(Jj\lambda'|\lambda i, \lambda_2 i_2)]\Big\}. \tag{4.34}$$

For quasiparticle ⊗ phonon configurations, which are strictly forbidden by Pauli exclusion principle, we have $\mathcal{L}(Jj\lambda i) = -1$, so that these configurations are automatically excluded from normalization and from the ensuing formulas. If we find maximum violation of the Pauli exclusion principle in two-phonon configurations, then $\mathcal{K}^J = -1$ and they are also eliminated. If the violation of the Pauli exclusion principle for the quasiparticle and one of the phonons in a quasiparticle ⊗ two phonons component is maximal, then $\mathcal{L}(Jj\lambda'|\lambda i, \lambda_2 i_2) = -1$ and this component disappears from the wave function. Therefore, a rigorous inclusion of commutation relations between the phonon and quasiparticle operators guarantees that the Pauli exclusion principle is satisfied.

We will now find the mean value of H_{QPNM} over state (4.4):

$$\begin{aligned}
&(\Psi_\nu^*(JM)H_{\text{QPNM}}\Psi_\nu(JM)) \\
&\quad = C_{J\nu}^2\Big\{\epsilon_J + \sum_{\lambda ij}\Big[(D_j^{\lambda i}(J\nu))^2 \\
&\quad\times [\epsilon_j + \omega_{\lambda i} + \Delta\omega(Jj\lambda i)] - 2D_j^{\lambda i}(J\nu)\Gamma(Jj\lambda i)\Big] \\
&\quad\times [1 + \mathcal{L}(Jj\lambda i)] + 2\sum_{\lambda i\lambda_2 i_2 j\lambda'}\Big[F_{j\lambda'}^{\lambda i\lambda_2 i_2}(J\nu)\Big]^2[\epsilon_j + \omega_{\lambda i} \\
&\quad + \omega_{\lambda_2 i_2} + S(Jj\lambda'|\lambda i, \lambda_2 i_2)][1 + \mathcal{K}^{\lambda'}(\lambda i, \lambda_2 i_2)] \\
&\quad\times [1 + \mathcal{L}(Jj\lambda'|\lambda i, \lambda_2 i_2)] + 2\sum_{\lambda ij\lambda_1 i_1\lambda_2 i_2} D_j^{\lambda i}(J\nu) \\
&\quad\times F_{j\lambda}^{\lambda_1 i_1 i_2}(J\nu)U_{\lambda_1 i_1}^{\lambda_2 i_2}(\lambda i)[1 + \mathcal{K}^{\lambda}(\lambda_1 i_1, \lambda_2 i_2)] \\
&\quad\times [1 + \mathcal{L}(Jj\lambda i)] - 4\sum_{\lambda i\lambda_1 i_1 jj'\lambda'} D_j^{\lambda i}(J\nu)F_{j'\lambda'}^{\lambda_1 i_1\lambda i}(J\nu) \\
&\quad\times (2j+1)^{1/2}(2\lambda'+1)^{1/2}(-)^{i'+\lambda_1+\lambda+J}\begin{Bmatrix}\lambda_1 & \lambda & \lambda' \\ J & j' & j\end{Bmatrix} \\
&\quad\times \Gamma(jj'\lambda_1 i_1)[1 + \mathcal{K}^{\lambda'}(\lambda i, \lambda_1 i_1)][1 + \mathcal{L}(Jj'\lambda_1 i_1)
\end{aligned}$$

$$+ \mathcal{L}(Jj\lambda i)]\Big\} \tag{4.35}$$

where

$$\Gamma(Jj\lambda i) = \sqrt{\frac{2\lambda+1}{2J+1}}\,\frac{1}{\sqrt{2}}\,\frac{f^{\lambda}(Jj)v^{(-)}_{Jj}}{(\mathcal{Y}^{\lambda i}_{\tau})^{1/2}}$$
$$= 2\sqrt{\frac{2\lambda+1}{2J+1}}\,\Gamma^{\lambda i}_{Jj}(\tau). \tag{4.36}$$

This vertex tends to zero as the phonon energy tends to the pole and $\mathcal{Y}^{\lambda i}_{\tau} \to \infty$:

$$\Delta\omega(Jj\lambda i) = -\sum_{\tau}\frac{X^{\lambda i}(\tau)}{\mathcal{Y}^{\lambda i}_{\tau}}\mathcal{L}(Jj\lambda i)$$

$$S(Jj\lambda|\lambda_1 i_1, \lambda_2 i_2) = \frac{1}{4}\sum_{\tau}\frac{X^{\lambda_1 i_1}(\tau) + X^{\lambda_2 i_2}(\tau)}{(\mathcal{Y}^{\lambda_1 i_1}_{\tau}\mathcal{Y}^{\lambda_2 i_2}_{\tau})^{1/2}}$$
$$\times [\mathcal{K}^{J}(\lambda_1 i_1, \lambda_2 i_2) - \mathcal{L}(Jj\lambda|\lambda_1 i_1, \lambda_2 i_2)]$$

$$U^{\lambda_1 i_1}_{\lambda_2 i_2}(\lambda i) = \langle Q_{\lambda\mu i} H_{vq}[Q^{+}_{\lambda_1\mu_1 i_1} Q^{+}_{\lambda_2\mu_2 i_2}]_{\lambda\mu}\rangle$$
$$= (-)^{\lambda_1+\lambda_2-\lambda}\frac{1}{\sqrt{2}}(2\lambda_1+1)^{1/2}(2\lambda_2+1)^{1/2}$$
$$\times \sum_{\tau}\sum_{j_1 j_2 j_3}{}^{\tau} v^{(\mp)}_{j_1 j_2}\left[\frac{f^{\lambda_2}(j_1 j_2)}{\sqrt{\mathcal{Y}^{\lambda_2 i_2}_{\tau}}}\begin{Bmatrix}\lambda_1 & \lambda_2 & \lambda\\ j_1 & j_3 & j_2\end{Bmatrix}\right.$$
$$\times \left(\psi^{\lambda i}_{j_3 j_1}\psi^{\lambda_1 i_1}_{j_2 j_3} \pm \varphi^{\lambda i}_{j_3 j_1}\varphi^{\lambda_1 i_1}_{j_2 j_3}\right)$$
$$+ \frac{f^{\lambda_1}(j_1 j_2)}{\sqrt{\mathcal{Y}^{\lambda_1 i_1}_{\tau}}}\begin{Bmatrix}\lambda_1 & \lambda_2 & \lambda\\ j_3 & j_2 & j_1\end{Bmatrix}\left(\varphi^{\lambda i}_{j_2 j_3}\varphi^{\lambda_2 i_2}_{j_3 j_1} \pm \psi^{\lambda i}_{j_2 j_3}\psi^{\lambda_2 i_2}_{j_3 j_1}\right)$$
$$\left. + \frac{f^{\lambda}(j_1 j_2)}{\sqrt{\mathcal{Y}^{\lambda i}_{\tau}}}\begin{Bmatrix}\lambda_1 & \lambda_2 & \lambda\\ j_1 & j_2 & j_3\end{Bmatrix}\left(\psi^{\lambda_1 i_1}_{j_3 j_1}\varphi^{\lambda_2 i_2}_{j_2 j_3} \pm \psi^{\lambda_2 i_2}_{j_2 j_3}\varphi^{\lambda_1 i_1}_{j_3 j_1}\right)\right] \tag{4.37}$$

where $v^{(-)}_{j_1 j_2}$ and the plus sign appear in combination with the multipole matrix elements $f^{\lambda}(j_1 j_2)$, while $v^{(+)}_{j_1 j_2}$ and the minus sign appear in combination with the spin-multipole matrix elements $f^{\lambda L}(jj_2)$. As we see from (4.35), the states that are strictly forbidden by the Pauli exclusion principle do not contribute to the mean value of the Hamiltonian.

Making use of the variational principle

$$\delta\{(\Psi^{*}_{\nu}(JM)H_{\text{QPNM}}\Psi_{\nu}(JM) \quad - \eta_{J\nu}[(\Psi^{*}_{\nu}(JM)\Psi_{\nu}(JM)) - 1]\} = 0 \tag{4.38}$$

we arrive at a set of three equations:

$$\epsilon_J - \eta_{J\nu} + \sum_{\lambda i j}\{(D_j^{\lambda i})^2[\epsilon_j + \omega_{\lambda i} + \Delta\omega(Jj\lambda i) - \eta_{J\nu}]$$
$$- 2D_j^{\lambda i}\Gamma(Jj\lambda i)\}[1 + \mathcal{L}(Jj\lambda i)]$$
$$+ 2 \sum_{\lambda_1 i_1 \lambda_2 i_2 j \lambda} (F_{j\lambda}^{\lambda_1 i_1 \lambda_2 i_2})^2[\epsilon_j + \omega_{\lambda_1 i_1} + \omega_{\lambda_2 i_2}$$
$$+ S(Jj\lambda|\lambda_1 i_1, \lambda_2 i_2) - \eta_{J\nu}][1 + \mathcal{K}^\lambda(\lambda_1 i_1, \lambda_2 i_2)]$$
$$\times [1 + \mathcal{L}(Jj\lambda|\lambda_1 i_1, \lambda_2 i_2)] + 2 \sum_{\lambda_1 i_1 \lambda_2 i_2 \lambda i j} D_j^{\lambda i} F_{j\lambda}^{\lambda_1 i_1 \lambda_2 i_2}$$
$$\times U_{\lambda_1 i_1}^{\lambda_2 i_2}(\lambda i)[1 + \mathcal{K}^\lambda(\lambda_1 i_1, \lambda_2 i_2)][1 + \mathcal{L}(Jj\lambda i)]$$
$$- 4 \sum_{\lambda i \lambda_1 i_1 j j' \lambda'} D_j^{\lambda i} F_{j'\lambda'}^{\lambda_1 i_1 \lambda i}(-)^{j' + \lambda_1 + \lambda + J}$$
$$\times (2j+1)^{1/2}(2\lambda'+1)^{1/2} \begin{Bmatrix} \lambda_1 & \lambda & \lambda' \\ J & j' & j \end{Bmatrix} \Gamma(jj'\lambda_1 i_1)$$
$$\times [1 + \mathcal{K}^{\lambda'}(\lambda i, \lambda_1 i_1)][1 + \mathcal{L}(Jj'\lambda_1 i_1) + \mathcal{L}(Jj\lambda i)] = 0 \qquad (4.39)$$

$$D_j^{\lambda i}[\epsilon_j + \omega_{\lambda i} + \Delta\omega(Jj\lambda i) - \eta_{J\nu}] - \Gamma(Jj\lambda i)]$$
$$\times [1 + \mathcal{L}(Jj\lambda i)] + \sum_{\lambda_1 i_1 \lambda_2 i_2} F_{j\lambda}^{\lambda_1 i_1 \lambda_2 i_2} U_{\lambda_1 i_1}^{\lambda_2 i_2}(\lambda i)$$
$$\times [1 + \mathcal{K}^\lambda(\lambda_1 i_1, \lambda_2 i_2)][1 + \mathcal{L}(Jj\lambda i)]$$
$$- 2 \sum_{\lambda_1 i_1 \lambda' j'} F_{j'\lambda'}^{\lambda_1 i_1 \lambda i}\Gamma(jj'\lambda_1 i_1)(-)^{j' + \lambda + \lambda_1 + J}$$
$$\times \begin{Bmatrix} \lambda_1 & \lambda & \lambda' \\ J & j' & j \end{Bmatrix} (2j+1)^{1/2}(2\lambda'+1)^{1/2}[1 + \mathcal{K}^{\lambda'}(\lambda i, \lambda_1 i_1)]$$
$$\times [1 + \mathcal{L}(Jj'\lambda_1 i_1) + \mathcal{L}(Jj\lambda i)] = 0 \qquad (4.39')$$

$$F_{j\lambda}^{\lambda_1 i_1 \lambda_2 i_2}[\epsilon_j + \omega_{\lambda_1 i_1} + \omega_{\lambda_2 i_2} + S(Jj\lambda|\lambda_1 i_1, \lambda_2 i_2)$$
$$- \eta_{J\nu}][1 + \mathcal{K}^\lambda(\lambda_1 i_1, \lambda_2 i_2)][1 + \mathcal{L}(Jj\lambda|\lambda_1 i_1, \lambda_2 i_2)]$$
$$+ \frac{1}{2}\sum_i D_j^{\lambda i} U_{\lambda_1 i_1}^{\lambda_2 i_2}(\lambda i)][1 + \mathcal{K}^\lambda(\lambda_1 i_1, \lambda_2 i_2)]$$
$$\times [1 + \mathcal{L}(Jj\lambda i)] - \frac{1}{2}\sum_{j_2}\Big\{D_{j_2}^{\lambda_2 i_2}\Gamma(j_2\lambda_1 i_1)$$
$$\times (2j_2+1)^{1/2}(2\lambda+1)^{1/2}(-)^{j+\lambda_1+\lambda_2+J} \begin{Bmatrix} \lambda_1 & \lambda_2 & \lambda \\ J & j & j_2 \end{Bmatrix}$$
$$\times [1 + \mathcal{K}^\lambda(\lambda_1 i_1, \lambda_2 i_2)][1 + \mathcal{L}(Jj\lambda_1 i_1)$$
$$+ \mathcal{L}(Jj_2\lambda_2 i_2)] + D_{j_2}^{\lambda_1 i_1}\Gamma(j_2 j\lambda_2 i_2)(2j_2+1)^{1/2}(2\lambda+1)^{1/2}$$

$$\times (-)^{J+j-\lambda} \begin{Bmatrix} \lambda_1 & \lambda_2 & \lambda \\ J & j & j_2 \end{Bmatrix} [1 + \mathcal{K}^{\lambda}(\lambda_1 i_1, \lambda_2 i_2)]$$
$$\times [1 + \mathcal{L}(Jj\lambda_2 i_2) + \mathcal{L}(Jj_2\lambda_1 i_1)]\Big\} = 0. \tag{4.39''}$$

Substituting the expression for $F_{j\lambda}^{\lambda_1 i_1 \lambda_2 i_2}(J\nu)$ from (4.39″) into (4.39), (4.39′), we obtain the equation

$$\mathcal{F}(\eta_{J\nu}) \equiv \epsilon_J - \eta_{J\nu} - \sum_{\lambda ij} \Gamma(Jj\lambda i) D_j^{\lambda i}(J\nu)$$
$$\times [1 + \mathcal{L}(Jj\lambda i)] = 0 \tag{4.40}$$

and an inhomogeneous equation for the functions $D_j^{\lambda i}(J\nu)$, whose explicit form is given in [145, 147]. With the functions $D_j^{\lambda i}(J\nu)$ determined and substituted into equation (4.40), we can calculate the energies $\eta_{J\nu}$ of the states with wave function (4.4).

4.2.3

Let us neglect the corrections caused by rigorously including the Pauli principle. Correspondingly, we set the functions $\mathcal{K}^J(\lambda i, \lambda_2 i_2)$, $\mathcal{L}(Jj\lambda i)$ and $\mathcal{L}(Jj\lambda|\lambda_1 i_1, \lambda_2 i_2)$ equal to zero. This operation results in the following set of equations:

$$\mathcal{F}(\eta_{J\nu}) \equiv \epsilon_J - \eta_{J\nu} - \sum_{\lambda ij} \Gamma(Jj\lambda i) D_j^{\lambda i}(J\nu) = 0 \tag{4.41}$$

$$\sum_{\lambda ij} D_j^{\lambda i} \Bigg\{ \Bigg[\epsilon_j + \omega_{\lambda i} - \eta_{j\nu} - \sum_{\lambda_2 i_2 j_2} \frac{\Gamma^2(jj_2\lambda_2 i_2)}{\epsilon_{j_2} + \omega_{\lambda i} + \omega_{\lambda_2 i_2} - \eta_{J\nu}} \Bigg]$$
$$\times \delta_{\lambda i, \lambda_1 i_1} \delta_{jj_1} - \sum_{j_3} \frac{\Gamma(j_3 j_1 \lambda i) \Gamma(j_3 j \lambda_1 i_1)}{\epsilon_{j_3} + \omega_{\lambda i} + \omega_{\lambda_1 i_1} - \eta_{J\nu}} (2j+1)^{1/2}$$
$$\times (2j_1+1)^{1/2} \begin{Bmatrix} \lambda_2 & j_3 & j_1 \\ \lambda_1 & J & j \end{Bmatrix}$$
$$- \frac{1}{2} \sum_{\lambda_2 i_2 \lambda_3 i_3} \frac{U_{\lambda_3 i_3}^{\lambda_2 i_2}(\lambda_1 i_1) U_{\lambda_3 i_3}^{\lambda_2 i_2}(\lambda_1 i)}{\epsilon_{j_1} + \omega_{\lambda_2 i_2} + \omega_{\lambda_3 i_3} - \eta_{J\nu}} \delta_{\lambda\lambda_1} \delta_{jj_1}$$
$$+ \sum_{\lambda_2 i_2} \Gamma(jj_1\lambda_2 i_2)(-)^{j+\lambda_1+J}(2j+1)^{1/2} \begin{Bmatrix} \lambda_2 & \lambda & \lambda_1 \\ J & j_1 & j \end{Bmatrix}$$
$$\times \left[\frac{U_{\lambda i}^{\lambda_2 i_2}(\lambda_1 i_1)(2\lambda_1+1)^{1/2}}{\epsilon_{j_1} + \omega_{\lambda i} + \omega_{\lambda_2 i_2} - \eta_{J\nu}} + \frac{U_{\lambda_2 i_2}^{\lambda_1 i_1}(\lambda i)(2\lambda+1)^{1/2}}{\epsilon_j + \omega_{\lambda_1 i_1} + \omega_{\lambda_2 i_2} - \eta_{J\nu}} \right] \Bigg\}$$
$$= \Gamma(Jj_1\lambda_1 i_1). \tag{4.42}$$

If the Pauli exclusion principle is rigorously taken into account, the poles get shifted and the matrix elements are renormalized.

If we drop the terms of (4.42) which contain $U_{\lambda_3 i_3}^{\lambda_2 i_2}(\lambda i)$ and are the most essential for the description of the fragmentation of quasiparticle⊗phonon-type states, we arrive at the set of equations which were first derived in [136]:

$$\sum_{\lambda ij} D_j^{\lambda i}(J\nu)\Bigg\{ \Bigg[\epsilon_j + \omega_{\lambda_1 i_1} - \eta_{J\nu} - \sum_{\lambda_2 i_2 j_2} \frac{\Gamma^2(jj_2\lambda_2 i_2)}{\epsilon_{j_2} + \omega_{\lambda i} + \omega_{\lambda_2 i_2} - \eta_{J\nu}}\Bigg] \delta_{\lambda i, \lambda_1 i_1}\delta_{jj_1} - \sum_{j_3} \frac{\Gamma(j_3 j_1 \lambda i)\Gamma(j_3 j \lambda_1 i_1)}{\epsilon_{j_2} + \omega_{\lambda_1 i_1} + \omega_{\lambda i} - \eta_{J\nu}} (2j_1+1)^{1/2}(2j+1)^{1/2} \times \begin{Bmatrix} \lambda & j_3 & j_1 \\ \lambda_1 & J & j \end{Bmatrix} \Bigg\} = \Gamma(Jj_1\lambda_1 i_1). \quad (4.43)$$

Equation (4.43) was used for the calculation of the strength distribution of deep hole states and high-lying particle states [142, 160–162] and of neutron strength functions [160, 163].

If we neglect the quasiparticle ⊗ two-phonon components in the wave functions of an excitation state of an odd nucleus, that is, if we consider only wave functions of the type

$$\Psi_\nu(JM) = C_{J\nu}\Big\{a_{JM}^+ + \sum_{\lambda ij} D_j^{\lambda i}(J\nu)\Big[a_{jm}^+ Q_{\lambda\mu i}^+\Big]_{JM}\Big\}\Psi_0 \quad (4.44)$$

then the diagonal approximation for the function $\mathcal{L}^J$ gives a simple expression for $D_j^{\lambda j}(J\nu)$ [143]:

$$D_j^{\lambda i}(J\nu) = \frac{\Gamma(Jj\lambda i)}{\epsilon_j + \omega_{\lambda i} + \Delta\omega(Jj\lambda i) - \eta_{J\nu}}. \quad (4.45)$$

Substituting (4.45) into (4.41), we obtain a secular equation for energy eigenvalues:

$$\mathcal{F}(\eta_{J\nu}) = \epsilon_J - \eta_{J\nu} - \sum_{\lambda ij} \frac{\Gamma^2(Jj\lambda i)[1 + \mathcal{L}(Jj\lambda i)]}{\epsilon_j + \omega_{\lambda i} + \Delta\omega(Jj\lambda i) - \eta_{J\nu}} = 0. \quad (4.46)$$

If we neglect the corrections due to including rigorously the Pauli principle and setting $\mathcal{L}$ and $\Delta\omega$ equal to zero, we arrive at the familiar equation for odd nuclei [6, 131]:

$$\mathcal{F}(\eta_{J\nu}) = \epsilon_J - \eta_{J\nu} - \sum_{\lambda ij} \frac{\Gamma^2(Jj\lambda i)}{\epsilon_j + \omega_{\lambda i} - \eta_{J\nu}} = 0. \quad (4.46')$$

4.2.4

Let us derive formulas for the strength functions that describe the fragmentation of single-quasiparticle and quasiparticle $\otimes$ phonon states. The rank of the set of equations for the function $D_j^{\lambda i}(J\nu)$ is from 100 to 1000. In view of the nonlinear dependence of the coefficients of the equations on energy η, the introduction of strength functions is unavoidable.

The strength function describing the fragmentation of a single-particle state can be chosen in the form

$$C_J^2(\eta) = \frac{\Delta}{2\pi} \sum_\nu \frac{C_{J\nu}^2}{(\eta - \eta_{J\nu})^2 + \Delta^2/4}. \tag{4.47}$$

Following [142], we can write the general solution of the set of linear equations using the Kramers formulas:

$$D_j^{\lambda i}(J\nu) = \frac{\Gamma(Jj\lambda i)}{\epsilon_j + \omega_{\lambda i} + \Delta\omega(Jj\lambda i) - \eta_{J\nu}} S_j^{\lambda i}(\eta_{J\nu}). \tag{4.48}$$

The function $S_j^{\lambda i}(\eta_{J\nu})$ is the ratio of two determinants; it includes all poles of the type

$$\epsilon_j + \omega_{\lambda_1 i_1} + \omega_{\lambda_2 i_2} + S(Jj\lambda|\lambda_1 i_1, \lambda_2 i_2).$$

If we take equation (4.48) into account, then (4.40) becomes

$$\begin{aligned} \mathcal{F}(\eta_{J\nu}) \equiv \epsilon_J - \eta_{J\nu} \\ - \sum_{\lambda ij} \frac{\Gamma^2(Jj\lambda i)[1 + \mathcal{L}(Jj\lambda i)]}{\epsilon_j + \omega_{\lambda i} + \Delta\omega(Jj\lambda i) - \eta_{J\nu}} S_j^{\lambda i}(\eta_{J\nu}) = 0. \end{aligned} \tag{4.49}$$

The following condition holds at the points where set (4.39)–(4.39″) has a solution:

$$\mathrm{d}\mathcal{F}(\eta)/\mathrm{d}\eta|_{\eta=\eta_{J\nu}} = -1/C_{J\nu}^2. \tag{4.50}$$

We make use of (4.50) and rewrite (4.47) in the form

$$C_J^2(\eta) = -\frac{\Delta}{2\pi} \sum_\nu \frac{1}{\mathrm{d}\mathcal{F}(\eta)/\mathrm{d}\eta|_{\eta=\eta_{J\nu}}} \frac{1}{(\eta - \eta_{J\nu})^2 + \Delta^2/4}.$$

Using the residues theory, we can carry out the same transformations as we did to derive formula (4.25′), to obtain

$$C_J^2(\eta) = \frac{1}{\pi} \operatorname{Im} \frac{1}{\mathcal{F}(\eta + \mathrm{i}\Delta/2)}. \tag{4.47'}$$

The explicit form of $C_J^2(\eta)$ was obtained in [142] from (4.47′) and (4.49):

$$C_J^2(\eta) = \frac{1}{\pi} \frac{(\Delta/2)[1+\Gamma_2(\eta)]}{[\epsilon_J - \gamma_2(\eta) - \eta]^2 + (\Delta/2)^2[1+\Gamma_2(\eta)]^2} \tag{4.51}$$

$$\gamma_2(\eta) = \sum_{j\lambda i} \frac{\Gamma^2(Jj\lambda i)[1+\mathfrak{L}(Jj\lambda i)]}{[\epsilon_j + \omega_{\lambda i} + \Delta\omega(Jj\lambda i) - \eta]^2 + \Delta^2/4} \times \left\{ \mathrm{Re}\, S_j^{\lambda i}(\eta)[\epsilon_j + \omega_{\lambda i} + \Delta\omega(Jj\lambda i) - \eta] \times -(\Delta/2)^2 \,\mathrm{Im}\, S_j^{\lambda i}(\eta) \right\} \tag{4.52}$$

$$\Gamma_2(\eta) = \sum_{\lambda ij} \frac{\Gamma^2(Jj\lambda i)[1+\mathfrak{L}(Jj\lambda i)]}{[\epsilon_j + \omega_{\lambda i} + \Delta\omega(Jj\lambda i) - \eta]^2 + \Delta^2/4} \times \left\{ \mathrm{Im}\, S_j^{\lambda i}(\eta)[\epsilon_j + \omega_{\lambda i} + \Delta\omega(Jj\lambda i) - \eta] + \mathrm{Re}\, S_j^{\lambda i}(\eta) \right\}. \tag{4.52′}$$

The strength function describing the fragmentation of the quasiparticle ⊗ phonon state will be defined by the formula

$$D^2(\eta) = \frac{\Delta}{2\pi} \sum_{\nu} \left[C_{J\nu} D_j^{\lambda i}(J\nu) \right]^2 \frac{1}{(\eta - \eta_{J\nu})^2 + \Delta^2/4}. \tag{4.53}$$

The term proportional to $[D_j^{\lambda i}(J\nu)]^2$ produces additional poles in the function $D^2(\eta)$; this creates difficulties for the application of the strength function method. However, the products $D_j^{\lambda i}(J\nu) D_{j'}^{\lambda' i'}(J\nu)$, where $\lambda ij \neq \lambda' i' j'$, do not contain additional poles. Consequently, we transform (4.53) as follows. Using (4.40), we express $D_j^{\lambda i}(J\nu)$ in terms of the other $D_{j'}^{\lambda' i'}(J\nu)$:

$$D_j^{\lambda i}(J\nu) = \Big\{ \epsilon_J - \eta_{J\nu} - \sum_{\lambda' i' j' \neq \lambda ij} \Gamma(Jj'\lambda' i') \times [1 + \mathfrak{L}(Jj'\lambda' i')] D_{j'}^{\lambda' i'}(J\nu) \Big\} \Big\{ \Gamma(Jj\lambda i)[1+\mathfrak{L}(Jj\lambda i)] \Big\}^{-1}$$

$$\left[D_j^{\lambda i}(J\nu) \right]^2 = \frac{D_j^{\lambda i}(J\nu)}{\Gamma(Jj\lambda i)[1+\mathfrak{L}(Jj\lambda i)]} \Big\{ \epsilon_j - \eta_{J\nu} - \sum_{\lambda' i' j' \neq \lambda ij} \Gamma(jj'\lambda' i')[1+\mathfrak{L}(Jj'\lambda' i')] D_{j'}^{\lambda' i'}(J\nu) \Big\}.$$

We now substitute this expression into (4.53) and obtain, after transformations,

$$D^2(\eta) = \frac{1}{\pi}\,\mathrm{Im}\,\frac{D_j^{\lambda i}(\eta + \mathrm{i}\Delta/2)}{\Gamma(Jj\lambda i)[1 + \mathfrak{L}(Jj\lambda i)]\mathfrak{F}(\eta + \mathrm{i}\Delta/2)}$$
$$\times \Big\{\epsilon_J - \eta - \mathrm{i}\Delta/2 - \sum_{\lambda' i' j'} \Gamma(jj'\lambda' i')[1 + \mathfrak{L}(jj'\lambda' i')]$$
$$\times\, D_{j'}^{\lambda' i'}(\eta + \mathrm{i}\Delta/2)\Big\}.$$

4.2.5

In order to calculate $C_J^2(\eta)$ and $D_J^2(\eta)$ and analyse the physical processes they determine, we need to calculate the functions $D_j^{\lambda i}(\eta + i\Delta/2)$ for different η and a fixed value of Δ. The method of solution was suggested in [164] and realized for set (4.43). Actually, this method is applicable to the general set of equations as well. It is based on the fact that the diagonal elements of the coefficients matrix of the unknown $D_j^{\lambda i}(J\nu)$ (we denote it by $\mathfrak{G}$, and its elements by $g_{kk'}$) are typically greater than its non-diagonal elements. This is caused, among other factors, by the non-diagonal matrix elements $g_{kk'}$ being formed by incoherent sums and the diagonal matrix elements g_{kk}, by coherent sums. The non-diagonal matrix elements become large only if the energy η is close to the pole. The role of these matrix elements is mostly to eliminate the redundant poles of the set of equations, while the roots $\eta_{J\nu}$ and the fragmentation are largely determined by the diagonal elements g_{kk}.

This property of set (4.43) was pointed out in [134]. Attempts have been reported in some papers to solve system (4.43) in the coherent approximation, that is, with only the diagonal terms retained. This procedure produced spurious, non-physical solutions. To eliminate them, an attempt was made to take into account the non-diagonal terms closest to the root that was sought [134]. Owing to the fluctuations of the values of $\Gamma(Jj\lambda i)$, however, it is difficult to formulate the general criterion of selecting such non-diagonal terms. The requirement for the poles of these non-diagonal terms to lie close to the root sought proved to be insufficient. Taking into account one or two non-diagonal poles is sufficient only to calculate the wave functions of solutions with a simple structure. If, however, a large number of components gave comparable contributions to the wave function of the solution, then an artificial limitation of the number of included non-diagonal poles produced considerable distortions in the solution structure. An increase in the number of retained non-diagonal poles resulted in unmanageable complications in the formulas, rendering them virtually useless for practical work.

Following [164], we use a matrix form for set (4.43),

$$\mathfrak{G}D = \Gamma \tag{4.54}$$

where $D_{\lambda i}$ is the vector formed by unknown coefficients $D_j^{\lambda i}(J\nu)$ and Γ is a vector formed by the right-hand sides of equations (4.43). The fact that the matrix elements $|g_{kk}|$ of the matrix $\mathfrak{G}$ are, as a rule, greater than $|g_{kk'}|$ $(k \neq k')$ makes it possible to use Jacobi's iterative method to solve equation (4.54). The nth approximation to the solution is then found from the relation

$$D^{(n)} = D^{(n-1)} + \mathcal{H}(\Gamma - \mathfrak{G}D^{(n-1)}).$$

The matrix elements of the diagonal matrix $\mathcal{H}$ are

$$h_{kk} = g_{kk}^{-1}.$$

The vector Γ is the initial approximation of $D^{(0)}$. The iterative process converges if the inequality

$$\tilde{r}_k = |g_{kk}|^{-1} \sum_{k \neq k'} |g_{k'k}| < 1 \qquad (4.54')$$

holds for each row of the matrix $\mathfrak{G}$. If condition (4.54') does not hold, a different form must be chosen for the matrix $\mathcal{H}$. We assume the following situation:

$$\tilde{r}_k > 1 \quad k = 1, 2, \ldots, m$$
$$\tilde{r}_k < 1 \quad k = m+1, m+2, \ldots, n$$

(n is the rank of the matrix $\mathfrak{G}$). We then define $\mathcal{H}$ as follows:

$$\mathcal{H} = \begin{cases} \mathfrak{G}'^{-1} & k = 1, 2, \ldots, m \\ g_{kk}^{-1} & k = m+1, m+2, \ldots, n. \end{cases} \qquad (4.54'')$$

The matrix $\mathfrak{G}'$ is formed by the first m rows and columns of the matrix $\mathfrak{G}$. Each value of $\eta + \mathrm{i}\Delta/2$ corresponds to a matrix $\mathfrak{G}'$, and its rank varies from point to point.

An analogy to the many-pole approximation method, mentioned above, will be useful here. The choice of $\mathcal{H}$ in the form (4.54'') means that in the first approximation, a submatrix is singled out of the matrix $\mathfrak{G}$ which corresponds to the poles most strongly coupled to the state whose fragmentation we calculate. Only the diagonal matrix elements are taken into account among the remaining $g_{kk'}$. This exactly corresponds to the many-pole approximation [134]. If, however, condition (4.54') holds for each row of the matrix $\mathcal{H}$, the first approximation to the exact solution is the result obtained with only the coherent terms retained. If the iterative method is used, we obtain a numerical criterion for finding such non-diagonal elements whose contribution is essential at a given point; the number of elements to be taken into account may vary from point to point. Furthermore, the subsequent iterations take into account the contribution of non-diagonal elements which are responsible for the largest corrections and eliminate redundant solutions.

In contrast to the direct inversion of the matrix $\mathfrak{G}$, the method is the more efficient the lower the rank of the matrix $\mathfrak{G}'$, that is, the closer the value of $(n-m)/n$ to unity. A certain role in improving the efficiency of the iterative method is played by calculating not the exact values of $C^2_{J\nu}$ or $\left[D^{\lambda i}_j(J\nu)\right]^2$ but the values averaged by using a weight function: an increase in the parameter Δ reduces the non-diagonal matrix elements and increases the rate of convergence of the iterative process.

As an example, the fragmentation of a hole neutron state of the nucleus $1g_{9/2}{}^{119}$Sn was treated in [164]. All the quasiparticle $\otimes$ phonon and quasiparticle $\otimes$ two phonons states in the interval $0 \leqslant E \leqslant 10$ MeV were taken into consideration. The rank of $\mathfrak{G}$ was found to be $n = 276$. The rank of $\mathfrak{G}'$ at each point $\eta + \mathrm{i}\Delta/2$ did not exceed 20 ($m \leqslant 20$), that is, it was by an order of magnitude less than n. The effect of the rate of convergence of the iterative process for Δ is seen from the following example. The degree of proximity of the lth approximation $D^{(l)}$ to the exact solution is characterized by the absolute value of the so-called residual vector. For $\Delta = 0.1$ MeV, the mismatch reached 10^{-4} after seven or eight iterations, but after two or three iterations for $\Delta = 0.5$ MeV.

4.3 EQUATIONS FOR EVEN MASS SPHERICAL NUCLEI

4.3.1

Let us derive QPNM equations for even–even spherical nuclei. The QPNM describes those properties of nuclei which are connected with the fragmentation of one-phonon state or two-quasiparticle states. The wave function of excited states is given by the formula

$$\Psi_\nu(JM) = \Big\{\sum_i R_i(J\nu)Q^+_{JMi} + \sum_{\lambda_1 i_1 \lambda_2 i_2} P^{\lambda_1 i_1}_{\lambda_2 i_2}(J\nu)[Q^+_{\lambda_1\mu_1 i_1}Q^+_{\lambda_1\mu_1 i_1}]_{JM}\Big\}\Psi_0 \tag{4.55}$$

where JMi or $\lambda\mu i$ are the quantum numbers of the multipole and spin-multipole phonons. In (4.55), we retained only two-phonon terms; three-phonon terms have to be added subsequently. The normalization condition is written as follows:

$$\sum_i [R_i(J\nu)]^2 + 2\sum_{\lambda_1 i_1 \lambda_2 i_2} [P^{\lambda_1 i_1}_{\lambda_2 i_2}(J\nu)]^2$$

$$+ \sum_{\substack{\lambda_1 i_1 \lambda_2 i_2 \\ \lambda'_1 i'_1 \lambda'_2 i'_2}} P^{\lambda'_1 i'_1}_{\lambda'_2 i'_2}(J\nu) P^{\lambda_1 i_1}_{\lambda_2 i_2}(J\nu)$$
$$\times \mathcal{K}^J(\lambda'_2 i'_2, \lambda'_1 i'_1 | \lambda_1 i_1, \lambda_2 i_2) = 1. \qquad (4.56)$$

The explicit form of the function $\mathcal{K}^J(\lambda'_2 i'_2, \lambda'_1 i'_1 | \lambda_1 i_1, \lambda_2 i_2)$ is given by (4.31) and (4.31′).

First we find the most general form of the equations of the model. We calculate the mean value of H_{QPNM} over (4.55):

$$(\Psi^*_\nu(JM) H_{\mathrm{QPNM}} \Psi_\nu(JM))$$
$$= \sum_i \omega_{Ji} R_i^2$$
$$+ \sum_{\substack{\lambda_1 i_1 \lambda_2 i_2 \\ \lambda'_1 i'_1 \lambda'_2 i'_2}} P^{\lambda'_1 i'_1}_{\lambda'_2 i'_2}(J\nu) P^{\lambda_1 i_1}_{\lambda_2 i_2}(J\nu) \{(\omega_{\lambda_1 i_1} + \omega_{\lambda_2 i_2})$$
$$\times [\delta_{\lambda_1 i_1, \lambda'_1 i'_1} \delta_{\lambda_2 i_2 \lambda'_2 i'_2} + \delta_{\lambda_1 i_1, \lambda'_2 i'_2} \delta_{\lambda_2 i_2 \lambda'_1 i'_1}$$
$$+ \mathcal{K}^J(\lambda'_2 i'_2, \lambda'_1 i'_1 | \lambda_1 i_1, \lambda_2 i_2)]$$
$$- \frac{1}{4} \sum_{i_3 \tau} \Big[\frac{X^{\lambda'_1 i'_1}(\tau) + X^{\lambda'_1 i_3}(\tau)}{(\mathcal{Y}^{\lambda'_1 i'_1}_\tau \mathcal{Y}^{\lambda'_1 i_3}_\tau)^{1/2}} \mathcal{K}^J(\lambda'_2 i'_2, \lambda'_1 i_3 | \lambda_1 i_1, \lambda_2 i_2)$$
$$+ \frac{X^{\lambda'_2 i'_2}(\tau) + X^{\lambda'_1 i_3}(\tau)}{(\mathcal{Y}^{\lambda'_2 i'_2}_\tau \mathcal{Y}^{\lambda'_2 i_3}_\tau)^{1/2}} \mathcal{K}^J(\lambda'_2 i_3, \lambda'_1 i'_1 | \lambda_1 i_1, \lambda_2 i_2) \Big]$$
$$- \frac{1}{4} \sum_{\substack{\lambda_4 i_4 \tau \\ \lambda_3 i_3 i'_3}} \frac{X^{\lambda_3 i_3}(\tau) + X^{\lambda_3 i'_3}(\tau)}{(\mathcal{Y}^{\lambda_3 i_3}_\tau \mathcal{Y}^{\lambda_3 i'_3}_\tau)^{1/2}}$$
$$\times \mathcal{K}^J(\lambda_4 i_4, \lambda_3 i'_3 | \lambda_1 i_1, \lambda_2 i_2) \mathcal{K}^J(\lambda'_2 i'_2, \lambda'_1 i'_1 | \lambda_3 i_3, \lambda_4 i_4)\}$$
$$+ 2 \sum_{\lambda_1 i_1 \lambda_2 i_2 i} R_i P^{\lambda_1 i_1}_{\lambda_2 i_2} \Big[U^{\lambda_1 i_1}_{\lambda_2 i_2}(Ji) + V^{\lambda_1 i_1}_{\lambda_2 i_2}(Ji) \Big] \qquad (4.57)$$

where $\delta_{\lambda i, \lambda' i'} = \delta_{\lambda \lambda'} \delta_{i i'}$. For spin-multipole forces, the function $X^{\lambda i}(\tau)$ of (2.106) is replaced by the function $X^{\lambda i}_L(\tau)$ which is defined by (2.113). The explicit form of the function $U^{\lambda_2 i_2}_{\lambda' i'}(Ji)$ is given by (4.37); it corresponds to the diagram shown in figure 4.5 (a). The function $V^{\lambda_1 i_1}_{\lambda_2 i_2}(Ji)$ corresponds to the diagrams of figures 4.5 (b,c); it vanishes at $\mathcal{K}^J = 0$.

Let us make use of the variational principle in the form

$$\delta\{(\Psi^*_\nu(JM) H_{\mathrm{QPNM}} \Psi_\nu(JM))$$
$$- \eta_\nu [(\Psi^*_\nu(JM) \Psi_\nu(JM)) - 1]\} = 0 \qquad (4.58)$$

which gives us a set of two equations. The first of these equations,

$$(\omega_{Ji} - \eta_\nu) R_i + \sum_{\lambda_1 i_1 \lambda_2 i_2} P^{\lambda_1 i_1}_{\lambda_2 i_2} \Big\{ U^{\lambda_1 i_1}_{\lambda_2 i_2}(Ji) + V^{\lambda_1 i_1}_{\lambda_2 i_2}(Ji) \Big\} = 0$$

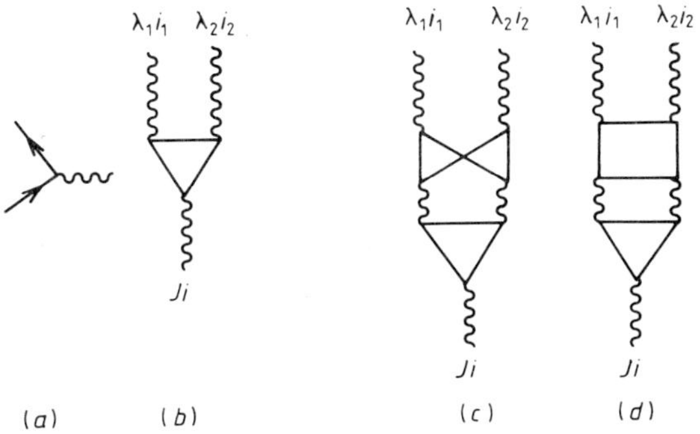

Figure 4.5 Diagrams taking into account (a) the quasiparticle–phonon interaction and (b–d) the coupling of the one-phonon and two-phonon configurations: wavy line—a phonon, straight line—a quasiparticle.

gives us R_i; substituting it into the second equation, we obtain a secular equation for determining the energies η_ν as a determinant in the space of two-phonon states [144]:

$$\begin{aligned}
\det \Bigg\| &(\omega_{\lambda_1 i_1} + \omega_{\lambda_2 i_2} - \eta_\nu)[\delta_{\lambda_1 i_1, \lambda_1' i_1'} \delta_{\lambda_2 i_2, \lambda_2' i_2'} \\
&+ \delta_{\lambda_1 i_1, \lambda_2' i_2'} \delta_{\lambda_2 i_2 \lambda_1' i_1'} + \mathcal{K}^J(\lambda_2' i_2', \lambda_1' i_1' | \lambda_1 i_1, \lambda_2 i_2)] \\
&- \frac{1}{4} \sum_{i_3 \tau} \Bigg[\frac{X^{\lambda_1' i_1'}(\tau) + X^{\lambda_1' i_3}(\tau)}{(\mathcal{Y}_\tau^{\lambda_1' i_1'} \mathcal{Y}_\tau^{\lambda_1' i_3})^{1/2}} \mathcal{K}^J(\lambda_2' i_2', \lambda_1' i_3 | \lambda_1 i_1, \lambda_2 i_2) \\
&+ \frac{X^{\lambda_2' i_2'}(\tau) + X^{\lambda_2' i_3}(\tau)}{(\mathcal{Y}_\tau^{\lambda_2' i_2'} \mathcal{Y}_\tau^{\lambda_2' i_3})^{1/2}} \mathcal{K}^J(\lambda_2' i_3, \lambda_1' i_1' | \lambda_1 i_1, \lambda_2 i_2) \Bigg] \\
&- \frac{1}{4} \sum_{\substack{\lambda_3 i_3 i_3' \\ \lambda_4 i_4 \tau}} \frac{X^{\lambda_3 i_3}(\tau) + X^{\lambda_3 i_3'}(\tau)}{(\mathcal{Y}_\tau^{\lambda_3 i_3} \mathcal{Y}_\tau^{\lambda_3 i_3'})^{1/2}} \\
&\times \mathcal{K}^J(\lambda_4 i_4, \lambda_3 i_3' | \lambda_1 i_1, \lambda_2 i_2) \mathcal{K}^J(\lambda_2' i_2', \lambda_1' i_1' | \lambda_3 i_3, \lambda_4 i_4) \\
&- \sum_i \frac{\left(U^{\lambda_1 i_1}_{\lambda_2 i_2}(Ji) + V^{\lambda_1 i_1}_{\lambda_2 i_2}(Ji)\right)\left(U^{\lambda_1' i_1'}_{\lambda_2' i_2'}(Ji) + V^{\lambda_1' i_1'}_{\lambda_2' i_2'}(Ji)\right)}{\omega_{Ji} - \eta_\nu} \Bigg\| = 0.
\end{aligned} \tag{4.59}$$

The rank of this determinant equals the number of two-phonon terms in wave function (4.55).

Let us illustrate equation (4.59) in terms of diagrams. The first term takes into account the diagrams shown in figures 4.6 (a, b), and subsequent terms cover the diagrams of figures 4.6 (c, d). The terms containing

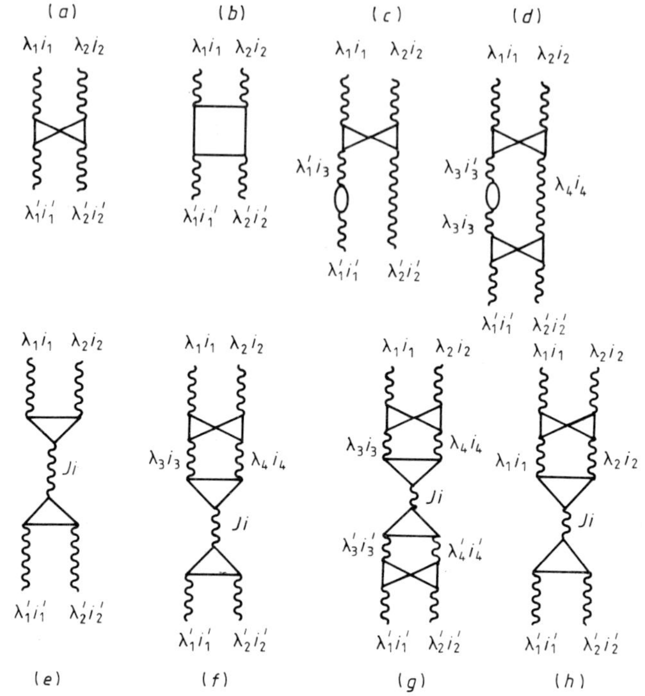

Figure 4.6 Diagrams in the space of two-phonon states. Notations are the same as in figure 4.5.

UU, *UV* and *VV* correspond to the diagrams of figures 4.6 (*e, f, g*), respectively. Furthermore, there exist diagrams which are obtained from those shown in figures 4.6 (*c, d, f, g*) by replacing (*a*) by a part of (*b*). Note the conditionality of this illustration, all the more so since the vertex parts are different in each specific case.

Secular equation (4.59) is very complex, the rank of the determinant being 10^3–10^5. It can be solved if the one-phonon state basis has been substantially truncated. Equation (4.59) takes into account a large number of graphs that contribute extremely little to the fragmentation of one-phonon states. Note that three-phonon terms of the wave function need be included in order to calculate the fragmentation of two-phonon states.

4.3.2

Let us consider an approximation of the set of equations. First, we retain only those terms of (4.59) which are proportional to $[P^{\lambda_1 i_1}_{\lambda_2 i_2}(J\nu)]^2$; secondly, we take one of $\mathcal{K}^J$ in the the term quadratic in $\mathcal{K}^J$ in the diagonal form.

This can be done because the absolute values of the diagonal terms $\mathcal{K}^J$ are much greater then those of the non-diagonal terms. We also limit the normalization condition (4.56) to the diagonal values of $\mathcal{K}^J$. As a result, we obtain

$$\sum_i R_i^2 + 2\sum_{\lambda_1 i_1 \lambda_2 i_2} \left(P_{\lambda_2 i_2}^{\lambda_1 i_1}\right)^2 [1 + \mathcal{K}^J(\lambda_1 i_1, \lambda_2 i_2)] = 1 \tag{4.60}$$

$$\begin{aligned}(\Psi_\nu^*(JM) H_{\mathrm{QPNM}} \Psi_\nu(JM)) &= \sum_i \omega_{Ji} R_i^2 + 2\sum_{\lambda_1 i_1 \lambda_2 i_2} \left(P_{\lambda_2 i_2}^{\lambda_1 i_1}\right)^2 \\ &\quad \times [\omega_{\lambda_1 i_1} + \omega_{\lambda_2 i_2} + \Delta\omega(\lambda_1 i_1, \lambda_2 i_2)][1 + \mathcal{K}^J(\lambda_1 i_1, \lambda_2 i_2)] \\ &\quad + 2\sum_{\lambda_1 i_1 \lambda_2 i_2} R_i P_{\lambda_2 i_2}^{\lambda_1 i_1} U_{\lambda_2 i_2}^{\lambda_1 i_1}(Ji)[1 + \mathcal{K}^J(\lambda_1 i_1, \lambda_2 i_2)]\end{aligned} \tag{4.61}$$

where

$$\begin{aligned}\Delta\omega(\lambda_1 i_1, \lambda_2 i_2) = -\frac{1}{4}\sum_{i_3 \tau} \Big[&\frac{X^{\lambda_1 i_1}(\tau) + X^{\lambda_1 i_3}(\tau)}{(\mathcal{Y}_\tau^{\lambda_1 i_1} \mathcal{Y}_\tau^{\lambda_1 i_3})^{1/2}} \\ &\times \mathcal{K}^J(\lambda_2 i_2, \lambda_1 i_3 | \lambda_1 i_1, \lambda_2 i_2) \\ &+ \frac{X^{\lambda_2 i_2}(\tau) + X^{\lambda_2 i_3}(\tau)}{(\mathcal{Y}_\tau^{\lambda_2 i_2} \mathcal{Y}_\tau^{\lambda_2 i_3})^{1/2}} \\ &\times \mathcal{K}^J(\lambda_2 i_3, \lambda_1 i_1 | \lambda_1 i_1, \lambda_2 i_2)\Big]\end{aligned} \tag{4.62}$$

and the summation over i_3 occurs because we went beyond the RPA framework.

Let us make use of the variational principle, which leads to the following set of equations:

$$\begin{aligned}(\omega_{Ji} - \eta_\nu) R_i + \sum_{\lambda_1 i_1 \lambda_2 i_2} & P_{\lambda_2 i_2}^{\lambda_1 i_1} U_{\lambda_2 i_2}^{\lambda_1 i_1}(Ji) \\ &\times [1 + \mathcal{K}^J(\lambda_1 i_1, \lambda_2 i_2)] = 0\end{aligned} \tag{4.63}$$

$$\begin{aligned}2(\omega_{\lambda_1 i_1} + \omega_{\lambda_2 i_2} + \Delta\omega(\lambda_1 i_1, \lambda_2 i_2) - \eta_\nu) & P_{\lambda_2 i_2}^{\lambda_1 i_1} \\ + \sum_i R_i U_{\lambda_2 i_2}^{\lambda_1 i_1}(Ji) = 0.\end{aligned} \tag{4.63'}$$

Now the secular equation can be derived both in the space of two-phonon states and in that of one-phonon states. The secular equation in the two-phonon space is

$$\mathcal{F}(\eta_\nu) \equiv \det \Big\| [\omega_{\lambda_1 i_1} + \omega_{\lambda_2 i_2} + \Delta\omega(\lambda_1 i_1, \lambda_2 i_2) - \eta_\nu] \\ \times (\delta_{\lambda_1 i_1, \lambda'_1 i'_1}\delta_{\lambda_2 i_2, \lambda'_2 i'_2} + \delta_{\lambda_1 i_1, \lambda'_2 i'_2}\delta_{\lambda_2 i_2, \lambda'_1 i'_1}) \\ - \frac{1}{2}\sum_i \frac{U^{\lambda_1 i_1}_{\lambda_2 i_2}(Ji) U^{\lambda'_1 i'_1}_{\lambda'_2 i'_2}(Ji)}{\omega_{Ji} - \eta_\nu}[1 + \mathcal{K}^J(\lambda_1 i_1, \lambda_2 i_2)] \Big\| = 0. \quad (4.64)$$

In this case, we take into account the diagrams shown in figures 4.6 (*a–c, e, h*); the diagram in figure 4.6 (*h*) differs from the one in figure 4.6 (*f*) in that there is no summation over intermediate two-phonon states, since only the diagonal value of $\mathcal{K}^J$ has been retained (i.e., a fewer number of diagrams are summed up than in equation (4.59)).

The secular equation in the one-phonon states space is written as follows:

$$\det \Big\| (\omega_{Ji} - \eta_\nu)\delta_{ii'} \\ - \frac{1}{2}\sum_{\lambda_1 i_1 \lambda_2 i_2} \frac{U^{\lambda_1 i_1}_{\lambda_2 i_2}(Ji) U^{\lambda_1 i_1}_{\lambda_2 i_2}(Ji')[1 + \mathcal{K}^J(\lambda_1 i_1, \lambda_2 i_2)]}{\omega_{\lambda_1 i_1} + \omega_{\lambda_2 i_2} + \Delta\omega(\lambda_1 i_1, \lambda_2 i_2) - \eta_\nu} \Big\| = 0. \quad (4.65)$$

In the one-phonon states space, we take into account the diagrams given in figures 4.7 (*a–c*). The diagrams in figures 4.7 (*b, c*) correspond to the corrections arising when the Pauli exclusion principle is taken into account in two-phonon terms of wave function (4.55).

Consider equation (4.65). The rank of the determinant equals the number of one-phonon states in the first term of wave function (4.55). It varies from 20 to 200 and thus happens to be less by two orders of magnitude than the orders of determinants (4.59) and (4.64). The factor $1+\mathcal{K}^J(\lambda_1 i_1, \lambda_2 i_2)$ stems from taking the Pauli exclusion principle into account in the two-phonon terms of wave function (4.55). In the case of the maximum violation of the Pauli exclusion principle, $\mathcal{K}^J = -1$ and the corresponding term is eliminated from the sum over $\lambda_1 i_1, \lambda_2 i_2$. The shift $\Delta\omega(\lambda_1 i_1, \lambda_2 i_2)$ of the two-phonon pole is caused by including diagrams similar to those of figures 4.6 (*c, d*). As shown in [140, 165], it is large for the first two-phonon collective states of deformed nuclei. The shift $\Delta\omega(\lambda_1 i_1, \lambda_2 i_2)$ is small for collective phonons forming giant resonances of various types. The $\Delta\omega(\lambda_1 i_1, \lambda_2 i_2)$ shift is zero for $\mathcal{K}^J = 0$. Using normalization condition (4.60), we find

$$R_i^2(J\nu) = A_{ii}^2 \Big\{ \sum_i A_{ii'}^2 \\ + \frac{1}{2}\sum_{\lambda_1 i_1 \lambda_2 i_2} \Big[\Big(\frac{\sum_{i'} U^{\lambda_1 i_1}_{\lambda_2 i_2}(Ji') A_{ii'}}{\omega_{\lambda_1 i_1} + \omega_{\lambda_2 i_2} + \Delta\omega(\lambda_1 i_1, \lambda_2 i_2) - \eta_\nu} \Big)^2$$

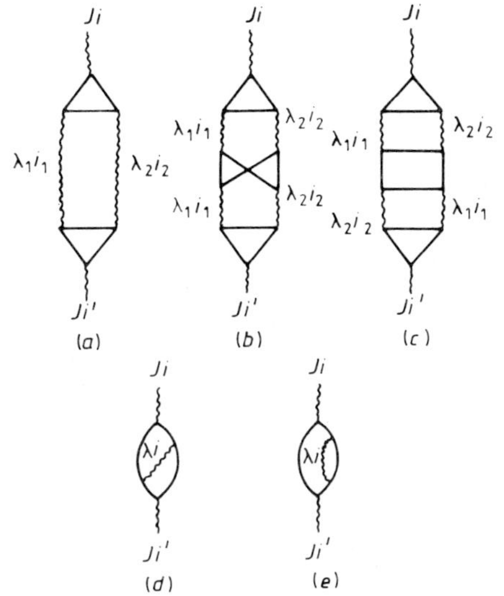

Figure 4.7 (a–c) Diagrams taken into account in the QPNM and (d, e) in the nuclear field theory. The notation is the same as in figure 4.5.

$$\times \left[1 + \mathcal{K}^J(\lambda_1 i_1, \lambda_2 i_2)\right]\Big]\Big\}^{-1} \tag{4.66}$$

where $A_{ii'}$ is the algebraic adjunction of determinant (4.65).

If the terms proportional to $\alpha^+\alpha$ in commutation relation (4.27) are dropped, we obtain from equations (4.63) and (4.63′) the expression

$$(\omega_{Ji} - \eta_\nu)R_i + \sum_{\lambda_1 i_1 \lambda_2 i_2} U^{\lambda_1 i_1}_{\lambda_2 i_2}(Ji) P^{\lambda_1 i_1}_{\lambda_2 i_2} = 0 \tag{4.67}$$

$$P^{\lambda_1 i_1}_{\lambda_2 i_2} = -\frac{1}{2}\sum_{i'} \frac{U^{\lambda_1 i_1}_{\lambda_2 i_2}(Ji')R_{i'}}{\omega_{\lambda_1 i_1} + \omega_{\lambda_2 i_2} - \eta_\nu}. \tag{4.67′}$$

Substituting (4.67′) into (4.67), we find the following set of equations:

$$\sum_{i'}\Big\{(\omega_{Ji} - \eta_\nu)\delta_{ii'} - \frac{1}{2}\sum_{\lambda_1 i_1 \lambda_2 i_2} \frac{U^{\lambda_1 i_1}_{\lambda_2 i_2}(Ji) U^{\lambda_1 i_1}_{\lambda_2 i_2}(Ji')}{\omega_{\lambda_1 i_1} + \omega_{\lambda_2 i_2} - \eta_\nu}\Big\} R_{i'} = 0. \tag{4.68}$$

The condition of existence of a non-trivial solution of set (4.68) yields a secular equation for determining the energies η_ν:

$$\mathcal{F}(\eta_\nu) \equiv \det\left\| (\omega_{Ji} - \eta_\nu)\delta_{ii'} - \frac{1}{2} \sum_{\lambda_1 i_1 \lambda_2 i_2} \frac{U^{\lambda_1 i_1}_{\lambda_2 i_2}(Ji) U^{\lambda_1 i_1}_{\lambda_2 i_2}(Ji')}{\Omega_{\lambda_1 i_1} + \omega_{\lambda_2 i_2} - \eta_\nu} \right\| = 0. \tag{4.69}$$

The determinant's rank equals the number of one-phonon states in wave function (4.55). Using the wave function normalization condition (4.55) in the form

$$\sum_i R_i^2 + 2 \sum_{\lambda_1 i_1 \lambda_2 i_2} \left(P^{\lambda_1 i_1}_{\lambda_2 i_2} \right)^2 = 1 \tag{4.70}$$

we obtain

$$R_i^2 = A_{ii}^2 \left\{ \sum_{i'} A_{ii'}^2 + \frac{1}{2} \sum_{\lambda_1 i_1 \lambda_2 i_2} \left(\frac{\sum_{i'} A_{ii'} U^{\lambda_1 i_1}_{\lambda_2 i_2}(i')}{\omega_{\lambda_1 i_1} + \omega_{\lambda_2 i_2} - \eta_\nu} \right)^2 \right\}^{-1} \tag{4.71}$$

where $A_{ii'}$ is the algebraic adjunction of determinant (4.69). The following relations hold for the solutions of equation (4.69):

$$R_i^{-2} = - \left[\frac{1}{A_{ii}} \frac{\partial \mathcal{F}(\eta)}{\partial \eta} \right]_{\eta = \eta_\nu} \tag{4.72}$$

$$R_i = (A_{ii'}/A_{ii}) R_{i'}. \tag{4.72'}$$

4.3.3

Let us compare the accuracy of the approximations in the QPNM and in the nuclear field theory (NFT) [78, 166, 167]; we will also find out the important role of the Pauli exclusion principle in the two-phonon terms of wave function (4.55).

In NFT calculations of the fragmentation of one-phonon states, one sums up the diagrams of the type shown in figures 4.7 (d, e). As shown in [166], they are particular cases of the diagram in figure 4.7 (a) when the $\lambda_2 i_2$ phonon is a collective phonon and the $\lambda_1 i_1$ phonon—a two-quasiparticle one. If the energy of a one-phonon state coincides with the appropriate two-quasiparticle pole, the vertex $\Gamma(Jj\lambda i)$ (4.36) vanishes and the wave function of the one-phonon state becomes the wave function of the two-quasiparticle state. This limit transition of a one-phonon $\lambda_1 i_1$ state into a two-quasiparticle state is shown in figure 4.8. The diagram shown in figure 4.7 (a) transforms either into that of figure 4.7 (e) (see figure 4.8 (a)) or

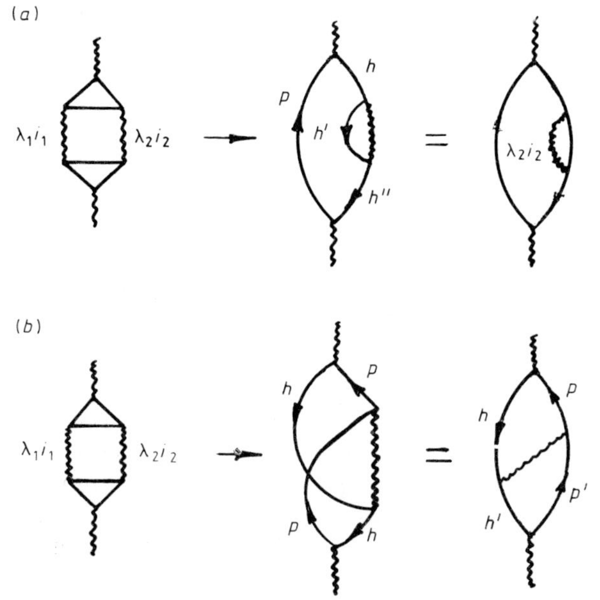

Figure 4.8 Transformation of diagrams summed up in the QPNM into diagrams taken into account in the NFT. The QPNM diagrams are on the left of the arrows. If the one-phonon state $\lambda_1 i_1$ transforms into a two-quasiparticle state, diagrams in the NFT are formed. The notation is the same as in figure 4.5.

into that of figure 4.7 (d) (see figure 4.8 (b)). This shows that the diagrams summed up in the NFT are a particular case of one of the diagrams that the QPNM takes into account. Therefore, a broader range of diagrams are summed up in the QPNM than in the NFT.

For a numerical comparison of the approximation accuracy in the QPNM and in the NFT, we will have to introduce the strength functions. Let $\Phi_{J\nu}$ be the excitation amplitude for the state $\Psi_\nu(JM)$ of some physical process; it is given by formula (4.55). Let us use the following definition of the strength function:

$$b(\Phi, \eta) = \sum_\nu |\Phi_{J\nu}|^2 \rho(\eta - \eta_\nu). \tag{4.73}$$

If the excitation of the state $\Psi_\nu(JM)$ goes via one-phonon state components of the wave function, then

$$\Phi_{J\nu} = \sum_i R_i(J\nu)\Phi_{Ji}. \tag{4.74}$$

Substituting (4.74) into (4.73) and using relations (4.72) and (4.72′), we rewrite (4.73) in the form

$$b(\Phi, \eta) = \sum_{\nu} \rho(\eta - \eta_{\nu}) \sum_{ii'} \frac{A_{ii'}}{A_{i'i'}} R_i^2 \Phi_{Ji} \Phi_{Ji'}$$
$$= \sum_{\nu} \rho(\eta - \eta_{\nu}) \frac{\Phi^2(\eta_{\nu})}{\partial \mathcal{F}/\partial \eta|_{\eta=\eta_{\nu}}}$$

where $\Phi^2(\eta_\nu) = -\sum_{ii'} A_{ii'}(\eta_\nu)\Phi_{Ji}\Phi_{Ji'}$. Now we consider the contour integral,

$$b(\Phi, \eta) = \frac{1}{2\pi i} \oint_{l_\eta} \frac{\Phi^2(z)}{\mathcal{F}(z)} \rho(\eta - z)\, \mathrm{d}z \tag{4.75}$$

over the contour l_η (see figure 4.3) which comprises the poles $z = \eta_\nu$ of the integrand $\mathcal{F}^{-1}(z)$; these poles are the roots of secular equation (4.69). If (4.75) is written in terms of the integral over a contour at infinity (the integral equals zero since the integrand falls off at infinity as $|z|^3$) and of the integrals over contours around the poles of the function $\rho(\eta - z)$, which equal $z = \eta \pm \mathrm{i}\Delta/2$, and if the residues theorem is then applied, we obtain

$$b(\Phi, \eta) = \frac{1}{\pi} \mathrm{Im} \left\{ \frac{\sum_{ii'} A_{ii'}(\eta + i\Delta/2)\Phi_{Ji}\Phi_{Ji'}}{\mathcal{F}(\eta + i\Delta/2)} \right\} \tag{4.76}$$

where $\mathcal{F}(\eta + \mathrm{i}\Delta/2)$ is the determinant of (4.69) for complex values of energy and $A_{ii'}$ are its algebraic adjunctions. Formula (4.76) holds in the case of the function $\Phi^2(z)$ having no singularities. Therefore, instead of solving the secular equations to find the mean characteristics, one calculates the determinant for the complex values of energy; this is a much simpler procedure. The energy averaging interval Δ determines the form in which the results of calculations are to be presented. If $\Delta \to 0$, the Lorentz function $\rho(\eta - \eta_\nu)$ transforms into the Dirac delta function and the strength function $b(\Phi, \eta)$ is non-zero only at the points $\eta = \eta_\nu$ of the solution to equation (4.64), where it equals $|\Phi_{J\nu}|^2$. The averaging interval Δ must be larger than the mean distance between the states that correspond to multiphonon components omitted in (4.55). Furthermore, Δ must be much smaller than the region of localization of the physical quantity of interest. In spherical nuclei, the region $0.1 \leqslant \Delta \leqslant 1$ MeV satisfies these criteria.

Let us consider, following [168], the differences between the results of the QPNM calculations, in which diagrams like those shown in figure 4.7 (*a*) are summed up, and of NFT calculations, in which diagrams shown in figure 4.7 (*d, e*) are summed up. In the NFT (see diagrams in figure 4.7 (*a*)), the collective phonon is taken for one of the intermediate phonons and the non-collective phonon (in fact, a two-quasiparticle state) is chosen as the other one. A certain arbitrariness is present in subsuming the phonons

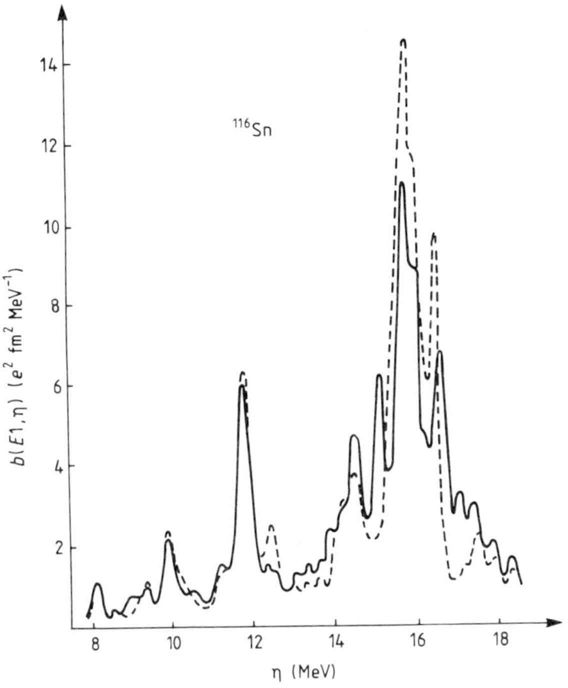

Figure 4.9 Strength function for the giant dipole resonance in ^{116}Sn: full curve—QPNM calculations taking into account the diagrams of figure 4.7 (*a*); broken curve—NFT calculations taking into account the diagrams of figure 4.7 (*d*, *e*); $\Delta = 0.2\,\text{MeV}$.

to be collective or non-collective or weakly collective. The results of QPNM and NFT calculations are given in figures 4.9 and 4.10. These figures show that the QPNM and NFT give not very different results, although the giant resonance strength in QPNM calculations is somewhat more fragmented. Even though the peak amplitudes obtained when the diagrams of figures 4.7 (*d*) and (*e*) are taken into account are somewhat greater, on the whole the two calculations give identical gross structure of strength distribution. This shows that the NFT does take into account all the more important diagrams.

In the QPNM solution of secular equation (4.69) (see [138, 161]) and in calculations of the corresponding strength functions (4.76), an approximate

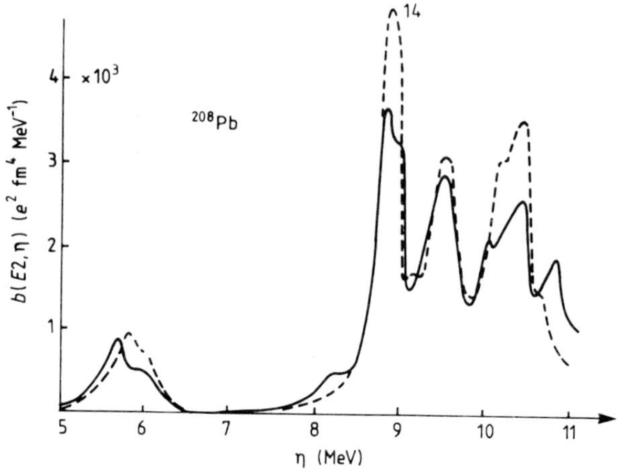

Figure 4.10 Strength function for the isoscalar quadrupole resonance in ^{208}Pb. The notation is the same as in figure 4.9; $\Delta = 0.2\,\text{MeV}$.

procedure is employed to take into account the Pauli exclusion principle. When equation (4.69) was numerically solved or (4.76) was calculated, two-phonon components with two non-collective phonons, violating the Pauli exclusion principle, were eliminated. This procedure ignored the shifting of two-phonon poles and eliminated that part of the two-phonon components which is allowed by the Pauli exclusion principle. Let us consider the degree of discrepancy between the results when the Pauli exclusion principle is taken into account exactly in the two-phonon components of wave function (4.55), that is, when equation (4.65) is solved and when an approximate procedure is employed. The results of calculating the strength function [168] in these two cases are plotted in figure 4.11. As we see from the figure, a more exact covering of the Pauli exclusion principle slightly reduces the value of $b(E2, \eta)$ in the neighbourhood of the maximum. The integral strength in the range from 5 to 14 MeV decreases by only 5%. On the whole, an approximate procedure of eliminating the states forbidden by the Pauli exclusion principle is practically equivalent to taking into account exactly the diagrams given in figures 4.7 (b, c), but is considerably simpler. If a state is not strictly forbidden by the Pauli exclusion principle, the renormalization of the interaction to include the corrections due to fermion structure of two-phonon states is found to be weak. If the demands implied by the Pauli exclusion principle are completely ignored, a large number of spurious two-phonon components may appear in the wave function.

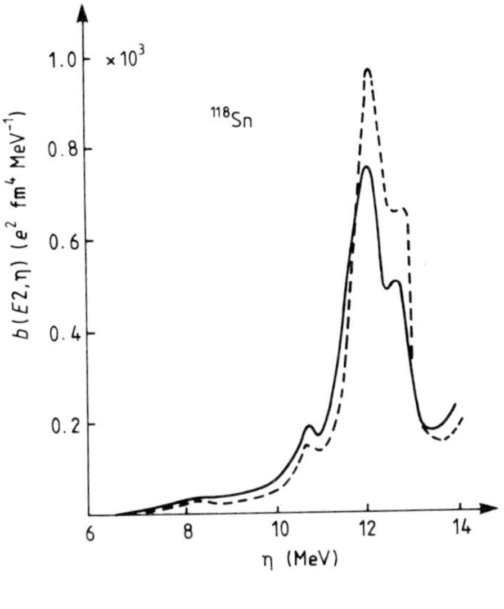

Figure 4.11 The strength function for the isoscalar quadrupole resonance in ^{118}Sn: full curve—the Pauli exclusion principle was exactly taken into account; broken curve—approximate procedure for taking the Pauli exclusion principle into account was used.

4.3.4

Let us derive a set of QPNM equations for a description of the strength distribution of collective charge-exchange states. We choose the following presentation of the wave function of an odd–odd spherical nucleus:

$$\Psi_\nu(JM) = \Big\{\sum_i \bar{R}_i(J\nu)\Omega^+_{JMi} + \sum_{\lambda_1 i_1 \lambda_2 i_2} \bar{P}^{\lambda_1 i_1}_{\lambda_2 i_2}\left[\Omega^+_{\lambda_1\mu_1 i_1} Q^+_{\lambda_2\mu_2 i_2}\right]_{JM}\Big\}\Psi_0 \qquad (4.77)$$

where Ψ_0 is the ground state wave function of the even–even nucleus. The normalization condition will be

$$1 = \sum_i \bar{R}_i^2 + \sum_{\lambda_1 i_1 \lambda_2 i_2}\left(\bar{P}^{\lambda_1 i_1}_{\lambda_2 i_2}\right)^2 + \sum_{\substack{\lambda_1 i_1 \lambda_2 i_2 \\ \lambda'_1 i'_1 \lambda'_2 i'_2}} \bar{P}^{\lambda'_1 i'_1}_{\lambda'_2 i'_2}\bar{P}^{\lambda_1 i_1}_{\lambda_2 i_2}\mathcal{K}^J_{\Omega Q}(\lambda'_2 i'_2, \lambda'_1 i'_1|\lambda_1 i_1, \lambda_2 i_2). \qquad (4.78)$$

Here

$$\begin{aligned}
&\mathcal{K}^J_{\Omega Q}(\lambda'_2 i'_2, \lambda' i' | \lambda i, \lambda_2 i_2) \\
&\quad = - \sum_{\mu\mu_2\mu'\mu'_2} \langle \lambda'\mu'\lambda'_2\mu'_2 | JM\rangle \langle \lambda\mu\lambda_2\mu_2 | JM\rangle \\
&\quad \times \sum_{j_m m_n j_p m_p} \Big\{ \sum_{j'_p m'_p j''_p m''_p} \Big[\langle j'_p m'_p j_n m_n | \lambda'\mu'\rangle \\
&\quad \times \langle j_p m_p j_n m_n | \lambda\mu\rangle \psi^{\lambda' i'}_{j'_p j_n} \psi^{\lambda i}_{j_p j_n} - (-)^{\lambda+\lambda'-\mu-\mu'} \\
&\quad \times \langle j'_p m'_p j_n m_n | \lambda' - \mu'\rangle \langle j_p m_p j_n m_n | \lambda - \mu\rangle \varphi^{\lambda' i'}_{j'_p j_n} \varphi^{\lambda i}_{j_p j_n} \Big] \\
&\quad \times \Big[\langle j'_p m'_p j''_p m''_p | \lambda_2\mu_2\rangle \langle j_p m_p j''_p m''_p | \lambda'_2\mu'_2\rangle \psi^{\lambda_2 i_2}_{j'_p j''_p} \psi^{\lambda'_2 i'_2}_{j_p j''_p} \\
&\quad \times -(-)^{\lambda_2+\lambda'_2-\mu_2-\mu'_2} \langle j''_p m''_p j_p m_p | \lambda_2 - \mu_2\rangle \\
&\quad \times \langle j'_p m'_p j''_p m''_p | \lambda'_2 - \mu'_2\rangle \varphi^{\lambda_2 i_2}_{j''_p j_p} \varphi^{\lambda'_2 i'_2}_{j'_p j''_p} \Big] \\
&\quad + \sum_{j'_n m'_n j''_n m''_n} \Big[\langle j_p m_p j'_n m'_n | \lambda'\mu'\rangle \langle j_p m_p j_n m_n | \lambda\mu\rangle \\
&\quad \times \psi^{\lambda' i'}_{j_p j'_n} \psi^{\lambda i}_{j_p j_n} - (-)^{\lambda+\lambda'-\mu-\mu'} \langle j_p m_p j'_n m'_n | \lambda' - \mu'\rangle \\
&\quad \times \langle j_p m_p j_n m_n | \lambda - \mu\rangle \varphi^{\lambda' i'}_{j_p j'_n} \varphi^{\lambda i}_{j_p j_n} \Big] \Big[\langle j'_n m'_n j''_n m''_n | \lambda_2\mu_2\rangle \\
&\quad \times \langle j_n m_n j''_n m''_n | \lambda'_2\mu'_2\rangle \psi^{\lambda_2 i_2}_{j'_n j''_n} \psi^{\lambda'_2 i'_2}_{j_n j''_n} - (-)^{\lambda_2+\lambda'_2-\mu_2+\mu'_2} \\
&\quad \times \langle j''_n m''_n j_n m_n | \lambda_2\mu_2\rangle \langle j'_n m'_n j''_n m''_n | \lambda'_2\mu'_2\rangle \varphi^{\lambda_2 i_2}_{j''_n j_n} \varphi^{\lambda'_2 i'_2}_{j'_n j_n} \Big] \Big\}.
\end{aligned} \tag{4.79}$$

We denote the diagonal value of this function by $\mathcal{K}^J_{\Omega Q}(\lambda_1 i_1, \lambda_2 i_2)$.

The next step is the calculation of the mean value of the QPNM Hamiltonian over state (4.77). We will use commutation relations (2.173) and (4.27) and also

$$\begin{aligned}
&\Big[\Omega_{\lambda\mu i}, Q^+_{\lambda'\mu' i'} \Big] \\
&\quad = \sum_{j_p m_p j_n m_n} \{ \alpha^+_{j_p m_p} \alpha_{j_n m_n} \\
&\quad \times \Big[\sum_{j'_p m'_p} \psi^{\lambda i}_{j'_p j_n} \psi^{\lambda' i'}_{j_p j'_p} \langle j'_p m'_p j_n m_n | \lambda\mu\rangle \langle j_p m_p j'_p m'_p | \lambda'\mu'\rangle \\
&\quad - (-)^{\lambda+\lambda'-\mu-\mu'} \sum_{j'_n m'_n} \varphi^{\lambda i}_{j_p j_n} \varphi^{\lambda' i'}_{j'_n j_n} \langle j_p m_p j'_n m'_n | \lambda - \mu\rangle \\
&\quad \times \langle j'_n m'_n j_n m_n | \lambda' - \mu'\rangle \Big] + \alpha^+_{j_n m_n} \alpha_{j_p m_p} \Big[\sum_{j'_n m'_n} \psi^{\lambda i}_{j_p j'_n} \psi^{\lambda' i'}_{j'_n j_n}
\end{aligned}$$

$$\times \langle j_p m_p j'_n m'_n | \lambda\mu \rangle \langle j'_n m'_n j_n m_n | \lambda'\mu' \rangle - (-)^{\lambda+\lambda'-\mu-\mu'}$$
$$\times \sum_{j'_p m'_p} \varphi^{\lambda i}_{j'_p j_n} \varphi^{\lambda' i'}_{j_p j'_p} \langle j'_p m'_p j_n m_n | \lambda - \mu \rangle \langle j_p m_p j'_p m'_p | \lambda' - \mu' \rangle \Big] \Big\}. \tag{4.80}$$

Using the variational principle, we now find the secular equation in the two-phonon space; its explicit form is given in [144].

The equations in two-phonon space are very cumbersome since they take into account a large number of diagrams producing small contributions to the fragmentation of np-phonons. Therefore, we perform the transformation to an approximate set of equations in the same way as we did in subsection 4.3.2. In this case, the normalization condition of wave function (4.77) and the mean QPNM Hamiltonian found from (4.77) become

$$\sum_i \bar{R}_i^2 + \sum_{\lambda_1 i_1 \lambda_2 i_2} \left(\bar{P}^{\lambda_1 i_1}_{\lambda_2 i_2}\right)^2 \left\{1 + \mathcal{K}^J_{\Omega Q}(\lambda_1 i_1, \lambda_2 i_2)\right\} = 1 \tag{4.81}$$

$$\begin{aligned}
&(\psi^*_\nu(JM) H_{\mathrm{QPNM}} \Psi_\nu(JM)) \\
&\quad = \sum_i \Omega_{\lambda i} \bar{R}_i^2 + \sum_{\lambda_1 i_1 \lambda_2 i_2} \left(\bar{P}^{\lambda_1 i_1}_{\lambda_2 i_2}\right)^2 \\
&\qquad \times [\Omega_{\lambda_1 i_1} + \omega_{\lambda_2 i_2} + \Delta\Omega_{\lambda_1 i_1} + \Delta\omega_{\lambda_2 i_2}][1 + \mathcal{K}^J_{\Omega Q}(\lambda_1 i_1, \lambda_2 i_2)] \\
&\qquad + 2 \sum_{\lambda_1 i_1 \lambda_2 i_2 i} \bar{R}_i \bar{P}^{\lambda_1 i_1}_{\lambda_2 i_2} W^{Ji}(\lambda_1 i_1, \lambda_2 i_2)[1 + \mathcal{K}^J_{\Omega Q}(\lambda_1 i_1, \lambda_2 i_2)]
\end{aligned} \tag{4.82}$$

where

$$\begin{aligned}
\Delta\Omega_{\lambda i} &= -\sum_{i'} \frac{1 + \mathsf{Y}^{\lambda i}\mathsf{Y}^{\lambda i'}}{(\mathcal{Y}^{\lambda i}\mathcal{Y}^{\lambda i'})^{1/2}} \mathcal{K}^J_{\Omega Q}(\lambda_2 i_2, \lambda i' | \lambda i, \lambda_2 i_2) \\
\Delta\omega_{\lambda_2 i_2} &= -\frac{1}{4} \sum_{i'_2 \tau} \frac{X^{\lambda_2 i_2}(\tau) + X^{\lambda_2 i'_2}(\tau)}{(\mathcal{Y}^{\lambda_2 i_2}_\tau \mathcal{Y}^{\lambda_2 i'_2}_\tau)^{1/2}} \\
&\quad \times \mathcal{K}^J_{\Omega Q}(\lambda_1 i_1, \lambda_2 i'_2 | \lambda_2 i_2, \lambda_1 i_1)
\end{aligned} \tag{4.83}$$

$$W^{Ji}(\lambda_1 i_1, \lambda_2 i_2) = \langle \Omega_{JMi} H_{vq} [\Omega^+_{\lambda_1 \mu_1 i_1} Q^+_{\lambda_2 \mu_2 i_2}]_{JM} \rangle. \tag{4.84}$$

Using the variational principle, we obtain the set of equations

$$\begin{aligned}
&(\Omega_{\lambda i} - \eta_\nu)\bar{R}_i + \sum_{\lambda_1 i_1 \lambda_2 i_2} \bar{P}^{\lambda_1 i_1}_{\lambda_2 i_2} W^{Ji}(\lambda_1 i_1, \lambda_2 i_2) \\
&\quad \times [1 + \mathcal{K}^J_{\Omega Q}(\lambda_1 i_1, \lambda_2 i_2)] = 0 \\
&(\omega_{\lambda_2 i_2} + \Omega_{\lambda_1 i_1} + \Delta\omega_{\lambda_2 i_2} + \Delta\Omega_{\lambda_1 i_1} - \eta_\nu)\bar{P}^{\lambda_1 i_1}_{\lambda_2 i_2} \\
&\quad + \sum_i \bar{R}_i W^{Ji}(\lambda_1 i_1, \lambda_2 i_2) = 0.
\end{aligned}$$

The second equation gives $\bar{P}^{\lambda_1 i_1}_{\lambda_2 i_2}$ which we substitute into the first and obtain a secular equation in the one-phonon state space:

$$\det\left\| (\omega_{Ji} - \eta_\nu)\delta_{ii'} - \sum_{\lambda_1 i_1 \lambda_2 i_2} \frac{W^{Ji}(\lambda_1 i_1, \lambda_2 i_2) W^{Ji'}(\lambda_1 i_1, \lambda_2 i_2)[1 + \mathcal{K}^J_{\Omega Q}(\lambda_1 i_1, \lambda_2 i_2)]}{\Omega_{\lambda_1 i_1} + \omega_{\lambda_2 i_2} + \Delta\Omega_{\lambda_1 i_1} + \Delta\omega_{\lambda_2 i_2} - \eta_\nu} \right\| = 0. \tag{4.85}$$

In this case we sum up the diagrams of figures 4.12 (*a–c*). Consider equation (4.85). The determinant's rank equals the number of one-phonon states in the first sum of wave function (4.77). It is hundreds of times less than the order of the corresponding determinant in the two-phonon state space. The factor $1 + \mathcal{K}^J_{\Omega Q}(\lambda_1 i_1, \lambda_2 i_2)$ and the shift $\Delta\Omega_{\lambda_1 i_1} + \Delta\omega_{\lambda_2 i_2}$ are due to the inclusion of the Pauli exclusion principle in the two-phonon terms of wave function (4.77).

The widths of giant charge-exchange resonances are a manifestation of the fragmentation of *np* phonons which are generated by the charge-exchange part of the isovector multipole and spin-multipole forces. Neither new forces nor new parameters need be introduced for the description. It is shown that the mathematical apparatus for the charge-exchange states can be so formulated that it becomes very similar to that for the description of low-lying collective states and collective states of the types of giant multipole and spin-multipole resonances in spherical nuclei. When the fragmentation of charge-exchange resonances is analysed (see [89]), one usually sets $\mathcal{K}^J_{\Omega Q} = 0$ and thereby $\Delta\Omega_{\lambda_1 i_1} + \Delta\omega_{\lambda_2 i_2} = 0$.

4.3.5

In the QPNM, the wave functions describing excitation states of even nuclei are taken as a sum of one- and two-phonon terms. It is necessary to improve the description of the fragmentation of collective one-phonon states. The general equations of the model [136], which takes into account three- and four-phonon terms of the wave functions, are extremely unwieldy. It was not possible to use them for developing an approximate method that would improve the description of the fragmentation of one-phonon states. A mathematical method for describing the fragmentation of one-phonon states more accurately was developed in [169]. The basic idea is to use two-phonon terms of the wave functions of the already fragmented one-phonon states. In this way we take into account certain three-phonon terms of the wave functions or a number of $3p - 3h$ configurations.

We will illustrate this method with the fragmentation of *np* phonons , choosing the following wave function of the odd–odd nucleus:

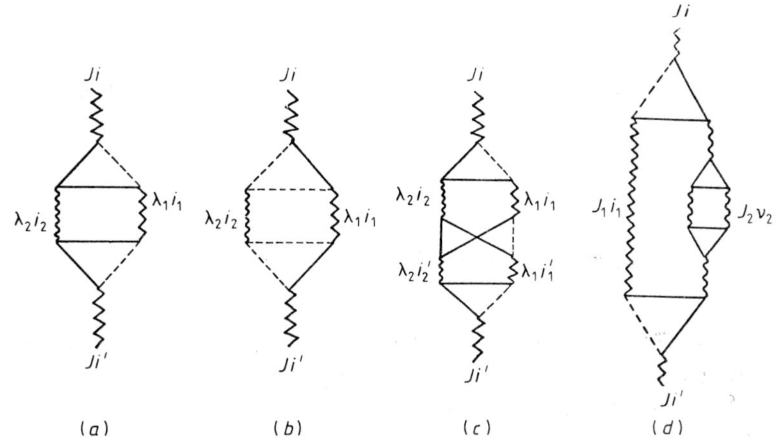

Figure 4.12 Diagrams in the one-phonon state space: wavy lines denote Q phonons, zig-zag lines—Ω phonons, full lines—neutron quasiparticles and broken lines stand for proton quasiparticles.

$$\Psi_n(JM) = \Big\{\sum_i \bar{C}_i(Jn)\Omega^+_{JMi} + \sum_{J_1 i_1 J_2 \nu_2} \bar{D}^{J_1 i_1}_{J_2\nu_2}(Jn)[\Omega^+_{J_1 M_1 i_1}\widetilde{Q}^+_{J_2 M_2 \nu_2}]_{JM}\Big\}\Psi_0 \quad (4.86)$$

where

$$\widetilde{Q}^+_{JM\nu} = \sum_i R_i(J\nu)Q^+_{JNi} + \sum_{\lambda_1 i_1 \lambda_2 i_2} P^{\lambda_1 i_1}_{\lambda_2 i_2}[Q^+_{\lambda_1\mu_1 i_1}Q^+_{\lambda_2\mu_2 i_2}]_{JM}.$$

We assume that we have determined the roots of equation (4.69) and thus found the functions $R_i(J\nu)$ and $P^{\lambda_1 i_1}_{\lambda_2 i_2}(j\nu)$. If the Pauli exclusion principle is ignored in multiphonon components, normalization condition (4.86) becomes

$$\sum_i [\bar{C}_i(Jn)]^2 + \sum_{J_1 i_1 J_2 \nu_2} [\bar{D}^{J_1 i_1}_{J_2\nu_2}(Jn)]^2 = 1. \quad (4.87)$$

The mean value of $H_{\rm QPNM}$ over (4.86) equals

$$(\Psi^*_n(JM)H_{\rm QPNM}\Psi_n(JM)) = \sum_i \Omega_{Ji}[\bar{C}_i(Jn)]^2 + \sum_{J_1 i_1 J_2 \nu_2}(\Omega_{J_1 i_1} + \eta_{J_2\nu_2})[\bar{D}^{J_1 i_1}_{J_2\nu_2}(Jn)]^2 + 2\sum_{iJ_1 i_1 J_2 \nu_2} \bar{C}_i(Jn)\bar{D}^{J_1 i_1}_{J_2\nu_2}(Jn)V^{J_1 i_1}_{J_2\nu_2}(Ji) \quad (4.88)$$

where

$$V^{J_1 i_1}_{J_2 \nu_2}(Ji) = \sum_{i_2} R_{i_2}(J_2\nu_2) W^{Ji}(J_1 i_1, J_2 i_2). \tag{4.88'}$$

Let us make use of the variational principle; we arrive at the following equations:

$$(\Omega_{Ji} - E_n)\bar{C}_i(Jn) + \sum_{J_1 i_1 J_2 \nu_2} V^{J_1 i_1}_{J_2 \nu_2}(Ji) \bar{D}^{J_1 i_1}_{J_2 \nu_2}(Jn) = 0 \tag{4.89}$$

$$(\Omega_{J_1 i_1} + \eta_{J_2 \nu_2} - E_n)\bar{D}^{J_1 i_1}_{J_2 \nu_2}(J_n) + \sum_{i'} V^{J_1 i_1}_{J_2 \nu_2}(Ji')\bar{C}_{i'}(Jn) = 0. \tag{4.89'}$$

Equations (4.87), (4.89) and (4.89′) form a complete set of equations for determining the functions $\bar{C}$, $\bar{D}$ and energies E_n of the states described by wave functions (4.86). If $\bar{D}$ from (4.89′) is substituted into (4.89), the secular equation changes to

$$\det \left\| (\Omega_{Ji} - E_n)\delta_{ii'} - \sum_{J_1 i_1 J_2 \nu_2} \frac{V^{J_1 i_1}_{J_2 \nu_2}(Ji) V^{J_1 i_1}_{J_2 \nu_2}(Ji')}{\Omega_{J_1 i_1} + \eta_{J_2 \nu_2} - E_n} \right\| = 0. \tag{4.90}$$

The rank of this determinant equals the number of one-phonon terms in wave function (4.86). In this case, we sum up diagrams of the type given in figure 4.12 (*d*). In numerical calculations, a part of the fragmented phonons can be replaced by one-phonon states.

As a result of solving equation (4.69) for states with fixed values of J^π, one calculates the energies $\eta_{J\nu}$; the number of these states equals the sum of the numbers of the one- and two-phonon poles. The strength of each *i*th one-phonon state is distributed over several levels ν. For instance, there are two one-phonon states with $J^\pi = 2^+$ and high values of $B(E2)$ in the energy range 7–17 MeV of ^{118}Sn. The fragmentation of these one-phonon states with energies $\omega_{2i=1}$ =11.97 MeV and $\omega_{2i=2}$ =12.38 MeV over the levels ν in ^{118}Sn is shown in figure 4.13. Instead of two poles in (4.69), $\Omega_{J_1 i_1} + \omega_{21}$ and $\Omega_{J_1 i_1} + \omega_{22}$, about 200 poles $\Omega_{J_1 i_1} + \eta_{J_2 \nu_2}$ have to be taken into account in secular equation (4.90). Figure 4.13 shows that there are solutions at the energies 10.4–10.8, around 11, 11.2–11.6, 11.8–12.9 MeV etc. The analysis can be restricted to taking into account a part of these levels, for example, to 60 levels over which more than 90% of the strength of these two one-phonon states is distributed. For each fixed state $J = 2$, $\nu = \nu_0$ with the energy $\eta_{2\nu_0}$, it is necessary to calculate the functions $R_i(2\nu_0)$ in (4.88′). In the case we discuss now, the summation in (4.88′) is carried out over two states: $i = 1$ and $i = 2$. The summation over $J_2\nu_2$ in the remaining terms in (4.90) is replaced by summation over $J_2 i_2$, as in (4.69).

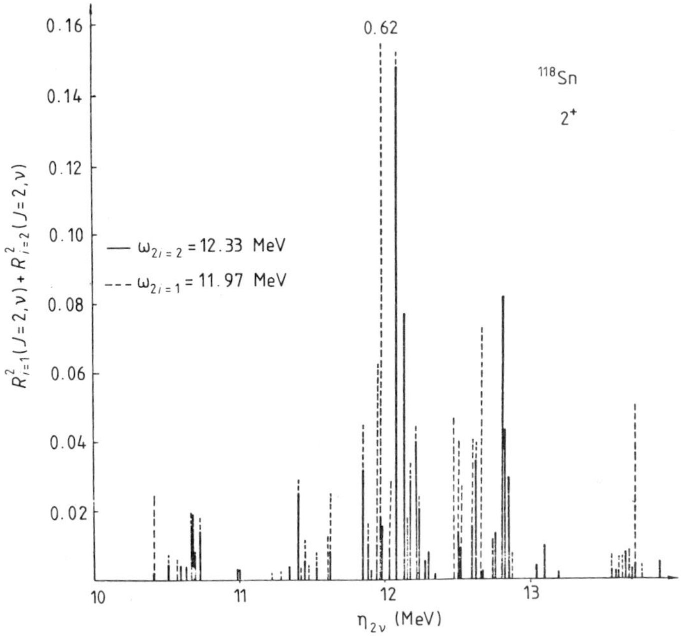

Figure 4.13 Strength distribution of single-phonon $R^2_{i=1}(2\nu)$ states with energy $\omega_{2i=1} = 11.97$ MeV (broken line) and $R^2_{i=2}(2\nu)$ states with energy $\omega_{2i=2} = 12.33$ MeV (solid line) in ^{118}Sn, obtained by solving secular equation (4.69) ($J^\pi = 2^+$).

The effect of taking into account the fragmented one-phonon states can be substantial because of a considerable increase in the number of levels over which fragmentation occurs, and because of the expansion of the energy range in which they are located. Taking these effects into account may reduce the strength of charge-exchange resonances close to their maximums and improve the description of the β-decay probabilities.

4.3.6

It is of interest to find the SU(6) limit of the QPNM, calculate the IBM parameters and check the extent to which the constraints imposed by the SU(6) approximation are satisfied. Let us consider quadrupole phonons and retain only the first root ($i = 1$), taking the index μ instead of $\lambda\mu i = 2\mu 1$. In this case we can obtain the SU(6) approximation; this was shown in section 3.2.

Following [170], we will find the SU(6) limit of the QPNM. To do this, we need to make sure that the operators Q_μ, Q^+_μ, $B(jj';2\mu)$ and their commutators form an SU(6) algebra. In order to determine the SU(6) algebra, it is necessary to satisfy the Jacobi identities, that is, the following equality must hold for any triplet of operators from the set under consideration:

$$[A,\,[B,C]]+[B,\,[C,A]]+[C,\,[A,B]]=0.$$

If the Jacobi identities are satisfied, we obtain the following conditions:

$$\tilde{t}^{\lambda\mu}_{\mu\mu_2}=0 \tag{4.91}$$

$$\widetilde{\mathcal{K}}(\mu_2',\mu'|\mu,\mu_2)=0 \tag{4.92}$$

$$[Q^+_\mu,Q^+_{\mu'}]=[Q_{\mu'},Q_\mu]=0. \tag{4.93}$$

We also obtain the equality

$$[B(jj';\lambda\mu),Q^+_{\mu'}]=\sum_{\mu_2}[t^{\lambda\mu}_{\mu'\mu_2}(jj')Q^+_{\mu_2}+\tilde{t}^{\lambda\mu}_{\mu'\mu_2}(jj')Q_{-\mu_2}] \tag{4.94}$$

and constraints on the function $\mathcal{K}(\mu'|\mu,\mu_2)$, so that we can rewrite (4.30):

$$[Q^+_{\mu_1},[Q_{\mu_2},Q^+_{\mu_3}]]=-\mathcal{K}(\delta_{\mu_1\mu_2}Q^+_{\mu_3}+\delta_{\mu_2\mu_3}Q^+_{\mu_1}). \tag{4.95}$$

As a result, only Q_μ, Q^+_μ and $[Q_{\mu'},\,Q^+_\mu]$ survive among the set of operators and their commutators, so that we obtain, as in section 3.2, an SU(6) algebra. Note that in the Tamm–Dancoff approximation, conditions (4.92) and (4.93) are satisfied automatically.

Let us use the Belyaev–Zelevinsky boson expansion method [92]. The expansion coefficients are found from the recurrent relations (3.7). The phonon operators will then be expressed, almost as in (3.34), in terms of the ideal boson operators d^+_μ and d_μ:

$$Q^+_\mu=d^+_\mu\sqrt{1+\gamma\hat{N}_d}\quad Q_\mu=\sqrt{1+\gamma\hat{N}_d}\,d_\mu$$

where

$$\hat{N}_d=\sum_\mu d^+_\mu d_\mu \tag{4.96}$$

and $N_d=-1/\gamma$ now means the total number of bosons.

We rewrite the Hamiltonian

$$H_2=H_{\rm av}+H_{\rm pair}+H_{\lambda=2}$$

in the form of (3.35), for the following values of the parameters:

$$h_0 = -\frac{5}{4}\sum_\tau (\mathcal{Y}_\tau^{21})^{-1}$$

$$h_1 = \omega_{21} - \frac{5}{2}\sum_{\tau\rho=\pm 1} (\kappa_0^{(2)} + \rho\kappa_1^{(2)})\mathfrak{G}_\tau \mathfrak{G}_{\rho\tau}$$

$$h_2 = -\frac{1}{4N_d}\sum_\tau (\mathcal{Y}_\tau^{21})^{-1}$$

$$h_3 = -\frac{5}{\sqrt{2N_d}}\sum_\tau \mathfrak{G}_\tau (\mathcal{Y}_\tau^{21})^{-1/2}$$

$$h_{4L} = -\frac{5}{2}\begin{Bmatrix} 2 & 2 & 2 \\ 2 & 2 & L \end{Bmatrix}\sum_{\tau\rho=\pm 1} (\kappa_0^{(2)} + \rho\kappa_1^{(2)})\mathfrak{G}_\tau \mathfrak{G}_{\rho\tau}$$

where

$$\mathfrak{G}_\tau = \sum_{j_1 j_2 j_3}^{\tau} f(j_1 j_2) v_{j_1 j_2}^{(-)} (-)^{j_1+j_3} \begin{Bmatrix} 2 & 2 & 2 \\ j_2 & j_1 & j_3 \end{Bmatrix} \times (\psi_{j_1 j_3}\psi_{j_2 j_3} + \varphi_{j_1 j_3}\varphi_{j_2 j_3}).$$

Therefore, a part of the Hamiltonian is represented in the SU(6) limit. It is immediately obvious that the term proportional to h_3 describes the quasiparticle–phonon interaction, while the terms with h_{4L} correspond to the terms, omitted in the QPNM, proportional to $B(j_1 j_2; 2\mu)B(j_3 j_4; 2\mu)$.

The expressions obtained for the parameters of the Hamiltonian were used in [170] for numerical calculations of the characteristics of zinc isotopes. The total number N_d of bosons was calculated microscopically; it must be equal to one half of the number of nucleons in open shells. Calculations demonstrated that conditions (4.91), (4.92), (4.93) and also those imposed on the function $\mathcal{K}(\mu_2', \mu_1' | \mu_1, \mu_2)$ are not well satisfied. This is an indication that the SU(6) approximation is not effective for spherical nuclei. It was shown in [171] that in the Tamm–Dancoff approximation, taking the interactions in the particle–particle channel into account does not improve the situation.

5

Fragmentation of single- and two-quasiparticle states of spherical nuclei

5.1 STRUCTURE OF LOW-LYING STATES

5.1.1

The most complete and accurate experimental data available refer to low-lying states. The wave functions of low-lying states have one dominant single- or two-quasiparticle component or a one-phonon component. As the excitation energy increases, the role of the dominant component decreases. The results of calculations for low-lying states strongly depend on the behaviour of single-particle levels at the Fermi surface. As a rule, the accuracy of calculations is limited by the rough description of energies and wave functions of single-particle states. An RPA description of the first, most collectivized, vibrational states in nuclei with open shells is not quite satisfactory. The description of the structure of low-lying states is improved if the projection is carried out in particle number and angular momentum; this was successfully done in [172].

In calculating the energies and wave functions of spherical nuclei, the mean field is calculated either by the Hartree–Fock method or by representing it in terms of the phenomenological Saxon–Woods potential (see (1.72)–(1.74)). The potential parameters are chosen to allow a correct description of the low-lying states of odd nuclei using equation (4.46′) which takes into account the quasiparticle–phonon interaction. The fitting of the parameters in the Saxon–Woods potential is carried out as follows. One chooses, for example, the parameters suggested in [45] and then calculates the RPA energies and wave functions of one-phonon 2^+ and 3^- states. Next, one solves equation (4.46′) and compares the obtained energies and spectroscopic factors with the experimental data. After this, the potential parameters are modified and the calculations are repeated. In many (but certainly not all) cases, this procedure leads to an acceptably good description of single-particle states. The results of this fitting for nuclei in the neighbourhood of ^{208}Pb ([145, 173]) are shown in figures 5.1 and 5.2. Figure 5.1 shows that taking the quasiparticle–phonon interaction into account, that is, solving equation (4.46′), considerably affects the spectrum. Figure 5.2 illustrates that a considerably good description of energies and spectroscopic factors is obtained for ^{207,209}Pb, ^{207}Tl and ^{209}Bi. The spectroscopic factors (equal to C^2_{Jv} from (4.44)) are nearly unity, so that the single-quasiparticle components of the wave function dominate. A conclusion was drawn, from the experimental data of [174] on the gap in proton single-particle states at $Z = 64$ and from specific features of the spectrum, that there must exist a new double-magic nucleus, ^{146}Gd. The calculations in [175] correctly reproduce the single-particle energies and the spectroscopic factors of the nuclei which differ from ^{146}Gd by one nucleon. An effective nucleon mass m^* was introduced in a number of papers, e.g., [55, 176]. The choice of QPNM parameters was described in detail in [74]. The limitation of the single-particle basis to discrete and quasi-discrete levels does not affect the description of the characteristics of the states of complex nuclei up to energies of about 15 MeV.

After the parameters of the Saxon–Woods potential and the pairing constants have been fixed, it is necessary to calculate the one-phonon states and fix the constants of the multipole and spin-multipole forces. The calculations of the one-phonon 2^+_1 and 3^-_1 states in RPA using formulas (2.105)–(2.108) appeared in a large number of papers [6, 62, 72–74]. These results are a basis for describing quadrupole and octupole vibrational states. Chapter 4 describes how to calculate the QPNM phonon basis and to fix the values of the constants (see also [74]).

If the constants $\kappa_0^{(2)}$, $\kappa_1^{(2)}$ are fixed using the condition $\omega_{21} = \omega_{21}^{\rm exp}$, that is, demanding the equality of the calculated and experimental energies of the first 2^+_1 states, then the calculated values of $B(E2)$ are found to be greater than the experimental ones. If the calculations use formulas (2.132)–(2.134) which take into account the interaction in the particle–particle channel, an increase in the quadrupole interaction constants in the particle–particle channel reduces the value of $B(E2)$. The effect of the interaction in the particle–particle channel on the vibrational states was analysed in several papers, such as [78, 79, 177, 178]. All even–even nuclei have low-lying collective octupole 3^-_1 states. As was observed in [179], the energies of 3^-_1 states do not vary smoothly as the mass number A grows. They are characterized by large values of $B(E3)$. An RPA description is a good first-order approximation.

5.1.2

Let us consider low-lying states of spherical even–even nuclei whose wave

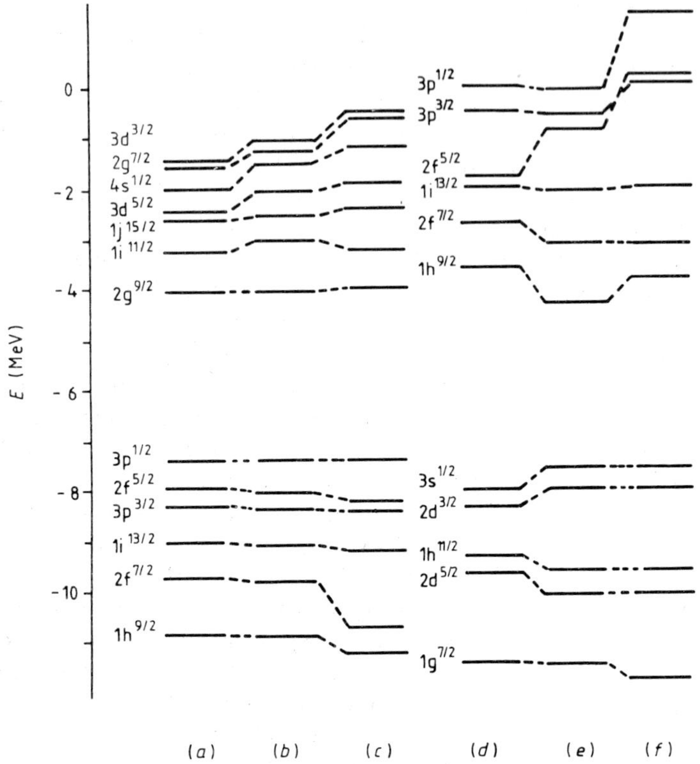

Figure 5.1 Single-particle states in ^{208}Pb: (a–c) neutron states, (d–f) proton states, (a, d) experimental values, (b, e) calculations taking into account the quasiparticle–phonon interactions; (c, f) energies of single-particle states of the Saxon–Woods potential with the parameters taken from [173].

functions have a dominating single- or two-phonon component. The experimental results reported in recent years indicate that the general pattern of spectra is more complex than the one implied by a simple slightly anharmonic vibrational model [4]. The simple vibrational model is elaborated by mixing in one- and two-phonon components, or by mixing one-phonon components with $2p$–$2h$ configurations (see [180, 181]).

We will analyse low-lying vibrational states in the framework of the QPNM. The wave function will be in the form of (4.55), with normalization condition (4.60). We will solve secular equation (4.65) and calculate the functions $R_i(J\nu)$ using (4.66), and $P^{\lambda_1 i_1}_{\lambda_2 i_2}(J\nu)$ using (4.63′).

Let us look at the effect of taking into account the Pauli exclusion principle on the two-phonon components of wave function (4.55). Correspond-

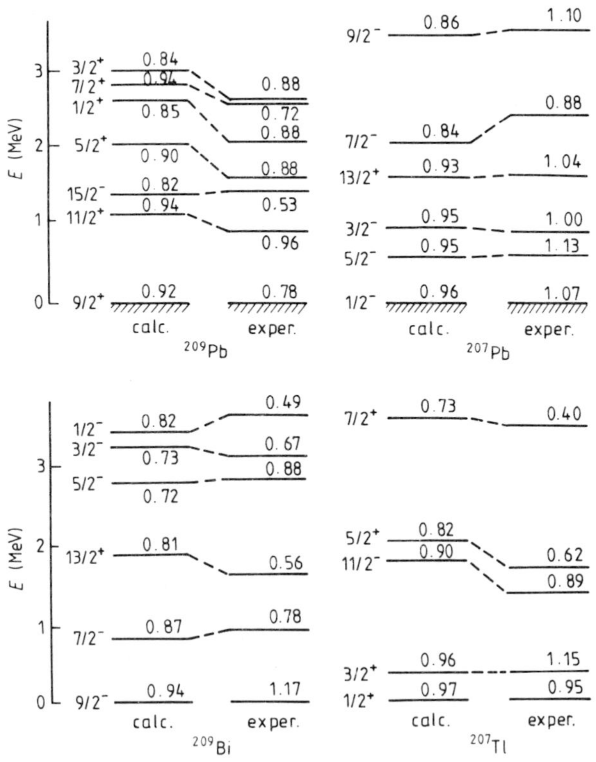

Figure 5.2 Experimental and QPNM-calculated energies and spectroscopic factors (shown by numbers over energy levels) of low-lying states of nuclei that differ from ^{208}Pb by one nucleon.

ingly, we compare the energies and the functions R_i^2 and $\left(P^{\lambda_1 i_1}_{\lambda_2 i_2}\right)^2$ calculated using formulas (4.65), (4.66) and (4.63′), with the values found using formulas (4.67), (4.67′), (4.69) and (4.71). The results calculated in [182] are listed in table 5.1. The table shows that the first 2_1^+ and 3_1^- states have a dominant one-phonon component and that the Pauli exclusion principle does not appreciably affect them. Taking the Pauli exclusion principle into account slightly changes the results for the 2^+ states with small two-phonon components. When two-phonon components in the second 2_2^+ states are large, taking the Pauli exclusion principle into account does not produce radical changes. The first 4_1^+ states of most nuclei have large two-phonon components $\{21, 21\}$, and sometimes large one-phonon components. The effect is somewhat greater for these states than for 2_2^+ states but is considerably smaller than for deformed nuclei. The effect of the Pauli exclusion principle is especially small in ^{208}Pb. The table also shows that the energies of the 2_1^+, 2_2^+ , 3_1^- and 4_2^+ states are described rather well.

In [183], three-phonon terms were added to wave function (4.55) and their role was analysed. It was shown that they appreciably influence the 2_1^+ states and strongly affect the 2_2^+, 0_1^+, 4_1^+ and higher-lying states in nuclei with open shells. Including three-phonon terms improves the description of the energies and of the structure of low-lying states in spherical nuclei.

Let us consider $E\lambda$ transitions between states that are described by wave functions (4.55). The $E\lambda$ transition operator will be in the form of (2.122′); after transformations, the reduced transition probability for the states $J\nu$ and $J'\nu'$ is given by the expression

$$B(E\lambda;J'\nu' \to J\nu) = \frac{1}{2J'+1}\Big|\sum_{\tau} e_{\text{eff}}^{(\lambda)}(\tau){\sum_{jj'}}^{\tau} p^{\lambda}(jj')$$
$$\times \Big\{\frac{1}{2\lambda+1}\sum_{ii'} u_{jj'}^{(+)} g_{jj'}^{\lambda i'}\Big[\sqrt{2J'+1}\,R_i(J\nu)P_{Ji}^{\lambda i'}(J'\nu')$$
$$+ (-)^{J+J'+\lambda}\sqrt{2J+1}\,R_i(J'\nu')P_{J'i}^{\lambda i'}(J\nu)\Big]$$
$$- (-)^{J+J'+\lambda}\sqrt{2J+1}\sqrt{2J'+1}\sum_{ii'} R_i(J\nu)R_{i'}(J'\nu')$$
$$\times \sum_{j''}\begin{Bmatrix}\lambda & J' & J\\ j'' & j' & j\end{Bmatrix} v_{jj'}^{(-)}\left(\psi_{j''j}^{J'i'}\psi_{j'j''}^{Ji} + \varphi_{j''j}^{J'i'}\varphi_{j'j''}^{Ji}\right)$$
$$- 4\sqrt{2J+1}\sqrt{2J'+1}\sum_{\substack{\lambda_1\lambda_2\lambda_3\\ i_1 i_2 i_3}} (-)^{J+\lambda_2+\lambda_3}\sqrt{2\lambda_1+1}$$
$$\times \sqrt{2\lambda_2+1}\,P_{\lambda_2 i_2}^{\lambda_3 i_3}(J\nu)P_{\lambda_1 i_1}^{\lambda_3 i_3}(J'\nu')v_{jj'}^{(-)}\begin{Bmatrix}\lambda_2 & \lambda_3 & J\\ J' & \lambda & \lambda_1\end{Bmatrix}$$
$$\times \sum_{j''}\begin{Bmatrix}\lambda & \lambda_1 & \lambda_2\\ j'' & j' & j\end{Bmatrix}\left(\psi_{j''j}^{\lambda_1 i_1}\psi_{j'j''}^{\lambda_2 i_2} + \varphi_{j''j}^{\lambda_1 i_1}\varphi_{j'j''}^{\lambda_2 i_2}\right)\Big\}\Big|^2. \tag{5.1}$$

The notation in (5.1) is the same as in formulas (2.124) and (2.124′). Note that all terms survive if $\varphi_{jj'}^{\lambda i} = 0$. If $p^{\lambda}(jj') = f^{\lambda}(jj')$, that is, if the radial dependence of the multipole forces is r^{λ}, then the term proportional to RP has a factor $X^{\lambda i}(\tau)(2\mathcal{Y}_{\tau}^{\lambda i})^{-1/2}$. Formula (5.1) is very useful since further progress in nuclear spectroscopy is largely connected with studying the γ transitions between excitation states.

The reduced probability of the $E\lambda$ transition from the ground state of an even–even nucleus to an excitation state, described by wave function (4.55), has the form

$$B(E\lambda;0_{\text{gs}}^+ \to J\nu) = \frac{1}{2\lambda+1}\Big|\sum_{\tau} e_{\text{eff}}^{(\lambda)}(\tau){\sum_{jj'}}^{\tau} p^{\lambda}(jj')$$

Table 5.1 Effect of the Pauli exclusion principle on the energies and structure of states with $J^\pi = 2^+$, 3^- and 4^+.

Nucleus	J^π_v	Energy (exper.) (MeV)	Calculation taking the Pauli exclusion principle into account						Calculation disregarding the Pauli exclusion principle					
			Energy (MeV)	λi	R_i^2	$\lambda_1 i_1$	$\lambda_2 i_2$	$(P^{\lambda_1 i_1}_{\lambda_2 i_2})^2$	Energy (MeV)	λi	R_i^2	$\lambda_1 i_1$	$\lambda_2 i_2$	$(P^{\lambda_1 i_1}_{\lambda_2 i_2})^2$
^{114}Sn	2_1^+	1.30	1.36	21	0.95	21	21	—	1.29	21	0.90	21	21	0.01
	2_2^+	2.17	2.63	22	0.27	21	21	0.53	2.53	22	0.21	21	21	0.58
	3_1^-	2.28	2.23	31	0.92	21	31	0.08	2.22	31	0.91	21	31	0.09
	4_1^+	2.19	2.71	41	0.29	21	21	0.48	2.60	41	0.22	21	21	0.57
^{142}Sm	2_1^+	0.77	0.68	21	0.82	21	21	0.03	0.64	21	0.79	21	21	0.05
	2_2^+	1.66	1.64	22	0.15	21	21	0.30	1.52	22	0.39	21	21	0.34
	3_1^-	1.78	1.80	31	0.94	21	31	0.06	1.76	31	0.91	21	31	0.08
	4_1^+	1.79	1.77	41	0.28	21	21	0.12	1.60	41	0.26	21	21	0.23
^{146}Gd	2_1^+	1.97	1.93	21	0.93	31	31	0.04	2.00	21	0.94	31	31	0.05
	2_2^+	3.37	3.47	22	0.04	31	31	0.95	3.43	22	0.05	31	31	0.94
	3_1^-	1.57	1.53	31	0.95	21	31	0.03	1.59	31	0.96	21	31	0.03
^{208}Pb	2_1^+	4.08	4.43	21	0.97	31	31	0.03	4.43	21	0.97	31	31	0.03
	3_1^-	2.61	2.59	31	0.99	—	—	—	2.60	31	0.99	—	—	—

Table 5.2 Reduced probabilities of $E\lambda$ transitions from the ground state to the excitation states.

Nucleus	Final state J_ν^π	Energy (MeV)	Transition type	$B(E\lambda)$ (s.p.u.) (exper.)	(calc.)
^{114}Sn	2_1^+	1.299	$E2$	16	19
	3_1^-	2.275	$E3$	11	28
^{116}Sn	2_1^+	1.294	$E2$	12.9	22
	2_2^+	2.112	$E2$	0.05	0.05
	3_1^-	2.266	$E3$	39	37
^{144}Sm	2_1^+	1.660	$E2$	17	20
^{146}Gd	2_1^+	1.97	$E2$	>0.4	15
	3_1^-	1.58	$E3$	37	36
^{208}Pb	2_1^+	4.086	$E2$	8.0	7.8
	3_1^-	2.614	$E3$	38	35

$$\times \Big\{ \sum_i R_i(J\nu) g_{jj}^{\lambda i} u_{jj'}^{(+)} + \sum_{\lambda_1 i_1 \lambda_2 i_2 j''} P_{\lambda_2 i_2}^{\lambda_1 i_1}(J\nu) f^\lambda(jj') v_{jj'}^{(-)} (2\lambda_1+1)^{1/2} \times (2\lambda_2+1)^{1/2} \begin{Bmatrix} \lambda_1 & \lambda_2 & \lambda \\ j & j' & j'' \end{Bmatrix} \times \left(\psi_{j'j''}^{\lambda_1 i_1} \varphi_{j''j}^{\lambda_2 i_2} + \varphi_{j'j''}^{\lambda_1 i_1} \psi_{j''j}^{\lambda_2 i_2} \right) \Big\} \Big|^2 . \quad (5.2)$$

Note that in the Tamm–Dancoff approximation, $\varphi_{jj'}^{\lambda i} = 0$ and the second term in (5.2) vanishes. The results of calculations with (5.2) for $e_n^{(\lambda)} = e_p^{(\lambda)} = 0$ (see (2.77′)) [160, 182] and the corresponding experimental data are shown in table 5.2. For most spherical (non-transition) nuclei, a description about as satisfactory as in table 5.2 was obtained for the probabilities of the $E2$ and $E3$ transitions to the 2_1^+, 2_2^+ and 3_1^- states. There is an agreement between the results of calculations and experimental data on transition densities for the 2_1^+ and 3_1^- states. If the radial dependence of multipole strengths differs from r^λ, then (5.2) includes a product of matrix elements, $p^\lambda(jj')f^\lambda(jj')$, instead of $[p^\lambda(jj')]^2$ which holds in the case of r^λ. No coherence appears explicitly in the sum over jj'; the $E\lambda$ transitions are then enhanced because the phases of the large matrix elements $p^\lambda(jj')$ and $f^\lambda(jj')$ are identical.

The lowest 1^- states in barium, cerium, neodimium and samarium offer an interesting possibility of observing a qualitative difference of the results of RPA calculations from those of Tamm–Dancoff calculations. These nu-

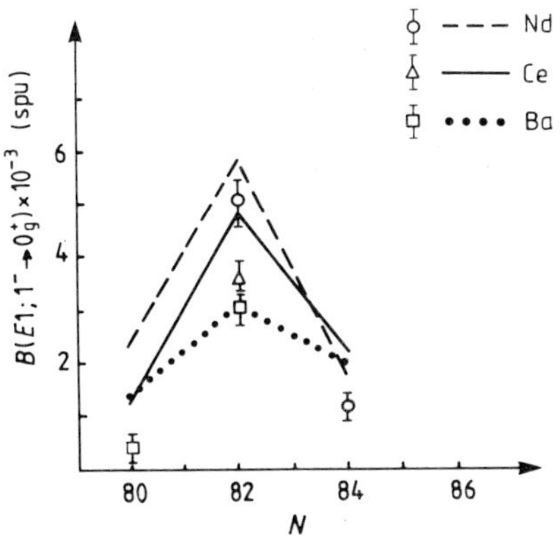

Figure 5.3 $B(E1)$ as a function of the number of neutrons, N, in barium, cerium and neodimium isotopes: the symbols show the experimental data; and the full, broken and dotted lines represent the calculated results.

clei show 1^- states [184] with energies somewhat smaller than those of the two-phonon poles $\{21, 31\}$ which are formed by the first quadrupole and octupole phonons. According to RPA calculations [160, 185], the energies of one-phonon states in these nuclei exceed 8 MeV. According to calculations using formulas (4.69) and (4.71), the energies of the lowest 1^- states with non-vanishing small one-phonon components lie 1–2 MeV above the energy of the 1^- states as measured in [184]. We can conclude that two-phonon states of the type $\{21,31\}$ are indeed observed in these nuclei. In this particular case, $R_i(1\nu) = 0$ in (5.2), so that the Tamm–Dancoff approximation gives $B(E1) = 0$. The reduced excitation probability for such states in RPA is non-zero.

The experimentally measured [184] and QPNM calculated [186] values of $B(E1)$ for the excitation of 1^- states in isotopes of some elements are given in figure 5.3. We see that the experimental and theoretical values of $B(E1)$ are not only non-zero but are rather close to one another. Besides, $B(E1)$ strongly depends on the number of neutrons, N, and $B(E1)$ reaches a maximum at $N = 82$. This behaviour arises because the contributions of neutrons and protons enter (5.2) with opposite signs and suppress each other. If $N = 82$, there is no pairing in the neutron system, so that $v_{jj'}^{(-)} = 0$ for the particle–hole states and the neutron contribution is small. The proton contribution is thus not compensated for, which increases $B(E1)$ for $N = 82$.

Only several of the features typical of the harmonic vibrator model have survived in the spectra of the 0^+, 2^+ and 4^+ states and the probabilities of the $E2$ transitions in spherical nuclei. The spectra are considerably more complex. The diagram of the levels of ^{116}Sn obtained in [187] is shown as

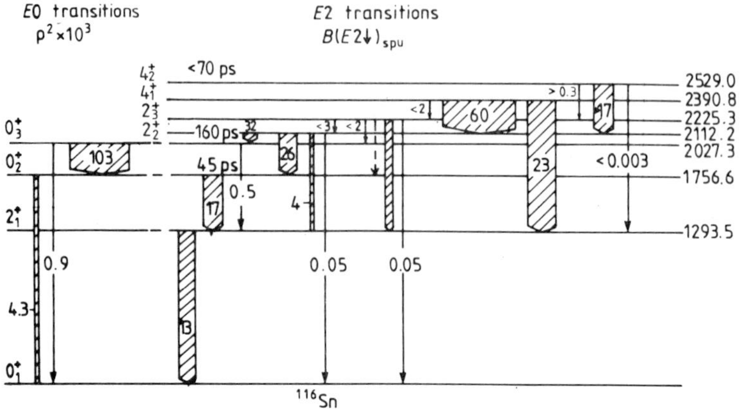

Figure 5.4 Energies in keV and the probabilities of the $E0$ and $E2$ transitions in ^{116}Sn.

an example in figure 5.4. Attributes of the harmonic vibrator are noticeable in the probabilities of the $E2$ transitions $2_1^+ \rightarrow 0_1^+$, $0_2^+ \rightarrow 2_1^+$ and $4_1^+ \rightarrow 2_1^+$. However, the energies of the 0_3^+, 2_3^+ and 4_2^+ states, the strong $E0$ transitions $0_2^+ \rightarrow 0_1^+$, $0_3^+ \rightarrow 0_2^+$ and very strong $E2$ transitions $4_1^+ \rightarrow 2_2^+$, $2_2^+ \rightarrow 0_2^+$ and $2_2^+ \rightarrow 0_3^+$ distort the simple relations of the harmonic vibrator model. This complex pattern of collective states is successfully described in the model which takes into account the coupling of the proton $2p-2h$ excitations to quadrupole vibrations, and in the model which takes into account the couplings between the one- and two-phonon states [188].

Numerous IBM calculations of quadrupole states in even–even spherical and transition nuclei have been carried out. As we have demonstrated in figure 3.5, the energies of the 2_1^+, 4_1^+ and 6_1^+ states in tellurium isotopes are adequately described in the IBM-2. The energies of such states were calculated for isotopes of ruthenium, palladium, xenon, barium, cerium, samarium, osmium and platinum in [110–112, 116, 189]. A satisfactory description of the probabilities of the $E2$ transitions and quadrupole moments of the 2^+ states was obtained when additional effective charges were introduced.

The discovery of a closed proton shell at $Z = 64$ led to difficulties in describing quadrupole states in neodimium, samarium and gadolinium isotopes in IBM-2, since the results of calculations strongly depend on the number of proton bosons. Thus the number of proton bosons in samarium isotopes with a closed shell is 6 for $Z = 50$, and equals unity for $Z = 64$. The behavior of the energy gap at $Z = 64$ was analysed in [190] as a function of the number of neutrons. It was shown that as neutrons fill up the subshell $1h_{9/2}$ at $N \geqslant 88$, a stable deformation appears and the gap vanishes. Correspondingly, it was suggested [190] to determine $N_{\rm p}$ for the neutron number $N < 88$ using the existence of the Z =50–64 shell, and for $N \geqslant 88$, using the existence of the Z =50–82 shell. The possibility of this approach was confirmed by the calculation of gyromagnetic ratios for 2_1^+ states of nuclei in the neighbourhood of $Z = 64$.

When calculating the energies and probabilities $B(E2)$ and $B(M1)$, a modified version of IBM-2 is used, in which new degrees of freedom are defined: additional s' and d' and also hexadecapole g bosons. The introduction of additional degrees of freedom into IBM-2 adds new states to the spectrum calculated and makes the model more flexible for the description of experimental data. Broadening of the model makes it possible to describe weakly-collective states.

5.1.3

Until recently, it was assumed that mixed-symmetry states, or isovector states, lie at energies above 10 MeV. The energies and secondary γ rays from the reaction ^{143}Nd$(n\gamma)^{144}$Nd were analysed in [191]. The analysis has demonstrated that the mixing ratio $\delta(E2/M1)$ is very low; this points to a strong $M1$ transition between the 2_3^+ and 2_1^+ states in ^{144}Nd. This behaviour is consistent with the γ transition from an isovector to isoscalar state. In [120], it was formulated that γ transitions in ^{140}Ba, ^{142}Ce and ^{144}Nd from 2_3^+ states with energies of about 2 MeV agree with their identification as the lowest-energy mixed-symmetry states in the U(5) limit of IBM-2. Note that the mixed-symmetry states $K^\pi = 1^+$ were discovered in [119] in ^{156}Gd and were later found in a considerable number of isotopes of Gd, Dy, Er and Yb.

Isovector states, or mixed-symmetry states, were found in ^{56}Fe [192, 193]. A QPNM analysis of how low-lying isovector states appear was described in [194], using ^{56}Fe as an example. The calculations in [194] were carried out with wave function (4.55) using formulas (4.67), (4.67′), (4.69), (4.70) and (4.71); the results are presented in tables 5.3 and 5.4, in the column denoted by $Q^+ + Q^+Q^+$. The ratio $B(IV, E2)/B(IS, E2)$ of the isoscalar and isovector transitions is a good qualitative characteristic of the isotopic structure of excitation states. The results listed in table 5.3 show that three out of four 2^+ states have isoscalar structure, and for them, $B(IV, E2)/B(IS, E2) \ll 1$. One of them is, however, an isovector state. The quasiparticle–phonon interaction results in strong mixing of 2^+ phonons in RPA and in changes in the structure of 2^+ states, so that it is the state 2_3^+ which is found to be isovector, not 2_2^+. Hence, the $M1$ transition $2_1^+ \to 2_3^+$ has the highest probability, as we see in table 5.4. Taking into account the quasiparticle–phonon interaction improved the description of energies and $B(E2)$ values. This example shows that states with large isovector components arise in a natural way in the excitation spectrum of ^{56}Fe.

Table 5.3 Energies, reduced probabilities of the electric $E2$ transitions and the probability ratio of isovector $B(\mathrm{IV})$ and isoscalar $B(\mathrm{IS})$ quadrupole transitions for some 2^+ states of ^{56}Fe.

J^π_ν	$\mathcal{E}_\nu$ (MeV)			$B(\mathrm{IV})/B(\mathrm{IS})$		$B(E2, 0^+_{gs} \to 2^+_\nu)e^2\mathrm{fm}^4$		
	(exper.) [193]	(theor.) RPA	(theor.) $Q^+ + Q^+Q^+$	(theor.) RPA	(theor.) $Q^+ + Q^+Q^+$	(exper.) [193]	(theor.) RPA	(theor.) $Q^+ + Q^+Q^+$
2^+_1	0.847	0.860	0.701	0.004	0.008	970±20	1341	678
2^+_2	2.650	2.790	2.520	14.60	0.730	12±1	4.7	10.4
2^+_3	2.960	3.802	2.995	0.028	9.150	10±4	57.4	1.2
2^+_4	3.370	4.079	3.330	0.007	0.150	40±3	30.4	5.0

Table 5.4 The probabilities of the $E2$ and $M1$ transitions and mixing coefficients δ for some 2^+ states of ^{56}Fe.

	$B(M1)\mu_0^2$			$B(E2)e^2$ fm^4		$\delta(E2/M1)$		
	(exper.) [193]	(theor.) RPA	(theor.) $Q^+ + Q^+Q^+$	(exper.) [193]	(theor.) $Q^+ + Q^+Q^+$	(exper.) [193]	(exper.) [192]	(theor.) $Q^+ + Q^+Q^+$
$2_1^+ \rightarrow 2_2^+$	0.22	1.30	0.16	34±11	4.7	−0.17	−0.18	−0.45
$2_1^+ \rightarrow 2_3^+$	0.14	0.16	0.95	36^{+14}_{-9}	36.4	−0.27	+0.19	−0.65
$2_1^+ \rightarrow 2_4^+$	0.12	0.19	0.38	6±2	6.6	−0.25	—	−0.36

5.1.4

Let us consider 0^+ excited states that cause quite a few of the difficulties that the theory faces. The low-lying 0^+ states have complex structure which is determined by the quadrupole and pairing-vibration branches of collective motion [4, 106]. The spin-orbital interactions were taken into account in [195]. So far we made use of only one term, $\sum_{jm}\epsilon_j\alpha^+_{jm}\alpha_{jm}$, from the part $H_{\rm av}+H_{\rm pair}$ of the Hamiltonian in the form (2.91). The following terms must be added to describe the 0^+ states:

$$H_0^{(1)} = -\frac{1}{4}\sum_\tau G_\tau {\sum_{jj'}}^\tau (2j+1)^{1/2}(2j'+1)^{1/2}[u_{j'}^2 A^+(j) - v_j^2 A(j)] \times [u_{j'}^2 A(j') - v_{j'}^2 A^+(j')] \tag{5.3}$$

$$H_0^{(2)} = -\frac{1}{2}\sum_\tau G_\tau {\sum_{jj'}}^\tau (2j+1)^{1/2}(u_j^2 - v_j^2) \times u_{j'}v_{j'}[A^+(j)B(j') + B(j')A^+(j)] \tag{5.4}$$

[the relevant notation is given in (2.96) and (2.97)]. By analogy with definition (2.101), we introduce pairing vibrational phonons $Q^+_{0i}(\tau)$, $Q_{0i}(\tau)$ for the proton and neutron systems. Calculating the average value

$$H_0' = \sum_{jm}\epsilon_j a^+_{jm}a_{jm} + H_0^{(1)} \tag{5.5}$$

over the state $Q^+_{0i}(\tau)\Psi_0$ and using the variational principle, we obtain secular equations separately for the proton and neutron systems, from which we calculate the energies ω_{0i} of the pairing vibrational phonons:

$$\Big\{{\sum_j}^\tau (2j+1)\frac{\epsilon_j(u_j^2-v_j^2)^2}{4\epsilon_j^2-\omega_{0i}^2} - \frac{1}{G_\tau}\Big\}\Big\{{\sum_j}^\tau (2j+1)\frac{\epsilon_j}{4\epsilon_j^2-\omega_{0i}^2} - \frac{1}{G_\tau}\Big\} - \frac{\omega_{0i}^2}{4}\Big\{{\sum_j}^\tau (2j+1)\frac{u_j^2-v_j^2}{4\epsilon_j^2-\omega_{0i}^2}\Big\}^2 = F_1(\omega_{0i})F_2(\omega_{0i}) - F_3^2(\omega_{0i}) \equiv F(\omega_{0i}) = 0. \tag{5.6}$$

The Hamiltonian containing all the interactions leading to pairing and the multipole interactions, taking into account the RPA secular equations including (5.6), can be written in the form

$$H = \sum_{jm}\epsilon_j a^+_{jm}a_{jm} + H_{0v} + H_{0vq} + H_{Mv} + H_{Mvq}. \tag{5.7}$$

If necessary, we can add to it the spin-multipole interactions. The expressions for H_{Mv} and H_{Mvq} are given by formulas (4.16) and (4.16′):

$$H_{0v} = -\sum_{\tau} \frac{2}{G_{\tau}} \sum_{ii'} \frac{F_1^{1/2}(\omega_{0i})F_1^{1/2}(\omega_{0i'}) + F_2^{1/2}(\omega_{0i})F_1^{1/2}(\omega_{0i'})}{(\mathcal{Y}_{\tau}^{0i}\mathcal{Y}_{\tau}^{0i'})^{1/2}} \times Q_{0i}^{+}(\tau)Q_{0i'}(\tau) \tag{5.8}$$

$$H_{0vq} = \frac{1}{2}\sum_{\tau i}\sum_{j}{}^{\tau} \frac{u_{jj}^{(+)}}{(\mathcal{Y}_{\tau}^{0i})^{1/2}} \{[(F_1^{1/2}(\omega_{0i}) - F_2^{1/2}(\omega_{0i}))Q_{0i}^{+}(\tau) + (F_1^{1/2}(\omega_{0i}) + F_2^{1/2}(\omega_{0i}))Q_{0i}(\tau)]B(j) + \text{h.c.}\}. \tag{5.9}$$

The wave function of the 0^+ excitation state can be given in the following form:

$$\Psi_{\nu}(J=0) = \Big\{\sum_{i\tau} R_i(\nu)Q_{0i}^{+}(\tau) + \sum_{\lambda\mu ii'} \frac{(-)^{\lambda-\mu}}{\sqrt{2\lambda+1}} P_{\lambda i'}^{\lambda i}(\nu) \times [Q_{\lambda\mu i}^{+}Q_{\lambda-\mu i}^{+}]_{00}\Big\}\Psi_0. \tag{5.10}$$

The normalization condition in a $\mathcal{K}$-diagonal approximation is

$$\sum_{i\tau}(R_i(\nu))^2 + 2\sum_{\lambda ii'}[P_{\lambda i'}^{\lambda i}(\nu)]^2\{1 + \mathcal{K}^{J=0}(\lambda i, \lambda i')\} = 1. \tag{5.11}$$

We will now find the mean value of H, over wave function (5.11). We make use of the variational principle and carry out the same transformations as in section 4.3; this gives us the secular equation of the type of (4.65). The function $U_{\lambda i_2}^{\lambda i_1}(i)$, which in this equation replaces $U_{\lambda_2 i_2}^{\lambda_1 i_1}(Ji)$, is

$$U_{\lambda i_2}^{\lambda i_1}(i) = \langle Q_{0i}(H_{0vq} + H_{Mvq})Q_{\lambda\mu i}^{+}Q_{\lambda-\mu i_2}\rangle. \tag{5.12}$$

The quasiparticle–phonon interaction results in mixing the pairing vibrational states $Q_{0i}^{+}\Psi_0$ with the two-phonon $[Q_{\lambda\mu i}^{+}Q_{\lambda-\mu i'}^{+}]_{00}\Psi_0$ states. A change in the structure of 0^+ states owing to mixing of this type is observed in reactions in which these states are excited and then decay. For example, in inelastic scattering reactions (see, e.g., [188]), one-phonon components of wave function (5.10) of 0^+ states are excited directly, while two-phonon components are excited by a two-stage process.

The 0^+ states are most efficiently analysed in two-nucleon transfer reactions. Let us look at the (p,t) reaction. The operator of two neutron transfer is given by

$$\Gamma(p,t) = \sum_{jm}(-)^{j-m}a_{jm}a_{j-m}$$

$$= -\sum_i \frac{2\sqrt{2}}{(\partial F/\partial \omega_0|_{\omega_0=\omega_{0i}})^{1/2}} \frac{1}{G_N}\{[F_2^{1/2}(\omega_{0i}) - F_1^{1/2}(\omega_{0i})]Q_{0i} + [F_2^{1/2}(\omega_{0i}) + F_1^{1/2}(\omega_{0i})]Q_{0i}^+\} - \sum_j u_{jj'}^{(+)} B(j) + 2\frac{C_n}{C_N}. \tag{5.13}$$

The probability of two-neutron transfer to the ground state of the nucleus is

$$S_0(p,t) = 4C_n^2/G_N^2. \tag{5.14}$$

The probabilities of (p,t) reactions between the ground states of even–even nuclei have been studied in sufficient detail [196]. They clearly display the effects of closed neutron shells. The maximum values of the cross sections are observed in spherical nuclei whose neutron number corresponds to a half-filled shell.

In the case of $\mathcal{K} = 0$, the probability for the 0^+ state (described by wave function (5.10)) to be excited in a (p,t) reaction is

$$S(p,t) = \left| \frac{\sqrt{2}}{G_N} \sum_i \frac{F_1^{1/2}(\omega_{0i}) + F_2^{1/2}(\omega_{0i})}{(\partial F/\partial \omega_0|_{\omega_0=\omega_{0i}})^{1/2}} R_i^n(\nu) + \sum_{\lambda i i'} \sqrt{2\lambda+1} \times P_{\lambda i'}^{\lambda i}(\nu) \sum_{jj'} u_{jj'}^{(+)} (\psi_{jj'}^{\lambda i} \varphi_{jj'}^{\lambda i'} + \psi_{jj'}^{\lambda i'} \varphi_{jj'}^{\lambda i}) \right|^2 \tag{5.15}$$

[197]. The second sum in (5.15) is not large and is often neglected. The strength distribution for such a transition is largely characterized by the fragmentation of the 0^+ one-phonon state.

A considerable number of papers (see, e.g., [187, 198]) are devoted to $E0$ transitions between 0^+ states. The probability of an $E0$ transition is determined by the dimensionless matrix element $\rho(E0)$. Its explicit form for the transition to the one-phonon 0^+ state of a deformed nucleus is given by (2.81′). The degree of the $E0/E2$ mixing is characterized by the quantity X, defined by (2.81″). An analysis of $E0$ transitions makes it possible to understand the degree of mixing of one-phonon pairing vibrational states and two-phonon quadrupole states.

Among low-lying 0^+ states, *'intruder'* states with large two-phonon components are encountered. It was shown in [199] that at a certain strength of neutron-proton interactions, $Q_{0i}^+(\mathrm{p})Q_{0i'}^+(\mathrm{n})\Psi_0$-type states may appear in spherical nuclei. It was concluded that the 0^+ *'intruder'* states are found in cadmium isotopes and other nuclei.

5.1.5

Let us consider spherical nuclei with an odd number of nucleons. The wave function is

$$\Psi_\nu(JM) = C_{j\nu}\{a^+_{JM} + \sum_{\lambda ij} D_j^{\lambda i}(J\nu)[a^+_{jm}Q^+_{\lambda\mu i}]_{JM}\}\Psi_0 \tag{5.16}$$

and the secular equation for the energies η_{J_ν} is given by (4.46). The reduced probabilities of $E\lambda$ and $M\lambda$ transitions between the states described by the wave functions $\Psi_\nu(JM)$ and $\Psi_{\nu'}(J'M')$ have the following form in odd proton ($\tau_0 = \mathrm{p}$) or odd neutron ($\tau_0 = \mathrm{n}$) nuclei:

$$\begin{aligned} B(E\lambda; J'\nu' \to J\nu) = {} & (2J'+1)^{-1} C^2_{J\nu} C^2_{J'\nu'} \Big\{ e^{(\lambda)}_{\text{eff}}(\tau_0) p^\lambda(J'J) v^{(-)}_{J'J} \\ & + \sum_\tau e^{(\lambda)}_{\text{eff}}(\tau) \sum_{jj'i}{}^{\tau} p^\lambda(jj') g^{\lambda i}_{jj'} u^{(+)}_{jj'} \Big[\Big(\frac{2J'+1}{2\lambda+1}\Big)^{1/2} \\ & \times D^{\lambda i}_J(J'\nu')(1 + \mathcal{L}(J'J\lambda i)) + (-)^{J'-J+\lambda} \Big(\frac{2J+1}{2\lambda+1}\Big)^{1/2} \\ & \times D^{\lambda i}_{J'}(J\nu)(1 + \mathcal{L}(J'J\lambda i)) \Big] \Big\}^2 \end{aligned} \tag{5.17}$$

$$\begin{aligned} B(M\lambda; J'\nu' \to J\nu) = {} & (2J'+1)^{-1} C^2_{J\nu} C^2_{J'\nu'} \Big\{ \langle J \,|\Gamma(M\lambda)| J'\rangle v^{(+)}_{J'J} \\ & + \sum_\tau \sum_{jj'i}{}^{\tau} \langle j|\Gamma(M\lambda)|j'\rangle u^{(-)}_{jj'} w^{\lambda i}_{jj'} \Big[\Big(\frac{2J'+1}{2\lambda+1}\Big)^{1/2} \\ & \times D^{\lambda i}_J(J'\nu')(1 + \mathcal{L}(J'J\lambda i)) + (-)^{J'-J+\lambda} \\ & \times \Big(\frac{2J+1}{2\lambda+1}\Big)^{1/2} D^{\lambda i}_{J'}(J\nu)(1 + \mathcal{L}(J'J\lambda i)) \Big] \Big\}^2 \end{aligned} \tag{5.18}$$

where $\Gamma(M\lambda)$ is defined by (2.123′).

A sufficiently good description of the energies and structures of low-lying states of odd spherical nuclei [201–203] was achieved in terms of the particle-vibrational model [4, 200], dynamic collective model [201], as well as the QPNM and some other models.

In some cases, taking into account one-phonon excitations or core polarization greatly affects the γ transition probability. Thus the probabilities of $M2$ transitions between the $11/2^-$ and $7/2^+$ states in $^{109-121}$Sn, $^{139-145}$Sb, $^{139-145}$Pr and $^{143-149}$Eu nuclei were calculated in [204]. The calculated values of $B(M2)$ between single-quasiparticle states $1h_{11/2}$ and $1g_{7/2}$ [the first term in (5.17)] were found to be greater than the experimental values by a factor of 15–20. Taking the one-phonon states with $\lambda^\pi = 2^+$, 3^- and 2^- into account reduced the discrepancy to a factor of 3–4. The role of the 2^- one-phonon states with energies up to 24 MeV is especially important (these are energies in the region of the giant $M2$ resonance). The $B(M2)$ probabilities in tin isotopes are determined to a large extent by the superfluidity factor $v^{(-)}_{h_{11/2}, g_{7/2}}$. An agreement of the calculated results and the experimental data was obtained for $g_s^{\text{eff}} = 0.5 g_s$. There is no doubt that the renormalization of the spin gyromagnetic factor is too high; typically, one takes $g_s^{\text{eff}} = (0.7-0.8)g_s$.

Let us consider the magnetic moments of odd nuclei. The magnetic moment μ_J in the state described by the wave function (5.16) is

$$\mu_J = g_J J \mu_0 = (\Psi_\nu^*(J, M = J)\mathfrak{M}(M;10)\Psi_\nu(J, M = J)) \tag{5.19}$$

where the operator $\mathfrak{M}(M;\lambda\mu)$ is given by (2.123′). Making use of the approximation

$$[\alpha_{jm}, Q_{\lambda\mu i}^+] = 0 \tag{5.20}$$

and dropping a number of small terms, we obtain

$$g_J = C_{J\nu}^2 (g_{\text{s.p.}} + g_1 + g_2) \tag{5.21}$$

where the single-quasiparticle part is

$$g_{\text{sp}} = (1/J)\langle \alpha_{JJ} \mathfrak{M}(M;10)\alpha_{JJ}^+ \rangle. \tag{5.22}$$

The coupling between quasiparticles and the one-phonon 2^+ and 3^- states leads to a correction:

$$g_1 = \left(\frac{4\pi}{3}\frac{2J+1}{J(J+1)}\right)^2 \sum_{j\lambda=2,3i} (-)^{\lambda+j+J+1}(D_j^{\lambda i}(J\nu))^2 \times \langle j|\Gamma(M1)|j\rangle \begin{Bmatrix} J & j & \lambda \\ j & J & \lambda \end{Bmatrix}. \tag{5.23}$$

The part due to the spin polarization of the core is

$$g_2 = \sqrt{\frac{4\pi}{2}}[3J(J+1)]^{-1/2} \sum_{jj'i} D_j^{1i} \langle j|\Gamma(M1)|j'\rangle u_{jj'}^{(-)} w_{jj'}^{1i}. \tag{5.24}$$

Calculations using formulas (5.19)–(5.24) for a large number of odd spherical nuclei have been performed in [205]. They provide a satisfactory description of magnetic moments. The spin-multipole interaction constants were chosen equal to those used in [156, 206] for the description of the $M1$ and $M2$ resonances; the g_s^{eff} factors were set equal to $0.9g_s$, $g_l(\text{p}) = 1$, $g_l(\text{n}) = 0$. In odd neutron nuclei, the contribution of core polarization, g_2, exceeds the contribution , g_1, of the collective motion, while these contributions are of similar values in odd proton nuclei. If the problem is formulated more rigorously, one has to clarify the role of meson exchange, non-nucleon degrees of freedom (e.g., excitation of Δ isobars) and core polarization by spin-orbital forces.

5.2 FRAGMENTATION OF SINGLE-QUASIPARTICLE STATES

5.2.1

Wave functions of the ground state and of low-lying states of odd nuclei contain a dominant single-quasiparticle component. If the excitation energy is slightly increased, the role of the quasiparticle ⊗ phonon components is enhanced, so that they are observed already in the ground states of nuclei with open shells. Quasiparticle ⊗ two-phonon components, quasiparticle ⊗ three-phonon components etc appear at high excitation energies. As the excitation energy grows, the nuclear state density increases and the state structure gets more complex. Not one but several components contribute to the normalization of the wave function. It is quite clear that it is absolutely impracticable to measure a large number of wave function components of each of many thousands of levels. Furthermore, it would be impossible to calculate the energies and wave functions of states with millions of components each. The difficulties and ambiguities involved were illustrated for light nuclei in [207], where diagonalization of high-order matrices was performed in the calculations of energies and wave functions. The problem is not to measure and calculate the energies and wave functions of each state but to find out the general trends in state structure evolution with increasing excitation energy [135, 208]. The first step in this direction is an investigation of the fragmentation of single-quasiparticle states. The attention is focused on studying the fragmentation of deep neutron and proton hole states and of high-lying particle proton states, since their widths $\Gamma\downarrow$ (caused by fragmentation due to the quasiparticle–phonon interaction) are considerably greater than the widths $\Gamma\uparrow$ due to the continuous spectrum ($\Gamma\uparrow$ are connected with the emission of particles).

First measurements of the fragmentation of low-lying states were performed in 1960 [209]. The fragmentation of deep hole states was studied much later [210, 211]. Single-nucleon transfer reactions proved to be efficient for analysing the strength distribution of high-lying particle and deep hole states in spherical nuclei. These reactions were used to study the fragmentation of single-quasiparticle states in many nuclei [212–219]. An important step in analysing fragmentation was the application of polarized particle beams; this technique made it possible to establish the spins of the excitation levels [220, 221]. Numerous calculations were published on the fragmentation of single-quasiparticle states in the QPNM [142, 143, 160–163, 173, 222–224], in the NFT [166], and also in other microscopic models [225, 226] and in other phenomenological approaches [227, 228].

5.2.2

Let us analyse the effect of the Pauli exclusion principle on the fragmentation of single-quasiparticle states. We begin with a wave function in the form (5.16). The condition of its normalization in a diagonal $\mathcal{L}$ approximation and the relevant secular equation are

$$C_{J\nu}^2\{1+\sum_{\lambda ij}[D_j^{\lambda i}(J\nu)]^2[1+\mathcal{L}(Jj\lambda i)]\}=1 \qquad (5.25)$$

$$\mathcal{F}(\eta_{J\nu})\equiv\epsilon_j-\eta_{J\nu}-\frac{1}{2}\sum_{\lambda ij}\frac{\Gamma^2(Jj\lambda i)[1+\mathcal{L}(Jj\lambda i)]}{\epsilon_j+\omega_{\lambda i}+\Delta\omega(Jj\lambda i)-\eta_{J\nu}}=0. \qquad (5.26)$$

With the exact commutation relations between the operators α_{jm}^+ and $Q_{\lambda\mu i}$ taken into account, new functions $\mathcal{L}(Jj\lambda i)$ and $\Delta\omega(Jj\lambda i)$ appear; they vanish in the approximation in which $[\alpha_{jm}, Q_{\lambda\mu i}^+]=0$. The function $\mathcal{L}(Jj\lambda i)$ determines the contribution of a given quasiparticle $\otimes$ phonon component to the normalization of the wave function and to the secular equation, depending on the degree of violation of the Pauli exclusion principle. If the state that violates the Pauli exclusion principle has a unity weight, that is, if the Pauli exclusion principle is maximally violated, then $\mathcal{L}=-1$ and this state is excluded automatically from both (5.25) and (5.26). The shift of the quasiparticle $\otimes$ phonon pole is determined by the function $\Delta\omega(Jj\lambda i)$; it vanishes at $\mathcal{L}=0$ and is the greater, the stronger the phonon λi is collectivized.

As a rule, the fragmentation of single-quasiparticle states is not appreciably dependent on whether the Pauli exclusion principle has been taken into account or not (figure 5.5). In certain infrequent cases, however, extra states appear in the approximation $[\alpha_{jm}, Q_{\lambda\mu i}^+]=0$. For example, a low-lying state, not observed experimentally, appears among the states with $J^\pi=7/2^-$ in ^{57}Ni. With the Pauli exclusion principle taken into account stringently, the extra states disappears and the experimental data fit better the results of calculations.

An empirical procedure of selecting correctly anti-symmetrized components was used to calculate the fragmentation of single-quasiparticle states with wave functions (5.16) and (4.4) [142]. A phonon 'collectivity' criterion was introduced (conditionally, to some extent) which separates all phonons into collective and non-collective ones. If the maximal two-quasiparticle component contributed to the normalization not more than a certain upper limit (40 to 50%, in most cases), phonons were classified as collective. All other phonons were treated as non-collective. The components of wave functions (5.16) and (4.4), which included only collective phonons, were retained in the wave function without additional analysis. In components with non-collective phonons, their structure was analysed and those not satisfying symmetry requirements were dropped. This procedure made it

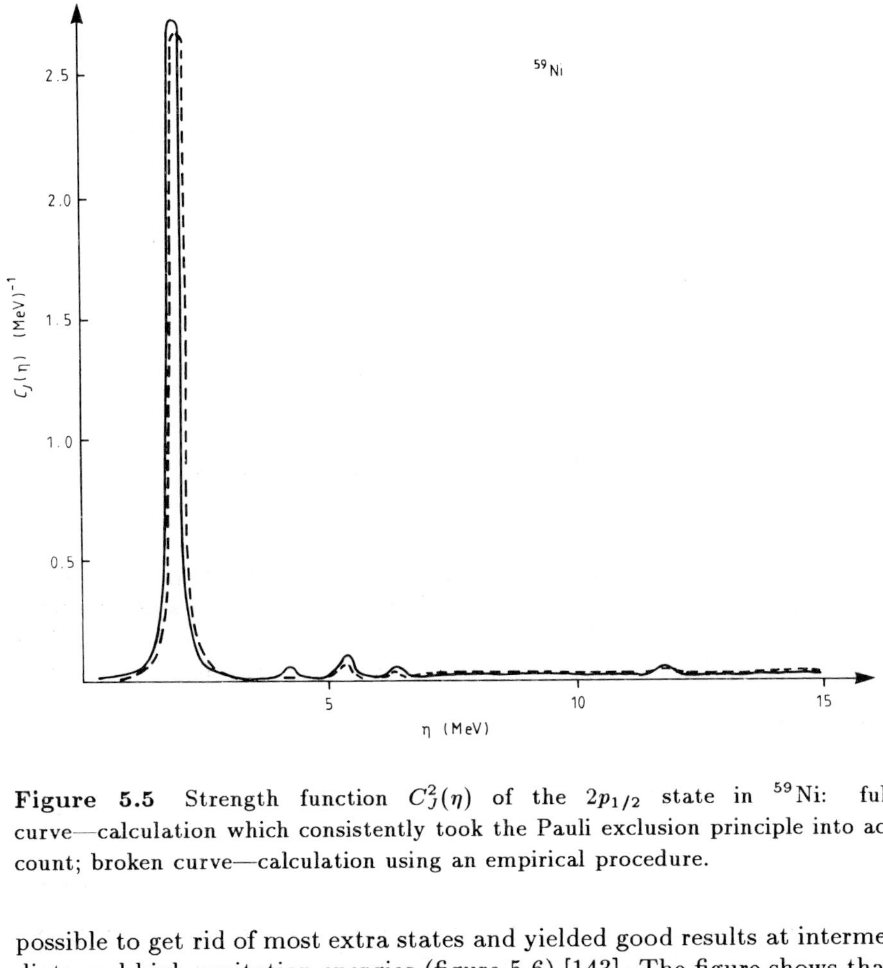

Figure 5.5 Strength function $C_J^2(\eta)$ of the $2p_{1/2}$ state in ^{59}Ni: full curve—calculation which consistently took the Pauli exclusion principle into account; broken curve—calculation using an empirical procedure.

possible to get rid of most extra states and yielded good results at intermediate and high excitation energies (figure 5.6) [143]. The figure shows that the strength functions of the $1h_{11/2}$ state in ^{205}Pb, calculated with the exact and with an empirical procedure of taking the Pauli exclusion principle into account, were practically identical. At the same time, calculations in the approximation of $[\alpha_{jm}, Q^+_{\lambda\mu i}] = 0$ (where the Pauli exclusion principle was ignored) evince very strong fragmentation of the $1h_{11/2}$ subshell; this fragmentation resulted from the presence of a large number of extra states.

5.2.3

When calculating the fragmentation of single-quasiparticle states in spherical nuclei, the wave function is written in the form

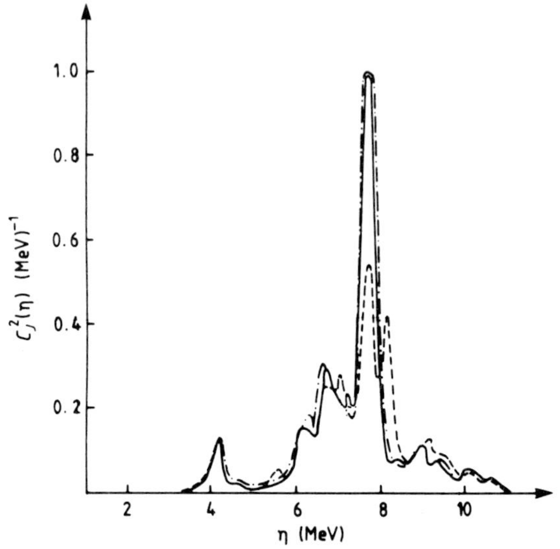

Figure 5.6 Strength function of the $1h_{11/2}$ state in ^{205}Pb: full curve—calculation using an empirical procedure of taking the Pauli exclusion principle into account; chain curve—calculation which consistently took the Pauli exclusion principle into account; broken curve—calculation ignoring the Pauli exclusion principle.

$$\Psi_\nu(JM) = C_{J\nu}\Big\{a_{JM}^+ + \sum_{\lambda ij} D_j^{\lambda i}(J\nu)[a_{jm}^+ Q_{\lambda\mu i}^+]_{JM} + \sum_{\lambda_1 i_1 \lambda_2 i_2 \lambda j} F_{j\lambda}^{\lambda_1 i_1 \lambda_2 i_2}(J\nu)\Big[a_{jm}^+[Q_{\lambda_1\mu_1 i_1}^+ Q_{\lambda_2\mu_2 i_2}]_{\lambda\mu}\Big]_{JM}\Big\}\Psi_0. \tag{5.27}$$

Let us use an empirical procedure of selecting correctly antisymmetrized components of this wave function, and hence, limit the analysis to solving equations (4.41) and (4.43). The strength function describing the fragmentation of a single-quasiparticle state is

$$C_J^2(\eta) = \frac{1}{\pi}\,\mathrm{Im}\,\frac{1}{\mathcal{F}(\eta + i\Delta/2)} \tag{5.28}$$

[see (4.47′)]. In order to describe the integral characteristics of fragmentation of a single-quasiparticle state in the interval $E_1 \leqslant \eta \leqslant E_2$, the energy centroid and the width $\Gamma\!\downarrow$ are introduced:

$$\bar{E} = \int_{E_1}^{E_2} C_j^2(\eta)\eta\,\mathrm{d}\eta \Big/ \int_{E_1}^{E_2} C_j^2(\eta)\,\mathrm{d}\eta \tag{5.29}$$

$$(\Gamma\downarrow)^2 = \int_{E_1}^{E_2} C^2(\eta)(\eta - \bar{E})\,\mathrm{d}\eta \Big/ \int_{E_1}^{E_2} C^2(\eta)\,\mathrm{d}\eta \qquad (5.30)$$

where η_{gs} is the ground state energy of an odd-mass nucleus.

In solving equation (4.43), one has to calculate a very large determinant whose order equals the number of α^+Q^+-type terms in (5.27). The rank of the matrix $\mathcal{H}$ of (4.54″) is also very high, and it is needed in the calculations of the strength function. Consequently, we need to select the most important components of the quasiparticle–phonon basis. Furthermore, it is of interest to find out the effect of the $a^+Q^+Q^+$-type components of wave function (5.27) on the fragmentation of single-quasiparticle states at excitation energies of 5–10 MeV. Additional interest in this problem arises because attempts are still made sometimes to describe the strength distribution for high-lying single-particle modes in the spectra of odd nuclei while taking into account only the interaction with the lowest quadrupole (and sometimes octupole) vibrational states; the effect of all other configurations is taken into account 'effectively', either by an artificial 'spreading' of the solutions obtained [225] or by introducing a complex phenomenological optical potential [227].

Let us consider first how the completeness of the phonon space affects the strength function $C_J^2(\eta)$ for states that are described by wave function (5.16) and secular equation (5.26). The results of calculations for a deep neutron hole subshell $1g_{9/2}$ in ^{119}Sn are shown in figures 5.7 (a–e). Evidently, if we consider only the first quadrupole and octupole phonons, then the predominant part of the subshell strength will be concentrated in two maximums. If the quadrupole and octupole phonons with energy up to 11 MeV are taken into account, calculations demonstrate a slight enhancement of fragmentation, including a splitting of the main peak. The important role of the large phonon basis is demonstrated in figure 5.7 (c).

Let us analyse the effect of $\alpha^+Q^+Q^+$-type components of wave function (5.27) on the fragmentation of a single-quasiparticle state. Note that taking these components into account enhances the fragmentation of the α^+Q^+ components of wave function (5.27) and thereby affects the fragmentation of the single-quasiparticle state. Figure 5.8 shows that the number of states calculated for a broad phonon basis is much greater than the number of states composed of quadrupole and octupole phonons. In calculations with a broad phonon basis at excitation energies of 6–8 MeV, the density of $\alpha^+Q^+Q^+$-type states exceeds that of α^+Q^+-type states by a factor of 3–5; this predominance grows as the excitation energy increases. As a result, the subshell strength shifts towards higher excitation energies. Hence, taking $\alpha^+Q^+Q^+$-type components into account affects the behaviour of the high-energy part of the corresponding strength functions.

A sufficiently large space of one-phonon states has to be used when cal-

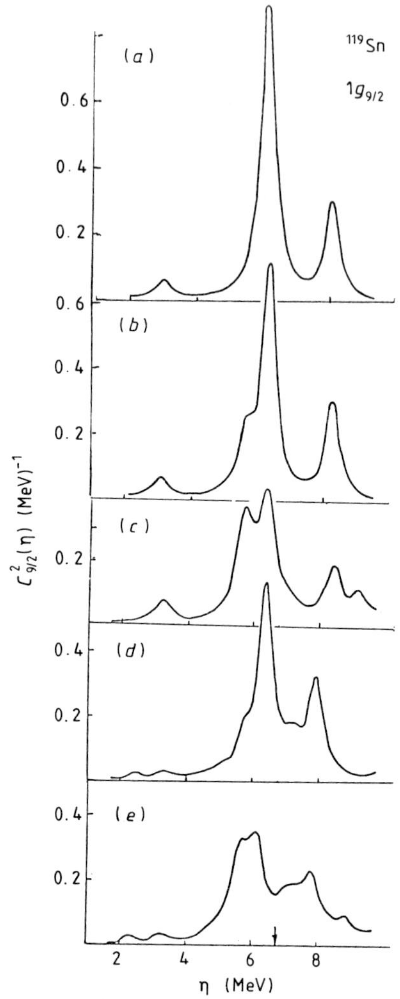

Figure 5.7 Strength function of the neutron hole state $1g_{9/2}$ in ^{119}Sn: (a) calculation for wave function (5.16) (containing α^+Q^+ terms) with only the first quadrupole and octupole phonons; (b) calculation for wave function (5.16) with quadrupole and octupole phonons at energies under 11 MeV; (c) calculation for wave function (5.16) with large phonon basis; (d) calculation for wave function (5.27) (containing $\alpha^+Q^+Q^+$ terms) with quadrupole and octupole phonons at energies under 11 MeV; (e) calculation for wave function (5.27) with large phonon basis.

culating the fragmentation of the single-quasiparticle state with wave function (5.27). The strength of interaction of a quasiparticle (whose quantum numbers are $nlJM$) with a configuration $[\alpha^+_{jm}Q^+_{\lambda\mu i}]_{JM}$ is determined by

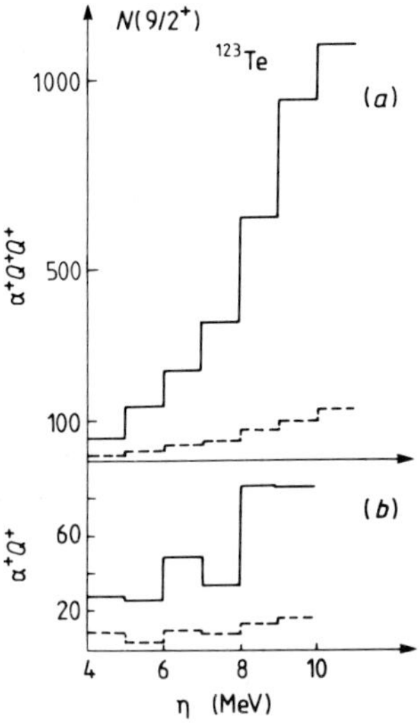

Figure 5.8 Histograms of (a) the number of $[\alpha^+Q^+Q^+]_{9/2^+}$-type states and (b) of $[\alpha^+Q^+]_{9/2^+}$-type states with $J^\pi = 9/2^2$ in the range $\Delta\eta = 1$ MeV, as functions of excitation energy η in ^{123}Te: full lines—calculation for broad phonon basis; broken lines—calculation for quadrupole and octupole phonons only.

the function $\Gamma(Jj\lambda i)$ in the form (4.36). The effect produced on $C^2_{9/2}(\eta)$ by truncating the space of complex configurations in strength $\Gamma(Jj\lambda i)$ was analysed in [145, 147] using as an example the fragmentation of the neutron subshell $1h_{9/2}$ in ^{117}Sn. If we limit the strength to $\Gamma(h_{9/2}j\lambda i) > 0.5\Gamma_{max}$, we find that the number of α^+Q^+-type terms is seven while the number of $\alpha^+Q^+Q^+$-type terms is 34. In this case, the subshell strength is concentrated at two levels. As the lower value for $\Gamma(Jj\lambda i)$ decreases, the number of components of wave function (5.27) grown considerably. For instance, if $\Gamma \geqslant 0.1\Gamma_{max}$, the number of α^+Q^+-type terms is 108 and that of $\alpha^+Q^+Q^+$-type terms is 4552. If $\Gamma \geqslant 0.005\Gamma_{max}$, the number of α^+Q^+-type terms is 238 and that of $\alpha^+Q^+Q^+$-type terms is 18 858. If $\Gamma \geqslant 0.1\Gamma_{max}$, the subshell fragmentation is stabilized and the transition to $\Gamma \geqslant 0.05\Gamma_{max}$ does not bring appreciable changes. Consequently, calculations are carried out with the constraint $\Gamma(Jj\lambda i) \geqslant 0.1\Gamma_{max}$. The importance of taking into account a broad phonon space is demonstrated in figure 5.7 (e), which shows how $C^2_{9/2}$ has changed in comparison with figure 5.7 (d). Note that the correct values of energy centroids $\bar{E}$ and widths $\Gamma\downarrow$ are obtained already for $\Gamma \geqslant 0.5\Gamma_{max}$.

For a description of the subshell strength distribution, it is necessary to carry out calculations with wave function (5.27) and a broad phonon basis.

5.2.4

It was typically assumed that the distribution of strength of a single-particle state far from the Fermi energy obeys the Breit–Wigner law, with the centre coinciding with the energy of the quasiparticle state. Even the pioneer calculations of the fragmentation of single-quasiparticle states demonstrated that their strength distribution has a complex form that differs from the Breit-Wigner one. Typical examples of fragmentation of hole states are given in figure 5.9. We see that a number of peaks are observed in the strength distribution of the states; they are caused by large matrix elements at the corresponding poles of the secular equation. When the fragmentation of the single-quasiparticle state is analysed, one typically discusses how the width $\Gamma\downarrow$ varies as $\bar{E}$ increases. It was assumed in some papers (see, e.g., [167]) that $\Gamma\downarrow\propto \bar{E}^2$. The available experimental and theoretical data on the fragmentation of single-quasiparticle states failed to establish a regular dependence of $\Gamma\downarrow$ on $\bar{E}$. There is no doubt that as $\bar{E}$ increases, the region of localization of the predominant part of the subshell strength increases. The width $\Gamma\downarrow$ defined by (5.30) characterizes the strength distribution if it has the Breit-Wigner shape or a single maximum. If the strength distribution has several peaks, then the widths $\Gamma\downarrow$ are not really informative.

The QPNM calculations of the fragmentation of single-quasiparticle states carry no free parameters, that is, the results are not fitted to experimental data. As a result, we cannot expect a complete (detailed) agreement of the theory with the experimental data. When the theoretical and experimental data are compared, one makes use of the spectroscopic factors, which are given by the formulas

$$C_J^2 S = (2J+1)v_J^2 C_J^2 \qquad C_J^2 S = (2J+1)u_J^2 C_J^2 \tag{5.31}$$

for reactions of the types (d,t) and (d,p), and also of the total spectroscopic factors, which contain integrals $\int_{\Delta E} C_J^2(\eta)\,\mathrm{d}\eta$ instead of C_J^2.

Let us consider the fragmentation of neutron hole subshells. The first to be studied experimentally was the fragmentation of the $1g_{9/2}$ subshell in tin and tellurium isotopes. A qualitatively correct description of fragmentation was obtained in [142]; this is illustrated in figure 5.9. The experimental data [217] for ^{143}Sm were obtained after the calculated results had been published. The data given in [142, 162] predict the existence of a second maximum in isotopes of tin and tellurium, at an energy 2 MeV higher than that of the first maximum. This effect was not found in the (^{3}He,α) and (d,t) reactions [221] (figure 5.10 (b)). The strength of the $1g_{9/2}$ subshell was singled out in the analysis of the ^{112}Sn(^{3}He, $\alpha\gamma$) reaction [229], in the 5–7 MeV excitation energy interval. As we see from figure 5.10 (a), a good agreement was obtained between the theoretical and experimental results. In [229], the experimental data were normalized to the results of [162] at

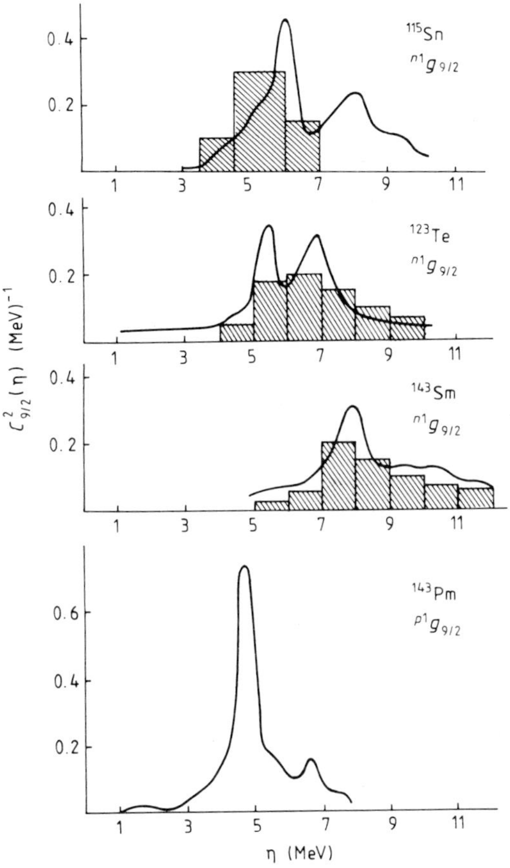

Figure 5.9 Fragmentation of the deep hole subshell $1g_{9/2}$ (neutron subshell in ^{115}Sn, ^{123}Te and ^{143}Sm and proton subshell in ^{143}Pm): curves plot QPNM calculations, shaded histograms are the experimental data.

the distribution maximum at 4.2 MeV. According to these results, 64% of the $1g_{9/2}$ subshell strength in ^{111}Sn is exhausted at excitation energies below 7 MeV. The results of calculations agree also with the experimental results reported in [230]. Note that (^{3}He, $\alpha\gamma$)-type reactions, for which both α particles and γ quanta are detected in the coincidence mode, offer new possibilities for studying both the fragmentation of single-quasiparticle states and their decay.

The calculated and the experimentally measured distributions of the sum of strength functions for the $2p_{1/2}$ and $2p_{3/2}$ subshells in ^{111}Sn are shown in figure 5.11. In this case as well, the experimental data reported in [229] for the ^{112}Sn(^{3}He, $\alpha\gamma$) reaction agree better with the results calculated in

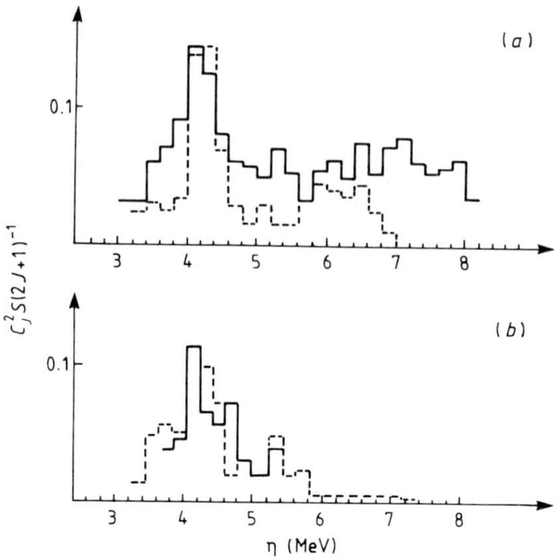

Figure 5.10 Fragmentation of the neutron hole $1g_{9/2}$ subshell in ^{111}Sn: (*a*) full line—experimental data for the (^{3}He, $\alpha\gamma$) reaction (1985); broken line—QPNM calculations (1980); (*b*) full line—experimental data for the (^{3}He, α) reaction (1980), broken line—for the (*d*, *t*) reaction (1982).

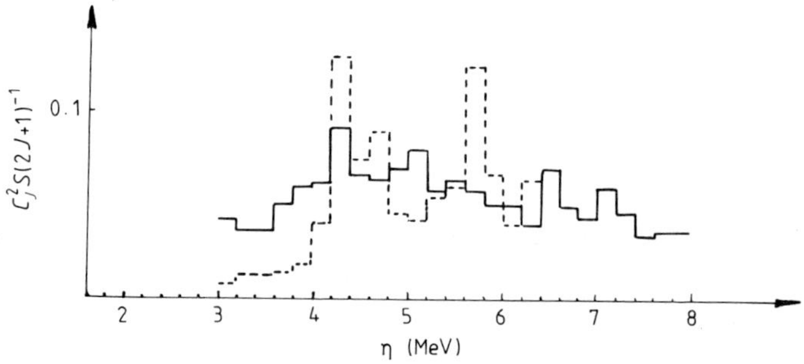

Figure 5.11 Fragmentation of the $2p$ neutron hole states (the sum of $2p_{1/2}$ and $2p_{3/2}$ subshells) in ^{111}Sn. The spectroscopic factors for the (^{3}He, $\alpha\gamma$) reaction (full line) are compared with the results of QPNM calculations (broken line).

[162] than with the measurement data for the (*dt*) reaction [221]. The figure shows that the strength is distributed almost uniformly in a wide energy interval.

A satisfactory agreement of the calculated results of [160] with the experimental data of [217] was found for the fragmentation of the $1h_{11/2}$ in ^{207}Pb (figure 5.12).

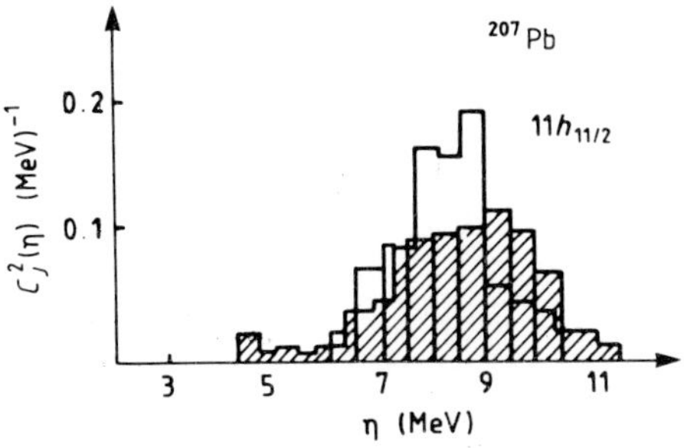

Figure 5.12 Strength distribution of the neutron hole $1h_{11/2}$ subshell in ^{207}Pb: experimental (shaded histogram) and QPNM-calculated (full line) distributions.

In the case of the fragmentation of deep proton state $1g_{9/2}$ in ^{143}Pm (see figure 5.9), the calculated results [142] are in qualitative agreement with experimental data obtained for the ^{144}Sm(d,^{3}He) reaction [231]. The fragmentation of proton hole states in ^{203,205,207}Tl was calculated in [160]; is does not differ too much from the fragmentation of neutron states. A certain difference between fragmentations of proton and neutron subshells in nuclei with a single closed shell is discussed in [145].

Let us consider the fragmentation of high-lying particle states, illustrated in figure 5.13. In ^{121}Sn, the maximum of strength distribution for $2f_{5/2}$ is at the neutron binding energy B_n, while in ^{209}Pb it is the tail of the subshell $2h_{11/2}$ we find at $\eta = B_n$. The results of calculations in [224] lead to a qualitatively correct description of the experimental data of [232]. When the fragmentation of neutron particle subshells with $\bar{E} > B_n$ is calculated, it is necessary to take into account the continuous spectrum, especially if l is small.

The theoretical predictions of the strength distribution in proton particle subshells of ^{145}Eu and ^{209}Bi had been made before the appropriate experiments were carried out. A comparison of the results of calculations with the experimental data, reported in [219], points to a qualitative agreement (figure 5.14).

The experimental information on the fragmentation of single-quasiparticle states in spherical nuclei is still rather scarce. Calculations are also limited to several states of a small number of nuclei.

So far, the following remarks can be made on the fragmentation of single-quasiparticle states in spherical nuclei.

(*i*) From 70 to 90% of the single-particle strength is concentrated at each of the ground or low-lying state of an odd nucleus. As the energy of a single-quasiparticle state grows, the fragmentation is enhanced and the single-particle strength is distributed over several levels. At high excitation energies, the subshell strength is distributed in a wide energy interval.

(*ii*) The shape of the strength distribution of a single-quasiparticle state

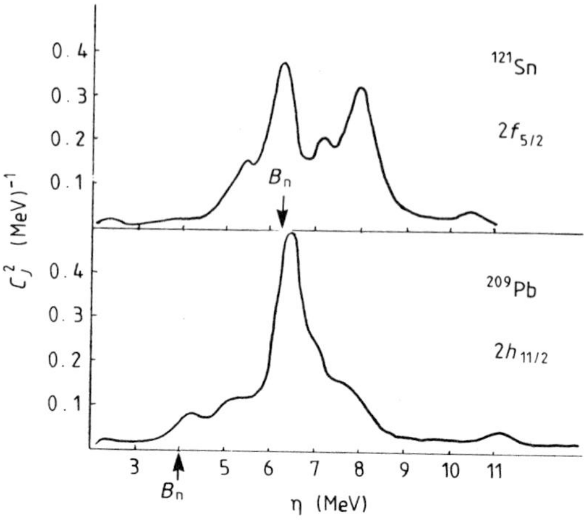

Figure 5.13 QPNM-calculated strength distribution in neutron particle subshells $2f_{5/2}$ in ^{121}Sn and $2h_{11/2}$ in ^{209}Pb.

differs from the Breit–Wigner curve; several peaks are found in the distribution.

(*iii*) The fragmentation is a function of the quantum numbers of the single-particle state and of the characteristics of the collective vibrational states of the corresponding even–even nucleus. Other conditions being equal, the fragmentation of low-l subshells is higher than that of high-l ones. The fragmentation of single-quasiparticle states is diminished if the nucleus has one, and especially two, closed shells.

The experimental data on fragmentation of single-quasiparticle states in spherical nuclei were classified in [233], where the results were also compared with QPNM calculations.

5.3 FRAGMENTATION OF ONE-PHONON AND TWO-QUASIPARTICLE STATES. GAMOW–TELLER TRANSITIONS

5.3.1

One of the processes in which the fragmentation of one-phonon states manifests itself very clearly is the excitation of giant resonances. If we only consider states below the nucleon binding energy, the fragmentation of one-phonon states must be taken into account when analysing the tail of

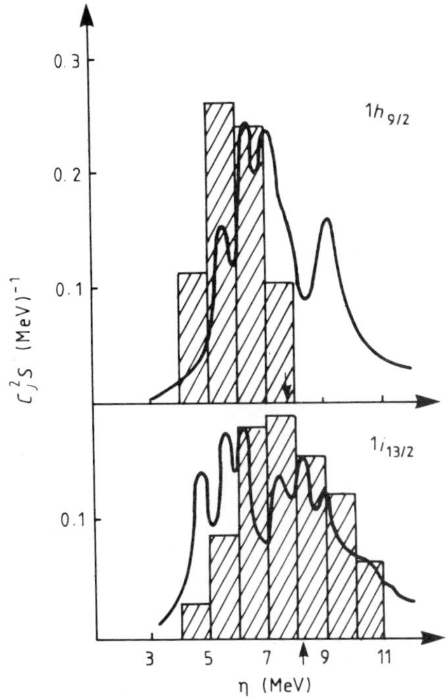

Figure 5.14 Comparison of QPNM-calculated fragmentation in protron particle subshells $1h_{9/2}$ and $1i_{13/2}$ in ^{145}Eu (full curves) with experimental data (shaded histograms). Arrows indicate the single-quasiparticle energy.

the giant dipole resonance, the low-energy part of the isoscalar octupole resonance, and other manifestations of collective states.

We choose the following form of the wave function of an excitation state of an even–even spherical nucleus:

$$\Psi_\nu(JM) = \Big\{\sum_i R_i(J\nu)Q^+_{Ji} + \sum_{\lambda_1 i_1 \lambda_2 i_2} P^{\lambda_1 i_1}_{\lambda_1 i_2}(J\nu)[Q^+_{\lambda_1\mu_1 i_1}Q^+_{\lambda_2\mu_2 i_2}]_{JM}\Big\}\Psi_0. \quad (5.32)$$

The reduced probability of the $E\lambda$ transition from the ground state to a state described by wave function (5.32) is given by formula (5.2). If collective states are excited, the second term in (5.2) that contains P is small and usually is neglected. In this case, the strength function of the $E\lambda$ transition is given by (4.76), where

$$\Phi_{Ji} = \sum_\tau e^{(\lambda)}_{\text{eff}}(\tau)\sum_{jj'}{}^{\tau} p^\lambda(jj')g^{\lambda i}_{jj'}u^{(+)}_{jj'}. \quad (5.33)$$

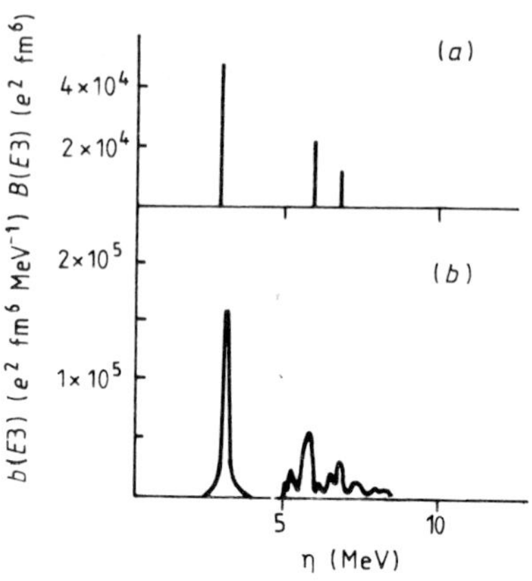

Figure 5.15 The characteristics of the 3_1^- state and of the low-energy octupole resonance in ^{90}Zr: (*a*) RPA calculation of $B(E3)$; (*b*) calculation of the strength function $b(E3, \eta)$, taking into account the quasiparticle–phonon interaction.

If $p^\lambda(jj') = f^\lambda(jj')$, then

$$\Phi_{Ji} = \sum_\tau e_{\text{eff}}^{(\lambda)}(\tau)\sqrt{2}\, X^{\lambda i}(\tau)(\mathcal{Y}_\tau^{\lambda i})^{-1}. \tag{5.33'}$$

Using this function, one calculates the distribution of the electromagnetic transition strength for various excitation energies.

One of the lowest-lying resonances is the isoscalar low-energy octupole resonance (LEOR). It is caused by transitions inside the valence shell between unequal-parity subshells and betwen the valence shell and the neighbouring shells. Such a resonance was observed in many spherical nuclei from ^{58}Ni to ^{208}Pb at energies of 5–7 MeV. Figure 5.15 plots the distribution of the $E3$ strength in ^{90}Zr, RPA-calculated using formula (2.124′), and the distribution calculated in [138] using (4.76) and taking into account the quasiparticle–phonon interaction with Φ_{Ji} in the form (5.33′). The figure refers to 3^- states with large values of $B(E3; 0_{\text{gs}}^+ \to 3^-)$ in the energy range from 0 to 10 MeV. Taking into account the quasiparticle–phonon interaction does not appreciably affect the first collective 3_1^- state; however, it results in the formation of a low-lying octupole resonance whose strength is distributed in the range $\Delta\eta \approx 2\,\text{MeV}$.

While the quasiparticle–phonon interaction does not greatly affect the low-lying octupole resonance in ^{90}Zr, it shifts this resonance and changes the strength distribution in, for example, ^{58}Ni. According to [160], taking into account the quasiparticle–phonon interaction in ^{208}Pb produced several 3^- states in the energy interval 4.7–5.7 MeV, which is in agreement with the experimental data [234].

5.3.2

It had been assumed for a long time [235] that the $E1$ strength distribution at excitation energies below the giant isovector electric dipole resonance is determined by its Lorentzian extrapolation. This is largely a true statement. Experimental studies of the resonance scattering of phonons by some spherical nuclei have demonstrated [236, 237] that the energy dependence of the cross section differs from that implied by the Lorentzian extrapolation of the giant dipole resonance. Substructures are observed in cross sections. An analysis of such substructures and of the effect produced by the giant dipole resonance on the radiative strength functions was performed in terms of QPNM [160, 173, 185] and in terms of other models [238].

The total photoabsorption cross section, which coincides with the cross section of dipole photoabsorption, is

$$\sigma_{\gamma t}(E) = 0.4025 \frac{E}{\Delta E} \int\limits_{E-\Delta E/2}^{E+\Delta E/2} b(E1, \eta)\, \mathrm{d}\eta. \tag{5.34}$$

where $\sigma_{\gamma t}$ is given in fm^2, E is given in MeV, and $b(E1, \eta)$ is determined by formula (4.76) with Φ_{Ji} given by (5.33′); the unit for b is e^2fm^2 MeV^{-1}. In the case of the Lorentzian extrapolation of the giant dipole resonance to the low-energy region, the formula for the dipole photoabsorption cross section is

$$\sigma_{\gamma t}(E) = \sigma_0 \frac{\Gamma_0^2 E^2}{(E^2 - E_0^2)^2 - E^2\Gamma^2} \tag{5.35}$$

where E_0 and Γ_0 are the energy and width of the giant dipole resonance.

Let us consider the behaviour of radiative strength functions $b(E1, \eta)$ in the 5–8 MeV energy range and analyse the effect of the giant dipole resonance on them. The behaviour of $b(E1, \eta)$ is determined not only by the giant dipole resonance but also by weak 1^- one-phonon states in the 5–8 MeV energy range. This feature is demonstrated in figure 5.16 for ^{206}Pb. The calculations were performed using wave function (5.32), whose one-phonon part first covered only the one-phonon states forming the giant resonance, but later took into account all one-phonon states, including those in the 5–8 MeV energy range. The figure shows that the quasiparticle–phonon interaction shifts part of the giant dipole resonance strength to the low-energy region. Taking the 1^- one-phonon states in the

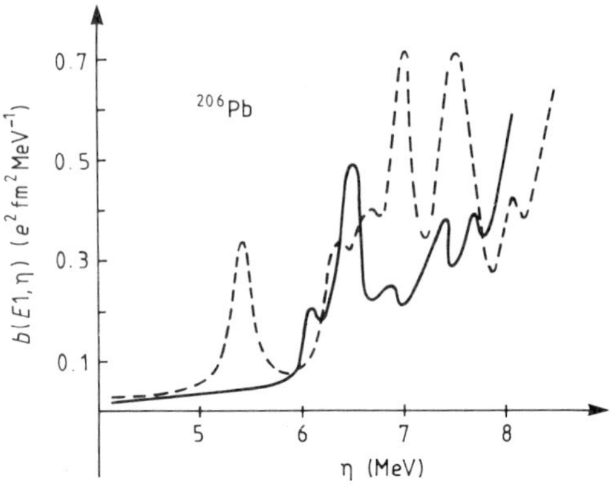

Figure 5.16 $E1$ strength function in ^{206}Pb calculated using (4.76) for Φ_{Ji} given by (5.33′): full curve—only the one-phonon states that form the giant dipole resonance are taken into account; broken curve—all 1^- one-phonon states are considered, including those in the 5–8 MeV energy range.

5–8 MeV energy range into account substantially modifies the strength function $b(E1,\eta)$, owing to the appearing substructures and the increase in the $E1$ strength in this range.

Irregularities in the energy dependence of $b(E1,\eta)$ manifest themselves in the photoexcitation cross sections. At energies below the nucleon emission threshold, the substructures in the dipole photoexcitation cross sections are caused by weak 1^- one-phonon states and by the effect of the giant dipole resonance. Let us consider the cross section of the dipole photoexcitation in ^{208}Pb (figure 5.17). The experimental data reveal substructures at the energies 5.5, 7.2 and 7.5 MeV (and perheps at other energies as well). This behaviour of the cross section does not agree with the Lorentzian extrapolation of the giant dipole resonance. According to calculations, the RPA manifests sub structures at the energies 6.0 and 7.5 MeV. Calculations with wave function (5.32), which contains all 1^- one-phonon states, give a qualitatively correct description of the experimental data. Although the effect of the giant dipole resonance on the $E1$ strength functions at 5–8 MeV is not large, taking it into account does improve the agreement of calculations with experimental data. A large effect of the giant dipole resonance in nuclei with open shells smoothes the substructures in the cross sections $\sigma_{\gamma t}(E)$ that are formed by weak 1^- states; hence, the $\sigma_{\gamma t}(E)$ cross section approaches the value given by (5.35).

The excitation states of spherical nuclei in the energy range from 3 MeV to the neutron binding energy B_n have not been carefully analysed. In view of this, the recently obtained results on the fragmentation of one-phonon and two-quasiparticle states seem to be quite valuable.

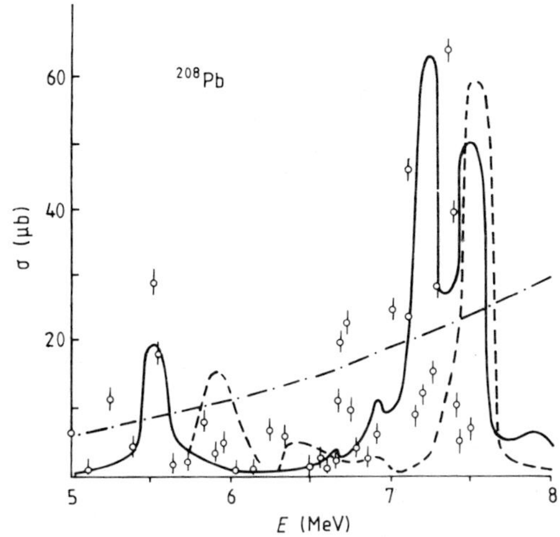

Figure 5.17 Dipole photoexcitation cross section in ^{208}Pb: full circles—experimental data [236]; full curve—calculation taking into account the quasiparticle–phonon interaction; broken curve—RPA calculations; dash-dot curve—Lorentzian extrapolation of the giant dipole resonance.

5.3.3

We will now consider the fragmentation of two-quasiparticle states in spherical nuclei. The experimental information on the fragmentation of two-quasiparticle states is obtained from the spectroscopic factors of one-nucleon transfer reactions. However, this is the information only about the two-quasiparticle states with one valence quasiparticle, that is, a quasiparticle at a single-particle level which corresponds to the ground state of the odd target nucleus. The two-nucleon transfer reactions, such as (p,t) and (t,p), give more complete (although less accurate) data on the fragmentation of two-quasiparticle states in spherical nuclei [213, 239–241].

I will list below the formulas describing the QPNM fragmentation of two-quasiparticle states. The two-quasiparticle component $\{j_1 j_2\}$ of the wave function $Q^+_{\lambda\mu i}\Psi_0$, described in the RPA, is determined by the phonon amplitude $(1/2)|\psi^{\lambda i}_{j_1 j_2}|^2$, where $\psi^{\lambda i}_{j_1 j_2}$ is given by (2.107), (2.114) or (2.119). The one-phonon states undergo fragmentation if the quasiparticle–phonon interactions are taken into account. The spin-J two-quasiparticle component $\{j_1 j_2\}$ of a state ν, represented by wave function (5.32), is

$$\Phi_{j_1 j_2}(J;\eta_\nu) = \frac{1}{2}\Big|\sum_i R_i(J\nu)\psi^{Ji}_{j_1 j_2}\Big|^2. \tag{5.36}$$

Let us look at the expressions for the spectroscopic factors of one-nucleon transfer in odd-mass target nuclei. The wave function of the target nucleon is

$$\Psi_{\nu_0}(j_0 m_0) = C_{j_0\nu_0} a^+_{j_0 m_0} \Psi_0.$$

The wave function of the final state with spin J_f is given by formula (5.32). Then the spectroscopic factors of nucleon transfer to the subshell j are [242]:

for (d,p)-type reactions,

$$\widetilde{S}_{jj_0}(J_f;\eta_\nu) = C^2_{j_0\nu_0} u_j^2 \Phi_{jj_0}(J_f;\eta_\nu) \tag{5.37}$$

for (d,t)-type reactions,

$$S_{jj_0}(J_f;\eta_\nu) = C^2_{j_0\nu_0} v_j^2 \Phi_{jj_0}(J_f;\eta_\nu). \tag{5.38}$$

If we sum up over all the spins of the final states J_f that form single-quasiparticle states j and j_0, we obtain

$$S_{jj_0}(\eta_\nu) = \sum_f S_{jj_0}(J_f;\eta_\nu). \tag{5.39}$$

The following expressions are often used:

$$S'_{jj_0}(J_f;\eta_\nu) = \frac{2J_f+1}{2j_0+1} S_{jj_0}(J_f;\eta_\nu) \tag{5.40}$$

$$S'_{jj_0}(\eta_\nu) = \sum_{J_f} \frac{2J_f+1}{2j_0+1} S_{jj_0}(J_f;\eta_\nu). \tag{5.41}$$

When the fragmentation of two-quasiparticle states at intermediate and high excitation energies is analysed, $\Phi_{j_1 j_2}(J;\eta_\nu)$ is replaced with the strength function

$$\Phi_{j_1 j_2}(J;\eta) = \frac{1}{\pi} \operatorname{Im} \left\{ \frac{\sum_{ii'} A_{ii'}(\eta + i\Delta/2) \psi^{Ji}_{j_1 j_2} \psi^{Ji'}_{j_1 j_2}}{\mathcal{F}(\eta + i\Delta/2)} \right\} \tag{5.42}$$

which is obtained from (4.76) for the case of $\Phi_{Ji} = \psi^{Ji}_{j_1 j_2}$. The strength functions for the spectroscopic factors are obtained from expressions (5.37)–(5.41) by substituting the strength function $\Phi_{j_1 j_2}(J;\eta)$ for $\Phi_{j_1 j_2}(J;\eta_\nu)$.

For the integral characteristics of the strength distribution of two-quasiparticle states in the energy interval ΔE, we will use the energy centroids

$$\bar{E}_{j_1 j_2} = \int_{\Delta E} S'_{j_1 j_2}(\eta)\eta \, \mathrm{d}\eta \Big/ \int_{\Delta E} S'(\eta) \, \mathrm{d}\eta \tag{5.43}$$

the total spectroscopic factors

$$\int_{\Delta E} S'_{j_1 j_2}(\eta)\, \mathrm{d}\eta \tag{5.44}$$

and the widths

$$\Gamma_{j_1 j_2} = 2.35 \left\{ \int_{\Delta E} (\bar{E}_{j_1 j_2} - \eta)^2 S'_{j_1 j_2}(\eta) \mathrm{d}\eta \Big/ \int_{\Delta E} S'(\eta) \mathrm{d}\eta \right\}^{1/2}. \tag{5.45}$$

Let us consider the fragmentation of particle–hole-type two-quasiparticle states. We will also take into account valence-particle–hole- and particle–valence particle two-quasiparticle states. To describe the fragmentation of two-quasiparticle particle–particle-type states, it is necessary to consider also the particle–particle residual interactions. The fragmentation of particle–hole-type two-quasiparticle states is caused, first, by the interaction between quasiparticles which generates the one-phonon states, and second, by the quasiparticle–phonon interaction.

Let us investigate the effect of each of these two factors on the fragmentation of two-quasiparticle states. The two-quasiparticle states $\{2d_{3/2}, 1g_{9/2}\}$ with $J^\pi = 3^+, 4^+, 5^+$ and 6^+ in ^{120}Sn (figure 5.18) manifest appreciable fragmentation already at the phonon generation stage. In figure 5.18, the fragmentation of two-quasiparticle states is shown as a strength function that takes into account the quasiparticle–phonon interaction. The total strengths of states in figures 5.18 (a) and (b) are identical. The figure shows that the quasiparticle–phonon interaction results in further, stronger fragmentation of two-quasiparticle states. The strength of the $\{2d_{3/2}, 1g_{9/2}\}$ states is mainly concentrated in the 7–9 MeV interval.

The analysis carried out in [242, 243] for a number of spherical nuclei has shown that the quasiparticle–phonon interaction substantially affects the strength distribution of two-quasiparticle states of spherical nuclei at excitation energies above 3 MeV. Very scanty experimental data are available so far on the fragmentation of two-quasiparticle states in one-nucleon transfer reactions [244–248].

We will consider the fragmentation of valence particle–hole-type states excited in one-nucleon transfer reactions, using ^{206}Pb as an example. The experimental data for the spectroscopic factors of the $^{207}\mathrm{Pb}(^3\mathrm{He}, \alpha)^{206}\mathrm{Pb}$ reactions were reported in [246] for a wide energy range. The analysis of cross sections in terms of the distorted waves method made possible the identification of the quantum numbers of the neutron hole subshells. Actually, this identification is not unambiguous. As a result, only a qualitative picture of the strength distribution in ^{206}Pb is known for the two-quasiparticle states at excitation energies above 4.3 MeV.

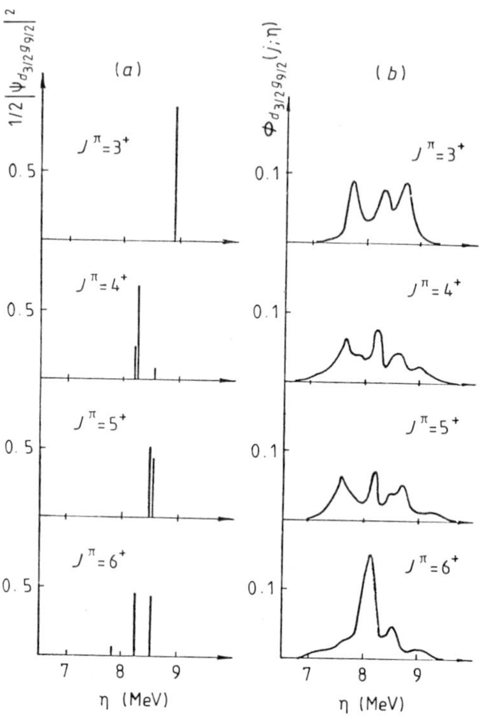

Figure 5.18 Fragmentation of the two-quasiparticle state $\{2d_{3/2}, 1g_{9/2}\}$ with $J^\pi = 3^+, 4^+, 5^+$ and 6^+ in ^{120}Sn: (*a*) RPA calculations; (*b*) strength functions calculated taking into account the quasiparticle–phonon interaction.

The low-lying states of ^{206}Pb were studied in great detail in various reactions. The experimental values of energies and spectroscopic factors are given in table 5.5. It also gives the results for wave function (5.32), which are obviously quite close to those of RPA calculations. Table 5.5 shows a considerable spread in the available experimental data. The calculated energies of low-lying states are in good agreement with the experiment. For states with large values of $S'_{jj_0}(J_f; \eta_v)$, the results calculated in [242] are close to those of [249]. An analysis of the reaction ^{207}Pb$(p,d)^{206}$Pb shows that the $\{2f_{7/2}, 3p_{1/2}\}$, $\{1i_{13/2}, 3p_{1/2}\}$ and $\{1h_{9/2}, 3p_{1/2}\}$ state strengths at energies below 4.2 MeV exhaust the total strength almost completely (to within several per cent). A smaller effect was observed in the reaction $(^3\text{He}, \alpha)$. The results of calculations in [242] fit best the data for the (d,t) reaction.

Table 5.5 Energies and spectroscopic factors $S'_{jj_0}(J_f;\eta_\nu)$ of low-lying states of ^{206}Pb.

$\{nlj, n_0l_0j_0\}$	J_f^π	η_ν (MeV) (exper.)	η_ν (MeV) (calc.)	$S'_{jj_0}(J_f;\eta_\nu)$ (exper.) [246] (^{3}He, α)	(exper.) [247] (d,t)	(exper.) [248] (p,d)	(calc.) [242]
$\{2f_{7/2}, 3p_{1/2}\}$	4^+	1.684	1.9	0.22	—	0.17	0.16
	4^+	1.998	2.2	0.2	—	0.14	0.23
	4^+	2.928	2.9	3.02	3.45	3.97	2.9
	3^+	3.122	3.0	2.60	2.69	3.37	2.6
	4^+	3.519	3.9	0.23	—	0.21	0.16
$\{1i_{13/2}, 3p_{1/2}\}$	7^-	2.200	1.8	4.25	5.5	7.05	6.4
	6^-	2.384	2.1	3.60	5.0	6.47	5.4
	7^-	2.865	3.0	0.2	—	0.32	0.24
$\{1h_{9/2}, 3p_{1/2}\}$	4^+	4.008	3.9	1.85	—	4.3	3.7
	5^+	4.116	4.0	2.55	—	5.00	4.8

5.3.4

Reactions of (p,t) and (t,p)-type two-nucleon transfer constitute an important tool for studying the fragmentation of two-quasiparticle states. The mechanism of these reactions is more complex than that of one-nucleon transfer reactions; consequently, we can only extract qualitative information on the fragmentation of two-quasiparticle states. While a (p,d)-type reaction provides information on the fragmentation of a valence particle–hole state, a (p,t)-type reaction gives additional information on the fragmentation of two-hole states.

A comparison of (p,d) and (p,t) reactions involving cadmium and tin isotopes was carried out in [213, 239]. An analysis of the (p,d) reaction of ^{111,113}Cd and ^{117,119}Sn demonstrates well-pronounced resonance-like structures in cross sections at $6.7-9.0$ MeV excitation energies. The energy and peak widths depend similarly on the mass number A in reaction cross sections of one- and two-nucleon transfers. The peak widths in (p,d) reactions are substantially smaller than those observed in (p,t) reactions. The fragmentation of two-quasiparticle states in tin isotopes was analysed in the QPNM framework in [242] (figure 5.19). Valence particle–hole-type two-quasiparticle states must be excited in (p,d)-type one-nucleon transfer reactions on odd target nuclei. Among such configurations, we find: $\{1g_{7/2}, 1g_{9/2}^{-1}\}$ in ^{110}Sn, $\{3s_{1/2}, 1g_{9/2}^{-1}\}$ in ^{112,114,116,118}Sn, $\{2d_{3/2}, 1g_{9/2}^{-1}\}$ in ^{120}Sn, and $\{1h_{11/2}, 1g_{9/2}^{-1}\}$ in ^{122,124}Sn. Two-hole states are additionally excited in the (p,t) reaction (e.g., $\{1g_{9/2}^{-1}, 1g_{9/2}^{-1}\}$); also excited are states

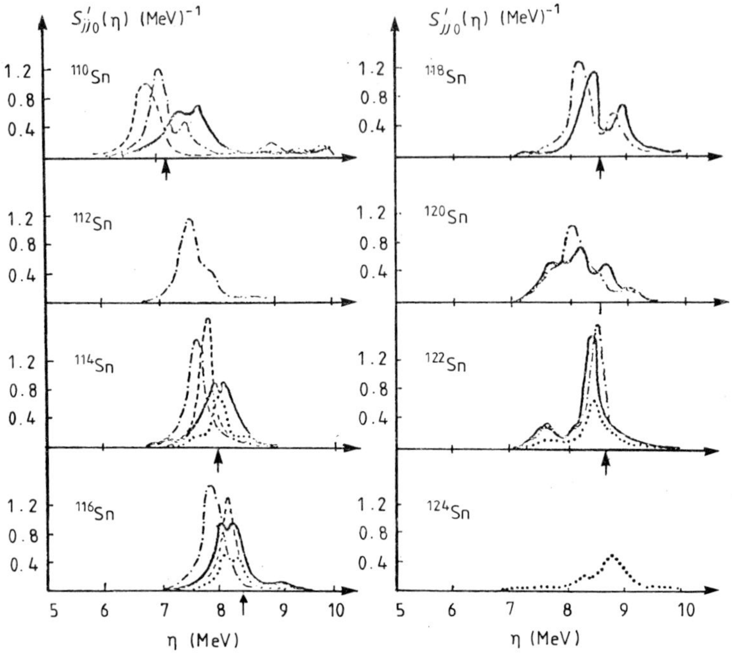

Figure 5.19 Strength functions $S_{jj_0}(\eta)$ for even–even tin isotopes: solid curves—for the $\{2d_{3/2}, 1g_{9/2}\}$ configurations; dashed curves—for the $\{1g_{7/2}, 1g_{9/2}\}$ configurations; dash-dot curves—for the $\{3s_{1/2}, 1g_{9/2}\}$ configurations; dotted curves—for the $\{1h_{11/2}, 1g_{9/2}\}$ configurations; arrows mark the peak energies observed in the (p, t) reaction.

which include a particle at a level close to the Fermi surface and a deep hole (e.g., $\{2d^{-1}_{5/2}, 1g^{-1}_{9/2}\}$ in ^{114}Sn).

Figure 5.20 shows, for a number of tin isotopes, the energy centroids $\bar{E}_{jj_0}$ and the widths Γ_{jj_0} for the states excited in (p, d) and (p, t) reactions which produce the same final even–even nuclei. The calculations were carried out for the (p, d) reaction and the configurations that include a valence particle and a hole $1g_{9/2}$ in the energy range $\Delta = 2\,\mathrm{MeV}$. The figure shows that the calculated energy centroids $\bar{E}_{jj_0}$ fit quite well the experimental data for the (p, d) reaction. The calculated values of $\bar{E}_{jj_0}$ lie slightly below the energies of the peaks excited in the (p, t) reaction. The energy centroid E_{jj_0} grows with increasing A since the hole state $1g^{-1}_{9/2}$ shifts downward with respect to the Fermi surface. The calculations make it possible to explain the difference between the energy centroids of the hole states in

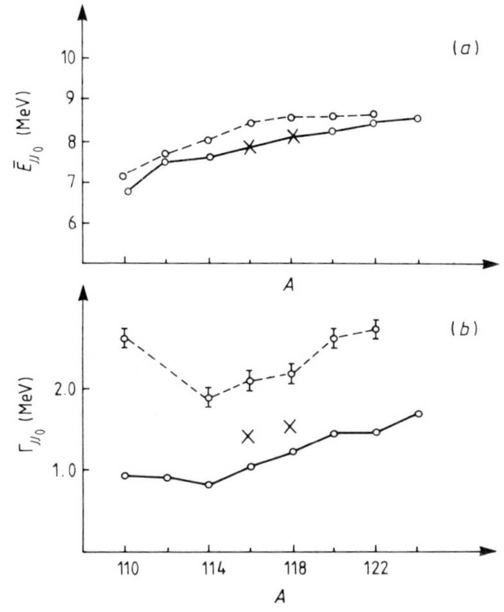

Figure 5.20 (a) Energy centroids $\bar{E}_{jj_0}$ and (b) widths Γ_{jj_0} as functions of A for even–even isotopes of Sn: A refers to the final nucleus; dots connected by a dashed line—experimental data for the (p,t) reaction [239]; crosses—experimental data for the (p,d) reaction [239]; dots connected by the solid line plot QPNM calculations for the (p,d) reaction.

odd and even–even isotopes of tin.

The calculated values of Γ_{jj_0} for (p,d) reactions are somewhat lower than the measured data. The dependence of Γ_{jj_0} on A is explained, first, by a change in the position of the $1g_{9/2}$ subshell with respect to the Fermi surface, and second, by the A-dependence of the spin j_0 which corresponds to the ground state of the odd target nucleus. If the spin j_0 is high, the two-quasiparticle configuration $\{jj_0\}$ is excited, with multiple spin values; this results, in its turn, in an increased width Γ_{jj_0}.

A resonance-like structure excited in a two-nucleon transfer reaction may have a sub structure. Indeed, a structure ($\Gamma_{jj_0} = 2.2\,\text{MeV}$) in the excitation energy range of 8 MeV was found in [240] in the reaction $^{116}\text{Sn}(\alpha, {}^{6}\text{He})^{114}\text{Sn}$, and two substructures at 7.45 and 8.3 MeV. An analysis of the angular distribution shows that mostly states with $J^{\pi} = 6^{+}$ are excited, and that one needs to assume, in order to obtain a good description of the angular distribution of a group of states at 7.45 MeV, that they also contain states with $J^{\pi} = 8^{+}$. QPNM calculations [243] shown in figure 5.21 give $\bar{E}_{jj_0} = 7.6\,\text{MeV}$ for the $\{1g_{7/2}^{-1}, 1g_{9/2}^{-1}\}$ configurations and $\bar{E}_{jj_0} = 8.2\,\text{MeV}$ for the $\{2d_{5/2}^{-1}, 1g_{9/2}^{-1}\}$ configurations. The states in-

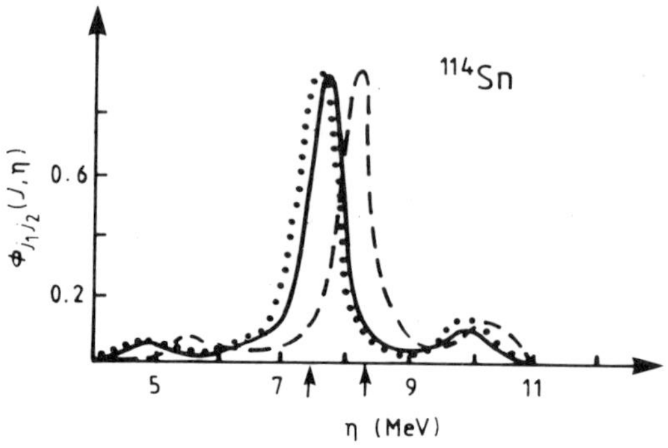

Figure 5.21 The strength functions for two-quasiparticle configurations in ^{114}Sn: solid curve—for the $\{1g^{-1}_{7/2}, 1g^{-1}_{9/2}\}_{6+}$ configuration; dotted curve—for the $\{1g^{-1}_{7/2}, 1g^{-1}_{9/2}\}_{8+}$ configuration; dashed curve—for the $\{2d^{-1}_{5/2}, 1g^{-1}_{9/2}\}_{6+}$ configuration; arrows mark the positions of the experimentally measured structures.

volving the transfer of two neutrons with $l = 6$ in the $(\alpha, {}^6\mathrm{He})$ reaction are excited to the highest extent. The lower group of states corresponds to the configuration $\{1g^{-1}_{7/2}, 1g^{-1}_{9/2}\}$, which may have $J^\pi_f = 6^+$, $J^\pi_f = 8^+$. The upper group of states correspond to the configurations $\{2d^{-1}_{5/2}, 1g^{-1}_{9/2}\}$, $\{1g^{-1}_{9/2}, 2d_{3/2}\}$ with $J^\pi_f = 6^+$. As a result, the observation of two groups of states in ^{114}Sn find a natural explanation in the framework of the QPNM.

The key role in the analysis of the structure of nuclear states in a wide range of excitation energy is played by identifying the regularities in the fragmentation of quasiparticle and phonon states. Important progress has been achieved recently in the studies of the fragmentation of single- and two-quasiparticle and also one-phonon configurations in spherical nuclei. The next stage is connected with the calculation of the fragmentation of the quasiparticle $\otimes$ phonon and quasiparticle $\otimes$ two phonons configurations. This approach makes it possible to obtain descriptions of γ-transitions and the channels in which particles are emitted in decays of deep holes and high-lying particle states, the giant resonances which undergo decay, and partial radiational strength functions for the transitions from neutron resonances.

5.3.5

One of the main objectives of nuclear physics is to find out how the properties of nucleons in nuclei change in comparison with those of free nucleons, and how non-nucleon degrees of freedom manifest themselves in reactions. The most interesting are those characteristics of the ground and excited states of nuclei and those features of nuclear reactions that cannot be described in the framework of the existing theories. One of the manifestations of the changes in the properties of nucleons in nuclei is the renormalization of the constant G_A of the axial-vector weak interaction. It can be stated that the Gamow–Teller GT interactions, which are responsible for the formation of the charge-exchange giant resonance and for fast β decays, play a special role in reaching the purpose formulated above. The double $\beta\beta$ decay is a process of considerable interest. An observation of the double neutrinoless $\beta\beta$ decay may lead to a conclusion on whether the neutrino is a Dirac or a Majorana particle.

As we know, an analysis of (pn) reactions yielded a conclusion on the quenching of approximately (40–50)% of the total GT strength determined by the model-independent sum rule,

$$S_- - S_+ = 3(N - Z) \tag{5.46}$$

(assuming $S_+ = 0$). Fragmentation owing to the coupling to $2p - 2h$ configurations and to mixing with Δ-isobar–neutron hole configurations was considered among other reasons for the quenching of a part of strength at the resonance peak. However, the quenching of only a part of the GT strength can be explained in this way. In our notation, $S_- = S(pn)$, $S_+ = S(np)$ are defined by (2.190) and (2.191). The need to measure the total strength S_+ of the GT branch of the resonance in the (np) reaction was recognized for quite some time already. However, the data on S_+ have appeared only recently, from a study of the reaction ${}^{54}Fe(np){}^{54}Mn$ in [250] and ${}^{90}Zr(np){}^{90}Y$ in [251].

Owing to a high decay energy, the study of β^+ decays of neutron-deficient short-lived isotopes provides experimental information on the distribution of the GT strength belonging to the S_+ branch. A conclusion on large renormalization of the constant G_A in nuclei was drawn from a comparison of the experimental data with the calculations in [252, 253].

We know that the calculations of the reduced probabilities of the allowed β decays in the independent quasiparticles model and in the RPA with the (ph) interaction give values substantially greater than the experimental ones. This difficulty exists for some years already. It was first shown in [254] that if the (ph) and (pp) interactions are taken into account in the RPA, it is possible to describe quite well the two-neutrinos double $\beta\beta 2\nu$ decay. Calculations in [255, 256] with more complex forces have confirmed the high effect of the (pp) interactions on the $\beta\beta 2\nu$ decay. It was shown in [257, 258] that the (pp) interactions do affect strongly the GT β^+ decay and the S_+ function, but leave the the GT β^- decay and the S_- function practically unchanged. Calculations of the reduced probabilities of GT β^+ decays in sperical nuclei were carried out in [257–259]; a satisfactory description of the appropriate experimental data was reported.

5.3.6

Let us investigate the effect of particle–particle interactions on the β^+

decays of neutron-deficient nuclei. A considerable part of the GT strength manifests itself in the β^+ decay of the ground 0^+_{gs} state of an even–even nucleus into a number of 1^+ states of an odd–odd nucleus, provided the energy of the β^+ decay is large. The matrix element of the β^+ decay from the ground 0^+_{gs} state of an even–even nucleus to a one-phonon 1^+i state with the wave function (2.174) is

$$(\Psi_0^*\Omega_i H_\beta \Psi_0) = \frac{1}{\sqrt{3}} \sum_{j_p j_n} f(j_p j_n)(\psi^i_{j_p j_n} u_{j_n} v_{j_p} + \varphi^i_{j_p j_n} u_{j_p} v_{j_n}). \tag{5.47}$$

The matrix element of the GT β^- decay for the ground 0^+_{gs} state and a one-phonon 1^+i state is

$$(\Psi_0^*\Omega_i H_\beta \Psi_0) = \frac{1}{\sqrt{3}} \sum_{j_p j_n} f(j_p j_n)(\psi^i_{j_p j_n} u_{j_p} v_{j_n} + \varphi^i_{j_p j_n} u_{j_n} v_{j_p}) \tag{5.48}$$

where

$$f(j_p j_n) = \frac{1}{2}\langle j_p \| \sigma\tau_\pm \| j_n \rangle.$$

A comparison of the matrix elements shows that owing to an excess of neutrons, the terms related with the direct amplitude $\psi^i_{j_p j_n}$ for the β^+ decay are smaller than the corresponding terms for the β^- decay. The amplitudes $\psi^i_{j_p j_n}$ and $\varphi^i_{j_p j_n}$ are found from the solutions of equations (2.181) and (2.181′). The GT (ph) constant is determined by the position of the Gamow–Teller giant resonance; it is taken to be $\kappa_1^{01} = -\frac{23}{A}$ MeV (the minus sign is connected with the sign of the Hamiltonian, (2.167)). Calculations are conducted for different values of the GT (pp) constant and for the ratio of the constants of the axial vector G_A and the vector G_V weak interactions $|G_A/G_V|$ equal to 1.26 and 1.0.

We define the GT strength of the β^+ decays from the ground state of an even–even nucleus to the 1^+ state with the wave function $\Omega^+_{1i}\Psi_0$ by the expression

$$B^+(1^+, i) = \Big| \sum_{j_p j_n} b_i^{(+)}(j_p j_n) \Big|^2 \tag{5.49}$$

where

$$b_i^+(j_p j_n) = f(j_p j_n)(\psi^i_{j_p j_n} v_{j_p} u_{j_n} + \varphi^i_{j_p j_n} u_{j_p} v_{j_n}) \tag{5.49′}$$

and $S_+ = \sum_i B^{(+)}(1^+, i)$. For such decays, ft values are given by

$$(ft)_i = \left(\frac{G_A}{G_V}\right)^2 \frac{6163.4}{B^+(1^+, i)}. \tag{5.50}$$

The reduced values ft_Σ are employed to compare the results of calculations with the experimental data:

$$(ft_\Sigma)^{-1} = \sum_i (ft_i)^{-1} \tag{5.50'}$$

where the summation over i runs over the nuclear states in a given energy range.

In calculating the reduced ft_Σ values for the GT β^+ decays of neutron-deficient even–even nuclei into 1^+i states of odd–odd nuclei, there is no need in conducting detailed calculations of the energies and structure of the low-lying 1^+ states that are being populated. The reasons for this are, first, that the total strength of transitions to all low-lying states is calculated, and second, that the low-lying states are separated from higher-lying 1^+ states by a gap of 3–4 MeV. Owing to this wide gap, the quasiparticle–phonon interactions do not mix the low-lying 1^+ states with the high-lying states of odd–odd nuclei. A tentative conclusion on the renormalization of the G_A constant in nuclei is based on these features and also on the fact that such β^+ decays exhaust a considerable part of the total strength S_+.

The effect of (pp) interactions on ft_Σ can be illustrated with table 5.6. It shows that ft_Σ increase by 0.2–0.4 if (ph) interactions are taken into account, and by 0.4–1.1 is (pp) interactions are additionally considered. From 30 to 60% of the total GT S_+ strength is involved in β_+ decays. Particle–particle interactions reduce the GT strength on the low-lying states to which β decays occur.

Table 5.7 demonstrates the splitting of a part of the GT S_+ strength from the region of low-lying states to higher excitation energies as a result of the (pp) interactions. If $G_1^{01} = 0$, the state $p1h_{11/2}, n1h_{9/2}$ contributes above 99.9% to the normalization of the 1^+ with the relative energy $\omega_{i_1} =$ 7.584 MeV and $|b_{i_1}^{(+)}(p1h_{11/2}, n1h_{9/2})|^2 = 6.57$; at the same time, $B(1^+i_1)$ is twice as small as this. If $G_1^{01} = 7.5/A$ MeV, this state splits into two, separated by 4.7 MeV; the GT S_+ strength of the upper state does not manifest itself in the β^+ decay. The role of ground states correlations is enhanced; as a result, in the lower state we have $B^{(+)}(1^+i_0) = 0.650$ which is considerably lower than $|b_{i_1}^{(+)}(p1h_{11/2}, n1h_{9/2})|^2 = 5.49$, since as G_1^{01} increases, the role of $b_{i_0}^{(+)}$ increases for other states that enter $B^{(+)}(1^+i_0)$ with the opposite sign relative to the dominating term.

The calculations of ft_Σ for the Gamow–Teller β^+ transitions $0^+_{\text{gs}} \to 1^+$ were carried out in [257, 258] for neutron-deficient nuclei with $G_1^{01} =$ 7.5/A MeV at $|G_A/G_V| = 1$ and with $G_1^{01} = 7.8/A$ MeV at $|G_A/G_V| =$ 1.25. If $G_1^{01} > 8.2/A$ MeV, the RPA becomes inapplicable to nuclei with open shells. The results of calculations in [257, 258] and the experimental data of [252, 253, 260–2] are shown in table 5.8. The table shows quite

Table 5.6 The effect of (pp) interactions on the GT β^+ decays $0^+_{gs} \to 1^+$.

β^+ transition	$\log ft_\Sigma$					Part of total strength	
	(exper.)	(calc.) $\lvert\frac{G_A}{G_V}\rvert = 1$			$\lvert\frac{G_A}{G_V}\rvert = 1.26$	$\lvert\frac{G_A}{G_V}\rvert = 1$	$\lvert\frac{G_A}{G_V}\rvert = 1.25$
		$\kappa_1^{01} = 0$ $G_1^{01} = 0$	$\kappa_1^{01} \cdot A = -23$ $G_1^{01} = 0$	$\kappa_1^{01} \cdot A = -23$ $G_1^{01} \cdot A = 7.5$	$\kappa_1^{01} \cdot A = -23$ $G_1^{01} \cdot A = 7.8$	$G_1^{01} \cdot A = 7.5$	$G_1^{01} \cdot A = 7.8$
$^{152}Yb \to {}^{152}Tm$	3.4	2.8	3.0	3.4	3.5	0.60	0.50
$^{150}Er \to {}^{150}Ho$	3.67	2.9	3.1	3.5	3.6	0.55	0.50
$^{148}Dy \to {}^{148}Tb$	3.9	3.0	3.3	3.8	3.9	0.50	0.43
$^{146}Dy \to {}^{146}Tb$	3.8	3.0	3.3	4.0	4.1	0.40	0.30
$^{96}Pd \to {}^{96}Rh$	3.3	2.7	3.1	3.4	3.3	0.35	0.32

Table 5.7 The structure of one-phonon 1^+i nuclear states in ^{148}Tb with a dominating component $p1h_{11/2}, n1h_{9/2}$, calculated for $\kappa_1^{01} = -23/A$ MeV, $G_1^{01} = 0$ and $G_1^{01} = 7.5/A$ MeV at $|G_A/G_V| = 1$.

		$G_1^{01} = 0$			$G_1^{01} = 7.5/A$ MeV			$G_1^{01} = 7.5/A$ MeV		
ω_i (MeV)		7.584			5.156			9.875		
$B^{(+)}(1^+, i)$		3.183			0.650			0.561		
j_p	j_n	$\psi^i_{j_p j_n}$	$\varphi^i_{j_p j_n}$	$b_i^{(+)}(j_p j_n)$	$\psi^i_{j_p j_n}$	$\varphi^i_{j_p j_n}$	$b_i^{(+)}(j_p j_n)$	$\psi^i_{j_p j_n}$	$\varphi^i_{j_p j_n}$	$b_i^{(+)}(j_p j_n)$
$3s_{1/2}$	$3s_{1/2}$	0.000	−0.038	−0.087	0.000	−0.074	−0.169	0.000	−0.004	−0.009
$2d_{5/2}$	$2d_{5/2}$	0.000	−0.020	−0.026	0.000	−0.138	−0.186	0.000	−0.037	−0.050
$1g_{7/2}$	$1g_{7/2}$	0.000	0.014	−0.013	0.000	0.118	−0.113	0.000	0.034	−0.033
$2d_{3/2}$	$2d_{5/2}$	0.006	0.040	−0.114	0.000	0.071	−0.204	0.000	0.003	−0.009
$1h_{11/2}$	$1h_{9/2}$	0.999	0.000	2.563	0.918	0.000	2.341	−0.440	0.000	−1.121
$1h_{9/2}$	$1h_{11/2}$	0.000	0.048	−0.220	0.000	0.039	−0.177	0.000	−0.013	0.058
$1g_{9/2}$	$1g_{9/2}$	0.000	0.016	−0.026	0.000	0.135	−0.215	0.000	0.043	−0.068
$1g_{9/2}$	$1g_{7/2}$	0.000	−0.004	−0.002	0.000	−0.140	−0.060	0.000	−0.052	−0.022
$1g_{7/2}$	$1g_{9/2}$	−0.023	0.000	−0.016	0.011	0.000	0.008	0.041	0.000	0.028
$1f_{5/2}$	$1f_{7/2}$	−0.017	0.000	−0.010	−0.003	0.000	−0.002	0.013	0.000	0.079
$1f_{7/2}$	$1f_{5/2}$	−0.010	0.000	−0.004	0.198	0.000	0.074	0.211	0.000	0.079
$1f_{7/2}$	$1f_{7/2}$	0.000	0.000	0.000	0.277	0.000	0.090	0.770	0.000	0.250

Table 5.8 Gamow–Teller β^+ decays $0^+_{gs} \to 1^+$ in spherical nuclei.

β^+ transition	$\log ft_\Sigma$ (exper.)	$\log ft_\Sigma$ (calc.)	
		$G_1^{01} \cdot A$ = 7.5 MeV	$G_1^{01} \cdot A$ = 7.8 MeV
^{154}Yb→^{154}Tm	3.6	3.5	3.8
^{152}Yb→^{152}Tm	3.5	3.4	3.5
^{152}Er→^{152}Ho	3.9	3.5	3.8
^{150}Er→^{150}Ho	3.7	3.5	3.6
^{148}Dy→^{148}Tb	3.9	3.8	3.9
^{146}Dy→^{146}Tb	3.8	4.0	4.1
^{108}Sn→^{108}In	3.4	3.5	3.3
^{106}Sn→^{106}In	3.2	3.3	3.1
^{104}Sn→^{104}In	3.2	3.3	3.2
^{104}Cd→^{104}Ag	3.5	3.8	3.7
^{102}Cd→^{102}Ag	3.4	3.6	3.5
^{100}Cd→^{100}Ag	3.2	3.2	3.1
^{98}Pd→ ^{98}Rh	3.5	3.5	3.4
^{96}Pd→ ^{96}Rh	3.2	3.4	3.3
^{94}Rh→ ^{94}Tc	3.6	3.6	3.5

satisfactory description of the experimental data with $G_1^{01} = 7.5/A$ MeV and $|G_A/G_V| = 1$ and with $G_1^{01} = 7.8/A$ MeV and $|G_A/G_V| = 1.25$.

The calculations of ft_Σ for the GT β^+ transitions with more complex forces were reported in [259]. The constants G_N, G_Z and κ_1^{01} were fixed using experimental data in the same way as in [257, 258], leaving G_1^{01} free. The magnitudes of the constants κ_1^{01} and G_1^{01} are quite close in [259] but differ considerably in [257, 258]. For example, if the matrix elements $\langle j_p \| \sigma\tau_\pm \| j_n \rangle$ are replaced by $\langle j_p \| \sigma\tau_\pm (\partial V(r)/\partial r) \| j_n \rangle$, the constant G_1^{01} has to be increased to obtain the same values of ft_Σ. If the space of single-particle states is truncated, the constant κ_1^{01} changes only slightly while G_1^{01} grows considerably.

5.3.7

Particle–particle interactions strongly affect the total GT strength S_+ of (n,p) transitions. The calculations in [258] were conducted with the same constants, including $G_1^{01} = 7.5/A$ MeV, as were the calculations of the β^+ decays with $|G_A/G_V| = 1$. The results are shown in table 5.9 which also gives the experimental data [250, 251]. The calculations with $G_1^{01} = 7.8/A$ MeV show a reduction in S_+ of 10%. The table demonstrates a good description of the experimental data; note that S_+ for ^{90}Zr(n,p)^{90}Y were calculated before they were measured. Calculations with smaller values

Table 5.9 Strength functions of Gamow–Teller (n, p) transitions.

Reaction	S_+ (exper.)	S_+ (calc.)	
		[258]	[263]
^{54}Fe(n,p)^{54}Mn	3.1 ± 0.6	4.2	—
^{90}Zr(n,p)^{90}Y	1.0 ± 0.3	1.2	1.6

of G_1^{01} are at variance with the experimental results. Table 5.9 lists the results of calculations in a model that includes the $2p - 2h$ excitations.

The calculated values of $\log ft_\Sigma$ agree with the experimental data if $|G_A/G_V| = 1.26$ and the renormalization is $|G_A/G_V| = 1$. If the renormalization of the constants G_A of the axial–vector weak interaction is increased, it is necessary to reduce G_1^{01} for the description of $\log ft_\Sigma$ of the GT β^+ decays, but this impairs the agreement of the calculated and the experimental values of S_+. We can conclude, on the basis of $\log ft_\Sigma$ calculations for GT β^+ decays of a large number of neutron-deficient spherical nuclei (given in table 5.8) and the total strength functions S_+ of the GT (n,p) transitions, that complex nuclei must obey the condition

$$|G_A/G_V| \geqslant 1. \tag{5.51}$$

Further experimental investigation of the β^+ decays of neutron-deficient spherical and deformed nuclei is required. The β^+ decay of ^{100}Sn must be measured in order to improve the accuracy of the renormalization of the constant G_A in nuclei.

6

Non-rotational states of deformed nuclei

6.1 EQUATIONS OF THE QUASIPARTICLE–PHONON MODEL FOR DEFORMED NUCLEI

6.1.1

The equilibrium shape of atomic nuclei is a sphere or an axially symmetric ellipsoid. Experimental evidence points to the existence of nuclei whose equilibrium shape is a three-axial ellipsoid. There also exist transition nuclei which are soft with respect to the β and γ vibrations. Their complex equilibrium shape depends on the excitation energy and angular momentum. Various calculation methods are used for the description of spherical and deformed nuclei; we will now formulate the QPNM for deformed nuclei.

A special feature of deformed nuclei is the explicit singling out of the degrees of freedom connected with the rotation of the nucleus as a whole. A phenomenological description of rotation makes it possible to write the wave function of a deformed nucleus in the form (see [4, 12])

$$\Psi^I_{MK}(\Theta_e, \nu) = \sqrt{\frac{2I+1}{16\pi^2}} \Big[D^I_{MK}(\Theta_e)\Psi_\nu(K+) + (-)^{I+K} D^I_{M-K}(\Theta_e)\Psi_\nu(K-) \Big] \tag{6.1}$$

where the generalized spherical functions $D^I_{M\sigma K}(\Theta_e)$ describe rotation. The total Hamiltonian is

$$H = T_{\text{rot}} + H_{\text{cor}} + H_{\text{QPNM}}. \tag{6.2}$$

The kinetic energy of rotation and the Coriolis interaction, coupling the internal motion with rotation, is usually written in the form

$$T_{\text{rot}} = \frac{I(I+1)}{2\mathfrak{J}} \tag{6.2'}$$

$$H_{\text{cor}} = -(I_+ J_- + I_- J_+)/2\mathfrak{J} \tag{6.3}$$

where I is the total angular momentum, J is the angular momentum of the internal motion and $\mathfrak{J}$ is the moment of inertia.

All kinds of non-rotational motion are described microscopically in many models, including the QPNM, while rotation is described phenomenologically. This chapter mostly gives the microscopic description of the internal motion with the wave function $\Psi_\nu(K\sigma)$, and the effects connected with rotation are discussed only when necessary. The rotation, the enhancement of the role played by the Coriolis interaction at high I and the high-spin states are described in [4, 12, 264–267] and in some other publications.

In the axially symmetric potential, the spherical subshell nlj splits into $j+1/2$ twice degenerate levels. The single-particle state is characterized by the parity π, the projection K of the angular momentum on the symmetry axis of the nucleus and the asymptotic quantum numbers $Nn_z\Lambda\uparrow$ (if $K = \Lambda + 1/2$) or $Nn_z\Lambda\downarrow$ (if $K = \Lambda - 1/2$). The set of the quantum numbers of single-particle states is denoted by $q\sigma$, $\sigma = \pm 1$ (see [12]). If the shape of the nucleus is a prolate ellipsoid of revolution, the levels with low K shift downwards and those with high K, upwards.

The mean field of a deformed nucleus is most often described using the Saxon–Woods potential in the form (1.76) and (1.77). The wave function φ_q of the single-particle state of the axially symmetric Saxon–Woods potential can be written as the following expansion in single-particle wave functions φ_{nljm} of the spherically symmetric Saxon–Woods potential:

$$\varphi_q = \sum_{nljm} a^q_{nlj}\varphi_{nljm}. \tag{6.4}$$

The normalization of the coefficients a^q_{nlj} is:

$$\sum_{nljm} \left(a^q_{nlj}\right)^2 = j + \frac{1}{2}. \tag{6.4'}$$

Figure 6.1 gives examples of distribution of neutron subshell strengths. A predominant part of the subshell strength is distributed over several states in the 5–7 MeV energy range. Actually, subshells with high j give an appreciable, even though not too large, contribution to states that are separated from them by 10–15 MeV. For example, around 4% of the strength of the neutron subshell $1h_{11/2}$ is concentrated on the states located in the vicinity of the neutron binding energy. If we take into account such 'tails', the strength of subshells with high l is found to be distributed over a larger number of states and a larger energy range than for subshells with low l. This feature is seen clearly in figure 6.1 which makes it possible to compare strength distributions of subshells with very different l. For the $3s_{1/2}$ subshell, 96% of strength is concentrated in the 400↑, 411↓, 420↑ and 431↓ states; the energy difference between the most distant states is 9.3 MeV.

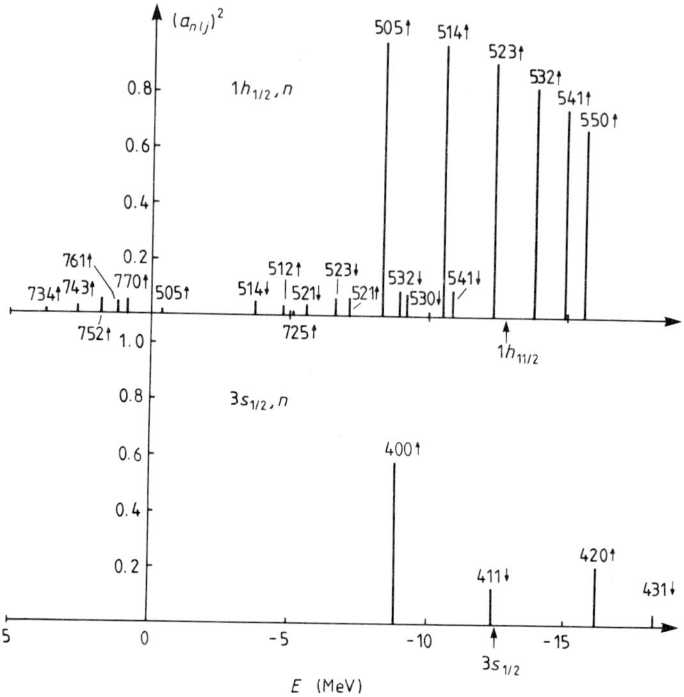

Figure 6.1 Distribution of the strength of neutron subshells $1h_{11/2}$ and $3s_{1/2}$ over single-particle states in the Saxon–Woods potential with axially-symmetric quadrupole and hexadecapole deformations. Arrows indicate the positions of subshells in the spherically symmetric potential with the same parameters.

The same 96% of the strength are distributed in the $1h_{11/2}$ subshell over 15 states in the energy range ΔE =11.8 MeV, even though 90% of its strength is concentrated in six strongest states.

Since single-particle energies and wave functions of the Saxon–Woods potential depend upon the mass number A, the region of deformed nuclei is split into zones so as not to perform calculations for each value of A. The rare-earth region and the actinides region are split in the calculations [6, 67, 71, 268–270] into zones around the following values of A: 155, 165, 173, 181, 229, 239, 247 and 255. The fitting of the Saxon–Woods potential parameters consists of the following four stages: (1) single-particle energies and wave functions are calculated for a specific set of parameters; (2) the equilibrium shape of the nucleus is calculated by the shell correction method [271]; this fixes the parameters of the quadrupole, β_2, and the hexadecapole, β_4, deformations (see (1.75)); (3) the phonon operators in the RPA are calculated; (4) the wave functions of odd nuclei, $\Psi_\nu(K\sigma)$, are taken as a sum of single-particle components and quasiparticle ⊗ phonon components, the quasiparticle–phonon interactions are taken into account and the energies and wave functions of non-rotational states of odd nuclei are found, after which they are compared with the available experimental data. In order to improve the agreement between the calculated results and the measured data, one varies the parameters of the Saxon–Woods potential and repeats the four calculation stages. The procedure is repeated until a sufficiently good description of the experimental data is achieved for the low-lying non-rotational levels of odd nuclei.

When the phonon basis is constructed for deformed nuclei, one uses multipole phonons with $\lambda\mu$ equal to 20, 22, 30, 31, 32, 33, 41, 43, 44, 54, 55, 65, 66, 76 and 77. For each multipolarity, one takes from 5 to 500 roots of the corresponding secular equation in the RPA, depending on the specific problem. The choice of pairing constants and isoscalar and isovector constants of multipole and spin-multipole forces was described in section 4.1.

6.1.2

Let us write the QPNM Hamiltonian in its most general form for separable interactions of rank $n_{\max} > 1$, using phonons (2.53) consisting of electric and magnetic parts. It has the following form:

$$H_{\mathrm{QPNM}} = \sum_{q\sigma} \varepsilon_q \alpha^+_{q\sigma}\alpha_{q\sigma} + H_v + H_{vq}$$

$$H_v = H_v^{00} + \sum_{\lambda} H_v^{\lambda 0} + \sum_K H_v^K \tag{6.5}$$

$$H_v^{00} = -\frac{1}{2}\sum_{\tau}\sum_{ii'} G_\tau \left[d^i_{y\tau} d^{i'}_{y\tau} + d^i_{\omega\tau} d^{i'}_{\omega\tau} \right] Q^+_{20i} Q_{20i'} \tag{6.6}$$

$$H_v^{\lambda 0} = -\sum_{ii'} W^{\lambda 0}_{ii'} Q^+_{\lambda 0 i} Q_{\lambda 0 i'} \tag{6.6$'$}$$

$$H_v^K = -\sum_{i_1 i_2 \sigma} W^K_{i_1 i_2} Q^+_{K i_1 \sigma} Q_{K i_2 \sigma} \tag{6.6$''$}$$

$$W^K_{i_1 i_2} = \sum_{\lambda}\left[W^{\lambda K}_{i_1 i_2} + W^{\lambda\lambda K}_{i_1 i_2} \right] + \sum_L \left[\sum_{\lambda' = L\pm 1} W^{\lambda' L K}_{i_1 i_2} + W^{TLK}_{i_1 i_2} \right]$$

$$W^{\lambda K}_{i_1 i_2} = \frac{1}{4}\sum_{n=1}^{n_{\max}} \sum_{\tau} \Big\{ \sum_{\rho=\pm 1} (\kappa_0^{\lambda K} + \rho\kappa_1^{\lambda K}) D^{\lambda K i_1}_{n\tau} D^{\lambda K i_2}_{n\rho\tau}$$
$$+ G^{\lambda K}\left[D^{\lambda K i_1}_{ng\tau} D^{\lambda K i_2}_{ng\tau} + D^{\lambda K i_1}_{n\omega\tau} D^{\lambda K i_2}_{n\omega\tau} \right] \Big\} (1 + \delta_{K0})^2$$

$$W^{\lambda' L K}_{i_1 i_2} = \frac{1}{4}\sum_{n=1}^{n_{\max}} \sum_{\tau} \Big\{ \sum_{\rho=\pm 1} (\kappa_0^{\lambda' L K} + \rho\kappa_1^{\lambda' L K}) D^{\lambda' L K i_1}_{n\tau} D^{\lambda' L K i_2}_{n\rho\tau}$$

$$+ G^{\lambda K}\Big[D^{\lambda' LKi_1}_{ng\tau} D^{\lambda' LKi_2}_{ng\tau} + D^{\lambda' LKi_1}_{n\omega\tau} D^{\lambda' LKi_2}_{n\omega\tau}\Big]\Big\}$$

$$W^{TLK}_{i_1 i_2} = -\frac{1}{2}\sum_{n=1}^{n_{\max}} \sum_{\tau,\rho=\pm 1} (\kappa^{LK}_{T0} + \rho\kappa^{LK}_{T1}) D^{L-1LKi_1}_{n\tau} D^{L+1LKi_2}_{n\rho\tau}$$

$$H_{vq} = H^{00}_{vq} + \sum_{\lambda} H^{\lambda 0}_{vq} + \sum_{K}\Big\{\sum_{\lambda}(H^{\lambda K}_{vq} + H^{\lambda\lambda K}_{vq})$$

$$+ \sum_{L}\Big[\sum_{\lambda'=L\pm 1} H^{\lambda' LK}_{vq} + H^{TLK}_{vq}\Big]\Big\} \tag{6.7}$$

$$H^{00}_{vq} = -\sum_{\tau i} G_\tau \sum_{qq'}^{\tau} (u_q^2 - v_q^2) u_{q'} v_{q'}\Big\{\Big[\psi^{20i}_{qq} Q^+_{20i} + \varphi^{20i}_{qq} Q_{20i}\Big]$$

$$\times \sum_{\sigma} \alpha^+_{q'\sigma}\alpha_{q'\sigma} + \text{h.c.}\Big\} \tag{6.8}$$

$$H^{\lambda K}_{vq} = -\frac{1}{4}\sum_{ni_2\tau\sigma} \sum_{qq'}^{\tau} f^{\lambda K}_n(qq') V^{\lambda K i_2}_{n\tau}(qq')\Big[(Q^+_{Ki_2\sigma} + Q_{Ki_2-\sigma})$$

$$\times B(qq'; K-\sigma) + B(qq'; K\sigma)(Q^+_{Ki_2-\sigma} + Q_{Ki_2\sigma})\Big] \tag{6.8$'$}$$

$$H^{\lambda 0}_{vq} = -\sum_{i\tau}\sum_{qq'}{}^{\tau}\, V^{\lambda 0 i}_{\tau}(qq') f^{\lambda 0}_n(qq')\Big\{(Q^+_{\lambda 0 i} + Q_{\lambda 0 i})$$

$$\times B(qq'; \mu = 0) + \text{h.c.}\Big\} \tag{6.8$''$}$$

$$H^{\lambda' LK}_{vq} = +\frac{\mathrm{i}}{4}\sum_{ni_2\tau\sigma}\sum_{qq'}{}^{\tau}\, f^{\lambda' LKi_2}_n(qq') V^{\lambda' LKi_2}_{n\tau}(qq')\Big[(Q^+_{Ki_2\sigma} - Q_{Ki_2-\sigma})$$

$$\times \mathfrak{B}(qq'; K-\sigma) + \mathfrak{B}(qq'; K\sigma)(Q^+_{Ki_2-\sigma} - Q_{Ki_2\sigma})\Big] \tag{6.9}$$

$$H^{TLK}_{vq} = -\frac{\mathrm{i}}{4}\sum_{ni_2\tau\rho}(\kappa^{LK}_{T0} + \rho\kappa^{LK}_{T1})\sum_{qq'}{}^{\tau}\Big[D^{L-1LKi_2}_{\rho\tau} f^{L+1LK}(qq')$$

$$+ D^{L+1LKi_2}_{\rho\tau} f^{L-1LK}(qq')\Big] v^{(+)}_{qq'}\Big[(Q^+_{Ki_2\sigma} - Q_{Ki_2-\sigma})$$

$$\times \mathfrak{B}(qq'; K-\sigma) + \text{h.c.}\Big] \tag{6.9$'$}$$

$$V^{\lambda K i_2}_{n\tau}(qq') = \sum_{\rho=\pm 1}(\kappa^{\lambda K}_0 + \rho\kappa^{\lambda K}_1) v^{(-)}_{qq'} D^{\lambda K i_2}_{n\rho\tau}$$

$$- G^{\lambda K} u^{(+)}_{qq'} D^{\lambda K i_2}_{ng\tau}$$

$$V^{\lambda' LKi_2}_{n\tau}(qq') = \sum_{\rho=\pm 1}(\kappa^{\lambda' LK}_0 + \rho\kappa^{\lambda' LK}_1) v^{(+)}_{qq'} D^{\lambda' LKi_2}_{n\rho\tau}.$$

Here ε_q is the quasiparticle energy for the monopole and quadrupole pairing, (2.73); the operator $Q_{\lambda 0 i}$ is given by (2.65)

$$
\begin{aligned}
d^i_{g\tau} &= \sum_q^\tau \frac{E(q)-\lambda_\tau}{\varepsilon_q} g^{20i}_{qq} \\
d^i_{n\tau} &= \sum_q^\tau w^{20i}_{qq} \\
D^{\lambda K i_2}_{n\tau} &= \sum_{qq'}^\tau f_n^{\lambda K}(qq') u^{(+)}_{qq'} g^{Ki_2}_{qq'} \\
D^{\lambda K i_2}_{ng\tau} &= \sum_{qq'}^\tau f_n^{\lambda K}(qq') v^{(-)}_{qq'} g^{Ki_2}_{qq'} \\
D^{\lambda K i_2}_{n\omega\tau} &= \sum_{qq'}^\tau f_n^{\lambda K}(qq') v^{(+)}_{qq'} w^{gi_2}_{qq'} \\
D^{\lambda' L K i_2}_{n\tau} &= \sum_{qq'}^\tau f_n^{\lambda' L K}(qq') u^{(-)}_{qq'} \chi(qq') w^{Ki_2}_{qq'} \\
D^{\lambda' L K i_2}_{ng\tau} &= \sum_{qq'}^\tau f_n^{\lambda' L K}(qq') v^{(-)}_{qq'} g^{Ki_2}_{qq'} \\
D^{\lambda' L K i_2}_{n\omega\tau} &= \sum_{qq'}^\tau f_n^{\lambda' L K}(qq') v^{(+)}_{qq'} w^{Ki_2}_{qq'} \\
g^{Ki_2}_{qq'} &= \psi^{Ki_2}_{qq'} + \varphi^{Ki_2}_{qq'} \qquad \omega^{Ki_2}_{qq'} = \psi^{Ki_2}_{qq'} - \varphi^{Ki_2}_{qq'} \\
f_n^{\lambda K}(qq') &= \langle q \| R_n^\lambda(r) Y_{\lambda K} \| q' \rangle \\
f_n^{\lambda' L K}(qq') &= \langle q \| R_n^{\lambda' L}(r) \{Y_\lambda \sigma\}_{LK} \| q' \rangle
\end{aligned}
\tag{6.10}
$$

$E(q)$ are the single-particle energies and λ_τ are the chemical potentials.

Simple separable interactions and phonon operators given by (2.35) are useful for many calculations. In this case, we take into account secular equations (2.42) and (2.68) for multipole phonons and (2.156) for neutron–proton phonons. We will use QPNM Hamiltonians (2.87), (2.90), (2.162), (2.163) and (2.163′), transformed by using the secular equations. The QPNM Hamiltonian can now be rewritten in the form

$$
\begin{aligned}
H_{\text{QPNM}} &= \sum_q \varepsilon_q B(q) + H_{Mv} + H_{Mvq} + H_{CMv} + H_{CMvq} \\
&\quad + H_{CSv} + H_{CSvq}
\end{aligned}
\tag{6.11}
$$

$$
H_{Mv} = -\frac{1}{4} \sum_{\lambda\mu i i'\sigma\tau} \frac{X^{\lambda\mu i}(\tau) + X^{\lambda\mu i'}(\tau)}{(\mathcal{Y}_\tau^{\lambda\mu i} \mathcal{Y}_\tau^{\lambda\mu i'})^{1/2}} Q^+_{\lambda\mu i\sigma} Q_{\lambda\mu i'\sigma} \tag{6.12}
$$

where $X^{\lambda\mu i}(\tau)$ and $\mathcal{Y}_\tau^{\lambda\mu i}$ for $\lambda\mu \neq 20$ are given by (2.42′) and (2.44), and for $\lambda\mu = 20$, by (2.68′) and (2.69″),

$$
H_{Mvq} = -\frac{1}{2} \sum_{\lambda\mu i i'\sigma\tau} \sum_{qq'}^\tau \Gamma^{\lambda\mu i}_{qq'}(\tau) \Big[(Q^+_{\lambda\mu i\sigma} + Q_{\lambda\mu i -\sigma})
$$

$$\times B(qq';\mu-\sigma)+\text{h.c.}\Big] \tag{6.12'}$$

$$H_{CMv}=-\sum_{\lambda\mu ii'\sigma}(1+\delta_{\mu 0})^{-2}\frac{1+\mathsf{Y}^{\lambda\mu i}\mathsf{Y}^{\lambda\mu i'}}{(\mathcal{Y}^{\lambda\mu i}\mathcal{Y}^{\lambda\mu i'})^{1/2}}\Omega^{+}_{\lambda\mu i\sigma}\Omega_{\lambda\mu i'\sigma} \tag{6.13}$$

$$H_{CMvq}=-\frac{1}{\sqrt{2}}\sum_{\lambda\mu i\sigma}(1+\delta_{\mu_0})^{-1}\left(\frac{\kappa_1^{(\lambda\mu)}}{\mathcal{Y}^{\lambda\mu i}}\right)^{1/2}\Big\{(\Omega^{+}_{\lambda\mu i\sigma}$$

$$\mathsf{Y}^{\lambda\mu i}\Omega_{\lambda\mu i-\sigma})\sum_{rs}\Big[f^{\lambda\mu}(rs)u_r u_s B(rs;\mu-\sigma)$$

$$-f^{\lambda\mu}(sr)v_r v_s B(sr;\mu-\sigma)\Big]+\text{h.c.}\Big\} \tag{6.13'}$$

$$H_{CSvq}=-\frac{1}{\sqrt{2}}\sum_{LKi\sigma}(1+\delta_{K_0})^{-1}\left(\frac{\kappa^{(L-1LK)}}{\mathcal{Y}^{LKi}}\right)^{1/2}\Big\{(\Omega^{+}_{LKi\sigma}$$

$$+\mathsf{Y}^{LKi}\Omega_{LKi-\sigma})\sum_{rs}f^{L-1LK}(rs)\Big[u_r u_s\mathfrak{B}(rs;K-\sigma)$$

$$-v_r v_s\mathfrak{B}(s,r;K-\sigma)\Big]+\text{h.c.}\Big\}. \tag{6.14}$$

The expression for H_{CSv} is obtained from (6.12) by the substitution of constants and the following matrix elements:

$$\Gamma^{\lambda\mu i}_{qq'}(\tau)=\frac{1}{\sqrt{2}}\frac{f^{\lambda\mu}(qq')v^{(-)}_{qq'}}{(\mathcal{Y}^{\lambda\mu i}_{\tau})^{1/2}}$$

$$\Gamma^{L-1LKi}_{qq'}(\tau)=\frac{1}{2\sqrt{2}}\frac{f^{L-1LK}(qq')v^{(+)}_{qq'}}{(\mathcal{Y}^{LKi}_{\tau})^{1/2}}. \tag{6.15}$$

6.1.3

I will now list QPNM formulas for odd mass deformed nuclei. When the fragmentation of single-particle states is calculated for the expansion of the wave function in the number of phonons, it is possible to retain only quasiparticle ⊗ phonon terms. The reason for this is that the stable quadrupole deformation has already resulted in the fragmentation of spherical subshells, and that additionally taking the quasiparticle ⊗ two phonons terms of the wave function into account cannot lead to appreciable changes. The formulas for the case of the wave function also containing quasiparticle ⊗ two phonons terms are given in [134]. The wave function of an odd nucleus can be chosen in the form

$$\Psi_\nu(K_0^{\pi_0}\sigma_0)=\Big\{\sum_{q_0}C^{\nu}_{q_0}\alpha^{+}_{q_0\sigma_0}$$

$$+ \sum_{qg_2\sigma\sigma_2} D^{\nu}_{qg_2}\delta_{\sigma K+\sigma_2\mu_2,\sigma_0 K_0}\alpha^+_{q\sigma}Q^+_{g_2\sigma_2}\Big\}\Psi_0 \qquad (6.16)$$

where $g = \lambda\mu i$; $\nu = 1, 2, 3$ number the states with a given $K_0^{\pi_0}$; Ψ_0 is the wave function of the ground state of an even–even nucleus; the remaining notation is given in section 2.1.

For taking into account the Pauli exclusion principle, we use the commutation relations:

$$[\alpha_{q\sigma}, Q^+_{g_2\sigma_2}] = \sum_{q_3}(\psi^{g_2}_{qq_3}\delta_{\sigma(K-K_3),\sigma_2\mu_2}\sigma\alpha^+_{q_3-\sigma} - \psi^{g_2}_{qq_3}\delta_{\sigma(K+K_3),\sigma_2\mu_2}\alpha^+_{q_3\sigma}) \qquad (6.17)$$

$$[\alpha^+_{q\sigma'}, Q^+_{g_2\sigma_2}] = \sum_{q_3}\{\varphi^{g_2}_{qq_3}\delta_{\sigma(K-K_3),\sigma_2\mu_2}\sigma\alpha_{q_3-\sigma} - \varphi^{g_2}_{qq_3}\delta_{\sigma(K+K_3),\sigma_2\mu_2}\alpha_{q_3\sigma}\}. \qquad (6.17')$$

Using them, we calculate

$$\sum_{\sigma\sigma_2\sigma'\sigma'_2}\delta_{\sigma'K'+\sigma'_2\mu'_2,\sigma_0K_0}\delta_{\sigma K+\sigma_2K_2,\sigma_0K_0}\langle Q_{g'_2\sigma'_2}\alpha_{q'\sigma'}\alpha^+_{q\sigma}Q^+_{g_2\sigma_2}\rangle$$
$$= \delta_{g_2g'_2}\delta_{qq'}(\delta_{K+\mu_2,K_0} + \delta_{|K-K_2|,K_0}) + \mathcal{L}^{K_0}(g'_2q'|qg_2). \qquad (6.18)$$

The function $\mathcal{L}^{K_0}(g'_2q'|qg_2)$ is sign-alternating; its diagonal values are much greater than the non-diagonal ones [272]. We will now need its diagonal and quasi-diagonal values. The quasi-diagonal value $\mathcal{L}^{K_0}(g'q'|qg)$ for which $q = q'$, $\lambda'\mu' = \lambda\mu$, $i' \neq i$ is

$$\mathcal{L}^{K_0}(qg, i') = -\sum_{q_2}\psi^{\lambda\mu i}_{qq_2}\psi^{\lambda\mu i'}_{qq_2}(\delta_{K+K_2,\mu}\delta_{K+\mu,K_0} + \delta_{K-K_2,\mu}\delta_{K+\mu,K_0} + \delta_{K_2-K,\mu}\delta_{|K-\mu|,K_0}). \qquad (6.19)$$

For the diagonal value $\mathcal{L}^{K_0}(g'q'|qg)$ (for $i = i'$), we use the notation $\mathcal{L}K_0(qg)$. In the approximation which is diagonal in the function $\mathcal{L}^{K_0}$, the normalization condition of wave function (6.16) takes the form

$$\sum_{q_0}(C^{\nu}_{q_0})^2 + \sum_{qg_2}(D^{\nu}_{qg_2})^2[1 + \mathcal{L}^{K_0}(qg_2)] = 1. \qquad (6.20)$$

Taking the Pauli principle into account adds the factor $1+\mathcal{L}^{K_0}(qg_2)$. If the violation of the Pauli exclusion principle is maximal, then $\mathcal{L}^{K_0}(q^0g^0_2) = -1$ and the corresponding term drops out of the sum over qg_2.

Let us only take into account the quasiparticles-multipole phonons interaction and take the following part of the QPNM Hamiltonian:

$$H_{\mathrm{QPNM}} = \sum_q \varepsilon_q B(q) + H_{Mv} + H_{Mvq}. \tag{6.21}$$

We average the value of H_{QPNM} over wave function (6.16) in the approximation diagonal in the function $\mathcal{L}^{K_0}$ and obtain

$$\begin{aligned}
[\Psi_\nu^*(K_0^{\pi_0}\sigma_0)H_{\mathrm{QPNM}}\Psi_\nu(K_0^{\pi_0}\sigma_0)] \\
&= \sum_{q_0}(C_{q_0}^\nu)^2\varepsilon_{q_0} \\
&+ \sum_{qg_2}(D_{qg_2}^\nu)^2[1+\mathcal{L}^{K_0}(qg_2)][\varepsilon_q+\omega_{g_2}+\Delta\omega(qg_2)] \\
&- 2\sum_{q_0qg_2} C_{q_0}^\nu D_{qg_2}^\nu \Gamma_{q_0q}^{g_2}[1+\mathcal{L}^{K_0}(qg_2)]
\end{aligned} \tag{6.22}$$

where

$$\Delta\omega(qg_2) = -\frac{1}{4}\sum_{i'\tau}\frac{X^{\lambda_2\mu_2i_2}(\tau)+X^{\lambda_2\mu_2i'}(\tau)}{(\mathcal{Y}_\tau^{\lambda_2\mu_2i_2}\mathcal{Y}_\tau^{\lambda_2\mu_2i'})^{1/2}}\mathcal{L}^{K_0}(qg_2,i'). \tag{6.23}$$

Let us make use of the variational principle and obtain, in the same manner as before, the following equations:

$$\sum_{q_0'} C_{q_0'}^\nu[(\varepsilon_{q_0}-\eta_\nu)\delta_{q_0q_0'} - V_{q_0q_0'}] = 0 \tag{6.24}$$

$$D_{qg_2}^\nu = \frac{\sum_{q_0} C_{q_0}^\nu \Gamma_{qq_0}^{g_2}}{\varepsilon_q+\omega_{g_2}+\Delta\omega(qg_2)-\eta_\nu} \tag{6.24'}$$

where

$$V_{q_0q_0'} = \sum_{qg_2}\frac{\Gamma_{q_0q}^{g_2}\Gamma_{q_0'q}^{g_2}[1+\mathcal{L}^{K_0}(qg_2)]}{\varepsilon_q+\omega_{g_2}+\Delta\omega(qg_2)-\eta_\nu}.$$

The following secular equation for calculating the energies η_ν of the states of an odd nucleus are obtained from the condition of existence of a non-trivial solution of set (6.24):

$$\Theta(\eta_\nu) = \det\|(\varepsilon_{q_0}-\eta_\nu)\delta_{q_0q_0'} - V_{q_0q_0'}\| = 0. \tag{6.25}$$

The rank of this determinant equals the number of single-particle states q_0 with a fixed value $K_0^{\pi_0}$, taken into account in (6.16).

Let us derive the functions $C^{\nu}_{q'_0}$ and $D^{\nu}_{q' g_2}$. We single out an arbitrary function $C^{\nu}_{q_0}$ and introduce the following notation for $q'_0 \neq q_0$:

$$\widetilde{C}^{\nu}_{q'_0} = C^{\nu}_{q'_0}/C^{\nu}_{q_0} \quad \widetilde{D}^{\nu}_{q g_2} = D^{\nu}_{q g_2}/C^{\nu}_{q_0}.$$

Equations (6.20), (6.24) and (6.24′) can be rewritten as

$$(C^{\nu}_{q_0})^2\Big\{1 + \sum_{q'_0 \neq q_0} (\widetilde{C}^{\nu}_{q'_0})^2 + \sum_{q g_2} (\widetilde{D}^{\nu}_{q g_2})^2 [1 + \mathfrak{L}^{K_0}(q g_2)]\Big\} = 1$$

$$\mathfrak{F}_{q_0}(\eta_\nu) = \varepsilon_{q_0} - \eta_\nu - \sum_{q g_2} \Gamma^{g_2}_{q_0 q} \widetilde{D}^{\nu}_{q g_2} = 0 \qquad (6.26)$$

$$(\varepsilon_{q'_0} - \eta_\nu)\widetilde{C}^{\nu}_{q'_0} - \sum_{q''_0 \neq q_0} V_{q'_0 q''_0} \widetilde{C}^{\nu}_{q'_0 q''_0} = V_{q'_0 q_0}$$

$$\widetilde{D}^{\nu}_{q g_2} = \frac{\Gamma^{g_2}_{q q_0} + \sum_{q'_0 \neq q_0} \Gamma^{g_2}_{q_0 q'_0} \widetilde{C}^{\nu}_{q'_0}}{\varepsilon_q + \omega_{g_2} + \Delta\omega(q g_2) - \eta_\nu}.$$

After some simple manipulation, we obtain

$$\widetilde{C}^{\nu}_{q'_0} = \Theta_{q_0}(q'_0; \eta_\nu)/\Theta_{q_0}(\eta_\nu) \qquad (6.27)$$

$$\widetilde{D}^{\nu}_{q g_2} = \frac{\Gamma^{g_2}_{q q_0} + \Theta^{-1}_{q_0}(\eta_\nu) \sum_{q'_0 \neq q_0} \Gamma^{g_2}_{q q'_0} \Theta_{q_0}(q'_0; \eta_\nu)}{\varepsilon_q + \omega_{g_2} + \Delta\omega(q g_2) - \eta_\nu} \qquad (6.27')$$

$$(C^{\nu}_{q_0})^{-1} = 1 + \Theta^{-2}_{q_0}(\eta_\nu)\Bigg\{\sum_{q'_0 \neq q_0} \Theta^2_{q_0}(q'_0; \eta_\nu) + \sum_{q g_2}\left[\frac{\Gamma^{g_2}_{q q_0}\Theta_{q_0}(\eta_\nu) + \sum_{q'_0 \neq q_0} \Gamma^{g_2}_{q q_0} \Theta_{q_0}(q'_0; \eta_\nu)}{\varepsilon_q + \omega_{g_2} + \Delta\omega(q g_2) - \eta_\nu}\right]^2 \times [1 + \mathfrak{L}^{K_0}(q g_2)]\Bigg\} \qquad (6.28)$$

where

$$\Theta_{q_0}(\eta_\nu) = \det \|(\epsilon_{q''_0} - \eta_\nu)\delta_{q'_0 q''_0} - V_{q'_0 q''_0}\|. \qquad (6.29)$$

The order of this determinant is less by unity than that of determinant (6.25); it is derived from (6.25) by eliminating the row and the column which contain q_0. The determinant $\theta_{q_0}(q'_0; \eta_\nu)$ is found from (6.29) by substituting the column $V_{q'_0 q_0}$ for the column q'_0. Note that

$$(C^{\nu}_{q_0})^{-2} = \partial \mathfrak{F}_{q_0}(\eta)/\partial \eta \qquad (6.28')$$

$$\mathfrak{F}_{q_0}(\eta) = \Theta(\eta)/\Theta_{q_0}(\eta). \qquad (6.28'')$$

If the Pauli exclusion principle in the quasiparticle ⊗ phonon states is not taken into account, we have

$$\mathcal{L}^{K_0}(qg, i) = \mathcal{L}^{K_0}(qg) = 0 \qquad \Delta\omega(qg_2) = 0 \tag{6.30}$$

and the secular equation changes to

$$\Theta(\eta_\nu) = \det \left\| (\epsilon_{q_0} - \eta_\nu)\delta_{q_0 q_0'} - \sum_{qg_2} \frac{\Gamma^{g_2}_{qq_0} \Gamma^{g_2}_{qq_0}}{\epsilon_q + \omega_{g_2} - \eta_\nu} \right\| = 0. \tag{6.31}$$

It is not difficult to obtain the remaining formulas by making use of conditions (6.30). They are identical to those given in [6].

The difference between deformed and spherical nuclei consists in that it is necessary to take into account several single-quasiparticle states with fixed values of $K_0^{\pi_0}$ in the single-quasiparticle part of the wave function. In the cases where it is possible to retain only one single-quasiparticle state with a given $K_0^{\pi_0}$ and ignore the Pauli exclusion principle, the formulas change to the following simple form:

$$\epsilon_{q_0} - \eta_\nu - \sum_{qg_2} \frac{(\Gamma^{g_2}_{qq_0})^2}{\epsilon_q + \omega_{g_2} - \eta_\nu} = 0 \tag{6.32}$$

$$D^\nu_{qg_2} = \frac{C^\nu_{q_0} \Gamma^{g_2}_{qq_0}}{\epsilon_q + \omega_{g_2} - \eta_\nu} \tag{6.32'}$$

$$(C^\nu_{q_0})^{-2} = 1 + \sum_{qg_2} \left[\frac{\Gamma^{g_2}_{qq_0}}{\epsilon_q + \omega_{g_2} - \eta_\nu} \right]^2 . \tag{6.32''}$$

The authors of some papers (see, e.g., [273–275]) take into account the quasiparticle–phonon interaction in the ground state of an even–even nucleus, that is, they redefine the vacuum. In this case, non-polar terms appear in secular equations (6.25), (6.31) and (6.32) in addition to the polar ones. As shown in [6, 139, 274], the effect of non-polar terms is minimal. They slightly decrease the energy of collective states of the quasiparticle ⊗ phonon type. Non-polar terms practically do not affect the energies and structures of nearly single-quasiparticle states. The reason for this is that in describing the energies of non-rotational states of odd nuclei, one has to calculate not the differences between the energies of systems with even and odd number of particles but the differences between the energies of the excited and the ground states. In these differences, the non-polar terms almost totally cancel out.

6.1.4

Let us derive QPNM equations for even–even deformed nuclei. The wave function of the excited state will be written in the form

$$\Psi_\nu(K_0^{\pi_0}\sigma_0) = \Big\{\sum_{i_0} R^\nu_{i_0} Q^+_{g_0\sigma_0} + \sum_{g_1\sigma_1 g_2\sigma_2} \frac{(1+\delta_{g_1g_2})^{1/2}\delta_{\sigma_1\mu_1+\sigma_2\mu_2,\sigma_0K_0}}{2[1+\delta_{K_0 0}(1-\delta_{\mu_1 0})]^{1/2}} \times P^\nu_{g_1g_2} Q^+_{g_1\sigma_1} Q^+_{g_2\sigma_2}\Big\}\Psi_0 \tag{6.33}$$

$$g_0 = \lambda_0\mu_0 i_0 \qquad (-)^{\lambda_0} = \pi_0;\ \mu_0 \equiv K_0.$$

In order to accurately take into account the Pauli exclusion principle in two-phonon terms of (6.33), consider the double commutator

$$\Big[[Q_{g'\sigma'}, Q^+_{g\sigma}], Q^+_{g_2\sigma_2}\Big] = \sum_{g_2'\sigma_2'}\{\mathcal{K}(g_2'\sigma_2', g'\sigma'|g\sigma, g_2\sigma_2)Q^+_{g_2'\sigma_2'} + \widetilde{\mathcal{K}}(g_2'\sigma_2', g'\sigma'|g\sigma, g_2\sigma_2)Q_{g_2'\sigma_2'}$$

and introduce the quantity

$$\mathcal{K}^{K_0}(g_2'g'|gg_2) = \sum_{\sigma\sigma_2\sigma'\sigma_2'} \delta_{\sigma'\mu'+\sigma_2'\mu_2',\sigma_0K_0}\delta_{\sigma\mu+\sigma_2\mu_2,\sigma_0K_0}\mathcal{K}(g_2'\sigma_2', g'\sigma'|g\sigma, g_2\sigma_2). \tag{6.34}$$

As we show in [140, 165], the absolute diagonal values of $\mathcal{K}(g_2g|gg_2)$ are much greater than the non-diagonal ones. As a rule, it is sufficient to use only the quasi-diagonal terms, which for $K_0 = \mu + \mu_2$ and $K_0 = \mu - \mu_2$, have the following form [276]:

$$\mathcal{K}^{K_0}(g_2\lambda\mu i'|\lambda\mu i g_2) = -\delta_{\mu+\mu_2,K_0}\frac{1}{1+\delta_{gg_2}} \times \sum_{q_1q_2q_3q_4} \delta_{K_2K_3}\Big[\psi^{\lambda\mu i'}_{q_1q_3}\psi^{\lambda\mu i}_{q_1q_2}\psi^{g_2}_{q_4q_2}\psi^{g_2}_{q_4q_3} - \varphi^{\lambda\mu i'}_{q_1q_3}\varphi^{\lambda\mu i}_{q_1q_2}\varphi^{g_2}_{q_4q_2}\varphi^{g_2}_{q_4q_3}\Big]\Big[\delta_{K_1-K_2,\mu}\delta_{K_4-K_2,\mu_2} + \delta_{K_2-K_1,\mu}\delta_{K_2-K_4,\mu_2} + \delta_{K_2-K_1,\mu}\delta_{K_2+K_4,\mu_2} + \delta_{K_2+K_1,\mu}\delta_{K_2-K_4,\mu_2} + \delta_{K_1+K_2,\mu}\delta_{K_2+K_4,\mu_2}\Big] \tag{6.35}$$

$$\begin{aligned}
&\mathcal{K}^{K_0}(g_2\lambda\mu i'|\lambda\mu i g_2)\\
&\quad = -\delta_{\mu-\mu_2,K_0}\frac{1}{1+\delta_{gg_2}}\\
&\quad \times \sum_{q_1q_2q_3q_4} \delta_{K_2K_3}\Big[\psi^{\lambda\mu i'}_{q_1q_3}\psi^{\lambda\mu i}_{q_1q_2}\psi^{g_2}_{q_4q_2}\psi^{g_2}_{q_4q_3}\\
&\quad - \varphi^{\lambda\mu i'}_{q_1q_3}\varphi^{\lambda\mu i}_{q_1q_2}\varphi^{g_2}_{q_4q_2}\varphi^{g_2}_{q_4q_3}\Big]\Big[\delta_{K_1-K_2,\mu}\delta_{K_2-K_4,\mu_2}\\
&\quad + \delta_{K_2-K_1,\mu}\delta_{K_4-K_2,\mu_2} + \delta_{K_1-K_2,\mu}\delta_{K_2+K_4,\mu_2}\\
&\quad + \delta_{K_1+K_2,\mu}\delta_{K_4-K_2,\mu_2}\Big]. \qquad (6.35')
\end{aligned}$$

The diagonal values of this function for $i' = i$ will be denoted by $\mathcal{K}^{K_0}(gg_2)$.

These calculations make use of the following commutation relation:

$$\begin{aligned}
&\Big[Q_{g'\sigma'},Q^+_{g\sigma}\Big]\\
&\quad = \delta_{gg'}\delta_{\sigma\sigma'} - \sum_{q_1q_2q_3\sigma_3}\Big\{\Big[\psi^{g'}_{q_1q_3}\psi^{g}_{q_1q_2}(\delta_{\sigma_3(K_2-K_1),\sigma\mu}\delta_{\sigma_3(K_3-K_1),\sigma'\mu'}\\
&\quad + \delta_{\sigma\sigma'}\delta_{\sigma_3\sigma}\delta_{K_1+K_3,\mu'}\delta_{K_1+K_2,\mu}) - \varphi^{g'}_{q_1q_2}\varphi^{g}_{q_1q_3}\\
&\quad \times (\delta_{\sigma_3(K_1-K_2),\sigma'\mu'}\delta_{\sigma_3(K_1-K_3),\sigma\mu}\\
&\quad + \delta_{\sigma\sigma'}\delta_{\sigma_3,-\sigma}\delta_{K_1+K_2,\mu}\delta_{K_1+K_3,\mu})\Big]\alpha^+_{q_2\sigma_3}\alpha_{q_3\sigma_3} + \Big[\psi^{g'}_{q_1q_3}\psi^{g}_{q_1q_2}\\
&\quad \times (\delta_{\sigma_3,-\sigma'}\delta_{\sigma'(K_2-K_1),\sigma\mu}\delta_{K_1+K_3,\mu'} - \delta_{\sigma_3\sigma}\delta_{\sigma(K_3-K_1),\sigma'\mu'}\\
&\quad \times \delta_{K_1+K_2,\mu}) + \varphi^{g'}_{q_1q_2}\varphi^{g}_{q_1q_3}(\delta_{\sigma_3,-\sigma'}\delta_{K_1+K_2),\mu'}\delta_{\sigma'(K_3-K_1),\sigma\mu}\\
&\quad - \delta_{\sigma_3\sigma}\delta_{K_1+K_3,\mu}\delta_{\sigma(K_2-K_1),\sigma'\mu'})\Big]\sigma_3\alpha^+_{q_2\sigma_3}\alpha_{q_3,-\sigma_3}\Big\}. \qquad (6.36)
\end{aligned}$$

In the approximation diagonal in $\mathcal{K}^{K_0}$, the condition of normalization of the wave function, (6.33), changes to

$$\sum_{i_0}(R^{\nu}_{i_0})^2 + \sum_{g\geqslant g_2}(P^{\nu}_{gg_2})^2[1+\mathcal{K}^{K_0}(gg_2)] = 1. \qquad (6.37)$$

The Hamiltonian of the model can be taken in the form (6.21). We will calculate the expectation value of H_{QPNM} over state (6.33). As a result, the approximation which is quasi-diagonal in $\mathcal{K}^{K_0}$ gives

$$\begin{aligned}
&(\Psi^*_\nu(K_0^{\pi_0}\sigma_0)H_{\text{QPNM}}\Psi_\nu(K_0^{\pi_0}\sigma_0))\\
&\quad = \sum_{i_0}\omega_{g_0}(R^{\nu}_{i_0})^2 + \sum_{g_1\geqslant g_2}[\omega_{g_1}+\omega_{g_2}\\
&\quad + \Delta\omega(g_1g_2)](P^{\nu}_{g_1g_2})^2[1+\mathcal{K}^{K_0}(g_1g_2)]
\end{aligned}$$

$$-2\sum_{i_0 g_1 \geqslant g_2} \frac{1+\mathcal{K}^{K_0}(g_1 g_2)}{(1+\delta_{g_1 g_2})^{1/2}[1+\delta_{K_0 0}(1-\delta_{\mu_1 0})]^{1/2}} \times R^{\nu}_{i_0} P^{\nu}_{g_1 g_2} U_{g_1 g_2}(g_0) \tag{6.38}$$

where

$$\Delta\omega(g_1 g_2) = -\frac{1}{4}\sum_{\tau i'}\Big\{\frac{X^{g_1}(\tau)+X^{\lambda_1\mu_1 i'}(\tau)}{(\mathcal{Y}^{g_1}_{\tau}\mathcal{Y}^{\lambda_1\mu_1 i'}_{\tau})^{1/2}}\mathcal{K}^{K_0}(g_1\lambda_1\mu_1 i'|g_1 g_2) + \frac{X^{g_2}(\tau)+X^{\lambda_2\mu_2 i'}(\tau)}{(\mathcal{Y}^{g_2}_{\tau}\mathcal{Y}^{\lambda_2\mu_2 i'}_{\tau})^{1/2}}\mathcal{K}^{K_0}(g_1\lambda_2\mu_2 i'|g_2 g_1)\Big\} \tag{6.39}$$

or, for the Hamiltonian H_{QPNM}

$$\Delta\omega(k_1 i_1, k_2 i_2) = -\sum_{i'}\Big\{\mathcal{K}^{K_0}(K_2 i_2, K_1 i'|K_1 i_1, K_2 i_2)W^{K_0}_{i,i'} + \mathcal{K}^{K_0}(K_2 i', K_1 i_1|K_1 i_1, K_2 i_2)W^{K_0}_{i_2 i'}\Big\} \tag{6.39$'$}$$

where $W^{K_0}_{i,i'}$ is given by formula (6.6″),

$$\sum_{\sigma_1\sigma_2}\delta_{\sigma_1\mu_1+\sigma_2\mu_2,\sigma_0 K_0}\{\langle Q_{g_0\sigma_0} H_{Mvq} Q^{+}_{g_1\sigma_1} Q^{+}_{g_2\sigma_2}\rangle + \langle Q_{g_2\sigma_2} Q_{g_1\sigma_1} H_{Mvq} Q^{+}_{g_0\sigma_0}\rangle\} = -2U_{g_1 g_2}(g_0) \times \left[1+\mathcal{K}^{K_0}(g_1 g_2)\right] \tag{6.40}$$

$$U_{g_1 g_2}(g_0) = \frac{1}{\sqrt{2}}\sum_{\tau}\left[\frac{S^{\lambda_1\mu_1}_{f\tau}(g_0 g_2)}{(\mathcal{Y}^{g_1}_{-\tau})^{1/2}} + \frac{S^{\lambda_2\mu_2}_{f\tau}(g_0 g_1)}{(\mathcal{Y}^{g_2}_{\tau})^{1/2}}\right] \tag{6.41}$$

$$S^{\lambda_1\mu_1}_{f\tau}(g_0 g_2) = \left(\frac{1+\delta_{K_0 0}}{1+\delta_{\mu_2 0}}\right)^{1/2}\sum_{qq'q_3}{}^{\tau} f^{\lambda_1\mu_1}(qq')v^{(-)}_{qq'} \times \left(\psi^{g_0}_{q_3 q'}\psi^{g_2}_{q_3 q} + \varphi^{g_0}_{q_3 q}\varphi^{g_2}_{q_3 q'}\right)\Theta^{\mu_1\mu_2\mu_0}_{KK'K_3} \tag{6.41$'$}$$

where $\theta^{\mu_1\mu_2 K_0}_{KK'K_3} = -1$ if $K+K'=\mu$ or $K_3+K'=\mu$ or $\mu_1+\mu_2=K_0$ or $\mu_1-\mu_2=K_0$, otherwise $\theta^{\mu_1\mu_2 K_0}_{KK'K_3}=1$ (see [276]).

Let us use the variational principle in the form

$$\delta\{(\Psi^{*}_{\nu}(K_0^{\pi_0}\sigma_0)H_{\text{QPNM}}\Psi_{\nu}(K_0^{\pi_0}\sigma_0)) - n_{\nu}[(\Psi^{*}_{\nu}\Psi_{\nu})-1]\} = 0 \tag{6.42}$$

and obtain the equations

$$(\omega_{g_0} - \eta_\nu)R^\nu_{i_0} - \sum_{g_1 \geqslant g_2} \frac{1 + \mathcal{K}^{K_0}(g_1 g_2)}{(1 + \delta_{g_1 g_2})^{1/2}[1 + \delta_{K_0 0}(1 - \delta_{\mu_1 0})]^{1/2}}$$
$$U_{g_1 g_2}(g_0) P^\nu_{g_1 g_2} = 0$$
$$[\omega_{g_1} + \omega_{g_2} + \Delta\omega(g_1 g_2) - \eta_\nu] P^\nu_{g_1 g_2} - (1 + \delta_{g_1 g_2})^{-1/2} \qquad (6.43)$$
$$\times [1 + \delta_{K_0 0}(1 - \delta_{\mu_1 0})]^{-1/2} \sum^{\nu}_{i'_0} R^\nu_{i'_0} U_{g_1 g_2}(\lambda_0 \mu_0 i'_0) = 0.$$

The second equation gives us $P^\nu_{g_1 g_2}$ which we substitute into the first equation and obtain a secular equation for calculating the energies η_ν of the states described by wave function (6.33):

$$\Theta(\eta_\nu) \equiv \det \left\| (\omega_{g_0} - \eta_\nu)\delta_{i_0 i'_0} - \sum_{g_1 \geqslant g_2} \frac{1 + \mathcal{K}^{K_0}(g_1 g_2)}{(1 + \delta_{g_1 g_2})[1 + \delta_{K_0 0}(1 - \delta_{\mu_1 0})]} \times \frac{U_{g_1 g_2}(\lambda_0 \mu_0 i_0) U_{g_1 g_2}(\lambda_0 \mu_0 i'_0)}{\omega_{g_1} + \omega_{g_2} + \Delta\omega(g_1 g_2) - \eta_\nu} \right\| = 0. \qquad (6.44)$$

The rank of this determinant equals the number of single-phonon terms in wave function (6.33). In this case, the diagrams given in figures 4.7 (*a–b*) are taken into account. If the approximation diagonal in $\mathcal{K}$ is not used, we obtain, just as for spherical nuclei (see (4.59)), a secular equation in the two-phonon states space. Taking the Pauli exclusion principle into account in the two-phonon terms of (6.33) leads to the appearance of the factor $1 + \mathcal{K}^{K_0}(g_1 g_2)$ in (6.44) and results in a shift of $\Delta\omega(g_1 g_2)$ of the two-phonon pole.

In order to find the explicit form of the functions $R^\nu_{i'_0}$ and $P^\nu_{g_1 g_2}$, we resort to the same procedure that was used in deriving the functions $C^\nu_{q'_0}$ and $D^\nu_{q g_2}$; this gives us the expressions

$$R^\nu_{i'_0}/R^\nu_{i_0} = \Theta_{i_0}(i'_0; \eta_\nu)/\Theta_{i_0}(\eta_\nu) \qquad (6.45)$$

$$(R^\nu_{i'_0})^2 = 1 + \Theta^{-2}_{i_0}(\eta_\nu) \Bigg\{ \sum_{i'_0 \neq i_0} \Theta^2_{i_0}(i'_0; \eta_\nu)$$
$$+ \sum_{g_1 \geqslant g_2} \Bigg[\frac{1 + \mathcal{K}^{K_0}(g_1 g_2)}{(1 + \delta_{g_1 g_2})^{1/2}[1 + \delta_{K_0 0}(1 - \delta_{\mu_1 0})]^{1/2}}$$

$$\times \frac{U_{g_1g_2}(\lambda_0\mu_0 i_0)\Theta_{i_0}(\eta_\nu) + \sum_{i'_0 \neq i_0} \Theta(i'_0;\eta_\nu) U_{g_1g_2}(\lambda_0\mu_0 i'_0)}{\omega_{g_1} + \omega_{g_2} + \Delta\omega(g_1g_2) - \eta_\nu}\Bigg]^2\Bigg\} \tag{6.45'}$$

where the rank of the determinant $\theta_{i_0}(\eta_\nu)$ is less by unity than that of the determinant $\theta(\eta_\nu)$, and $\theta_{i_0}(i'_0 : \eta_\nu)$ is obtained from $\theta_{i_0}(\eta_\nu)$ by replacing the i'_0 column by the column of constant terms. The function $P^\nu_{g_1g_2}$ is found from the second equation of (6.43).

Equations (6.44), (6.45) and (6.45′) coincide in form with the equations given in [69, 70] in which pn and pp multipole, spin-multipole and tensor separable interactions of rank $n_{\max} > 1$ are considered. Taking into account ph and pp separable $n_{\max} > 1$ interactions of electric and magnetic types leads to more complex expressions for the shift $\Delta\omega(g_1g_2)$ and the function $U_{g_1g_2}(g_0)$ as compared with formulas (6.39) and (6.41). The QPNM is so formulated that all complications due to the complex form of interactions between quasiparticles get reduced to additional complexity in RPA equations. Fortunately, the solution of RPA equations with arbitrarily complex interactions never meets with excessive difficulties.

Therefore, the form of equations (6.43), (6.44), (6.45) and (6.45′), including the rank of determinant (6.44), is independent of which multipole, spin-multipole and tensor interactions have been taken into account; neither does it depend on the rank $n_{\max}$ of separable interactions. This means that calculations in QPNM can be performed for arbitrarily complex interactions given in a separable form. No additional difficulties, caused by the complexity of interactions, arise when three-phonon terms are added to wave function (6.33).

If $\mathcal{K}^{K_0}(g_1g_2) = 0$, $\Delta\omega(g_1g_2) = 0$, equations (6.43) and (6.44) take the form

$$\begin{aligned}
&(\omega_{g_0} - \eta_\nu)R^\nu_{i_0} - \sum_{g_1 \geqslant g_2} (1+\delta_{g_1g_2})^{-1/2} \\
&\qquad \times [1+\delta_{K_0 0}(1-\delta_{\mu_1 0})]^{-1/2} U_{g_1g_2}(g_0) P^\nu_{g_1g_2} = 0 \\
&(\omega_{g_1} + \omega_{g_2} - \eta_\nu)P^\nu_{g_1g_2} - (1+\delta_{g_1g_2})^{-1/2} \\
&\qquad \times [1+\delta_{K_0 0}(1-\delta_{\mu_1 0})]^{-1/2} \sum_{i'_0} U_{g_1g_2}(\lambda_0\mu_0 i'_0) R^\nu_{i'_0} = 0
\end{aligned} \tag{6.46}$$

$$\det \left\| (\omega_{g_0} - \eta_\nu)\delta_{i_0 i'_0} - \sum_{g_1 \geqslant g_2} (1+\delta_{g_1g_2})^{-1} \times [1+\delta_{K_0 0}(1-\delta_{\mu_1 0})]^{-1} \frac{U_{g_1g_2}(\lambda_0\mu_0 i_0) U_{g_1g_2}(\lambda_0\mu_0 i'_0)}{\omega_{g_1} + \omega_{g_2} - \eta_\nu} \right\| = 0. \tag{6.47}$$

These formulas are used to calculate the properties of non-rotational states of deformed nuclei.

6.2 FRAGMENTATION OF SINGLE-PARTICLE STATES

6.2.1

It is possible to choose the mean field in strongly deformed nuclei in such manner that the density matrix $\rho(q, q')$ is diagonal for the ground states of nuclei in the stability region. This is the case of validity of the independent quasiparticles model. Owing to the superfluid pairing correlations, the low-lying state can be characterized by quantum numbers of single-particle level occupied by the quasiparticles. If there were no such interaction in nuclei, weak residual forces would smear the single-particle structure. The independent particles model gives a correct first-approximation picture of spectra of odd deformed nuclei. Data tables, such as given in [12], characterize the sequence of filling of single-particle states as the number of neutrons and protons increases.

The structure of the states of deformed nuclei is dictated by the internal motion and rotation, and also by the Coriolis interaction coupling them. A simple model, postulating that a nucleus is a rotor coupled to a particle or a quasiparticle [4] played an important role for understanding the general features of spectra of odd deformed nuclei. A large number of nuclear spectra were analysed in the framework of this model [277]. This model needs, for the calculation of spectra and probabilities of electromagnetic transitions, an artificial attenuation of the Coriolis interaction; hence, attempts to improve the model continue (see, e.g., [278]). The simple picture of nuclear spectra given by the quasiparticle $\otimes$ rotor model becomes more complicated when coupling of several rotational bands is taken into account. It was shown in [279], using the three rotational bands of ^{171}Yb as an example, that the analysis of experimental data is often very difficult.

As a rule, single-particle energies and wave functions of the axially symmetric Saxon–Woods potential are used for a microscopic description of deformed nuclei. Calculations presented in [268, 269] and in some others used single-particle energies and wave functions of the Saxon–Woods potential and the monopole pairing. The pairing constants were found from the nuclear mass difference (see [6, 12]). In the fitting of parameters of the Saxon–Woods potential, carried out in 1968–1974, the region of deformed nuclei was divided into the following zones of the mass number A: 155, 165, 173, 181, 229, 239, 247 and 255. The Saxon–Woods potential parameters for the mentioned bands are given in [12, 66, 268, 269].

In describing odd deformed nuclei, the wave function is taken in the form (6.1) with $\Psi_\nu(K^\pi\sigma)$ given by (6.16), and the quasiparticle–phonon interaction is taken into account. In [71, 280, 281] and in some other papers, non-rotational states of odd deformed nuclei were calculated in the rare-earth region and in the actinide region, using wave function (6.16) and formulas (6.27)–(6.32″). A fairly satisfactory description of the energies and structure of low-lying states was obtained. There is an agreement with the experimental data [282, 283] (a large part of these data were obtained after the calculations were published).

6.2.2

Let us consider electromagnetic transitions in odd nuclei. The reduced probability of $E\lambda$ transitions between states described by wave functions (6.1) and (6.16) equals

$$\begin{aligned}B(E\lambda; I_0^{\pi_0} K_0 \nu_0 \to I_f^{\pi_f} K_f \nu_f) &= \frac{2\lambda+1}{4\pi} \\ &\times |\langle I_0 K_0 \lambda K_f - K_i | I_f K_f\rangle [\Psi^*_{\nu_f}(K_f^{\pi_f}) \mathfrak{M}(E\lambda; K_f - K_i) \\ &\times \Psi_{\nu_0}(K_0^{\pi_0})] + (-)^{I_0+K_0} \langle I_0 - K_0 \lambda K_f + K_0 | I_f K_f\rangle \\ &\times (\Psi^*_{\nu_f}(K_f^{\pi_f}) \mathfrak{M}(E\lambda; K_f + K_i) \Psi_0(K_0^{\pi_0}))|^2 \qquad (6.48)\end{aligned}$$

where

$$\begin{aligned}&(\Psi^*_{\nu_f}(K_f^{\pi_f}) \mathfrak{M}(E\lambda; \mu) \Psi_{\nu_0}(K_0^{\pi_0})) \\ &= \sum_{q_0 q_f} C^{\nu_f}_{q_f} C^{\nu_0}_{q_0} v^{(-)}_{q_f q_0} \\ &\times e^{(\lambda\mu)}_{\text{eff}}(\tau) p^{\lambda\mu}(q_0 q_f) + \sum_i \mathfrak{M}^E_{\lambda\mu i} \sum_{q g_2} \left\{ \sum_{q_0 q_f} C^{\nu_f}_{q_f} D^{\nu_0}_{q g_2} \right. \\ &\times [\delta_{q_f q} \delta_{g g_2} - \mathcal{L}^{K_f}(g q_f | q g_2)] + \sum_{q_0} D^{\nu_f}_{q g_2} C^{\nu_0}_{q_0} [\delta_{q q_0} \delta_{g g_2} \\ &\left. - \mathcal{L}^{K_0}(g q_0 | q g_2)] \right\} + \sum_{q q' g_2} D^{\nu_f}_{q' g_2} D^{\nu_0}_{q g_2} v^{(-)}_{q q'} e^{(\lambda\mu)}_{\text{eff}} p^{\lambda\mu}(q q') \\ &\times [1 - \mathcal{L}(q g_2)]. \qquad (6.49)\end{aligned}$$

Here $e^{(\lambda\mu)}_{\text{eff}}(\tau)$ is defined by (2.77″); for an odd neutron (proton) nucleus, neutron (proton) single-particle states are used; $\mathcal{M}^E_{\lambda\mu i}$ and $\mathcal{L}^K(g q_0 | q g_2)$ are given by formulas (2.79′), (6.18) and (6.19), and $\mathcal{L}(q g_2)$ is the sum of $(\psi^{g_2}_{q q_2})^2$ over q_2, with the appropriate Kroneker symbols (see [285]). If the radial dependence of multipole forces is taken in the form $R_\lambda(r) = r^\lambda$, the function $\mathcal{M}_{\lambda\mu i}$ is given by (2.79″).

The terms of matrix element (6.49) which contain $C^{\nu_f}_{q_0 f} C^{\nu_0}_{q_0}$ and $D^{\nu_f}_{q' g_2} D^{\nu_0}_{q g_2}$ describe the $E\lambda$ transitions between single-quasiparticle components and between quasiparticle ⊗ phonon components. The terms containing CD describe transitions between single-quasiparticle components and quasiparticle ⊗ phonon components, which are enhanced if the $\lambda\mu i$ phonon is collective. The formulas for reduced probabilities of the $M\lambda$ transitions are obtained from (6.48) and (6.49) by the same substitution as they are obtained from (5.19) and (5.18) in the case of odd spherical nuclei.

The experimental data given on γ transitions and magnetic momenta s are listed in systematic form in [282, 283, 284–287] and in some other publications. An analysis of experimental data on electromagnetic transitions in odd deformed nuclei, calculations of reduced probabilities and a comparison of the theory with experimental results can be found in a considerable number of papers, for instance, [4, 6, 12, 284–292]. The reduced probabilities of electromagnetic transitions greatly depend on the structure of the states involved in the transition. Hence, the experimental information on the absolute values of the probabilities of electromagnetic transitions plays an important part in improving the accuracy of the Saxon–Woods potential parameters and of the effective coupling constants, and also in clarifying the role of 'admixtures' to the dominating components of the wave functions. For example, the probabilities of K-forbidden transitions are determined by the components with different values of K. The probabilities of $E1$ transitions are strongly influenced by the admixtures in the wave functions which yield matrix elements allowed in the asymptotic quantum numbers. Small admixtures to the quasiparticle ⊗ $\lambda\mu i$ =221 phonon considerably increase the $E2$ transition probability. Obviously, the probabilities of a large number of $E\lambda$ transitions are determined by single-quasiparticle components of wave functions.

The above arguments can be illustrated by the calculations of [285] using formula (6.48) for the probabilities of $E2$ transitions with $\Delta K = \pm 2$. The results of calculations are given in table 6.1 in the form of hindrance factors

$$F = \frac{B(E2)_{\text{calc.}}}{B(E2)_{\text{exper.}}}$$

The calculations were performed ignoring (if $S = 0$), or taking into account (if $S \neq 0$), the Pauli exclusion principle in quasiparticle ⊗ phonon configurations, for the effective charges $e^{(2)}_{\text{eff}}(p) = 1.2e$, $e^{(2)}_{\text{eff}}(n) = 0.2e$, $e^{(2)}_{\text{eff}}(p) = e$, $e^{(2)}_{\text{eff}}(n) = 0$. The experimental values of $B(E2)$ are given in units of $e^2 b^2$ ($b = 10^{-24}$ cm^2). The dominating components, the single-quasiparticle or the phonon $\lambda\mu i$ =221, are given for the initial, i, and the final, f, states. The table shows that a sufficiently satisfactory description of experimental data was obtained. A small admixture of the quasiparticle ⊗ {221} phonon to the wave function results in a considerable enhancement of the $E2$ transition.

When the reduced probabilities of electromagnetic transitions between levels of various rotational bands are calculated, it is necessary to take into account the Coriolis interaction (see [284, 289, 293]). The importance of taking into account the Coriolis interaction and the quasiparticle–phonons interaction can be illustrated with $E1$ transitions between the levels of the bands constructed of states with dominating one-quasi-proton components 532↑, 413↓ and 411↑ in ^{153,155}Eu and ^{155,157}Tb. Calculations in [292] were carried out taking into account the Coriolis interaction in the independent quasiparticles model (IQM+Cor) and in QPNM (QPNM+Cor); together with the experimental data, these results are shown in figure 6.2. The figure shows that IQM+Cor calculations do not give a correct description: deviations from the experimental data reach two to three orders of magnitude. At the same time, a qualitatively correct description of experimental data for the $E1$ transitions was obtained in the framework of the QPNM+Cor model. Hence, the calculations of the probabilities of the $E\lambda$ transitions in odd deformed nuclei must be carried out in the QPNM, with the Coriolis interaction taken into account.

The correct description of the low-lying excitations in odd-mass nuclei in IQM+Cor requires the attenuation of certain Coriolis-mixing matrix elements. The Coriolis attenuation problems are discussed in many papers (e.g., [289, 292]). Typically, in particle-rotor model calculations, in IQM+Cor and in other models the fitted attenuation coefficients are used. In the QPNM calculations with wave function (6.16), the coefficient $\alpha_{\mathrm{QPNM}} = C^i_{q_i} C^f_{q_f}$ appears. It is connected with the decrease in the contribution of single-particle components to the wave functions of the initial and final states. The fitted coefficients α_{fit} are compared in table 6.2 with the coefficients α_{QPNM}. The table shows that the quasiparticle–phonon interaction is mainly responsible for the attenuation of the Coriolis-mixing matrix elements.

6.2.3

Let us consider the vibrational states in odd deformed nuclei. The existence of vibrational states in them and the important role of the quasiparticle–phonon interaction have been known for some time already [294]. Let us clarify whether a set of vibrational states must exist at each single-quasiparticle level. To do this, we need to take into account the Pauli exclusion principle in quasiparticle ⊗ phonon configurations; correspondingly, we will make use of formulas (6.16), (6.20) and (6.24)–(6.29).

Let us consider the states of ^{159}Gd and ^{159}Tb where the quasiparticle ⊗ phonon dominate, and analyse how taking the Pauli exclusion principle into account affects them. We will operate with phonons whose struc-

Table 6.1 Hindrance factors F for $E2$ (ΔK =2) transitions calculated without ($S = 0$) and with ($S \neq 0$) the Pauli exclusion principle taken into account.

Nucleus	$I^{\pi}K[Nn_z\Lambda]$		$B(E2)_{exp}$	F			
	i	f	e^2b^2	$e^{(2)}_{eff}(p) = e$, $e^{(2)}_{eff}(n) = 0$		$e^{(2)}_{eff}(p) = 1.2e$, $e^{(2)}_{eff}(n) = 0.2e$	
				$S = 0$	$S \neq 0$	$S = 0$	$S \neq 0$
^{163}Dy	$1/2^- 1/2$ [521]	$5/2^- 5/2$ [523]	1.25×10^{-2}	0.88	0.41	2.2	1.1
	$3/2^- 3/2$ {221}	$7/2^- 5/2$ [523]	4.8×10^{-2}	1.2	1.2	3.1	3.1
^{165}Ho	$11/2^- 11/2$ {221}	$7/2^- 7/2$ [523]	3.8×10^{-2}	1.0	1.0	2.6	2.6
^{165}Er	$1/2^- 1/2$ [521]	$5/2^- 5/2$ [523]	1.05×10^{-2}	2.3	1.3	5.5	3.1
^{165}Tm	$5/2^+ 5/2$ {221}	$3/2^+ 1/2$ [411]	4.1×10^{-2}	0.80	0.37	1.9	0.85
	$3/2^+ 3/2$ [651]	$7/2^+ 7/2$ [633]	3.8×10^{-2}	0.95	0.92	2.6	2.6
^{167}Er	$11/2^+ 11/2$ {221}	$7/2^+ 7/2$ [633]	4.8×10^{-2}	0.77	0.70	2.1	2.0
^{169}Yb	$5/2^- 5/2$ [512]	$1/2^- 1/2$ [521]	9.4×10^{-5}	0.85	0.35	3.0	1.4
^{177}Lu	$1/2^+ 1/2$ [411]	$5/2^+ 5/2$ [402]	6.3×10^{-6}	2.2	2.9	3.7	4.9
^{179}Ta	$1/2^+ 1/2$ [411]	$5/2^+ 5/2$ [402]	8.24×10^{-5}	2.02	0.03	0.03	0.06

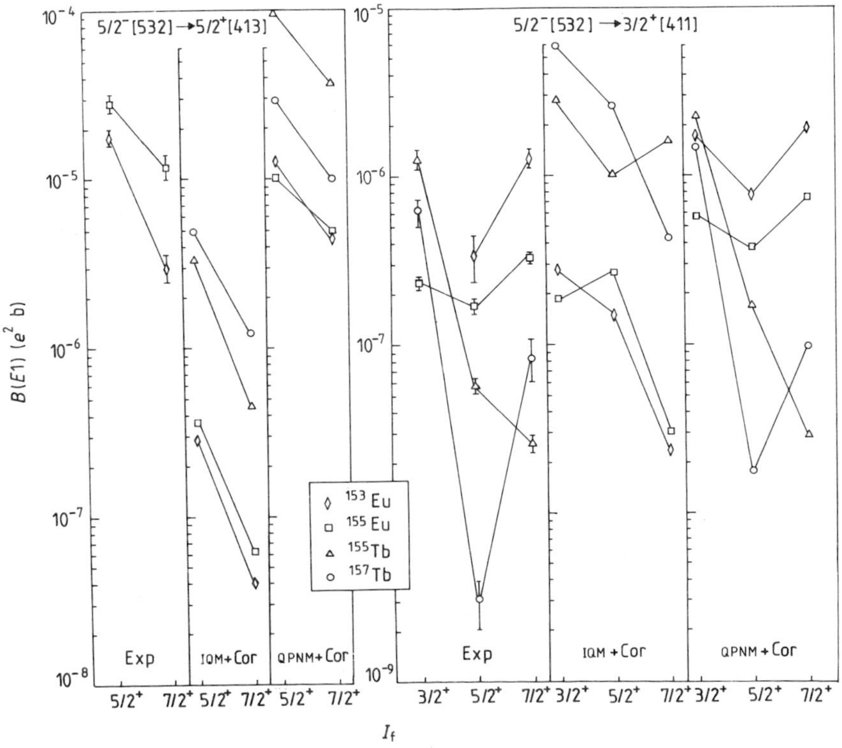

Figure 6.2 The reduced probabilities of $E1$ transitions between the members of the $5/2^-[532]$ band and the $5/2^+[413]$ and $3/2^+[411]$ bands in ^{153,155}Eu and ^{155,157}Tb, calculated in the IQM+Cor and QPNM+Cor.

Table 6.2 Attenuation coefficient of the Coriolis-mixing matrix elements.

f	i	^{153}Eu		^{155}Eu		^{155}Tb		^{157}Tb	
		fit	QPNM	fit	QPNM	fit	QPNM	fit	QPNM
523↑	532 ↑	0.7	0.9	0.8	0.9	0.7	0.9	0.8	0.9
532↑	541 ↑	0.6	0.9	0.6	0.9	0.6	0.9	0.7	0.9
541↑	550 ↑	0.8	0.7	0.7	0.7	0.8	0.7	0.8	0.6
404↓	413 ↓	0.8	0.9	0.8	0.9	0.8	0.9	0.8	0.9
402↑	411 ↑	0.4	0.4	0.5	0.7	0.7	0.8	0.6	0.7
411↑	420 ↑	0.5	0.8	0.6	0.7	0.9	0.8	0.8	0.8

ture is given in table 6.3. The calculated values of $\mathcal{L}^{K_0}(qg)$ and $\Delta\omega(qg)$ are given in table 6.4.

The states $7/2^-$ and $1/2^-$ in ^{159}Gd consist of a 521↑ quasiparticle and

Table 6.3 Characteristics of one-phonon states in ^{158}Gd.

$g = \lambda\mu i$	ω_g MeV	Phonon structure	
221	1.18	nn 521↑ + 521↓ 29%	nn 642↑ − 660↑ 16%
		nn 523↓ − 521↓ 11%	nn 411↑ + 411↓ 7%
222	2.54	nn 642↑ − 660↑ 72%	nn 521↑ + 521↓ 27%
301	1.26	nn 642↑ − 523↓ 39%	nn 651↑ − 521↑ 20%
		nn 413↓ − 532↑ 4%	nn 411↑ − 541↑ 3%

Table 6.4 The functions $\mathcal{L}^{K_0}(qg)$ and shifts of the poles $\Delta\omega(qg)$ of the secular equation.

Nucleus	K_0^π	$q\otimes\lambda\mu i \equiv g$	$\mathcal{L}^{K_0}(qg)$	$\Delta\omega(qg)$
^{159}Gd	7/2⁻	521↑⊗221	−0.34	1.8
	1/2⁻	521↑⊗221	−0.01	0.07
	7/2⁻	521↑⊗222	−0.27	0.1
	9/2⁺	642↑⊗221	−0.2	1.0
	1/2⁺	642↑⊗221	0	0
	9/2⁺	642↑⊗222	−0.72	0.07
	3/2⁺	521↑⊗301	−0.22	1.3
^{159}Tb	7/2⁺	411↑⊗221	−0.1	0.12
	1/2⁺	411↑⊗221	−0.01	0.01
	3/2⁻	411↑⊗301	-0.6×10^{-3}	0.08

a 221 phonon whose wave function contains a neutron nn 521↑ + 521↓ component. Let us add up the projections

$$K_0 = 7/2 = K_{521\uparrow} + \mu = K_{521\uparrow} + \ldots(K_{521\uparrow} + K_{521\downarrow})$$
$$K_0 = -1/2 = K_{521\uparrow} - \mu = K_{521\uparrow} - \ldots(K_{521\uparrow} + K_{521\downarrow}).$$

This expression shows that for $K_0 = 7/2$, the projections $K_{521\uparrow}$ of a free quasiparticle and quasiparticle in a phonon have identical signs. Such components are forbidden by the Pauli exclusion principle, so that $\mathcal{L}^{K_0=7/2}(521\uparrow, 221) = -0.34$ and the shift is large. The large shift is also caused by the strongly collective nature of the γ-vibrational state. For $K_0 = 1/2$, the projections $K_{521\uparrow}$ enter with opposite signs, the Pauli exclusion principle is not violated and the values of $\mathcal{L}^{K_0=1/2}(521\uparrow, 221)$ and $\Delta\omega(521\uparrow, 221)$ are small. The deviation from zero occurs owing to the violation of the Pauli exclusion principle in the small components of the 221 phonon and 521↑ quasiparticle. The values of $\mathcal{Y}^g_\tau$ are large for the non-collective phonon and a strong violation of the Pauli exclusion principle does not produce a large shift; an example of this is the $\lambda\mu i$ =222 phonon (see table 6.4).

Table 6.5 Vibrational states in odd nuclei.

Nucleus	K_0^π	Energy (MeV)	Configuration	Calculation: Contribution of configuration (%)	Calculation: $\Delta\eta_\nu$ (MeV)	Effect of Pauli principle
^{159}Ho	$3/2^-$	0.312	523↑⊗221	97	0.001	very weak
^{165}Ho	$3/2^-$	0.514	523↑⊗221	98	0.001	very weak
	$11/2^-$	0.689	523↑⊗221	99	0.01	very weak
	$5/2^+$	0.995	411↓⊗221	31	0.2	small
^{167}Er	$3/2^+$	0.532	633↑⊗221	12	0.01	very weak
^{169}Yb	$3/2^+$	0.720	633↑⊗221	11	0.01	very weak
	$3/2^-$	0.660	521↓⊗221	90	1.5	strong
^{233}Th	$1/2^+$	0.584	622↑⊗221	54	0.001	very weak
	$1/2^+$	0.842	633↓⊗221	17	0.001	very weak
	$3/2^+$	0.815	631↓⊗221	90	1.0	strong
^{233}U	$5/2^-$	0.748	633↓⊗301	100	2	strong
^{235}U	$7/2^+$	0.444	743↓⊗301	6	1	strong
	$3/2^-$	0.636	743↑⊗221	100	0.01	very weak
	$1/2^-$	0.761	631↑⊗301	100	1.2	strong
	$1/2^+$	0.844	622↑⊗221	40	0.001	very weak
^{237}Np	$5/2^-$	0.721	642↑⊗301	14	0.02	very weak
^{239}U	$1/2^+$	0.687	622↑⊗221	40	0.001	very weak
	$1/2^-$	0.814	631↑⊗301	40	0.2	small

Let us consider γ-vibrational and octupole states in odd nuclei. Table 6.5 lists the experimental data for specific quasiparticle ⊗ phonon components, the shifts of the energy centroid and the values of the wave function components. The last column of the table gives conclusions on the effect of taking into account the Pauli exclusion principle. Its effect is defined as very low if the Pauli exclusion principle can be ignored, as moderate if the vibrational state can exist, and as strong if the shift is large and the quasiparticle⊗phonon state must be fragmented over several nuclear levels.

Let us discuss results shown in table 6.5. γ-vibrational states, based on the proton single-particle state 523↑ are observed in ^{159,165}Ho. Taking the Pauli exclusion principle into account hardly affects the states with $K^\pi = 3/2^-$ and $K^\pi = 11/2^-$ and the configuration 523↑⊗221; such vibrational states must exist in accordance with the available experimental data. The Pauli exclusion principle leaves the 411↓⊗221 state with $K^\pi = 5/2^+$ almost unaffected. In ^{167}Er, we find the 633↑⊗221 state with $K^\pi = 3/2^+$ and the energy 0.532 MeV. According to calculations, the Pauli exclusion principle is only slightly violated in the 633↑⊗221 state with $K^\pi = 3/2^+$. The first state with $K^\pi = 3/2^+$ has the following structure: 651↑—76%, 633↑⊗221—12%, 660↑⊗221—5%. A large part of the strength of the 633↑⊗221 configuration is concentrated on the next level with $K^\pi = 3/2^+$. The state with $K^\pi = 3/2^+$ and the energy 0.532 MeV is strongly excited in the (d,d') reaction. In the state $521\downarrow\otimes 221$ with $K^\pi = 3/2^-$ in ^{169}Yb, the Pauli exclusion principle is strongly violated, so that the state must be fragmented over several levels. According to calculations, ^{169}Yb has a state with energy 0.660 MeV and a large component $521\downarrow$. The $\langle 521\uparrow|E2|521\downarrow\rangle$ matrix element is large, possibly explaining the strong $E2$ transition to the $521\downarrow$ state.

In ^{233}Th, the Pauli exclusion principle hardly influences the $622\uparrow\otimes 221$ and $633\downarrow\otimes 221$ states with $K^\pi = 1/2^+$ at all. In the $631\downarrow\otimes 221$ component with $K^\pi = 3/2^+$, the Pauli exclusion principle is strongly violated and $\mathcal{L}^{3/2}(631\downarrow 221) = -0.66$. Only a small part of strength is in the region of 1 MeV. Taking the Pauli exclusion principle into account strongly affects the $633\downarrow\otimes 301$ state with $K^\pi = 5/2^-$ in ^{233}U; its energy centroid is sufficiently large.

In ^{235}U, the structure of the first state with $K^\pi = 7/2^+$ is: $624\uparrow\otimes 221$—80%, $743\uparrow\otimes 301$—6%. A large part of the strength of $743\uparrow\otimes 301$ is concentrated at the second $7/2^+$ state with $K^\pi = 7/2^+$ and an energy above 1 MeV. The violation of the Pauli exclusion principle in the $743\uparrow\otimes 301$ state is small. In the $631\downarrow\otimes 301$ state with $K^\pi = 1/2^-$ it is slightly higher and its energy centroid lies above 2 MeV. The state with $K^\pi = 1/2^-$ and the energy 0.671 MeV in ^{235}U is excited in the (d,p) and (d,t) reactions: we have no direct experimental indications for the existence of a large $631\downarrow\otimes 301$ component. The $743\uparrow\otimes 221$ with $K^\pi = 3/2^-$, $622\uparrow\otimes 221$ with $K^\pi = 1/2^+$ and $633\downarrow\otimes 221$ with $K^\pi = 1/2^+$ states are unaffected by the Pauli exclusion principle. The energies of these states are about 1 MeV, and they must exist as γ-vibrational states. The $622\uparrow\otimes 221$ state with $K^\pi = 1/2^+$ in ^{239}U is not influenced by the Pauli exclusion principle; such a γ-vibrational state is observed experimentally.

Table 6.5 demonstrates that vibrational states must be expected to exist at many single-quasiparticle levels in odd deformed nuclei. In those cases, however, when the Pauli exclusion principle is strongly violated in the quasiparticle $\otimes$ phonon configuration, the corresponding vibrational states cannot exist. Further experimental study of vibrational states in deformed nuclei would be of considerable interest.

6.2.4

Let us consider the fragmentation of single-quasiparticle states. Using expression (6.28′), we obtain in the same manner as for spherical nuclei a strength function which describes the fragmentation of the single-quasiparticle state:

$$C^2_{q_0}(\eta) = \frac{1}{\pi}\,\mathrm{Im}\,\frac{1}{\mathcal{F}_{q_0}(\eta + i\Delta/2)} = \frac{1}{\pi}\,\mathrm{Im}\,\frac{\Theta_{q_0}(\eta + i\Delta/2)}{\Theta(\eta + i\Delta/2)}. \tag{6.50}$$

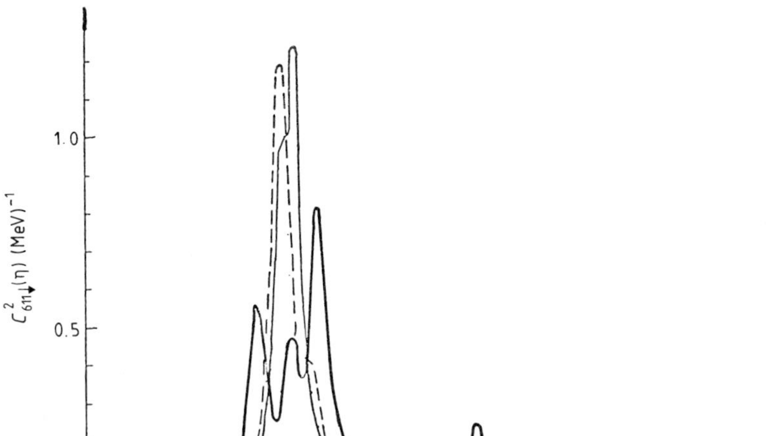

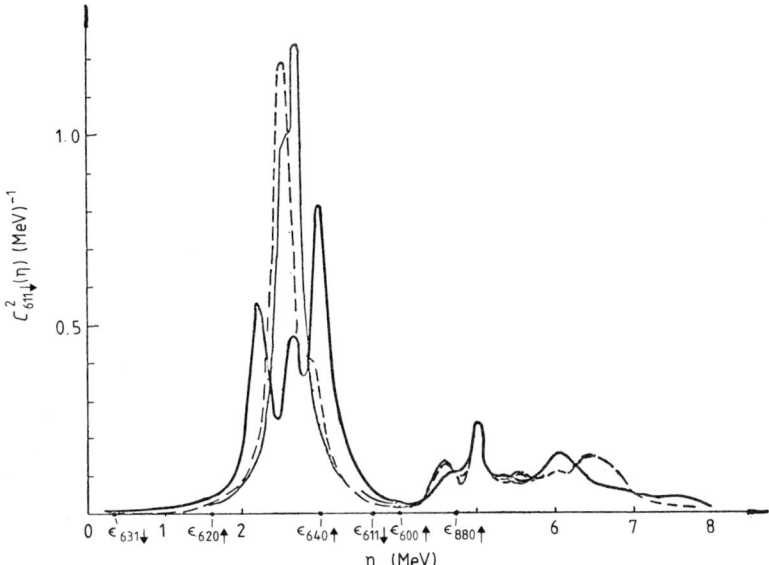

Figure 6.3 Effect of adding other states q_0' with $K^\pi = 1/2^+$ on the fragmentation of the $q_0 = 611\downarrow$ in ^{239}U: full curve with fine line—fragmentation of $611\downarrow$ with the effect of other states ignored; broken curve—fragmentation of $611\downarrow$ with the effect of $631\downarrow$ and $620\uparrow$ taken into account; full curve with bold line—fragmentation of $611\downarrow$ with the effect of the $631\downarrow$, $620\uparrow$, $640\uparrow$, $600\uparrow$ and $880\uparrow$ taken into account.

The role of the Pauli exclusion principle in the study of the fragmentation of single-quasiparticle states in the quasiparticle $\otimes$ phonon components is insignificant and approximation (6.30) can be used.

As the fragmentation regions of various single-quasiparticle states with identical values of K^π overlap, several other single-particle states q_0' with the same values of K^π have to be considered in analysing the fragmentation of a specific q_0 state. An example of strong influence of adding other states on the fragmentation of the $611\downarrow$ with $K^\pi = 1/2^+$ in ^{239}U is shown in figure 6.3. In other cases, the effect of such adding is small. When the fragmentation of single-quasiparticle states in odd deformed nuclei is studied, one has to take into account the influence of the nearest single-particle states with the same value of K^π.

Let us analyse how the fragmentation of a single-particle state changes, depending on the position of the level relative to the Fermi level. If the single-particle state lies close to the Fermi level, about 90% of its strength is concentrated at the lowest level with a given value of K^π, while the remaining 10% is distributed in a wide energy interval. The maximum

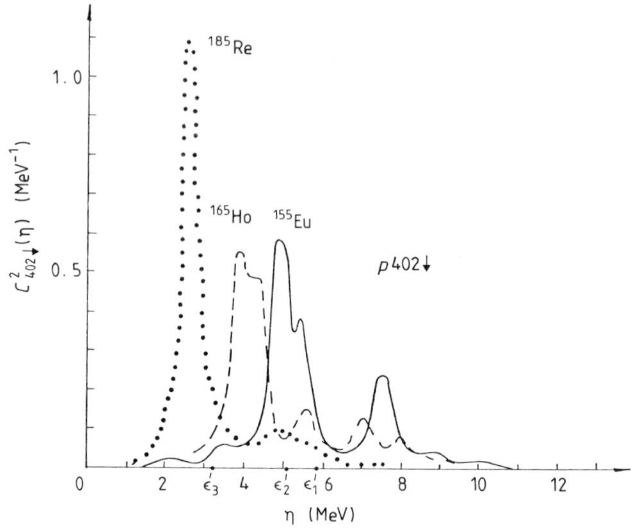

Figure 6.4 Fragmentation of the single-particle proton state with $K^\pi = 3/2^+$, 402↓ in ^{185}Re (dotted curve), in ^{165}Ho (broken curve) and in ^{155}Eu (full curve). The quasiparticle energies are denoted by ϵ_3, ϵ_2 and ϵ_1, respectively.

of the distribution is shifted by 0.5–1.5 MeV towards lower energies with respect to the quasiparticle energy ϵ_{q_0}. As the single-particle state moves away from the Fermi level, the strength concentrated at one level decreases and the distribution broadens [137]. If the energy of a single-particle state is separated from the Fermi energy by more than 2 MeV, the distribution maximum fluctuates around ϵ_{q_0}. An example of changes in the fragmentation of the proton state 402↓ with $K^\pi = 3/2^+$ in the sequence from ^{185}Re to ^{165}Ho and then to ^{155}Eu is given in figure 6.4. For ^{165}Ho and especially for ^{155}Eu, the 402↓ state is a high-lying particle state. In these nuclei, it is fragmented in a wide energy range; furthermore, other local maxima appear in addition to the large one.

The fragmentation essentially depends not only on the position but also on the quantum numbers of the single-particle states and on the degree of collectivity of the lowest vibrational states of a given nucleus. The shape of the distribution differs from the Breit–Wigner distribution. As a rule, there are several small peaks in addition to the main maximum. The height of these peaks is determined by the asymptotic quantum numbers of single-particle states and their relation to the vibration states. The distribution function is non-symmetric with respect to the maximum, owing to a less steep decrease with increasing excitation energies. As the quantum number K of a single-particle state increases, the fragmentation typically goes down. Crude estimates of the effect of the terms of the quasiparticle ⊗ two phonons wave function on the fragmentation of single-particle states show that the effect is small and only somewhat smoothes the highest maxima and the deepest minima.

6.2.5

One-nucleon transfer reactions manifest not the fragmentation of the single-particle state but the fragmentation of a specific nlj subshell of the spherical basis in a deformed nucleus. Section 6.1 demonstrated the effect of stable quadrupole deformation on the fragmentation of spherical subshells. Now we can calculate the subsequent fragmentation owing to the quasiparticle–phonons interaction.

We will use the expansion of a single-particle wave function of the axially symmetric Saxon–Woods potential in the wave function of the spherically symmetric Saxon–Woods potential in the form of (6.4). The strength function describing the fragmentation of the nlj subshell which enters the single-quasiparticle q_0 state with $K = K_0$ then takes the form

$$S_{nlj}^{q_0 K_0}(\eta) = \sum_{\nu} |a_{nlj}^{q_0 K_0} C_{q_0}^{\nu}|^2 \rho(\eta - \eta_\nu) = C_{q_0}^2(\eta)(a_{nlj}^{q_0 K_0})^2. \tag{6.51}$$

In order to calculate the fragmentation of the nlj subshell, we need to sum up over all single-particle states with a given value of $K_0^{\pi_0}$ and then over all values of K_0:

$$S_{nlj}^{K_0}(\eta) = \sum_{q_0} S_{nlj}^{q_0 K_0}(\eta) \tag{6.52}$$

$$S_{nlj}(\eta) = \sum_{q_0 K_0} S_{nlj}^{q_0 K_0}(\eta). \tag{6.53}$$

In calculating the spectroscopic strength functions which are to be compared with the functions extracted from the experimental data, one uses for (d,p)-type reactions, instead of $S_{nlj}^{q_0 K_0}$, the functions

$$\tilde{S}_{nlj}^{q_0 K_0}(\eta) = C_{q_0}^2(\eta)(a_{nlj}^{q_0 K_0} u_{q_0})^2 \tag{6.54}$$

and for (d,t)-type reactions,

$$\approx{S}_{nlj}^{q_0 K_0}(\eta) = C_{q_0}^2(\eta)(a_{nlj}^{q_0 K_0} v_{q_0})^2 \tag{6.55}$$

where u_q, v_q are the Bogoliubov transformation coefficients. Strength functions (6.54) and (6.55) are then substituted in formulas (6.52) and (6.53);

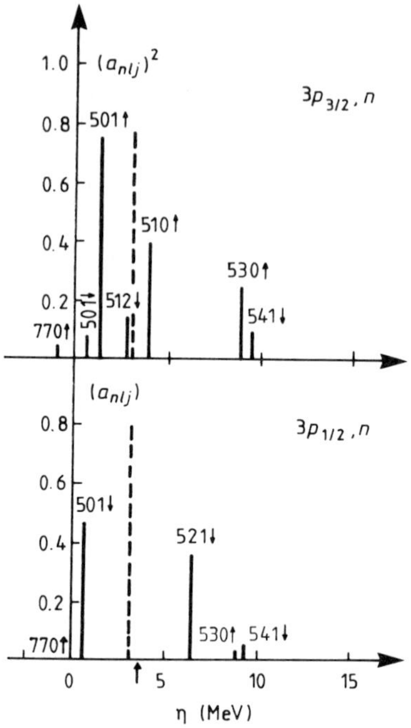

Figure 6.5 Strength distribution for $3p_{1/2}$ and $3p_{3/2}$ subshells over single-particle states of the Saxon–Woods potential, with β_2 =0.24 and β_4 =0.03: arrows indicate the position of subshells in the spherical basis; broken lines mark the Fermi level corresponding to the ^{183}W nucleus.

the corresponding functions are denoted by $\widetilde{S}^{K_0}_{nlj}(\eta)$, $\approx\!\!\!\!\!S^{K_0}_{nlj}(\eta)$ and $\tilde{S}_{nlj}(\eta)$, $\approx\!\!\!\!\!S_{nlj}(\eta)$.

Let us consider the fragmentation of neutron subshells $3p_{1/2}$ and $3p_{3/2}$, which manifest themselves in (d,p) reactions on tungsten isotopes. The fragmentation of the strength of these subshells in response to stable deformation is given in figure 6.5. A large part of the strength of the $3p_{3/2}$ neutron subshell is concentrated in the 501↑, 510↑ and 530↑ single-particle states. The strength of the $3p_{1/2}$ subshell is mostly concentrated on two levels, 501↓ and 521↓. The greatest contribution to the normalization of the 770↑ and 761↑ states comes from the $1j_{15/2}$ subshell; hence, they provide a very small contribution to the cross section of (d,p) reactions in which low-spin states are excited. Both in the case of the $3p_{1/2}$ and $3p_{3/2}$ subshells, and in the cases calculated in [295], the fragmentation of the spherical basis subshells was found to be substantial, owing to the stable deformation.

Figure 6.6 plots the strength functions $S_{nlj}(\eta)$ and the spectroscopic

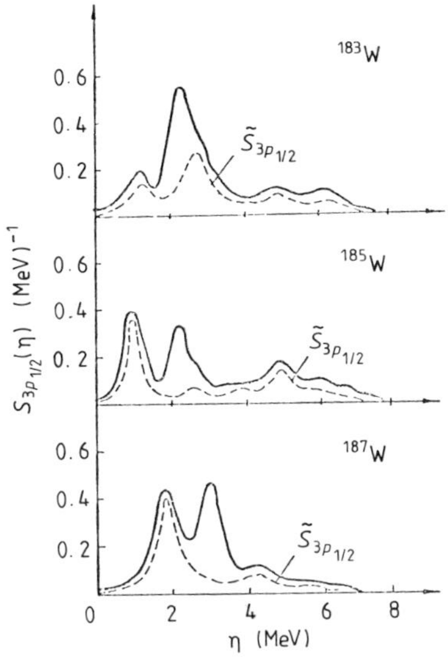

Figure 6.6 Fragmentation of the $3p_{1/2}$ neutron subshell in $^{183,185,187}W$ and the spectroscopic strength functions, $\tilde{S}_{3p_{1/2}}(\eta)$, (broken curves) in these nuclei.

strength functions $\tilde{S}_{nlj}(\eta)$ for the neutron $3p_{1/2}$ subshell in the $^{183,185,187}W$ isotopes (calculated in [296], with the quasiparticle–phonon interaction taken into account). The functions $\tilde{S}_{nlj}(\eta)$ coincide, for all practical purposes, with the spectroscopic strength functions of the (d,p) reaction for the 501↓ and 501↑ states. The regions of localization of the strength of the $3p_{1/2}$ subshell in thee nuclei are quite close to one another: 6 to 7 MeV. Within this region, the strength distribution is different for different isotopes. While about 30% of the subshell strength in ^{185}W and ^{187}W lie in the region of the excitation energy below 2 MeV, only about 20% of the subshell strength is found to lie in the same region in ^{183}W. The region of localiztion of the strength of the $3p_{3/2}$ subshell is slightly wider than that of the $3p_{1/2}$ subshell, reaching 7–8 MeV. The strength distribution inside the localization region is sufficiently uniform in ^{183}W, while in ^{185}W and ^{187}W the strength is mostly concentrated at low excitation energies.

When the $3p_{1/2}$ and $3p_{3/2}$ subshells were calculated, it became possible to interpret the experimental data [297] on the cross sections of (d,p) reactions that excite the $1/2^-$ and $3/1^-$ states in $^{183,185,187}W$; the interpretation did not require high hexadecapole deformation. Calculations of for many neutron and proton subshells were performed in [295, 296], for a number of isotopes of rare-earth elements and actinide nuclei. Experiments on the study of fragmentation of single-quasiparticle states in deformed nuclei are being carried out in a number of laboratories [297, 298].

6.2.6

Let us consider the distribution of subshell strength in the transition from spherical to deformed nucleus. In the spherical nucleus, the entire strength of an nlj subshell is concentrated on one single-quasiparticle state; its strength distribution over the spectrum of nuclear excitations due to the interaction with phonons is described by the strength function $C_J^2(\eta)$, given by (5.28). Owing to the static deformation in the deformed nucleus, the subshell strength is distributed over a large number of single-quasiparticle states with different values of K, and the strength function of the $S_{nlj}(\eta)$ subshell is obtained by adding up the strength functions which describe the strength distribution of these single-quasiparticle states, with the weight factors $(a_{nlj}^{qK})^2$.

We will now compare the behaviour of the strength functions $C_J^2(\eta)$ and $S_{nlj}(\eta)$ for the same subshells. The strength functions describing the fragmentation of the neutron subshell $1g_{7/2}$ in ^{145}Sm and ^{151}Sm are plotted in figure 6.7 (see [295]). For the ^{151}Sm nucleus, individual terms $(a_{nlj}^{qK})^2 C^2 qK(\eta)$ to which the $1g_{7/2}$ subshell gives the main contribution are shown in addition to the total strength function $S_{nlj}(\eta)$ (these are the states 402↓, 404↓, 413↓, 422↓ and 431↓). The single-quasiparticle states which carry the main part of the strength of the $1g_{7/2}$ subshell are distributed in the energy range $\Delta\eta \approx 9$ MeV; the interaction with phonons distributes the strength of each of them over an energy interval of 4–5 MeV. The strength distributions of individual quasiparticle states overlap rather considerably, but nevertheless the region in which most of the strength of the subshell is localized widens appreciably and reaches 13 MeV. In ^{145}Sm, the strength of the $1g_{7/2}$ subshell is concentrated in a narrower interval, $\Delta\eta \approx 6$ MeV. Note, however, that this interval is wider than the one in which the strength of the single-quasiparticle state is distributed in the deformed nucleus.

For subshells with small l and j, the regions of localization in spherical and deformed nuclei are not very different. One example is the strength distribution of the proton hole subshell $2p_{3/2}$ in ^{151}Pm and ^{145}Pm (figure 6.8). The splitting of the $2p_{3/2}$ subshell due to deformation is quite small and the interaction with phonons increases the region of localization of the subshells up to 8 MeV. Figure 6.8 implies that the strength of the $2p_{3/2}$ subshell in ^{145}Pm is distributed in the same region.

The examples discussed above referred to deep hole subshells with energies of about 6–7 MeV. A study of the subshells close to the Fermi level of deformed nuclei (e.g., $1i_{3/2}$ in the neutron system and $1h_{11/2}$ in the proton system) has demonstrated that their strength distribution manifests a clearly pronounced maximum at low excitation energy, owing to a weaker fragmentation of single-quasiparticle states with low excitation energy.

The mechanisms of fragmentation of the strength of single-particle subshells in spherical and deformed nuclei are different. As in the case of

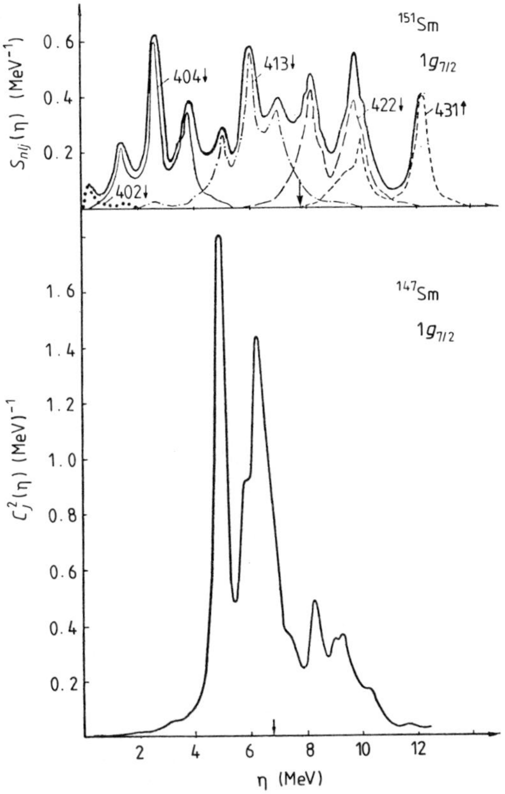

Figure 6.7 The strength functions of the neutron $1g_{7/2}$ subshell in ^{151}Sm and ^{145}Sm (bold full curves): the fine full, broken and chain curves plot the strength functions $C^2_{q0}(\eta)$ of individual single-quasiparticle states on which the strength of the $1g_{7/2}$ subshell is mostly concentrated; the functions are multiplied by the appropriate factors $(a^{qK}_{nlj})^2$. Arrows indicate the position of the subshell in the two nuclei.

giant resonances, the splitting of a spherically symmetric subshell in the deformed mean field mostly determines the size of the region in which its main strength is concentrated; in this respect, the interaction with phonons plays a less significant role than in spherical nuclei. In spherical nuclei, relatively narrow regions (2–3 MeV), in which most of the subshell strength is concentrated, are typically found. In fact, these are the regions that we observe in the cross sections of one-nucleon transfer reactions as resonance-like structures. The absence of such regions in deformed nuclei, a considerable overlapping of strength distributions for subshells make is difficult to single out the contributions of individual subshells in the cross sections

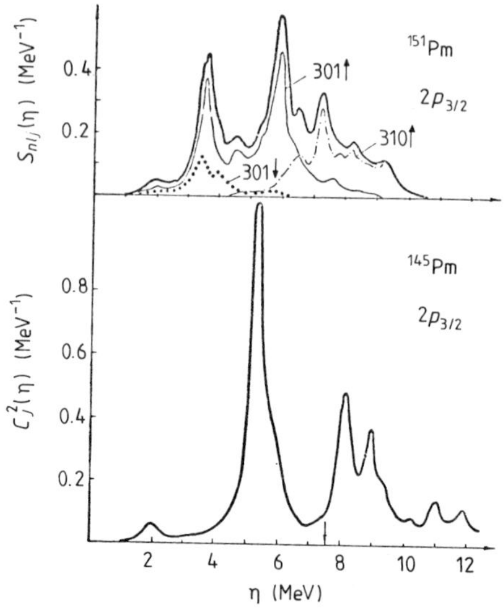

Figure 6.8 The strength functions of the proton hole $2p_{3/2}$ subshell in ^{151}Pm and ^{145}Pm. The notation is the same as in figure 6.7.

of one-nucleon transfer reactions.

6.3 STRUCTURE OF NON-ROTATIONAL STATES IN EVEN DEFORMED NUCLEI

6.3.1

The low-lying states of even–even deformed nuclei are the collective vibrational and two-quasiparticle states. Each of them contains a rotational band. The treatment of rotation and its relation to the internal motion in axially symmetric nuclei is presented in detail in monograph [4]. A large number of papers were devoted to the study of rotational bands and high-spin states, so that this aspect will not be discussed here. A cranking model and the HFB method were successfully used to analyse the states on the YRAST-line [299]. This model, plus the subsequent application of the RPA, was used in [300] for the description of quadrupole and octupole vibrational states in a number of deformed nuclei. The presentation below is limited to non-rotational states and the internal wave functions described here can be used in calculations that take into account the coupling between different rotational bands.

Deformed nuclei contain collective vibrational quadrupole, octupole and, in exceptional cases, hexadecapole states. They are typically described in terms of the RPA. The remaining non-rotational low-lying states can be interpreted, as a first approximation, as two-quasiparticle states. Tables listing the energies, structure and $B(E\lambda)$ values for low-lying non-rotational states can be found in [6, 165, 268, 269, 301–303].

The reduced probabilities of $E\lambda$ transitions from the ground states of even–even nuclei to their excited states described by wave function (6.33) are

$$B(E\lambda; 0^+_{\text{gs}} \to I^\pi K\nu) = \langle 00\lambda\mu|IK\rangle^2(2-\delta_{\mu 0})\Big(\sum_{i_0} R^\nu_{i_0}\mathcal{M}^E_{\lambda\mu i_0}\Big)^2 \qquad (6.56)$$

where $\mathcal{M}^E_{\lambda\mu i}$ is given by (2.79′) or (2.82). The reduced probabilities of isoscalar $E\lambda$ transitions in (pp'), (d,d'), (α,α') and some other reactions are given by the formula

$$B(ISE\lambda; 0^+_{\text{gs}} \to I^\pi K\nu) = \langle 00\lambda\mu|IK\rangle^2(2-\delta_{\mu 0})\Big(\sum_{i_0} R^\nu_{i_0}\frac{e}{\sqrt{2}}\frac{Z}{A}$$

$$\times \sum_{qq'} p^{\lambda\mu}(qq')g^{\lambda\mu i_0}_{qq'}u^{(+)}_{qq'}\Big)^2. \qquad (6.56')$$

These probabilities are chosen for $B(ISE\lambda)$ to be close to the values of $B(E\lambda)$ calculated using (6.56) for $e^{(\lambda)}_{\text{eff}}(p) = e$, $e^{(\lambda)}_{\text{eff}}(n) = 0$. In some papers, for instance, in [302, 303], the authors choose $e^{(\lambda)}_{\text{eff}}(p) = 1.2e$ and $e^{(\lambda)}_{\text{eff}}(n) = 0.2e$, since they make use of the truncated space of single-particle states from the bottom of the well up to the energy +5 MeV. The values of $B(E\lambda)$ calculated in this manner practically coincide with the values calculated for the single-particle states in the interval from the well bottom to the energy of +50 MeV, for $e^{(\lambda)}_{\text{eff}}(p) = e$ and $e^{(\lambda)}_{\text{eff}}(n) = 0$. The use of the radial dependence $\partial V(r)/\partial r$ instead of r^λ for multipole interactions in calculations somewhat decreases the value of $B(E\lambda)$. When $B(E\lambda)$ is calculated, it is often necessary to take into account the Coriolis interaction. The calculation results are given by many authors in single-particle units

$$B(E\lambda)_{\text{s.p.u.}} = \frac{2\lambda+1}{4\pi}\Big(\frac{3}{\lambda+3}\Big)^2\Big(0.12A^{1/3}\Big)^{2\lambda}e^2 10^{-24\lambda}.$$

When the structure of states is studied, an important role is played by the electromagnetic transitions between excited states. A matrix element of the $E\lambda$ transition between excited states described by wave functions (6.33) is given by the expression [276]

$$(\Psi^*_{\nu'}(K_0'^{\pi_0'})\mathfrak{M}(E\lambda\mu)\Psi_\nu(K_0^{\pi_0}))$$

$$
\begin{aligned}
&= \sum_{i_0 i_0'} R^{\nu'}_{i_0'} R^{\nu}_{i_0} \mathcal{M}^E_{\lambda\mu}(K_0^{\pi_0} i_0 \to K'^{\pi_0'}_0 i_0') + \sum_i \mathcal{M}^E_{\lambda\mu i} \\
&\times \Bigg\{ \sum_{i_0'} R^{\nu'}_{i_0'} P^{\nu}_{g_0' g} (1+\delta_{g_0' g})^{1/2} [1+\delta_{K_0' 0}(1-\delta_{\mu' 0})]^{-1/2} \\
&\times [1 + \mathcal{K}^{K_0}(g_0' g)] + \sum_{i_0} R^{\nu}_{i_0} P^{\nu'}_{g_0 g} (1+\delta_{g_0 g})^{1/2} [1 + \delta_{K_0' 0} \\
&\times (1-\delta_{\mu 0})]^{-1/2} [1 + \mathcal{K}^{K_0'}(g_0 g)] \Bigg\} + \sum_{g_1 g_2 g_3} P^{\nu}_{g_1 g_3} P^{\nu'}_{g_2 g_3} \\
&\times \mathcal{M}^E_{\lambda\mu}(g_1 \to g_2)(1+\delta_{g_1 g_3})^{1/2}(1+\delta_{g_2 g_3})^{1/2}[1+\delta_{K_0' 0}(1-\delta_{\mu' 0})]^{-1/2} \\
&\times [1+\delta_{K_0' 0}(1-\delta_{\mu 0})]^{-1/2}[1 + \frac{1}{2}\mathcal{K}^{K_0}(g_1 g_3) + \frac{1}{2}\mathcal{K}^{K_0'}(g_2 g_3)]
\end{aligned}
\qquad (6.57)
$$

where $\mathcal{M}^E_{\lambda\mu}(K_0^{\pi_0} i_0 \to K'^{\pi_0'}_0 i_0')$ is given by (2.83).

6.3.2

Calculations of the energies and wave functions of two-quasiparticle and one-phonon states of even–even deformed nuclei were performed in 1960–1975. The quadrupole and octupole states were calculated in the RPA. The results were summarized in [61, 68, 268, 269]. A sufficiently good description of the data available at the time had been obtained and predictions formulated, a considerable part of which was later confirmed by new experiments. In the subsequent years, a great deal of new and more complete experimental data were obtained on the energies, $B(E\lambda)$ values and the spectroscopic factors of one- and two-nucleon transfer reactions. We can expect that considerable amounts of experimental data will appear from the new generation of accelerators and detectors. New QPNM calculations were therefore carried out with wave function (6.33) and equations (6.43), (6.44) taking into account ph or ph and pp isoscalar and isovector multipole interactions. I will outline now some results of these calculations.

The results of [302, 303] were obtained for single-particle energies and the wave functions of the Saxon–Woods potential. In order to reduce the number of arbitrary constants, the constants of the isovector ph interactions, $\kappa_1^{\lambda\mu} = -1.5\kappa_0^{\lambda\mu}$, and the constants of the pp interactions, $G^{\lambda\mu} = 0.9\kappa_0^{\lambda\mu}$. The constants of isoscalar ph interactions were taken in the interval $\kappa_0^{\lambda\mu} = 0.012$–$0.020\,\mathrm{fm}^2\,\mathrm{MeV}^{-1}$ and only κ_0^{44} was taken above $0.020\,\mathrm{fm}^2\,\mathrm{MeV}^{-1}$. The constants $\kappa_0^{\lambda\mu}$ increase with increasing λ and decrease as the mass number A increases. The monopole pairing constants G_τ were calculated using the pairing energies at fixed values of G^{20}. The energies of two-quasiparticle poles were calculated taking into account the monopole and quadrupole pairing, the blocking effect and the Gallagher–Moszkowski corrections.

The results of the calculations are presented in tables 6.6, 6.7 and 6.8. They give the energies (in MeV) and $B(E\lambda)$ (in single-particle units) found when the Coriolis interaction was ignored. The resulting structure is represented by contributions (in per cent) of one-phonon $\lambda\mu i$ and two-phonon $\{\lambda_1\mu_1 i_1, \lambda_2\mu_2 i_2\}$ components to normalization (6.37) of wave function (6.33). The contribution of two-phonon components takes into account the factor $\{1 + \mathcal{K}^{K_0}(\lambda_1\mu_1 i_1, \lambda_2\mu_2 i_2\}$. Also given in the tables (in per cent) are the largest two-quasi-neutron nn and two-quasi-proton pp components of the wave functions of the one-phonon states $\lambda\mu i$. To denote the single-particle states, we use the asymptotic quantum numbers $Nn_z\Lambda\uparrow$ for $K = \Lambda+1/2$ and $Nn_z\Lambda\downarrow$ for $K = \Lambda-1/2$.

In the case of ^{174}Yb, the energy of the first $K^{\pi}_{\nu} = 2^{+}_{1}$ state has increased in comparison with that of ^{170,172}Yb, which is correctly described by the QPNM. The experimental value $B(E2) = 1.7$ single-particle units does not contradict the calculated result. According to the experimental data [304], the contribution of configuration nn $512\uparrow - 510\uparrow$ to the 2^{+}_{1} state is about 75%, which is much greater than the calculated value. According to the calculations, the most collective among the $K^{\pi} = 0^{-}$ states is the fourth 0^{-}_{4} state with $B(E3) = 3.6$ s.p.u. and energy of about 3 MeV; for the 0^{-}_{3} state, $B(E3) = 1.0$ s.p.u. There are no experimental data on $K^{\pi} = 1^{-}$ states; calculations indicate that the most collective are the states with energies about 3 MeV and $B(E3) \approx 1$ s.p.u. The calculated value of $B(E3)$ for the excitation of $I^{\pi}K_{\nu} = 3^{-}2_{1}$ is considerably lower than the value $B(E3) = 4.1$ s.p.u. given in [304].

The experimental results on the structure of the states ^{178}Hf given in table 6.7 are extracted from the data on the (dp) and (dt) reactions and the *au* β decays. In some cases, the contribution of the two-quasiparticle component is determined, and in others a specific two-quasiparticle component is indicated through which the reaction goes. The first 2^{+}_{1} state of ^{178}Hf is collective; the two-quasi-neutron nn $514\downarrow - 512\downarrow$ configuration is contained in the wave functions of the 2^{+}_{1} and 2^{+}_{2} states. Consequently, they are easily excited in (dp) reactions. These properties of the 2^{+}_{1} and 2^{+}_{2} states are reproduced quite well in calculations. At the same time, the calculated energy of the 2^{+}_{2} state equals 2 MeV, which is explained by the absence of two-quasiparticle poles with energies below 2.2 MeV. According to calculations, the 3^{+}_{1} state of ^{178}Hf is collective and must be easily excited in a (dp) reaction; the second 3^{+}_{2} state must be weakly collectivized. The energies of the 3^{+}_{3}, 3^{+}_{4} and 3^{+}_{5} states are 2.2, 2.3 and 2.4 MeV, respectively. The wave functions of these states contain one dominating one-phonon component. The 4^{+}_{1} state of ^{178}Hf is collective, with dominating 441 component in the wave function; the contribution of the two-phonon component {221,221}

Table 6.6 Non-rotational states of ^{174}Yb.

	Experiment	QPNM calculation					
K^{π}_{ν}	$\mathcal{E}$ (MeV)	$\mathcal{E}$ (MeV)	$B(E\lambda)$ (s.p.u.)	Structure (%)			
2^+_1	1.634	1.6	2.2	221 92	{201,221} 6	221: nn 512↑ − 510↑ 36 nn 512↑ − 521↓ 16	
2^+_2	2.172	2.0	0.2	222 99			
0^-_1	1.710	1.7	1.0	301 96		301: nn 514↓ − 633↑ 44	pp 404↓ − 523↑ 1
0^-_2		2.2	0.5	302 94		302: pp 404↓ − 523↑ 43	
1^-_1		1.8	0.1	311 97			
1^-_2		1.9	0.8	312 98			
2^-_1	1.318	1.2	1.6	321 94		321: nn 624↑ − 512↑ 92 {204,321} 4	pp 514↑ − 402↑ 1
2^-_2		2.7	1.7	322 85	{201,321} 3		
3^-_1	1.851	2.0	4.0	331 98		331: nn 615↑ − 512↑ 25	pp 514↑ − 411↑ 25
3^-_2	(2.050)	2.2	0.2	332 98		332: nn 633↑ − 521↓ 92	
3^+_1	1.606	1.5	1.4	431 97		431: pp 404↓ − 411↓ 61 nn 514↓ − 521↓ 10	nn 512↑ + 521↓ 23 nn 505↑ − 512↑ 1
3^+_2	(2.284)	1.9	0.1	432 99		432: pp 404↓ − 411↓ 34	nn 512↑ + 521↓ 60
4^+_1		1.8	0.01	441 92			
4^+_2		1.9	0.001	442 90			
6^+_1	1.518	1.5				nn 512↑ + 514↓ 100	
7^-_1		1.6				nn 512↑ + 624↑ 100	
7^-_2		1.7				nn 514↓ + 633↑ 100	
8^-_2		1.8				nn 514↓ + 624↑ 100	
5^-_1	1.885	1.9				pp 411↓ + 514↑ 78	nn 521↓ + 624↑ 21
5^-_2	2.379	2.2				pp 411↓ + 514↑ 21	nn 521↓ + 624↑ 77
6^-_1		2.0				nn 633↑ + 512↑ 100	

equals 3%. The experimental data [305] do not give evidence of the collective behaviour of the $E2$ transition from the 4_1^+ to the 2_1^+ state.

The experimental data on octupole states of ^{178}Hf are scarce. For example, the $K_\nu^\pi = 0^-$ state has not been found. Among the 1_1^-, 2_1^- and 3_1^- states, only the $B(E3)$ value for the excitation of the 3^-2_1 state has been measured. According to our calculations, the energy of the 0_1^- state is 2 MeV and $B(E3) = 4$ s.p.u.. The calculated $B(E3)$ values for the 1_1^-, 2_2^- and 2_3^- states are small, so that it is not surprising that they have not yet been experimentally measured. The calculated energies of the 2_2^- and 2_3^- states lie above the experimental data.

Let us discuss the description of non-rotational states of ^{240}Pu reported in [303] and presented in table 6.8. A sufficiently good description of the energies and $B(E\lambda)$ values is obtained for $K_\nu^\pi = 2_1^+$ and for the first octupole states. According to calculations for the wave functions 0_1^+ and 0_2^+, there is a considerable contribution of the nn $631\downarrow - 631\downarrow$ configuration, so that they must be excited in the (dp) reaction; this conclusion agrees with the experimental data of [306].

The RPA and QPNM descriptions of several first 0^+ states of deformed nuclei including ph interactions, and descriptions in other models cannot be regarded as satisfactory. The calculations of energy and spectroscopic factors of (pt) and (tp) reactions with excitation of 0^+ states are given in [307] for a number of deformed nuclei in the rare-earth region. The calculations employed exact diagonalization and took into account the effective pairing and neutron–proton quadrupole interactions. The observed enhancement of the (tp) transitions to the excited 0^+ states has been reproduced in calculations for a number of nuclei.

The role of the pp interactions is very significant in the description of the 0^+ excited states. As the constant G^{20} increases, the RPA energies of several poles of the secular equation and the structure of low-lying 0^+ states change. In calculations the densities of low-lying 0^+ states are increased and the the values of $B(E2)$ decrease for the excitation of the $I^\pi 0_\nu = 2^+0_1$ state and increase for the the $I^\pi 0_\nu = 2^+0_2$ state. Calculations of [303] led to a conclusion that the inclusion of the pp interaction improves the description of 0^+ states.

Tables 6.6, 6.7 and 6.8 show that a sufficiently good description of non-rotational states was obtained and predictions could be made. An analysis of the vibrational states with $K^\pi \neq 0^+$ in well-deformed even–even nuclei demonstrated that the energy and structure of each state are mostly determined by single-particle energies and wave functions of the Saxon–Woods potential, by the multipole pairing and the ph isoscalar multipole interaction. The multipole ph isovector interaction, quadrupole pairing and multipole pp interaction play a secondary role. Taking the pp interaction into account improves the description of the vibrational states. Furthermore, it makes the applicability of the RPA more conclusive for the description of

Table 6.7 Non-rotational states of ^{178}Hf.

K^{π}_{ν}	Experiment			QPNM calculation		
	$\mathcal{E}_{\nu}$ (MeV) $[B(E\lambda)_{\text{s.p.u.}}]$	Structure (%)		$\mathcal{E}_{\nu}$ (MeV) $[B(E\lambda)_{\text{s.p.u.}}]$	Structure (%)	
8^-_1	1.147	log ft =4.7:	pp 404↓ + 514↑ 34	1.11	981 100	981: nn 514↓ + 624↑ 75
			nn 514↓ + 624↑ 66			pp 514↑ + 404↓ 25
2^+_1	1.174	(dp):	nn 514↓ − 512↓	1.12	221 92 {221,441} 4	221: nn 514↓ − 512↓ 29
	[3.9]			[4.1]		nn 512↑ − 510↑ 32
2^-_1	1.260	(dt):	nn 624↑ − 512↑	1.2	321 96	321: nn 624↑ − 512↑ 86
	[4.0]			[2.0]		pp 514↑ − 402↑ 8
1^-_1	1.310			1.4	311 89	311: nn 514↓ − 624↑ 96
				[0.5]		
8^-_2	1.479	log ft =4.7:	pp 404↓ + 514↑ 66	1.42	982 100	982: pp 514↑ + 404↓ 75
			nn 514↓ + 624↑ 34			nn 514↓ + 624↑ 25
4^+_1	1.514	(dp):	nn 514↓ + 510↑	1.5	441 94 {221,221} 3	
				[2.0]		441: nn 514↓ − 510↑ 60
						nn 512↑ + 512↓ 23
6^+_1	1.554			1.7		nn 512↑ + 514↓ 100
2^-_2	1.568			1.9	322 96	322: pp 514↑ − 402↑ 88
				[0.8]		nn 624↑ − 512↑ 10

Table 6.7 (continued)

K^{π}_{ν}	Experiment		QPNM calculation		
	$\mathcal{E}_{\nu}$ (MeV) $[B(E\lambda)_{\text{s.p.u.}}]$	Structure (%)	$\mathcal{E}_{\nu}$ (MeV) $[B(E\lambda)_{\text{s.p.u.}}]$	Structure (%)	
5^-_1	1.637		1.8	551 100	551: pp 514↑ + 411↓ 73 nn 624↑ + 521↓ 22
3^+_1	1.758	(dp): nn 514↓ − 510↑	1.8 [1.7]	431 99	431: pp 404↓ − 411↓ 37 nn 514↓ − 510↑ 21
3^-_1	1.803		1.9 [4.0]	331 98	331: nn 615↑ − 512↑ 45 pp 514↑ − 411↑ 24
2^+_2	1.808	(dp): nn 514↓ − 512↓	2.0 [0.01]	222 90 {202, 221} 8	222: nn 514↓ − 512↓ 56 nn 624↑ − 642↑ 40
2^-_3	1.857		2.6 [0.02]	223 98	223: nn 615↑ − 514↓ 90
4^+_2	(2.007)		2.0 [0.0004]	442 90	442: pp 404↓ + 411↓92

Table 6.8 Non-rotational states of ^{240}Pu.

K^π_ν	Experiment $\mathcal{E}$ (MeV)	Experiment $B(E\lambda)$ (s.p.u.)	QPNM calculation $\mathcal{E}$ (MeV)	$B(E\lambda)$ (s.p.u.)				Structure (%)		
0^+_1	0.861		0.8	0.8	201	90	{221,221} 5	201:	nn $622\uparrow - 622\uparrow$ 31	nn $624\downarrow - 624\downarrow$ 30
									pp $523\downarrow - 523\downarrow$ 20	nn $631\downarrow - 631\downarrow$ 7
0^+_2	1.091		1.1	0.03	202	96		202:	nn $622\uparrow - 622\uparrow$ 48	nn $624\downarrow - 624\downarrow$ 36
									nn $631\downarrow - 631\downarrow$ 7	
0^+_3	(1.525)		1.3	0.07	203	76	204 15	203:	nn $631\downarrow - 631\downarrow$ 38	nn $523\downarrow - 523\downarrow$ 36
2^+_1	1.137	2.3	1.2	2.6	221	96	{201,221} 3	221:	nn $622\uparrow - 631\downarrow$ 30	nn $633\downarrow - 631\downarrow$ 21
2^+_2	(1.223)		1.3	1.0	222	96		222:	nn $622\uparrow - 631\downarrow$ 69	nn $633\downarrow - 631\downarrow$ 10
0^-_1	0.597	17	0.6	13	301	94		301:	nn $743\uparrow - 624\downarrow$ 17	pp $523\downarrow - 642\uparrow$ 14
0^-_2	(1.411)		1.2	0.3	302	88		302:	pp $523\downarrow - 642\uparrow$ 34	nn $733\uparrow - 624\downarrow$ 15
1^-_1	0.938	9.6	0.94	7.0	311	96		311:	nn $743\uparrow - 622\uparrow$ 65	pp $642\uparrow - 521\uparrow$ 17
1^-_2	(1.488)		1.4	2.2	312	94		312:	pp $642\uparrow - 521\uparrow$ 58	nn $743\uparrow - 622\uparrow$ 31
2^-_1	1.241	10.8	1.24	8.0	321	98		321:	nn $734\uparrow - 622\uparrow$ 45	pp $642\uparrow - 530\uparrow$ 12
3^+_1	1.036		1.0	1.0	431	99		431:	nn $622\uparrow + 631\downarrow$ 90	
4^+_1			1.35	0.3	441	90		441:	nn $624\downarrow + 631\downarrow$ 90	pp $523\downarrow + 521\uparrow$ 6
5^-_1	1.308		1.2		551	100		551:	pp $642\uparrow + 523\downarrow$ 100	

states with energies below 1 MeV.

6.3.3

Let us consider hexadecapole states and multipole states for $\lambda \geqslant 5$. The calculation results of [301] and later publications are given in table 6.9. The notation is the same as in tables 6.6–6.8. The table shows experimental data on energies, two-particle components (information on them is obtained from nuclear reactions and the β decay) and the collectivity of states which manifested itself in (d, d') and (α, α') reactions. Hexadecapole non-rotational states with $K^\pi = 3^+, 4^+$ are generated by isoscalar and isovector forces with $\lambda = 4$. QPNM calculations are carried out using formulas (6.33), (6.43)–(6.45) and (6.56), with the ph interaction taken into account.

An analysis of the experimental data and their comparison with the calculation results made it possible to identify the following regularities in the behaviour of the hexadecapole 3_1^+ and 4_1^+ states in the region $158 \leqslant A \leqslant 188$. Gadolinium and dysprosium isotopes have practically two-quasiparticle 4_1^+ states with energies in the interval 1.1–1.9 MeV. According to the experimental data of [308] and some others, there are collective 3_1^+ states in ^{168,170}Er, ^{168,170,172,174}Yb and ^{174,176,178}Hf. According to calculations, the 4_1^+ states in Os isotopes are collective; calculations seem to give a too high value $B(E4; 0_1^+ \rightarrow 4^+4_1) \approx 2$–4 s.p.u. [309].

QPNM calculations with fixed values of $\kappa_0^{(43)}$ and $\kappa_0^{(44)}$ (the isovector constants play an insignificant role) give a qualitatively correct description of hexadecapole states with $K^\pi = 3^+, 4^+$. The results of these calculations reproduce the main two-quasiparticle components extracted from experimental data and explain the experimentally observed [308] mixing of neutron and proton two-quasiparticle components in ytterbium isotopes in states with $K^\pi = 3^+$. The two-phonon components $\{221, 221\}$ in all calculated 4_1^+ and 4_2^+ states do not contribute more than 20%. The calculated energies of the lowest two-quasiparticle states with $K^\pi = 3^+, 4^+$ in those nuclei where low-lying states with these values of K^π were not found experimentally, are typically found to exceed 2 MeV.

Hexadecapole forces with $\lambda\mu = 42$ affect the 2_1^+ states. It was recently shown [310] that the experimental data on the excitation of a state with $I^\pi K = 4^+2$ γ-vibrational band of ^{168}Er in a (p, p') reaction can be explained only if large hexadecapole forces are introduced. Calculations in which quadrupole forces with $\lambda\mu = 22$ and hexadecapole forces with $\lambda\mu = 42$ are simultaneously taken into account have been reported in [301]. One-phonon energies have been found in the solution of fourth-order secular equation (2.75); the amplitudes of the two-quasiparticle components of such one-phonon states contain a quadrupole and a hexadecapole parts. With the appropriate values of $\kappa_0^{(22)}$ and $\kappa_0^{(42)}$, it is possible to obtain

Table 6.9 Experimental data and results of calculations for low-lying states with $K^{\pi} = 3^+, 4^+$.

Nucleus	K^{π}_{ν}	Experiment: $\mathcal{E}$ (MeV)		Experiment: Structure (%)	QPNM calculation [301]: $\mathcal{E}$ (MeV) $[B(E4)_{\text{s.p.u.}}]$	QPNM calculation [301]: Structure (%)	
^{158}Gd	4^+_1	1.380	(t,α)	pp $411\uparrow + 413\downarrow$ large	1.2	441 90 {221, 221}	8
					[0.4]	441: pp $413\downarrow + 411\uparrow$	85
	4^+_2	1.920	(d,p)	nn $521\uparrow + 523\downarrow$ large	1.7	442 91	
					[0.01]	442: nn $523\downarrow + 521\uparrow$	87
^{158}Dy	4^+_1	1.895	$\log ft = 4.9$	nn $521\uparrow + 523\downarrow$ large	2.2	441 92	442 6
					[0.4]	441: pp $413\downarrow + 411\uparrow$	36
						nn $523\downarrow + 521\uparrow$	28
^{160}Dy	4^+_1	1.694	$\log ft = 4.7$	nn $521\uparrow + 523\downarrow$ large	1.7	441 91 {221, 221}	6
					[0.1]	441: nn $521\uparrow + 523\downarrow$	85
	4^+_2	2.096	$\log ft = 5.8$		2.0	442 85 {201, 442}	6
					[0.1]	442: nn $642\uparrow + 651\uparrow$	74
						nn $523\downarrow + 521\uparrow$	8
^{162}Dy	4^+_1	1.535	(d,t)	nn $521\uparrow + 523\downarrow$ large	1.7	441 89	
					[0.2]	441: nn $523\downarrow + 521\uparrow$	92
^{170}Er	3^+_1	1.217	$\sigma(d,d')$	for 4^+3_1 large	1.2	431 99	
					[0.6]	431: nn $512\uparrow + 521\downarrow$	94
^{168}Yb	3^+_1	1.452	$\sigma(d,d')$	for 4^+3_1 large	1.6	431 98	
			$(\alpha,2n)$	nn 20	[1.6]	431: pp $404\downarrow - 411\downarrow$	61
			g_K − factor	pp 80		nn $512\uparrow + 521\downarrow$	13
^{170}Yb	3^+_1	1.470	$\sigma(d,d')$	for 4^+3_1 large	1.3	431 99	
					[1.4]	431: nn $512\uparrow - 521\downarrow$	66
						pp $404\downarrow - 411\downarrow$	24

Table 6.9 (continued)

Nucleus	K^{π}_{ν}	Experiment $\mathcal{E}$ (MeV)		Experiment Structure (%)	QPNM calculation [301] $\mathcal{E}$ (MeV) $[B(E4)_{\text{s.p.u.}}]$		QPNM calculation [301] Structure (%)	
^{172}Yb	3^+_1	1.172	$\sigma(d,d')$	for 4^+3_1 large	1.3	431	99	
			(d,t) (d,p)	nn 512↑ + 521↓ 75	[1.3]	431	nn 512↑ + 521↓	68
			(p,α)	nn 512↓ + 521↓ 74				
			μ	pp 404↓ − 411↓ 25			pp 404↓ − 411↓	20
	3^+_2	1.663	(d,t) (d,p)	nn 512↑ + 521↓	1.6	432		100
				appreciable	[0.4]	432:	pp 404↓ − 411↓	52
			(p, α)	pp 404↓ − 411↓ 26			nn 512↑ + 521↓	30
^{174}Yb	3^+_1	1.606	$\sigma(d,d')$	for 4^+3_1 large	1.4	431		99
					[1.3]	431:	nn 514↓ − 521↓	45
							pp 404↓ − 411↓	39
^{174}Hf	3^+_1	1.303	$\sigma(d,d')$	for 4^+3_1 large	1.3	431		99
					[1.0]	431:	nn 512↑ + 521↓	75
^{176}Hf	3^+_1	1.578	$\sigma(d,d')$	for 4^+3_1 large	1.3	431		99
			(d,t)	nn 514↓ − 521↑ large	[1.2]	431:	nn 514↓ − 521↓	72
							pp 404↓ − 411↓	12
	4^+_1	1.888	(d,t)	nn 514↓ + 521↑ large	1.7	431:		93
					[0.01]	441:	nn 514↓ + 521↓	99
^{186}Os	4^+_1	1.352	$B(E4)$ large	for (α,α')	1.2	441	90{201, 441}	3
							{221, 221}	4
					[4.0]	441:	pp 402↑ + 402↓	56
							nn 514↓ + 510↓	13

a large contribution of hexadecapole forces to the state with $K^\pi_\nu = 2^+_1$ and the energy and $B(E2)$ which agree with the experimental data. The results of calculation, in which quadrupole and hexadecapole forces are simultaneously taken into account, do agree with the experimental data of [310], pointing to a considerable contribution of hexadecapole forces to the 2^+_1 state.

Some experimental results point to a strong mixing of two-quasi-proton and two-quasi-neutron configurations with large values of K in even–even deformed nuclei; an analysis of these data is given in [311]. In [312], this mixing was calculated taking into account the additional neutron–proton interaction. Paper [313] treated the role of high-multipole interactions; it calculated the states with $K \geqslant 4$ of a number of rare-earth nuclei in the framework of the QPNM with ph interactions for $\kappa_0^{\lambda\mu} = (0.020$–$0.024)\,\mathrm{fm}^2\,\mathrm{MeV}^{-1}$. The results of calculations of [313] and the experimental data from [314–318] are systematized in table 6.10.

The mixing of two-quasi-neutron nn $514\downarrow + 624\uparrow$ and two-quasi-proton pp $404\downarrow + 514\uparrow$ states in ^{178}Hf is described by taking into account the multipole interaction $\lambda\mu = 98$. The table shows a good description of experimental data on the energy and structure of the $K^\pi_\nu = 8^-_1$ and 8^-_2 states for $\kappa_0^{98} = 0.024\,\mathrm{fm}^2\,\mathrm{MeV}^{-1}$. Note that if $\kappa_0^{98} = 0.020\,\mathrm{fm}^2\,\mathrm{MeV}^{-1}$, we obtain a somewhat smaller mixing, namely, 80% and 20%. The corresponding neutron and proton matrix elements are large. What is relatively unexpected is the important role of an interaction with such a high multipolarity as $\lambda = 9$, with the projection $\mu = 8$ and a constant κ_0^{98} whose value is close to those of κ_0^{22}, κ_0^{33} and κ_0^{44} (which give a good description of quadrupole, octupole and hexadecapole states of ^{178}Hf). Note that even though the 8^-_1 and 8^-_2 states are not pure two-quasiparticle states, the isomer state with $K^\pi_\nu = 16^+_1$ and energy 2.447 MeV of ^{178}Hf is a pure four-quasiparticle state p $514\uparrow$ + p $404\downarrow$ + n $514\downarrow$ + n $624\uparrow$. The reason is that the total strength of the p $514\uparrow$ + p $404\downarrow$ and n $514\downarrow$ + n $624\uparrow$ states is concentrated on two levels, 8^-_1 and 8^-_2.

According to the experimental data [314], two $K^\pi = 6^+$ configurations, nn $514\downarrow + 512\uparrow$ and pp $404\downarrow + 402\uparrow$ in ^{176}Hf, are mixed. The calculated mixing and the energies of the 6^+_1 and 6^+_2 states are close to the experimental data. However, the calculated structure of the first state is close to that of the experimentally observed second state, and vice versa. This discrepancy is caused by the single-particle states scheme.

According to the calculations [313], the energies of the $K^\pi_\nu = 8^-_1$ and 8^-_2 states of ^{176}Hf are 1.52 and 1.84 MeV. In comparison with ^{178}Hf, the mixing of the pp $404\downarrow$+$514\uparrow$ and nn $514\downarrow$+$624\uparrow$ configurations is somewhat lower. The description of the structure of the first 8^-_1 state agrees with the experimental data which show that the contribution of the configuration pp $404\downarrow$+$514\uparrow$ is (86±6)%. It would be interesting to experimentally detect the second 8^-_2 state. According to calculations, ^{176}Hf possesses 7^-_1 and

7_2^- states with energies 1.76 and 1.81 MeV. Even though the energies of these states are close to each other, the two-quasi-neutron configurations nn 514↓+633↑ and nn 512↑+624↑ are practically not mixed. The two respective neutron matrix elements are not large.

Table 6.10 Mixing of two-quasiproton and two-quasineutron configurations in deformed nuclei.

Nucleus	$\lambda\mu$	$\mathcal{E}$ (MeV)		1 configuration(%)		2 configuration(%)	
	K_ν^π	exper.	calc.	exper.	calc.	exper.	calc.
^{178}Hf	98			pp 404↓ + 514↑		nn 514↓ + 624↑	
	8_1^-	1.147	1.11	34±4	25	(66)	75
	8_2^-	1.479	1.42	(66)	75	34±4	24
^{176}Hf	66			pp 404↓ + 402↑		nn 514↓ + 512↑	
	6_1^+	1.333	1.35	62	26	38	73
	6_2^+	1.762	1.75	38	71	62	27
	98			pp 404↓ + 514↑		nn 514↓ + 624↑	
	8_1^-	1.559	1.52	95	86		14
	8_2^-		1.84		14		86
	77			nn 514↓ + 633↑		nn 512↑ + 624↑	
	7_1^-	1.860	1.76	∼ 100	99.7		0.04
	7_2^-		1.81		0.2		99.8
^{174}Yb	55			pp 411↓ + 514↑		nn 521↓ + 624↑	
	5_1^-	1.885	1.9	46±2	78		21
	5_2^-	2.379	2.2	54±2	21		77
^{168}Er	54			nn 633↑ + 521↓		pp 411↓ + 523↑	
	4_1^-	1.094	1.0	70	81	25	18
	4_2^-	1.905	1.6	30	18	60	80
^{158}Gd	44			pp 413↓ + 411↑		nn 523↓ + 521↑	
	4_1^+	1.380	1.32	large	94	appreciable	4
	4_2^+	1.920	1.9		4	75	95
	54			nn 521↑ + 642↑		pp 532↑ + 411↑	
	4_1^-	1.636	1.66	72	92		6
	4_2^-		1.86		7		87

A study of the β decay of ^{174}Tm has revealed in ^{174}Yb a mixing of the two-quasi-proton pp 411↓ + 514↑ and two-quasi-neutron nn 521↓ + 624↑ configurations in the 5_1^- and 5_2^- states with the energies 1.885 and 2.379 MeV. Calculations with $\lambda\mu = 55$ interactions show less mixing than found from experimental data. If we take $\kappa_0^{55} = 0.026\,\text{fm}^2\text{MeV}^{-1}$, then the mixing of these configurations increases to 72% and 27%. The mixing of the nn 633↑ + 521↓ and pp 411↓ + 523↑ configurations in the $K_\nu^\pi = 4_1^-$ and 4_2^- states in ^{168}Er was first calculated in [302].

We can conclude from such calculation results that taking into account high multipole interactions led to a qualitatively correct description of experimental data on the mixing of the two-quasi-neutron and two-quasi-proton configurations with high values of K. In those cases where the energies of two-quasi-proton and two-quasi-neutron states with identical values of K^π are not very different and the corresponding matrix elements are not small, high multipole interactions with $\lambda = 5$–9 do play an important role in the mixing of such states. These are the situations where high multipole interactions have to be taken into account.

Table 6.11 lists the predictions, made in [303], for the mixture of two-quasi-proton and two-quasi-neutron states in ^{238}U, ^{250}Cf, ^{256}Fm and ^{260}104.

Table 6.11 Calculated mixing of two-quasi-proton and two-quasi-neutron configurations.

Nuclei	K^π_ν	$\lambda\mu$ / $\kappa_0^{\lambda\mu}$ ($\mathrm{fm^2\,MeV^{-1}}$)	$\mathcal{E}$ (MeV)	Structure (%): nn two-quasi-neutron		pp two-quasi-proton	
^{238}U	5^-_1	55	1.61	$631\downarrow + 734\uparrow$	93	$642\uparrow + 523\downarrow$	2.2
	5^-_2	0.02	1.99	$622\uparrow + 752\uparrow$	93	$642\uparrow + 523\downarrow$	5.8
	5^-_3		2.01	$622\uparrow + 752\uparrow$	3.4	$642\uparrow + 523\downarrow$	87
^{250}Cf	4^-_1	54	1.2	$734\uparrow - 620\uparrow$	72	$660\uparrow + 514\downarrow$	9
	4^-_2	0.018	1.8	$734\uparrow - 620\uparrow$	26	$660\uparrow + 514\downarrow$	32
^{256}Fm	5^+_1	65	1.1	$613\uparrow + 622\downarrow$	69	$521\uparrow + 514\downarrow$	21
	5^+_2	0.02	1.4	$613\uparrow + 622\downarrow$	27	$521\uparrow + 514\downarrow$	70
	7^-_1	77	1.4	$622\uparrow + 725\downarrow$	88	$633\uparrow + 514\downarrow$	10
	7^-_2	0.02	1.5	$622\uparrow + 725\downarrow$	10	$633\uparrow + 514\downarrow$	88
^{260}104	5^-_1	55	1.0	$620\uparrow - 725\uparrow$	74	$624\uparrow + 521\downarrow$	23
	5^-_2	0.02	1.2	$620\uparrow - 725\uparrow$	24	$624\uparrow + 521\downarrow$	76

Probably, the need to consider high multipole interactions is relevant to taking into account high multipole deformations with $\lambda = 5$, 6 and 7 [319], which were found to be important in the barium and radium regions.

6.3.4

Let us discuss the role of two-phonon states in deformed nuclei. According to the accepted interpretation (see [4]), even–even deformed nuclei must possess one-, two- and three-phonon states. We will carry out the analysis with wave function (6.33). In order to find the energies η_ν, we need to solve equations (6.43) or (6.44). The shift $\Delta\omega(g_1 g_2)$ of the two-phonon pole is determined by formula (6.39); it corresponds to taking into account the diagrams given in figure 4.6 (*c, d*). This shows that the shift $\Delta\omega(g_1 g_2)$ appears only if the fermion structure of phonons is taken into account. The calculations [140, 165] demonstrated that the shift of the two-phonon pole in a number of deformed nuclei is 0.5–1.5 MeV for both collective phonons. It was shown [165] that the same shift takes place for the corresponding root of secular equation (6.44). For the two weakly collective phonons, the shifts of two-phonon poles $\Delta\omega(g_1 g_2)$ are small, owing to the large values of $\mathcal{Y}^g_\tau$ in (6.39). The energies of such states are close to the sum of energies of the constituent phonons.

As a result of taking into account the Pauli exclusion principle in the two-phonon components of wave function (6.33), the energies of the collective two-phonon states of even–even deformed nuclei increased by 1–2 MeV and fell into the 2.5–4 MeV range of excitation energies. At 3–4 MeV energies, collective two-phonon states are fragmented over many levels. A conclusion was drawn on this basis in [165] that collective two-phonon states cannot exist in deformed nuclei.

The fragmentation of two-phonon states cannot be described in calculations using wave function (6.33). To achieve the description, (6.33) must be expanded by adding three-phonon terms. Consequently, conclusions are made on the energy centroids of two-phonon states.

The increase in the energy centroids of the two-phonon $\{g_1, g_2\}$ states in relation to the sum of the energies of the vibrational states with dominating one-phonon components of their wave functions is a result of two factors. The first is the anharmonicity of vibrations, since the energy of one-phonon states is greater than the energies of states dominated by a one-phonon component described by wave function (6.33). The second is the shift of the two-phonon pole $\Delta\omega(g_1 g_2)$ resulting from taking into account the Pauli exclusion principle in the two-phonon states of wave function (6.33).

Considering the three-phonon terms in the diagonal approximation in the two-phonon states space reduces the shift $\Delta\omega(g_1 g_2)$. The pp interactions taken into account in [302, 303] in addition to the ph interactions also reduce the shift $\Delta\omega(g_1 g_2)$. As a result, for collective states this shift takes on values in the range 0.5–1.5 MeV. The function $U_{g_1 g_2}(g_0)$ defined by (6.40) is a sum of non-coherent terms. When the total number of terms, including those resulting from the introduction of pp interactions, are taken into account, the values of $U_{g_1 g_2}(g_0)$ increase in comparison with the calculations of [165, 301–303]. As a result, the contribution of two-phonon configurations to normalization (6.37) of wave function (6.33) for the states with $K^\pi = 0^+$, 2^+ and 4^+ and energy 1.8–2.4 MeV reaches 20%.

Let us define the two-phonon state. We will call a state two-phonon if the contribution of the two-phonon component to the normalization of its wave function exceeds 50%. With this definition, not more than one two-phonon state with a fixed value of K^π can be constructed out of two phonons g_1 and g_2. It appears that the experimental measurement and a theoretical calculation of the contribution of two-phonon configurations to the wave functions of the low-lying states of deformed nuclei become important tasks.

The following statements can be formulated about the general features of vibrational states of even–even deformed nuclei with excitation energies below 5 MeV.

(1) The anharmonicity of nuclear vibrations for states with energies up to 2.5–3.0 MeV is small. This means that the contribution of the one-phonon component to the normalization of the wave function exceeds 90%. States with fixed values of K^π are mostly one-phonon states corresponding to the first, second, etc solutions of the RPA secular equation. The small anharmonicity constitutes the decisive difference between the vibrational states in deformed nuclei and those in open shell spherical nuclei.
(2) Strong fragmentation of one-phonon states begins at excitation energies of 2.5–3.0 MeV and above, that is, in the energy range of two-phonon poles.

The small anharmonicity of low-lying vibrational states of deformed nuclei is caused by two factors:

(a) numerical values of the function $U_{g_1 g_2}(g_0)$ are from 0.01 to 0.20 MeV, that is, are less by one to two orders of magnitude than in open shell spherical nuclei. Only rarely does $U_{g_1 g_2}(g_0)$ take on values above 0.1 MeV. The function $U_{g_1 g_2}(g_0)$ is a sum of a large number of terms with various signs. In deformed nuclei, opposite-sign terms suppress each other. In spherical nuclei, the largest terms in $U_{g_1 g_2}(g_0)$ have identical signs for the first roots of the RPA secular equations;
(b) the shift $\Delta\omega(g_1 g_2)$ of the two-phonon pole, which increases the energy of two-phonon poles to values above 2.5–3.0 MeV.

The strong fragmentation of one-phonon states at excitation energies above 3 MeV will manifest itself in the experimental measurement of $B(E2)$, $B(E3)$ and $B(E4)$, and also in measuring $B(M2)$ and $B(M3)$ values.

The basic contradiction in the description of the structure of non-rotational states of deformed nuclei in the QPNM, on one hand, and the IBM, the Bohr–Mottelson model and its microscopic analogue, and other phenomenological models, on the other, lies in whether low-lying two-phonon collective states exist. I support the conclusions made in [320] and some other publications that no really reliable experimental data are available on the existence of two-phonon collective states in deformed nuclei. Attempts to explain their absence were made for ^{168}Er and for thorium and uranium isotopes. According to [124, 321, 322], the absence of quadrupole two-phonon states with energies below 2 MeV in ^{168}Er occurs because of the high anharmonicity of the γ-vibrational mode (the ^{168}Er nucleus is the three-axial ellipsoid of revolution with $\gamma \neq 0$). Nevertheless, calculations [323] indicate that although ^{168}Er is a soft nucleus with respect to a non-axial deformation, the minimum of energy is reached at $\gamma_0 = 0$.

An important stage in the experimental study of the contribution of two-phonon configurations to the wave functions of states with energies about 2 MeV was the development of the technique of identifying short-lived nuclear states using gamma-ray induced Doppler broadening (GRIDB) [324–326]. The measurement of the lifetime of the $K^{\pi}_{\nu} = 4^+_1$ state of ^{168}Er [324, 325] made it possible to find the value of $B(E2)$ for the $4^+_1 \rightarrow 2^+_1$ transition in the 0.5–1.5 s.p.u. interval. If the strength of the two-phonon configuration {221,221} is totally concentrated on the 4^+_1 state, then the ratio

$$\frac{B(E2; 4^+_1 \rightarrow 2^+_1)}{B(E2; 2^+_1 \rightarrow 0^+_{gs})} \tag{6.58}$$

equals 2.78. According to [324], this ratio changes to 0.5–1.6, which means that the contribution of the two-phonon configuration to the wave function of the 4^+_1 state of ^{168}Er is (20–50)%. This was the first experimental measurement of states with a large two-phonon contribution in deformed nuclei.

According to the estimates of [327], the contribution of the configuration {221,221} to the 4^+_1 state of ^{168}Er, taking into account the decrease in $\Delta\omega(221, 221)$ due to the three-phonon terms and a more accurate calculation of $U_{221,221}(441)$, is 17%, and ratio (6.58) is 0.46, which does not contradict the experimental data [324, 325]. According to the multiphonon method calculations of [328], the 4^+_1 state of ^{168}Er is a two-phonon one and ratio (6.58) is 1.1, which is in accord with [324, 325]. However, the absolute values of $B(E2)$ and $B(E4)$ calculated in [328] are much smaller than the experimental data.

Even–even isotopes of radium, thorium and uranium are shown to have low-lying states with $I^{\pi}K = 1^-0$. Ever since the publication of [329], they have been treated as octupole vibrational states $\lambda\mu i = 301$. Octupole two-phonon 0^+ states of the type {301, 301} have not been found experimentally in thorium and uranium isotopes. Their absence is explained by the existence of a stable octupole deformation that has been identified in nuclei with $A < 228$. According to the calculations of [330], a stable octupole deformation may exist in isotopes of radium and thorium with $A < 228$. Consequently, the absence of two-phonon {301, 301}-type 0^+ states in ^{228,230}Th, ^{232,234}U and some other nuclei still awaits its explanation in the framework of the Bohr–Mottelson, IBM and other phenomenological models.

Table 6.12 $\log ft_{\Sigma}$ for the β^+ decay $0^+_{gs} \to 1^+$.

Nucleus	Exper.	QPNM calculations			Part of total strength manifested in the β^+ decay
		$G_1^{11}=0$ $\lvert G_A/G_V\rvert=1$	$G_1^{11}=-8.3/A$ $\lvert G_A/G_V\rvert=1$	$G_1^{11}=-8.5/A$ $\lvert G_A/G_V\rvert=1.26$	
^{162}Yb	4.72	4.1	4.7	4.8	0.24
^{164}Yb	4.8	4.2	4.7	4.7	0.14
^{166}Yb	4.9	4.3	4.9	4.9	0.12
^{166}Hf	4.76	4.2	4.8	5.0	0.34
^{164}Hf	—	4.1	4.6	4.5	0.17

The shift of two-phonon poles in deformed nuclei was investigated in [331] using the boson expansion method. New phonons, expressed via boson operators, were constructed. The QPNM Hamiltonian is written in terms of the operators of the new bosons up to terms of order four. The shift of the two-phonon pole was also calculated. It coincided with the shift calculated in [165] to the accuracy of the main terms. This result corroborates the conclusion on the absence of collective two-phonon states in even–even deformed nuclei.

6.3.5

Let us consider the β^+ decays in deformed nuclei. It was shown in [257–259] that pp interactions affect the Gamow–Teller β^+ decays of spherical and deformed nuclei. Let us consider, following [332], the β^+ transitions from the ground states 0^+_{gs} even–even nuclei to the one-phonon $\Omega^+_{\mu i\sigma}\Psi_0$ states (see (2.152)), for which the equations are given by (2.160) and (2.160′). The results of calculations and the experimental data of [333] are shown in table 6.12. Taking into account the pp interactions brought a decrease in the reduced probabilities of the Gamow–Teller β^+ transitions by a factor of 5 to 6. The table clearly shows that the β^+ decay reflects a part of the total Gamow–Teller strength. Since there are no states with large Gamow–Teller strength, available for the β^+ decay, at energies above 3 MeV, we are justified in neglecting the fragmentation of one-phonon states caused by the quasiparticle–phonon interaction.

6.3.6

Let us compare the descriptions of non-rotational states in the QPNM and IBM [334, 335].

A comparison of a description of non-rotational states of even–even deformed nuclei in the phenomenological IBM with those in microscopic models is possible because ideal bosons in the IBM can be presented as series in fermion pairs, that is, bosons are related to the microscopic fermion structure. An important role is played in the IBM by the number N of valent nucleons: owing to this, the calculation results are connected with a specific nucleus and a comparison of spectra of different nuclei becomes a natural procedure. The IBM gives a description of (tp)- or (pt)-type two-nucleon transfer reaction, that is, a description of the relation between two nuclei. It is also important that the same non-rotational excited states are described both in the phenomenological IBM and in microscopic models.

Two cardinal differences can be indicated between the IBM and many microscopic models, such as the RPA and its modifications, the theory of finite Fermi systems [55], the theory of nuclear fields [78], the QPNM and some others.

(1) The IBM singles out a subspace of collective states which constitutes a small fraction of the total space of two-quasiparticle states. Namely, only that part is included which is contained in the s-, d- and f-bosons, or in s_p-, d_p- and s_n-, d_n- bosons. When the g-boson is introduced, the space of two-quasiparticle states with $K^\pi = 0+$ and 2^+ is enlarged and the subspace with $K^\pi = 1+$, $3+$ and 4^+ is additionally taken into account. The introduction of the s'- and d'-bosons also enlarges the space of two-quasiparticle states.

Microscopic models make use of a large, almost entire space of two-quasiparticle particle–hole-type states. Some of these models also consider the space of particle–particle and hole–hole-type states. The completeness of the employed space of two-quasiparticle states, required to describe giant resonances, is monitored by checking the model-independent and model-dependent sum rules.

The strong feature of the IBM lies in that it partially takes into account the multi-quasiparticle components of the wave functions, since the boson operator is written as a series in fermion pairs. In microscopic models, taking into account the 2p–2h, and even more so 3p–3h configurations entails considerable difficulties.

(2) When the subspace of collective states is singled out in the IBM, all coupling to other collective states is cut off, for instance, coupling to the states that form giant resonances and with a large number of weakly collective and two-quasiparticle states. One should bear in mind that this separation of collective states is not unambiguous, owing to the absence of a clear distinction between strongly collective and somewhat less collective states. The addition of a number of weakly collective states to the IBM when the g-, s'- and d'-bosons are introduced diminishes the attractiveness of the model, since the number of not-included weakly

collective states is considerable.

As the excitation energy increases, the state density goes up and the state structure becomes more complicated. The strength of simple single-quasiparticle or one-phonon states gets distributed (fragmented) over many nuclear levels. The responsibility for fragmentation lies in the coupling between collective and non-collective motions or the quasiparticle–phonon interaction. This coupling is to some extent taken into account in microscopic models but is typically absent from the IBM. The IBM is therefore powerless to describe one of the most important properties of nuclei, namely, that the structure of states gets progressively more complex as the excitation energy increases. Furthermore, the way the IBM takes into account weakly collective and two-quasiparticle states is artificial and therefore generates a large number of new parameters.

The cardinal differences between the QPNM and IBM in their description of a number of non-rotational states of even–even deformed nuclei can be expressed in a fairly specific way and thus an approach to testing this difference experimentally can be formulated. The possibility of formulating the specific differences stems from the fact that the wave functions of non-rotational states with excitation energies up to 2.5 MeV have one dominating component in both models. Thus the principal differences are the different dominating components in the QPNM and IBM; this was first pointed out in [334]. Figure 6.9 schematically presents the dominating components of the wave functions in the QPNM, in the *sd* IBM and the *sdg* IBM for the states $K^{\pi}_{\nu} = 0^+_1$, 0^+_2, 0^+_3, 0^+_4, 2^+_1, 2^+_2, 2^+_3, 2^+_4, 2^+_5, 4^+_1, and 4^+_2. In the QPNM, wave function (6.33) has a single dominating one-phonon component $\lambda\mu i$. In the *sd* IBM the dominating components are one-boson with $n_\gamma = 1$ and $n_\beta = 1$, two-boson with $\{n_\gamma = 1, n_\beta = 1\}$, $n_\gamma = 2$ and $n_\beta = 2$, and three-boson with $\{n_\gamma = 2, n_\beta = 1\}$, $n_\gamma = 3$ and $n_\beta = 3$, and other components as well.

The *sdg* IBM wave functions in the harmonic approximation are:
for the ground state

$$|g) = \frac{1}{\sqrt{N!}}(\Sigma^+)^N|0)$$

for the one-boson states

$$\left.\begin{aligned} |\gamma) &\approx \Pi^+_2(\Sigma^+)^{N-1}|0) \\ |\beta) &\approx \Pi^+_0(\Sigma^+)^{N-1}|0) \\ |2^+_2) &\approx \Gamma^+_2(\Sigma^+)^{N-1}|0) \\ |0^+_2) &\approx \Gamma^+_0(\Sigma^+)^{N-1}|0) \\ |4^+_1) &\approx \Gamma^+_4(\Sigma^+)^{N-1}|0) \end{aligned}\right\}$$

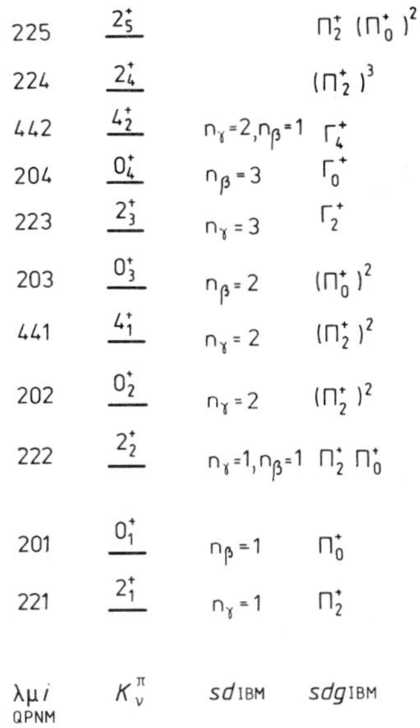

Figure 6.9 Schematic diagram of the dominating components of the wave functions of non-rotational states with $K^\pi = 0^+$, 2^+ and 4^+ in the QPNM, *sd* IBM and *sdg* IBM.

for the two-boson states

$$
\left.\begin{aligned}
|2^+) &\approx \Pi_2^+\Pi_0^+(\Sigma^+)^{N-2}|0) \\
|0^+) &\approx (\Pi_0^+)^2(\Sigma^+)^{N-2}|0) \\
|4^+) &\approx (\Pi_2^+)^2(\Sigma^+)^{N-2}|0) \\
|0^+) &\approx (\Pi_2^+)^2(\Sigma^+)^{N-2}|0)
\end{aligned}\right\}
$$

where

$$
\left.\begin{aligned}
\Sigma^+ &= 5^{-1/2}\,s^+ + 2\cdot 7^{-1/2}\,d_0^+ + 2\sqrt{2}\cdot 35^{-1/2}\,g_0^+ \\
\Pi_0^+ &= 2\cdot 15^{-1/2}\,s^+ + 21^{-1/2}\,d_0^+ - 2\sqrt{6}\cdot 35^{-1/2}\,g_0^+ \\
\Pi_2^+ &= 7^{-1/2}\,d_2^+ + \sqrt{6}\cdot 7^{-1/2}\,g_2^+ \\
\Gamma_0^+ &= 2\sqrt{\frac{2}{15}}\,s^+ - 2\sqrt{\frac{2}{21}}\,d_0^+ + \sqrt{\frac{3}{35}}\,g_0^+ \\
\Gamma_2^+ &= \sqrt{\frac{6}{7}}\,d_2^+ - \sqrt{\frac{1}{7}}\,g_2^+ \\
\Gamma_4^+ &= g_4^+ .
\end{aligned}\right\}
$$

The wave functions of the γ vibrational state with $K_\nu^\pi = 2_1^+$ and β-vibrational state with $K_\nu^\pi = 0_1^+$ state possess a dominating one-boson or one-phonon component. Hence, there are no essential differences in the description of these states in the QPNM, *sd* IBM and *sdg* IBM.

The wave functions of states with $K^{\pi}_{\nu} = 2^+_2, 0^+_2, 4^+_1$ and 0^+_3 in the QPNM have one dominating one-phonon component each, and in the *sd* IBM and *sdg* IBM each has one dominating two-boson component. Here QPNM and IBM manifest a principal difference in the structure of these states. The same difference between the QPNM and *sd* IBM is found for the $K^{\pi}_{\nu} = 2^+_3$, 0^+_4 and 4^+_2 states, since their wave functions have dominating one-phonon components in the QPNM and three-boson components in the *sd* IBM.

In the *sdg* IBM, the wave function of the $K^{\pi}_{\nu} = 4^+_2$ state has a dominating one-boson component, while the wave functions of the $K^{\pi}_{\nu} = 2^+_3$ and 0^+_4 states consist of a combination of one-boson operators. There is no fundamental difference with the QPNM. An appreciable difference is that the wave function of a one-phonon state (for example, $\lambda\mu i = 223$) may greatly differ in the contribution of various two-quasiparticle components to the normalization from the wave functions of one-phonon states 221 and 222.

We need to answer the question of how to distinguish between the states whose wave functions possess dominating one-phonon components from states with dominating one-boson components. Let us consider inelastic scattering of protons, α-particles and heavy ions, one-nucleon transfer reactions and allowed unhindered *au* β decay.

If the $K^{\pi}_{\nu} = 2^+_2, 0^+_2$ and 4^+_1 states are collective and have large values of $B(E2)$, they can be distinguished from two-boson states. Two-boson states are excited in two stages and their excitation cross sections are small. In many cases, the $K^{\pi}_{\nu} = 2^+_2, 0^+_2$ and 4^+_1 states are not collective and their $B(E2)$ and $B(E4)$ values are not high. As a result, the inelastic scattering allows the identification of only sufficiently collective one-phonon states, separating them from states with dominating two-boson components.

One-nucleon transfer reactions manifest well-defined large two-quasiparticle components of one-phonon wave functions. In their turn, they determine in an unambiguous manner the existence of a large one-phonon component of the wave function. The cross sections of one-nucleon transfer reactions with the excitation of two-boson states are small. It is with one-nucleon transfer reactions that we can distinguish between states with dominating one-phonon components and states with dominating two-boson components in the cases of large two-quasiparticle components.

There are two *au* β transitions between p 523↑⇆n 523↓ and p 514↑⇆n 514↓. In the case of *au* β decay, it is possible to distinguish between a transition to one-phonon states and a transition to two-phonon states. Owing to the interaction between bosons, the wave functions of the $K^{\pi}_{\nu} = 2^+_2, 0^+_2, 4^+_1$ and 0^+_3 states, shown in figure 6.9, contain an admixture of one-boson components. In the *sdg* IBM, the states with $K^{\pi}_{\nu} = 2^+_3$, 0^+_4 and 4^+_2 have dominating one-boson components. It was not clear whether it were possible to show experimentally which of the two models (QPNM or *sdg* IBM) gave a correct description of the structure of these states. In some cases, the results of studying the one-nucleon transfer reactions and the *au* β decays did give an affirmative answer to this question.

The wave functions of $\lambda\mu i$ one-phonon states consist of different two-quasiparticle components. Some states with fixed $\lambda\mu$ or K^π may be excited, for example, in the $(t\alpha)$ reactions, others in (dp), still others in both types of reactions, and so on. As a result of QPNM calculations, the contribution of the one-phonon component to the normalization of wave function (6.33) is found, and also the contribution of the two-phonon components to the normalization of the wave function of the one-phonon state. Calculations give the spectroscopic factors of one-nucleon transfer reactions and the intensities of the *au* β decays.

In the *sdg* IBM, the operators Π_0^+, Π_2^+, Γ_0^+, Γ_2^+ and Γ_4^+ are combinations of the s^+, d^+ and g^+ bosons with quite definite weights. These weights impose strict selection rules for the one-nucleon transfer reactions and *au* β decays.

Let us illustrate the possibility of exciting a state in one-nucleon transfer reactions and *au* β decays in the *sdg* IBM and QPNM, using as an example five $K^\pi = 2^+$ states as shown in figure 6.10. The dominating components of the wave functions of these five states are given in figure 6.9. The possible excitations of these states in the *sdg* IBM and QPNM are shown schematically in figure 6.10. Consider the $(t\alpha)$ reaction. If the 2_1^+ reaction is strongly excited, then the *sdg* IBM predicts that the 2_3^+ state must be excited strongly and the 2_2^+ state, weakly. According to the QPNM, the 2_1^+, 2_4^+ and 2_5^+ states can be excited strongly but 2_2^+ and 2_3^+ will not be excited. In the framework of the QPNM, for example, in a (dp) reaction, the 2_5^+ state may be excited but the first four 2^+ states may not be, while in a (dt) reaction, all five 2^+ states may be excited. If, in the framework of the *sdg* IBM, the 2_1^+ state is not excited in a (dt) reaction, no other 2^+ states will be excited. In a (dp) reaction, the 2_5^+ state cannot be excited if the 2_1^+ and 2_3^+ states are not. Similar selection rules hold for the *au* β transitions. For instance, if $\log ft$ is small for the transition to the 2_1^+ state, in the *sdg* IBM it must be small for transition to the 2_3^+ state. In the QPNM, $\log ft$ can be small for the *au* β transitions, for example, to the 2_1^+, 2_4^+ and 2_5^+ states.

The possibility of describing in the QPNM such excitation of five 2^+ states, illustrated in figure 6.10, is realized in the case of ^{168}Er. The results of calculations of the energy and structure of five 2^+ states performed in [302], and the experimental data of [336, 337] are listed in table 6.13. The table gives the calculated contributions (in %) of the one-phonon components $22i$, $i = 1, 2, 3, 4, 5$ to the normalization of wave functions (6.33) for the first five 2^+ states. For each $22i$ phonon, the table gives the contribution (in %) of the largest two-quasiparticle components to the normalization of their wave functions. The experimental data [336] on the $(\vec{t}\alpha)$ reaction indicate that the two-quasi-proton state pp $411\uparrow + 411\downarrow$ enters with large weight the wave

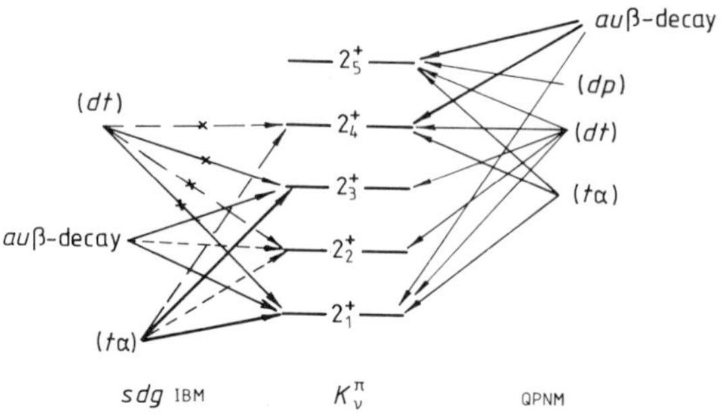

Figure 6.10 The schematic diagram of excitation of the $K^\pi = 2^+$ states in one-nucleon transfer reactions and *au β* decay in the QPNM and *sdg* IBM. Notation: bold full lines—high-intensity transitions, fine and broken lines—weak and very weak transitions, crossed arrows—forbidden transitions.

functions of the 2_1^+ and 2_4^+ states and with an appreciable weight, the 2_5^+ state. According to the calculations of [302], the contribution of pp $411\uparrow + 411\downarrow$ to the 2_1^+ and 2_4^+ states is large and to the 2_3^+ and 2_5^+ states is also considerable. In the *au β* decay from the $K^\pi = 3^+$ state with the p $523\uparrow - n521\downarrow$ configuration of ^{168}Ho we find a manifestation of the strength of the two-quasi-neutron state nn $523\downarrow - 521\downarrow$ which is observed in the 2_1^+ state; its contribution to the 2_4^+ and 2_5^+ states is large but to 2_2^+ and 2_3^+ is very small. The distribution of the strength of the nn $523\downarrow - 521\downarrow$ state among the first five 2^+ states is described sufficiently well in the QPNM. The table shows that the peculiarities of the excitation of the 2^+ states in a $(\vec{t}\alpha)$ reactions and in the *au β* decay, illustrated in figure 6.10, are given qualitatively correct QPNM descriptions. According to the calculations, all five 2^+ states can be excited in the ^{169}Er$(dt)^{168}$Er reaction. Unfortunately, this reaction is very difficult to observe experimentally because of the insufficient lifetime of ^{169}Er. The 2_5^+ state can be observed in a (dp) reaction.

It is impossible to obtain a qualitatively correct description of the experimental data on the excitation of the first five 2^+ states of ^{168}Er in a $(\vec{t}\alpha)$ reaction and in the *au β* decay in the framework of the *sdg* IBM. This model is absolutely powerless in the description of the structure of these 2^+ states of ^{168}Er. It is not clear at all how the *sdg* IBM could be modified to remove the contradiction with the experimental data on the 2^+ states of ^{168}Er.

When the g-boson is introduced into the IBM [126, 338], states with $K^\pi = 1^+$, 3^+ and 4^+, containing two-quasiparticle components, appear in deformed nuclei, and the space of two-quasiparticle states with $K^\pi = 0^+, 2^+$ broadens. Also, one normally takes into account some weakly collectivized states. These states are not very different from other weakly collectivized

Table 6.13 $K^{\pi} = 2^{+}$ states in ^{168}Er.

	Experiment				QPNM calculation			
K^{π}_{ν}	$\mathcal{E}$ (MeV)		Structure (%)		$\mathcal{E}$ (MeV)		Structure (%)	
2^{+}_{1}	0.821	$(\vec{t}\alpha)$:	pp 413↓ − 411↓	50	0.8	221:	96	
			pp 411↑ + 411↓	37		221	pp 413↓ − 411↓	36
		log ft =5.2:	nn 523↓ − 521↓				pp 411↑ + 411↓	30
			considerable				nn 523↓ − 521↓	18
							nn 521↑ + 521↓	18
2^{+}_{2}	1.848	log ft =6.1:	nn 523↓ − 521↓		1.8	222:	98	
			small			222	nn 512↑ − 521↓	97
							pp 411↑ + 411↓	2
2^{+}_{3}	1.930	log ft =6.2:	nn 523↓ − 521↓		1.9	223:	94	
			small			223	nn 521↑ + 521↓	60
							pp 411↑ + 411↓	13
							nn 523↓ − 521↓	3
2^{+}_{4}	2.193	$(\vec{t}\alpha)$:	pp 411↑ + 411↓	20–30	2.2	224:	98	
		log ft =4.8:	nn 523↓ − 521↓	224			nn 523↓ − 521↓	60
			large					
							pp 411↑ + 411↓	28
							nn 521↑ + 521↓	4
2^{+}_{5}	2.425	$(\vec{t}\alpha)$:	pp 411↑ + 411↓		2.5	225:	97	
			considerable		225		nn 633↑ − 651↑	36
		log ft =4.6:	nn 523↓ − 521↓				pp 411↑ + 411↓	15
			large				nn 523↓ − 521↓	15
							nn 521↑ + 521↓	6

states which are ignored. When the g-boson is introduced, the 4^+_1 state has a large two-boson component [126, 338]. Among the 0^+ and 2^+ states that we consider now, there remain states with large two-boson configurations. A renormalization of s- and d-bosons using the g-boson without explicitly introducing the g-boson degrees of freedom does not result in expanding the space of two-quasiparticle states considered in the *sd* IBM; hence, it does not remove the fundamental difference between this model and the QPNM for the description of the 0^+_3, 0^+_4, 2^+_2, 2^+_3, 4^+_1 and 4^+_2 states.

Different models must be compared for a large number of deformed nuclei in the rare earth and actinide regions, so as to avoid distorting the general picture by the specifics of a selected nucleus. For example, the contradiction between the *sdg* IBM and the experimental data is not removed in the description of the 4^+_1 and 4^+_2 states in ^{156,158}Gd and ^{160,164}Dy and of the 3^+_1 and 3^+_2 states in ^{172,174}Yb. Is the introduction of the g' boson the only way out? The difficulties of the *sdg* IBM are connected with the description of the 2^+_2 states with large values of $B(E2)$ that we find in many nuclei.

States of negative parity are described in the *sdf* IBM and in the *spdf* IBM [339, 340]. It is clear that collective states of positive and negative parity in deformed nuclei should be described in the framework of one model. Owing to the need of introducing the g-boson, the model required may be the *spdfg* IBM. It will undoubtedly be possible to describe in its terms the energies and probabilities $B(E\lambda)$ of the rotational bands constructed on the states with a dominating one-boson component. Since no two-phonon or two-boson states were experimentally found in deformed nuclei, it is not clear whether the *spdfg* IBM will help to remove the contradiction with the experimental data.

Further investigation of the structure of even–even deformed nuclei calls for experiments measuring the contribution of two-quasiparticle components to the wave functions of the rotational bands constructed on states with $K^\pi_\nu = 0^+_3$, 0^+_4, 0^+_5, 2^+_2, 2^+_3, 3^+_1, 3^+_2, 4^+_1, 4^+_2 and on other states at the energies about 1.5–2.5 MeV. Also needed are experiments searching for two-phonon collective states.

6.3.7

Let us consider the distribution of the $E2$ and $E3$ strength over low-lying states. It is typically assumed that the quadrupole β and γ states are collective while other collective states form giant isoscalar and isovector quadrupole resonances. In the case of octupoles, the collective states are, in addition to the first octupole states, those states that form the low-lying (LEOR) and high-lying (HEOR) isoscalar and isovector octupole giant resonances. Low-lying collective hexadecapole states are observed in a number of nuclei. Phenomenological models, including the IBM, are based on the collective nature of the first quadrupole and octupole states and on the

Table 6.14 Distribution of the $E2$ strength.

Nucleus	K^{π}_{ν}	$\mathcal{E}$ (MeV) or $\Delta\mathcal{E}$ (MeV)	$B(E2)_{(\text{s.p.u.})}$ (exper.) [341]	(calc.) QPNM [302]	IBM [341]
^{168}Er	2^+_1	0.821	4.7	4.6	4.7
	0^+, 2^+	1.2–2.5	—	1.5	0.02
^{172}Yb	2^+_1	1.47	1.4	1.5	1.0
	2^+_2	1.61	0.42	0.7	0.007
	0^+	1.0–2.0	0.66	0.9	0.2
	0^+, 2^+	2.0–3.0	2.4	1.0	0.2
^{178}Hf	2^+_1	1.174	3.9	4.1	—
	2^+	3.0–3.2	—	0.6	—

absence of collectivity in higher-lying states, up to the giant resonances. A study of the $E\lambda$ strength distribution became possible after experiments [341] on inelastic scattering of α particles.

We will discuss the distribution of the quadrupole strength. The experimental data and the results of QPNM [302] and IBM [341] calculations are listed in table 6.14. In ^{168}Er and ^{178}Hf, we find the standard situation: the main part of the $E2$ strength is concentrated on the γ vibrational state. According to calculations [302], the remaining states in ^{168}Er share about 30%, and in ^{178}Hf, about 15% of the $E2$ strength of the γ vibrational state.

The distribution of the $E2$ strength in ^{172}Yb is qualitatively different from that in ^{168}Er, ^{178}Hf and a number of other nuclei. The peculiarity of the distribution of the $E2$ strength in ^{172}Yb manifests itself, first of all, in that the second 2^+_2 state, in addition to the first 2^+_1 state, is also a collective one. Second, a considerable part of the $E2$ strength is concentrated in the energy range 2–3 MeV. Cases in which the first 2^+_1 and second 2^+_2 states are both collective constitute great difficulties for a phenomenological description.

Let us look at the $E3$ distribution in ^{168}Er. According to [341], ^{168}Er has the first collective states with $K^{\pi}_{\nu} = 0^-_1$, 1^-_1 and 2^-_1. Six collective $K^{\pi}_{\nu} = 3^-_1$ states were observed. The first three 3^-_1, 3^-_2 and 3^-_3 states concentrate 1.3 s.p.u. while the fourth 3^- state has 4.68 s.p.u., that is, three times as much as the first three. Besides, 7.9 s.p.u. is found in the interval 2.25–2.50 MeV. This distribution of the $E3$ strength is very different from standard cases.

The reason why the first most collective state with a given value of K^{π} is not the first to appear is shown in figure 6.11 for the $K^{\pi} = 3^-$ state in ^{168}Er. The figure shows the behaviour of $\mathcal{F}^{M}_{33}(\omega)$ defined by formula (2.42) as a function of the energy ω. The calculations of $\mathcal{F}^{M}_{33}(\omega)$ were performed

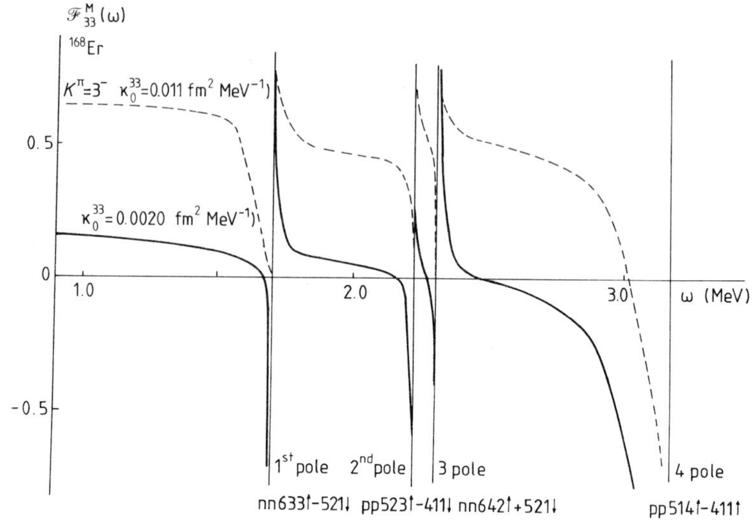

Figure 6.11 Function $\mathcal{F}^{M}_{33}(\omega)$ for $K^{\pi}=3^{-}$ state in ^{168}Er, for $\kappa_0^{33}=0.020$ fm^{2}MeV^{-1} (full curve) and $\kappa_0^{33}=0.011$ fm^{2}MeV^{-1} (broken curve). Vertical lines mark the poles of equation (2.42). The roots of (2.42) are calculated for $\mathcal{F}^{M}_{33}(\omega)=0$.

for the values $\kappa_0^{33}=0.020$ fm^2 MeV^{-1} and $\kappa_0^{33}=0.011$ fm^2 MeV^{-1}, that is, for values much lower than those used in the calculations of [302]. A state is the more collective, the greater the shift of the energy root with respect to the respective pole towards lower energies. The fourth root is considerably lowered with respect to the fourth pole, so that the fourth 3^-_4 state is collective. It remains collective if $\kappa_0^{33}=0.011$ fm^2 MeV^{-1} while 3^-_1, 3^-_2 and 3^-_3 states are nearly two-quasiparticle: the values of $B(E3)$ for them are 0.01, 0.04 and 0.1 s.p.u., that is, are less by a factor of 30–60 than those for $\kappa_0^{33}=0.020$ fm^{2}MeV^{-1}. In the *sdf* IBM calculations of $K^{\pi}=3^{-}$ states of ^{168}Er [340], the first three $K^{\pi}=3^{-}$ levels are omitted. The main part of the $E3$ strength is concentrated on the 3^-_4 states which the authors were regarding as the first collective $K^{\pi}=3^{-}$ state. Is it really permissible to assume as collective the states whose reduced excitation probabilities are greater than the two-quasiparticle values by factors of 30–60?

Examples of a non-standard distribution of the $E3$ strength calculated in [302, 303] are given in table 6.15. It would be of considerable interest to obtain experimental data to test this distribution.

Let us demonstrate the $E2$ and $E3$ strength distributions over low-lying vibrational states using ^{238}U [303]. To study the RPA distributions of the $E2$ strength, $B(E2;i)$ values were calculated for the $K^{\pi}_i=2^+_i$ and 0^+_i states. The low-lying 1^+ states are not very different from the two-quasiparticle

Table 6.15 Non-standard distribution of the $E3$ strength: QPNM calculations.

Nuclei	$I=3$ K^{π}_{ν}	$\mathcal{E}$ or $\Delta\mathcal{E}$ (MeV)	$B(E3)$ (s.p.u.)	Nuclei	$I=3$ K^{π}_{ν}	$\mathcal{E}$ or $\Delta\mathcal{E}$ (MeV)	$B(E3)$ (s.p.u.)
^{168}Er	3^-_1	1.6	0.14	^{178}Hf	0^-_1	2.0	2.0
	3^-_2	2.1	0.6		0^-_2	2.4	4.0
	3^-_3	2.2	0.3		1^-_1	1.4	0.5
	3^-_4	2.4	2.0		1^-_2	1.5	0.3
					1^-	2.0–3.0	9.5
^{170}Yb	1^-_1	1.5	0.7				
	1^-_2	2.2	1.2	^{238}U	3^-_1	1.6	0.2
	2^-_1	1.6	2.4		3^-_2	1.9	2.0
	2^-_2	1.9	2.0				
	3^-_1	1.6	0.1	^{250}Cf	0^-_1	1.4	3
	3^-	2.0–2.6	3.3		0^-_2	1.8	4
					1^-_1	1.15	8
^{172}Yb	1^-_1	1.2	1.4		1^-_2	1.5	2
	1^-_2	2.2	2.2		1^-_3	1.8	2
^{174}Yb	0^-_1	1.7	1.0				
	0^-	2.0–3.0	5.0	^{256}Fm	2^-_1	0.9	5
	1^-_1	1.8	0.1		2^-	1.5–2.5	5
	1^-	1.9–3.2	4.2				
	2^-_1	1.2	1.6				
	2^-_2	2.7	1.7				

states, so their contribution can be ignored. The results of calculations are shown in figure 6.12 as the sum $\sum_i B(E2;i)$ containing a part corresponding to $K^{\pi}=2^+$ and 0^+ and in table 6.16 as the sums $\sum_{\mu i} B(E\lambda,\mu;i)$ and $\sum_{\mu i}\omega_{\lambda\mu i}B(E\lambda,\mu;i)$, for energy intervals of 0.5 MeV. The $K^{\pi}=2^+$ states take about 2/3 of the total sum $\sum_i B(E2;i)$. The figure shows that the total sum increases up to 5.5 MeV and then reaches a plateau. For states with energies above 5.5 MeV up to the isoscalar quadrupole resonance, the $B(E2;i)$ values are very small. The contribution of the γ-vibrational $K^{\pi}=2^+_1$ state to the sum $\sum_i B(E2;K=2,i)$ is 45%, and the contribution of the β-vibrational $K^{\pi}=0^+_1$ state to the sum $\sum_i B(E2;K=0,i)$ is 15%.

For all states with energies up to 6.5 MeV, the contribution of the sums $\sum_i \omega_{22i}B(E2;K=2,i)+\sum_i\omega_{20i}B(E2;K=0,i)$ to the energy-weighted isoscalar sum rule is 15%, with the contribution of the β and γ vibrational states being only 2.2%. The results of these analyses imply that a substantial part of the $E2$ strength falls on the states above the β and γ vibrational states. Note that none of the phenomenological models, including the IBM, reproduce that part of the $E2$ strength which falls on the states above the β and γ vibrational states.

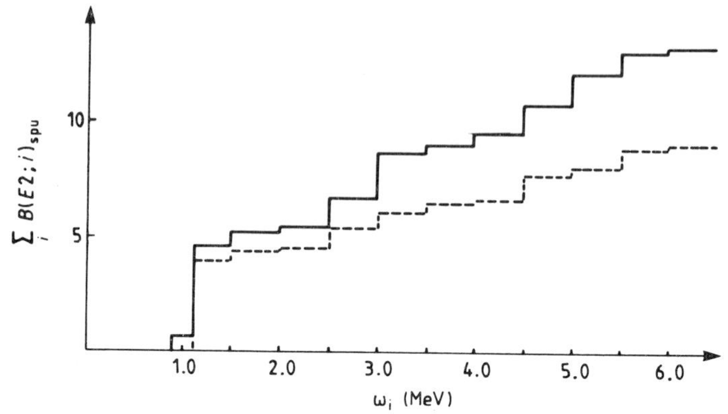

Figure 6.12 Distribution of $E2$ strength in ^{238}U plotted as the total sum of $B(E2;i)$ values for energy intervals of 0.5 MeV. Notation: the broken histogram is the contribution of the states with $K^\pi = 2^+$ to $\sum_i B(E2;i)$, and the full line is the contribution of the states with $K^\pi = 0^+$ and 2^+.

Table 6.16 Distribution of the $E2$ and $E3$ strength in ^{238}U.

	$\Delta\mathcal{E}$ (MeV)					
	1–2	2–3	3–4	4–5	5–6	1–6
$\sum_{\mu=0,2}\sum_i B(E2,\mu;i)_{\text{s.p.u.}}$	5.4	2.7	0.9	2.4	1.5	12.9
$\sum_{\mu=0,2}\sum_i \omega_{2\mu i} B(E2,\mu;i)_{\text{s.p.u.}}$	6.1	6.4	3.0	10.5	7.8	33.8
$\sum_{\mu=0,1,2,3}\sum_i B(E3,\mu;i)_{\text{s.p.u.}}$	45.2	10.2	9.3	4.6	1.7	71.0
$\sum_{\mu=0,1,2,3}\sum_i \omega_{3\mu i} B(E3,\mu;i)_{\text{s.p.u.}}$	46.0	25.6	33.7	21.0	7.3	134.0

The $E3$ strength distribution in ^{238}U, RPA-calculated in [303] in the form of $B(E3,i)$ values for each state with $K^\pi = 0^-$, 1^-, 2^- and 3^-, is plotted in figure 6.13. It shows that a large number of collective octupole states are present in addition to the first states. As the excitation energy increases to 5.5 MeV, the $B(E3,i)$ decrease; they are very small above 6 MeV. In the interval 3.5–4.0 MeV, there is a local maximum which can hardly be regarded as a low-lying isoscalar octupole resonance. According to calculations, there is no gap in the distribution of the $B(E3,i)$ values between the first and higher-lying octupole states. We can assume that all octupole states with energies up to 5.0–5.5 MeV, form a low-energy isoscalar octupole resonance. For ^{238}U, the contribution of these states to

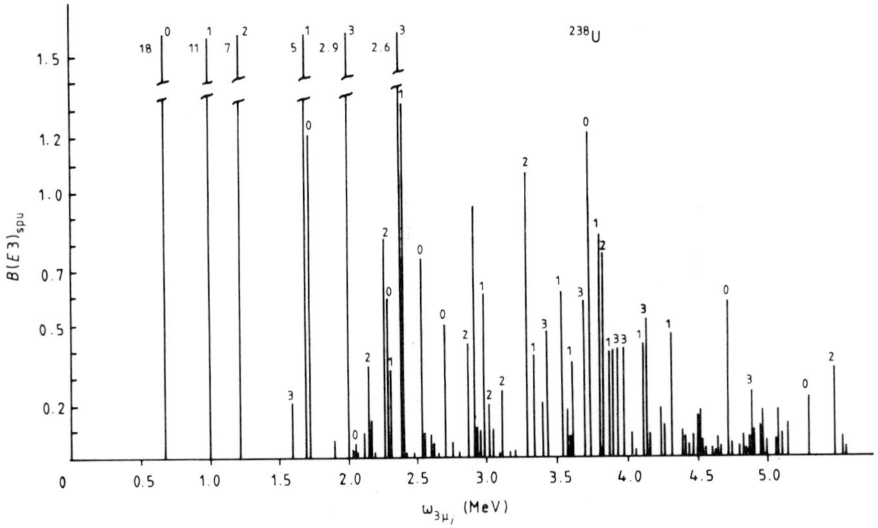

Figure 6.13 Distribution of the $E3$ strength among the low-lying states in ^{238}U. Notation: $B(E3)$ are given in single-particle units, large $B(E3,i)$ values are given on the left of the straight lines, and K^π = 0, 1, 2, 3 are given on the right for $B(E3) > 0.2$ s.p.u.

the isoscalar energy-weighted sum rule equals 28%.

As we see from figure 6.13, collective octupole states with large values of $B(E3,i)$ are found at energies of 3.5, 4.8 and even 5.4 MeV. The calculation of the $K^\pi = 0^-, 1^-, 2^-, 3^-$ states with wave function (6.33) demonstrated that a considerable number of states in the 2–5 MeV energy range have complex structure. In many cases, states with small values of $B(E3,i)$ are not two-quasiparticle states and their wave functions contain a large number of two-quasiparticle and two-phonon components. A minor part of the results are listed in table 6.17, with the same notation as in table 6.6. The wave functions of many states possess a dominating one-phonon component, and relatively large two-phonon components in some states.

An analysis of the $E2$ and $E3$ distributions in ^{238}U showed that a considerable part of the strength concentrating on low-lying states lies above the β and γ vibrational and the first octupole states. Experimental study of the distribution of the $E\lambda$ strength in even–even deformed nuclei is definitely of substantial scientific interest.

Table 6.17 Structure of selected states in ^{238}U.

K^{π}_{ν}	$\mathcal{E}$ (MeV)	$B(E\lambda)$ (s.p.u.)		Structure (%)			
0^{-}_{16}	3.73	1.50	30 16	24; {201, 302}	21; {331, 434} 12;	{205, 301}	8
			30 16:	pp 521↑ − 651↑ 8	pp 512↑ − 642↑ 7;		
					nn 633↓ − 503↓ 4		
0^{-}_{28}	4.80	0.29	30 28	82; {334, 434}	8 {201, 301}	4;	
			30 28:	nn 620↑ − 501↓ 36	nn 761↑ − 622↓ 3;		
1^{-}_{27}	3.49	10^{-4}	31 27	97; {201, 312}	1		
			31 27:	nn 501↓ − 631↑	98		
2^{-}_{33}	4.14	9.03	32 33	92 {321,434}	3; {322,442}	2	
			32 33:	nn 615↓ − 502↓ 42;	pp 514↓ − 402↓ 41;		
3^{-}_{4}	2.40	2.5	33 4	86 {221, 311}	5		
			33 4:	pp 642↑ + 530↑ 29;	nn 725↑ − 622↑ 23;		
				pp 633↑ − 530↑ 10			
3^{-}_{9}	2.91	0.1	33 9	94; {201, 332}	2		
			33 9:	nn 725↑ − 622↑ 48;	nn 734↑ − 631↑ 44;		
				pp 633↑ − 530↑	6		

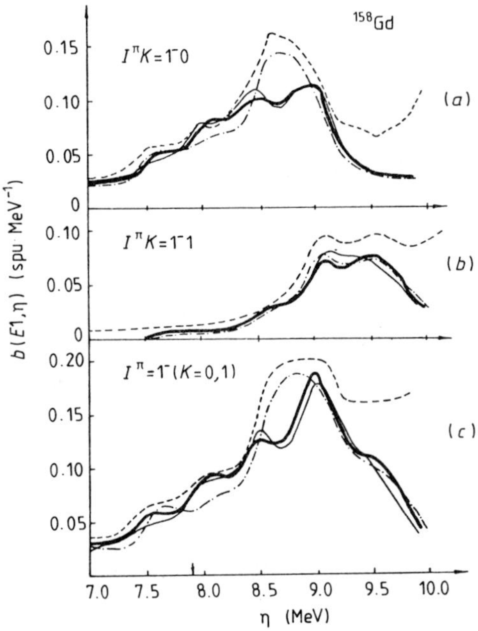

Figure 6.14 Strength functions $b(E1, \eta)$ for the $E1$ transitions from the ground state to states with $I^\pi = 1^-$ in ^{158}Gd: (a) transitions to states with $K = 0$; (b) transitions to states with $K = 1$; (c) transitions to states with $K = 0$ and $K = 1$; broken curves—RPA calculations taking into account all one-phonon states; chain curves—calculations on a restricted basis; bold full curve—calculations taking into account the quasiparticle–phonon interactions on a restricted basis; fine full curve—calculations on a restricted basis, neglecting the Pauli exclusion principle in two-phonon components of the wave function; $\Delta = 0.4$ MeV.

6.3.8

The structure of states of deformed nuclei in the energy range from 2 MeV up to the neutron binding energy B_n has not been studied well so far. The high-spin states are the only exception. An important role in the study of the properties of the states in the intermediate range of excitation energy is played by the electromagnetic transitions. The strength function of the γ transition from the ground state to the state described by wave function (6.33) can be written in the form

$$b(\lambda; \eta) = \frac{1}{\pi} \operatorname{Im} \left\{ \frac{\begin{vmatrix} 0 & \mathcal{M}_{\lambda\mu i} \\ \mathcal{M}_{\lambda\mu i} & \|\theta(\eta)\| \end{vmatrix}}{\theta(\eta)} \right\} \tag{6.59}$$

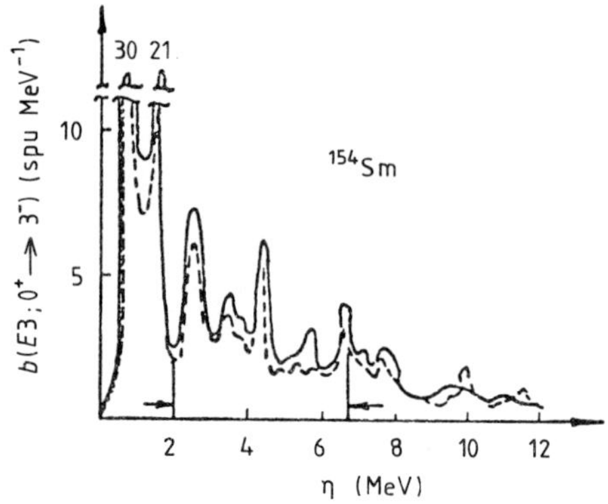

Figure 6.15 Strength functions $b(E3, \eta)$ in ^{152}Sm for the $E3$ transitions from the ground state to 3^- states (full curve) and for transitions to isoscalar components of the 3^- states (broken curve).

(we follow here [342]). Here $\theta(\eta)$ is the determinant of the matrix $||\theta(\eta)||$ given by (6.44). The numerator of (6.59) is the determinant of a bordered matrix whose rank is greater by unity than that of the matrix $||\theta(\eta)||$. The function $\mathcal{M}^E_{\lambda\mu i}$ for electric transitions is given by formulas (2.79′) or (2.82), and the function $\mathcal{M}^M_{LKi}$ for magnetic transitions is given by (2.85).

An example of calculation of strength functions $b(E1, \eta)$ is given in figure 6.14. The calculations were carried out with the total and a restricted number of one-phonon states. A limitation of the number of one-phonon states to 40 is needed to be able to use formula (6.59) when taking into account the quasiparticle–phonon interactions. The calculations in which the Pauli exclusion principle is taken into account strictly or approximately in the two-phonon components of (6.33) give not very different values of the strength functions. Taking into account the fragmentation of one-phonon states affects only slightly the radiation strength functions. For this reason, it is possible to calculate the strength functions of many deformed nuclei in the framework of the RPA.

Among the collective states in the 2–7 MeV excitation energy interval, the low-lying isoscalar octupole resonance is especially well pronounced. An example of the results of RPA calculations of the strength functions $b(E3; 0^+_{gs} \to 3^-; \eta)$ [343] for the transitions to the $3^- K$ states of ^{152}Sm with $K = 0, 1, 2, 3$ is given in figure 6.15. The region of the low-lying octupole resonance is shown by arrows. The results for ^{152}Sm are given for the $E3$ transitions to the isoscalar components of a state which are directly connected with the cross sections of (α, α') reactions. The figure shows that the isoscalar contribution is prevalent. The existence of a low-lying isoscalar octupole resonance in deformed nuclei has been confirmed experimentally [343].

Conclusion

Most of the known nuclear characteristics are determined by the behaviour of few-quasiparticle components of the wave functions of excited states. This monograph demonstrates the possibility of a unified microscopic description of few-nucleon components and of the nuclear characteristics they imply at low, intermediate and high excitation energies. The mathematical apparatus presented in the book is also suitable for describing the electric and magnetic isoscalar and isovector giant resonances, analogue states and charge-exchange resonances, neutron strength functions and partial probabilities of radiational transitions from neutron resonances. The introduction of neutron resonances into the microscopic description together with low-lying states and giant resonances constitutes one of the successes of the QPNM. There is no doubt that as we shift the study to high excited states, we encounter increased complexities and the mathematical apparatus has to be further developed.

The analysis of the fragmentation of few-quasiparticle states, that is, single-quasiparticle, two-quasiparticle and one-phonon states, is a first step in the study of the structure of atomic nuclei. The next step should be the fragmentation of three-and four-quasiparticle states or quasiparticle$\otimes$phonon and quasiparticle$\otimes$two phonons states. First experiments [229, 230] on the γ decay of deep hole states and on giant resonance decay [344–346] are evidence that the second step has begun. Experiments with giant resonances on excited states appear to be very promising.

The field of multi-quasiparticle or multi-phonon configurations is enormously large and rather poorly known. One should expect that many-nucleon transfer reactions on spherical and deformed nuclei will play an important role in its experimental investigation. What we need is a broad range of research into the behaviour of the fragmentation of multi-quasiparticle configurations. Without this knowledge, we will not be able to build up the general picture of nuclear structure.

References

[1] Blatt J M and Weisskopf V F 1952 *Theoretical Nuclear Physics* (New York: Wiley)

[2] Davydov A S 1958 *Theory of Atomic Nucleus* (Moscow: Fizmatgiz) (in Russian)

[3] Preston M A 1962 *Physics of the Nucleus* (Reading, MA: Addison–Wesley)

[4] Bohr A and Mottelson B 1969 *Nuclear Structure* vol **1**, 1975 vol **2** (New York: Benjamin)

[5] Brown G E 1967 *Unified Theory of Nuclear Models and Forces* (Amsterdam: North-Holland)

[6] Soloviev V G 1976 *Theory of Complex Nuclei* (Oxford: Pergamon)

[7] Eisenberg J and Greiner W 1970 *Nuclear Theory* vol **1**, vol **2**, 1972 vol **3** (Amsterdam: North-Holland)

[8] Wildermuth K and Tang Y C 1977 *Unified Theory of the Nucleus* (Braunschweig: Vieweg)

[9] Sitenko A G and Tartakovsky V K 1972 *Lectures on the Theory of the Nucleus* (Moscow: Atomizdat) (in Russian)

Krainov V P 1973 *Lectures on the Microscopic Theory of Atomic Nucleus* (Moscow: Atomizdat) (in Russian)

[10] De Shalit A and Feshbach H 1974 *Theoretical Nuclear Physics* vol **1** (New York: Wiley)

[11] Ring P and Schuk P 1980 *Nuclear Many Body Problem* (New York: Springer)

[12] Soloviev V G 1981 *Theory of the Atomic Nucleus: Nuclear Models* (Moscow: Energoizdat) (in Russian)

[13] *Symbols, Units and Nomenclature in Physics Documents* 1978, U.I.P. 20 (International Union of Pure and Applied Physics, S.U.N. Commission)

[14] Condon E V and Shortley G G 1935 *Theory of Atomic Spectra* (London: Cambridge University Press)

Varshalovich D A, Moskalev A N, Khersonsky V K 1975 *Quantum Theory of Angular Momentum* (Moscow: Nauka) (in Russian)

[15] Baldin A M 1980 *Prog. Part. Nucl. Phys.* **4** 95–132

Burov V V, Lukyanov V K and Titov A I 1984 *Part. Nucl.* **15** 1249–95

[16] *10th Int. Conf. on Particles and Nuclei.* 1985 (Amsterdam: North-Holland)

Proceedings of the 7th Int. Seminar on High Energy Physics 1984, D1, 2-84-599 (Dubna: JINR)

[17] Bogoliubov N N 1949 *Lectures on Quantum Statistics* (Kiev: Sovetskaya Shkola) (in Russian)

[18] Bardeen J, Cooper L and Schrieffer J 1957 *Phys. Rev.* **108** 1175–206

Bogoliubov N N 1958 *ZhETF* **36** 58–65, 73–9
Bogoliubov N N, Tolmachev V V and Shirkov D V 1958 *A New Method in Superconductivity Theory* (Moscow: AN SSSR) (in Russian)
[19] Landau L D 1958 *ZhETF* **35** 97–103
[20] Vanagas V 1971 *Algebraic Methods in Nuclear Theory* (Vilnus: Mintas)
Filippov G F, Ovcharenko V I and Smirnov Yu F 1981 *Microscopic Theory of Collective Excitations of Atomic Nuclei* (Kiev: Naukova Dumka) (in Russian)
[21] Bogoliubov N N 1958 *Dokl. AN SSSR* **119** 52–5
[22] Bogoliubov N N, Soloviev V G 1959 *Dokl. AN SSSR* **124** 1011–4
[23] Kerman A K and Klein A 1963 *Phys. Rev* **132** 1326–42
Klein A and Kerman A K 1965 *Phys. Rev* **138** 1323–32
[24] Klein A, Gelenza L and Kerman A K 1965 *Phys. Rev* **140** 245–63
Dreizler R M, Klein A, Chi-Chiang Wu and Do Dang G 1967 *Phys. Rev* **156** 1167–72
[25] Belyaev S T and Zelevinsky V G 1972 *Yad. Fiz.* **16** 1195–208; 1973 **17** 525–39
[26] Bogoliubov N N 1959 *Uspekhi Fiz. Nauk* **67** 549–80
[27] Tyablikov S V 1959 *Dokl. AN SSSR* **121** 250–2; 1958 *Nauchn. Dokl. Vyssh. Shkoly* **3** 183–91
[28] Soloviev V G 1963 *Effect of Pairing Correlations on the Properties of Atomic Nuclei* (Moscow: Gosatomizdat) (in Russian); *Selected Topics in Nuclear Theory* (Vienna: IAEA) p 233
[29] Bogoliubov N N 1960 *JINR Preprint* P-511, Dubna; *Selected Works* Kiev: Naukova Dumka vol **3** p. 110 (in Russian)
[30] Belyaev S T 1959 *Mat.-fiz. medd. Kgl. danske vid. selskab.* **31** No. 11
[31] Soloviev V G 1958 *Zh. Eksp. Theor. Fiz.* **35** 823–5; 1958 *Dokl. AN SSSR* **123** 652–4; *Nucl. Phys.* **9** 655–64
[32] Soloviev V G 1959 *Zh. Eksp. Theor. Fiz.* **36** 1869–74; 1960 *Dokl. AN SSSR* **133** 325–8
[33] Lein A 1967 *Nuclear Structure* (New York: Benjamin)
[34] Soloviev V G 1961 *Zh. Eksp. Teor. Fiz.* **40** 654–65
[35] Soloviev V G 1961 *Izv. AN SSSR* **25** 1198–216; 1961 *Mat.-Fiz. Medd. Dan. Vidensk. Selsk.* **1** No. 11
[36] Wahlborn S 1962 *Nucl. Phys.* **37** 554–84
[37] Nilsson S G and Prior O 1961 *Mat.-Fiz. Medd. Dan. Vidensk. Selsk.* **32** No. 16
[38] Volkov M K, Pavlinovsky A, Rybarska V and Soloviev V G 1963 *Izv. AN SSSR* **27** 878–90
[39] Mikhailov I N 1963 *Acta. Phys. Polon.* **23** 745–50
[40] Mang H J, Poggenburg J K and Rasmussen I O 1965 *Nucl. Phys.* **64** 353–78
[41] Kuzmenko N K and Mikhailov V M 1973 *Izv. AN SSSR* **37** 1911–9; 1980 **44** 942–8
[42] Brown G E and Jackson A D 1976 *The nucleon–nucleon interaction* (Amsterdam: North-Holland)
HaidenbauerJ, Koike Y amd Plessas W 1986 *Phys. Rev.* **33** 439–46
[43] Vauterin D and Brink D M 1973 *Phys. Rev.* **5** 626–47
Beiner M, Flocard H, Nguyen van Giai and Quentin P 1975 *Nucl. Phys.* A **238** 29–69
[44] Barts B I, Bolotin Yu L, Inopin E V and Gonchar V Yu 1982 *The Hartree–Fock Method in the Theory of the Nucleus* (Kiev: Naukova Dumka)

[45] Nemirovsky P E and Chepurnov V A 1966 *Yad. Fiz.* **3** 998–1010
Chepurnov V A 1967 *Yad. Fiz.* **6** 955–60
[46] Gareev F A, Ivanova S P and Kalinkin V N 1967 *Acta. Phys. Polon.* **32** 461–89
[47] Pashkevich V V 1971 *Nucl. Phys.* **169** 275–93
[48] Dudek J, Majhofer A and Skalski J, Werner T, Čwiok S and Nazarewicz W 1979 *J. Phys. G: Nucl. Phys.* **5** 1359–81
[49] Belyaev S T 1965 *Nucl. Phys.* **64** 17–29
Belyaev S T 1968 *Collective Excitations in Nuclei* (New York: Gordon and Breach)
[50] Dobaczewski J and Skalski J 1981 *Nucl. Phys.* **369** 123–40
[51] Baranger M and Veneroni M 1978 *Ann. Phys.* **114** 123–200
[52] Goeke K and Reinhard P G 19 *Ann. Phys.* **112** 444–71
Goeke K, Grummer F and Reinhard P G 19 *Ann. Phys.* **150** 504–51
[53] Madler P 1971 *Part. Nucl.* **15** 418–82
[54] Jolos R V and Soloviev V G 1971 *Part. Nucl.* **1** 365–90
[55] Migdal A B 1983 *Theory of Finite Fermi Systems and Properties of Atomic Nuclei* (Moscow: Nauka)
[56] Migdal A B 1967 *Method of Quasiparticles in the Theory of the Nucleus* (Moscow: Nauka)
[57] Krainov V P 1966 *Yad. Fiz.* **3** 804–12
[58] Werner E 1966 *Z. Phys.* **191** 381–94
[59] Birbrair B L 1968 *Nucl. Phys.* A **108** 449–62; 1973 **212** 27–44
[60] Liu K F and Brown C E 1976 *Nucl. Phys.* A **265** 385–415
Kamerdzhiev S P 1972 *Yad. Fiz.* **15** 676–89
[61] Soloviev V G 1965 *Nucl. Phys.* **69** 1–36
[62] Kisslinger L S and Sorensen R A 1963 *Rev. Mod. Phys.* **35** 853–915; 1960 *Mat.-Fiz. Medd. Dan. Vidensk. Selsk.* **32** No. 9
[63] Bës D R and Sorensen R A 1969 *Advances Nucl. Phys.* **2** 129–222
Rowe D J 1967 *Phys. Rev.* **162** 866-871
[64] Malov L A, Nesterenko V O and Soloviev V G 1977 *Teor. Matem. Fiz.* **32** 134–44
[65] Soloviev V G 1978 *Part. Nucl.* **9** 580–622
Soloviev V G 1978 *Nucleonica* **23** 1149–78; 1987 *Prog. Part. Nucl. Phys.* **19** 107–65
[66] Malov L A and Soloviev V G 1980 *Part. Nucl.* **11** 301–41
[67] Soloviev V G 1982 *Teor. Matem. Fiz.* **53** 399–405
[68] Soloviev V G 1965 *Atomic Energy Review* **3** No. 2 117–93
[69] Soloviev V G 1991 *Z. Phys.* A **338** 271–9
[70] Soloviev V G 1989 *Z. Phys.* A **334** 143–8
[71] Gareev F A, Ivanova S P, Soloviev V G and Fedotov S I 1973 *Part. Nucl.* **4** 357–455
[72] Tamura T and Udagava T 1961 *Prog. Theor. Phys.* **25** 1051–82
Yoshida S 1962 *Nucl. Phys.* **38** 380–419
[73] Birbrair B L, Erokhina K I and Lemberg I Kh 1963 *Izv. AN SSSR* **27** 150–71
[74] Vdovin A I and Soloviev V G 1983 *Part. Nucl.* **14** 237–85
[75] Voronov V V and Soloviev V G 1983 *Part. Nucl.* **14** 1380–442
[76] Ponomarev V Yu, Soloviev V G, Stoyanov Ch and Vdovin A I 1979 *Nucl. Phys.* **323** 446-460
Vdovin A I, Stoyanov Ch and Yudin I P 1978 *Izv. AN SSSR* **42** 2004–9
[77] Soloviev V G 1989 *Yad. Fiz.* **50** 40

[78] Bortignon P F, Broglia R A, Bes D R and Lotta R 1977 *Phys. Repts* C **30** 306–60

[79] Vdovin A I, Dambasuren D, Soloviev V G and Stoyanov Ch 1976 *Izv. AN SSSR* **40** 2183–8
Dambasuren D 1977 *Izv. AN SSSR* **41** 1281–6

[80] Soloviev V G, Stoyanova O and Stoyanov Ch 19 *Izv. AN SSSR* **44** 1938–46
Nesterenko V O, Soloviev V G and Khalkin A V 1980 *Yad. Fiz.* **32** 1209–17

[81] Ken-ji Hara 1964 *Prog. Theor. Phys.* **32** 88–105
Ikeda K, Udagawa T and Yamamura H 1965 *Prog. Theor. Phys.* **33** 22–37

[82] Rowe D J 1968 *Phys. Rev.* **175** 1283–92
Johnson R E, Dreizler R M and Klein A 1969 *Phys. Rev.* **176** 1289–92

[83] Hernandes E S and Plastino A 1972 *Phys. Lett.* B **39** 163–5; 1974 *Z. Phys* A **268** 377–46; 1975 **273** 253–8

[84] Navrotska-Rybarska V, Stoyanova O and Stoyanov Ch 1981 *Yad. Fiz.* **33** 253–8

[85] Nguen Dinh Dang and Soloviev V G 1983 *JINR Communication* R4–84–325 (Dubna: JINR)

[86] Bainum D E, Rapaport J, Goodman G D, Horen D J, Foster C C, Greenfield M B and Goulding C A 1980 *Phys. Rev. Lett.* **44** 1751–4
Vinogradov A A *et al* 1981 *Pisma Zh. Eksp. Teor. Fiz.* **33** 233–6
Horen D J *et al* 1981 *Phys. Lett.* **99** 383–6

[87] Gaarde C, Rapaport J, Taddeucci T N, Goodman C D, Foster C C, Beninm D E, Goulding S A, Greenfield M B, Horen D J and Sugarbaker E 1981 *Nucl. Phys.* A **369** 258–80
Gaarde C 1985 *Proc. of the Int. School on Nuclear Structure* D4–85–851 ed V G Soloviev and Yu P Popov (Dubna: JINR) pp 104–23

[88] Soloviev V G, Sushkov A V and Shirikova N Yu 1983 *Pisma Zh. Eksp. Teor. Fiz.* **38** 151–3; *Z. Phys* A **316** 65–74

[89] Kuzmin V A and Soloviev V G 1982 *Yad. Fiz.* **35** 620–7; 1984 *J. Phys. G: Nucl. Phys.* **10** 1507–22; 1985 **11** 603–12

[90] Grotz K, Klapdor H V and Metzinger J 1983 *Phys. Lett.* **132** 22–6
Klapdor H V and Grotz K 1984 *Phys. Lett.* **142** 323–8
Vogel P and Fisher P 1985 *Phys. Rev.* **32** 1362–8

[91] Davydov A S 1967 *Excited States of Atomic Nuclei* (Moscow: Atomizdat) (in Russian)

[92] Belyaev S T and Zelevinsky V G 1962 *Zh. Eksp. Theor. Fiz.* **42** 1590–603; 1962 *Nucl. Phys.* **39** 582–604

[93] Marumori T, Yamamura M and Tokunaga A 1964 *Prog. Theor. Phys.* **31** 1009–25

[94] Janssen D, Dönau F, Frauendorf S and Jolos R V 1971 *Nucl. Phys.* A **172** 145–65
Jolos R V and Rybarska V 1972 *Part. Nucl.* **3** 739–69

[95] Marshalek E R and Weneser J 1972 *Part. Nucl.* **3** 739–69
Marshalek E R 1971 *Nucl. Phys.* A **161** 401–9

[96] Dyson F 1956 *Phys. Rev.* **102** 1217–30, 1230–44

[97] Holzwarth G, Janssen D and Jolos R V 1976 *Nucl. Phys.* A **261** 1–12

[98] Iwasaki S, Sakata F and Takada K 1977 *Prog. Theor. Phys.* **57** 1289–302
Tamura T and Kishimoto T 1983 *Prog. Theor. Phys.* **74** 282–95

[99] Pedrochi V G and Tamura T 1983 *Phys. Rev.* **28** 410–27; 1982 **29** 1461–74
Tamura T, Li C T and Pedrochi V G 1985 *Phys. Rev.* **32** 2129–40

[100] Maglione E and Vitturi A 1984 *Nucl. Phys.* A **430** 158–74

Faessler A and Morrison I 1984 *Nucl. Phys.* A **423** 320–32
[101] Janssen D, Jolos R V and Dönau F 1984 *Nucl. Phys.* A **224** 93–115
Jolos R V and Janssen D 1977 *Part. Nucl.* **8** 330–73
[102] Holstein T and Primakoff H 1940 *Phys. Rev.* **58** 1098–107
[103] Arima A and Iachello F 1975 *Phys. Rev. Lett.* **35** 1069–72; 1978 *Phys. Rev. Lett.*. **40** 385–7; 1976 *Ann. Phys.*. **99** 253–317
[104] Kuriyama A, Marumori T and Matsuyanagi K 1971 *Prog. Theor. Phys.* **45** 784–809
Suzuki T and Matsuyanagi K 1976 *Prog. Theor. Phys.* **56** 1156–73
[105] Bes D R and Broglia R A 1966 *Nucl. Phys.* **80** 289–304
Bohr A 1968 *Nuclear Structure Dubna Symp.* 179–89
[106] Soloviev V G 1962 *Phys. Lett.* **1** 202–5
[107] Blaizot J P and Marshalek E R 1978 *Nucl. Phys.* A **309** 422–52
Kurchev G 1980 *Nucl. Phys.* A **349** 416–32
[108] Sakai M 1984 *Atomic Data and Nuclear Data Tables* **31** 399–432
[109] Zelevinsky V G 1985 *Proc. Int. School on Nuclear Structure* D4–85–851 ed V G Soloviev and Yu P Popov (Dubna: JINR) pp 173–96
[110] 1979 *Interacting Bosons in Nuclear Physics* ed F Iachello (New York: Plenum)
1981 *Interacting Bose–Fermi Systems in Nuclear Physics* ed F Iachello (New York: Plenum)
[111] Casten R F and Warner D D 1988 *Rev. Mod. Phys.* **60** 389
[112] Jolos R V, Lemberg I Kh and Mikhailov B M 1985 *Part. Nucl.* **16** 280–348
Golovkov N A, Dzhelepov B S, Ivanov R B and Mikhailova M A 1985 *Izv. AN SSSR* ser. fiz. **50** 2–8
[113] Arima A, Otsuka T, Iachello F and Talmi I 1977 *Phys. Lett.* B **66** 205–8; 1978 **76** 139–43
[114] Elliot J P 1985 *Rep. Prog. Phys.* **48** 171–221
Dieperink A E 1984 *Nucl. Phys.* A **421** 189c–204c; 1987 *Proc. Int. Nucl. Phys. Conf. (Inst. Phys. Conf. Ser. 86)* ed J Durrel, J Irvine, G Morrison (Bristol: Institute of Physics) pp 139–53
[115] Duval P D and Barrett B R 1981 *Phys. Lett.* B **100** 223–7; 1982 *Nucl. Phys.* A **376** 213–28
van Isacker P, Pittel S, Frank A and Duval P D 1986 A *Nucl. Phys.* **451** 201–18
[116] Sambataro M and Molnar G 1982 *Nucl. Phys.* A **376** 201–12
Sambataro M 1982 *Nucl. Phys.* A **380** 365–82
[117] Otsuka T, Arima A and Iachello F 1978 *Nucl. Phys.* A **309** 1–39
[118] Gambhir Y K, Ring P and Schuk P 1982 *Phys. Rev.* C **25** 2858–61
Pittel S, Duval P D and Barrett B R 1982 *Phys. Rev.* C **25** 2834–6; 1982 *Ann. Phys. (N.Y.)* **144** 168–89
Barrett B R 1985 *Proc. Int. School on Nuclear Structure* ed V G Soloviev and Yu P Popov D4–85–891 (Dubna: JINR) pp 153–72
Sambataro M, Schaaser H and Brink D M 1986 *Phys. Lett.* **167** 145–9
[119] Bohle D, Richter A, Steffen W, Dieperink A E L, Lo Iudice H, Palumbo F and Scholten O 1984 *Phys. Lett.* B **137** 27–31
[120] Hamilton W D, Irbäck A and Elliot J P 1984 *Phys. Rev. Lett.* **53** 2469–72
Otsuka T and Ginocchio J N 1985 *Phys. Rev. Lett.* **54** 777–80
[121] Harter H, von Brentano P, Gelberg A and Casten R F 1985 *Phys. Rev.* C **32** 631–3
[122] Warner D D, Casten R F and Davidson W F 1980 *Phys. Rev. Lett.* **45** 1761–5; 1981 *Phys. Rev.* C **24** 1713–33

Warner D D and Casten R F 1982 *Phys. Rev.* **25** 2025–28; 1982 *Phys. Rev. Lett.* **48** 666–9

[123] Casten R F and Aprahamian A 1984 *Phys. Rev.* C **25** 1919–21

Casten R F, von Brentano P and Haque A M 1985 *Phys. Rev.* C **31** 1991–4

Zhang M, Valliers M, Gilmore R, Da Hsuan Feng, Hoff R W and Hong–Zhou Sun 1985 *Phys. Rev.* C **32** 1076–9

Casten R F, Frank W and von Brentano P 1985 *Nucl. Phys.* A **444** 133–53

[124] Bohr A and Mottelson B R 1980 *Phys. Scripta* **22** 468–74; 1982 **25** 28–36

[125] Davidson W F *et al* 1981 *J. Phys. G: Nucl. Phys.* **7** 455–528

Burke D G, Davidson W F, Cizewski J A, Brown R E, Flynn E R and Sunier J W 1985 *Can. J. Phys.* **63** 1309–19

[126] Wu Hua-Chuan and Zhou Xiao–Qian 1984 *Nucl. Phys.* A **417** 67–76

[127] Arima A 1985 *Nuclear Structure 1985* ed R Broglia, G B Hagemann and B Herskind (Amsterdam: Elsevier) pp 147–59

[128] Lipas P O and Helimaki K 1985 *Phys. Rev. Lett.* B **165** 244–6

Zimmerman M and Dobeš J 1985 *Phys. Rev. Lett.* B **156** 7–10

Scholten O, Heyde K and van Isaker 1985 *Phys. Rev. Lett.* **55** 1866–9

Dobeš J 1985 *Proc. Int. School on Nuclear Structure* D4–85–851 ed V G Soloviev and Yu P Popov (Dubna: JINR) pp 197–207

[129] Faessler A 1983 *Nucl. Phys.* A **396** 291c–305c

[130] Brant S, Paar V and Vretenar D 1984 *Z. Phys.* A **319** 355–6

[131] Soloviev V G 1965 *Phys. Lett.* **16** 308–311; 1966 **21** 320–2

[132] Soloviev V G 1968 *Nuclear structure Dubna Symp.* (Vienna: IAEA) pp 101–18; 1972 *Proc. Int. School on Nuclear Structure* D4–6465 (Dubna: JINR) pp 77–123

[133] Soloviev V G 1971 *Izv. AN SSSR* **35** 666–77; 1977 **38** 1580–7; 1973 *Teor. Matem. Fiz.* **17** 90–102

[134] Soloviev V G and Malov L A 1972 *Nucl. Phys.* A **196** 433–51

Malov L A and Soloviev V G 1975 *Yad. Fiz.* **21** 502–9; 1975 *Teor. Matem. Fiz.* **25** 265

[135] Soloviev V G 1975 *Neutron Capture Gamma-Ray Spectroscopy* (Petten: Reactor Centrum Nederland) pp 99–117; 1976 *Selected Topics in Nuclear Structure* D–9920 (Dubna: JINR) pp 146–75

[136] Vdovin A I and Soloviev V G 1973 *Teor. Matem. Fiz.* **17** 90–102

Krychev G and Soloviev V G 1976 *Teor. Matem. Fiz.* A **22** 244–52

[137] Malov L A and Soloviev V G 1976 *Nucl. Phys.* A **270** 87–107; 1977 *Yad. Fiz.* **26** 729–39

[138] Soloviev V G, Stoyanov Ch and Vdovin A I 1977 *Nucl. Phys.* A **288** 376–96

[139] Soloviev V G 1978 *Izv. AN SSSR* **42** 1991–2003

[140] Jolos R V, Molina Kh L and Soloviev V G 1979 *Teor. Matem. Fiz.* **40** 245–50;1980 *Z. Phys* **295** 147–52

[141] Soloviev V G 1979 *Fundamental Problems in Theoretical and Mathematical Physics* D-12831 (Dubna: JINR) pp 424–38; 1980 *Proc. Int. School on Nuclear Structure* D4–80–385 (Dubna: JINR) pp 57–88

[142] Soloviev V G, Stoyanov Ch and Vdovin A I 1980 *Nucl. Phys.* A **342** 261–82

[143] Chan Zuy Khuong, Soloviev V G and Voronov V V 1981 *J. Phys. G: Nucl. Phys.* **7** 151–63

Voronov V V and Chan Zuy Khuong 1981 *Izv. AN SSSR* ser. Fiz. **45** 1909–15

[144] Voronov V V and Soloviev V G 1983 *Teor. Matem. Fiz.* **57**75–84
Soloviev V G 1983 *Teor. Matem. Fiz.* **57** 438–47
[145] Vdovin A I, Voronov V V and Dao Tien Khoa 1985 *Teor. Matem. Fiz.* **64** 259–68
Stoyanov Ch, Vdovin A I and Voronov V V 1985 *Proc. Int. School on Nuclear Structure* D4–85–851 ed V G Soloviev and Yu P Popov (Dubna: JINR) pp 27–50
[146] Soloviev V G 1983 *Proc. Int. School on Nuclear Structure* ed V G Soloviev and Yu P Popov (Dubna: JINR) pp 8–26
[147] Vdovin A I, Voronov V V, Soloviev V G and Stoyanov Ch 1985 *Part. Nucl.* **16** 245–79
[148] Tabakin F 1964 *Ann. Phys.* **14** 51–94
Schmid E W and Zigelmann H 1975 *The Quantum Mechanical Three-Body Problem* (London: Pergamon)
[149] Knüpfer W and Huber M G 1976 *Phys. Rev.* C **14** 2254–68
[150] Speth J, Klemt V, Wambach J and Brown G E 1980 *Nucl. Phys.* A **343** 382–416
Nakayama K, Krewald S, Speth J and Love W G 1984 *Nucl. Phys.* A **431** 419–60
[151] Dehesa J S, Krewald S, Lallen A and Donnelly T A 1985 *Nucl. Phys.* A **436** 573–92
[152] Bës D R, Broglia R A, Dussel G G, Liotta R and Perazzo D 1976 *Nucl. Phys.* A **260** 77–94
[153] Vdovin A I, Voronov V V and Malov 1976, Soloviev V G and Stoyanov Ch 1976 *Part. Nucl.* **7** 952–88
[154] Peterson D F and Veje C J 1967 *Phys. Lett.* B **24** 444-463
[155] Bes D R, Broglia R A and Nillson B S 1975 *Phys. Repts* **16** 1–56
[156] Castel B and Hamamoto I 1976 *Phys. Lett.* B **65** 27–31
[157] Malov L A 1981 *JINR Communication* P4–81–228 (Dubna: JINR)
[158] Bellman R 1960 *Introduction to Matrix Analysis* (New York: McGraw Hill)
[159] Kuliev A A and Salamov D I 1984 *Izv. AN AzSSR Ser. Fiz.-Tekh. Mat.* No 2 60–69
[160] Soloviev V G, Stoyanov Ch and Voronov V V 1983 *Nucl. Phys.* A **399** 141–62
[161] Stoyanov Ch and Vdovin A I 1983 *Phys. Lett.* B **130** 134–8
[162] Vdovin A I, Nguen Dinh Thao, Soloviev V G and Stoyanov Ch 1983 *Yad. Fiz.* **37** 43–51
[163] Dambasuren D, Soloviev V G, Stoyanov Ch, Vdovin A I 1976 *J. Phys. G: Nucl. Phys.* **2** 25–31
[164] Stoyanov Ch 1979 *Teor. Matem. Fiz.* **40** 422–8
[165] Soloviev V G and Shirikova N Yu 1981 *Z. Phys.* A **301** 263–9; 1982 *Yad. Fiz.* **36** 1376–86
[166] Bortignon P F and Broglia R A 1981 *Nucl. Phys.* A **371** 405–29
[167] Bertsch G F, Bortignon P F and Broglia R A 1983 *Rev. Mod. Phys.* **55** 287–314
[168] Voronov V V, Nguen Dinh Dang, Ponomarev V Yu, Soloviev V G and Stoyanov Ch 1984 *Yad. Fiz.* **40** 683
[169] Soloviev V G 1984 *Yad. Fiz.* **40** 1163–70
[170] Voronov V V and Kyrchev G 1986 *Teor. Matem. Fiz.* **69** 236–44
[171] Suzuki T, Fuyuki M and Matsuyangi K 1979 *Prog. Theor. Phys.* **61** 1082–92

[172] Schmid K W, Grümmer F and Faessler A 1984 *Phys. Rev.* C **29** 291–323
Faessler A 1985 *Proc. Int. School on Nuclear Structure* D4–85–851 ed V G Soloviev and Yu P Popov (Dubna: JINR) pp 208–19
[173] Voronov V V, Dao Thien Khoa 1984 *Izv. AN SSSR Ser. Fiz.* **48** 2008–15
[174] Kleinheinz P, Broda R, Daly P J, Lunardi S, Ogana M and Blomqvist J 1979 *Z. Phys.* A **290** 279–96
Nagai Y, Styczen J, Püparinen M, Kleinheinz P, Bazzacco D, von Brentano P, Zell K O and Blomqvist J 1981 *Phys. Rev. Lett.* **47** 1259–62
[175] Artamonov S A, Isakov B I, Oglobin S G, Sliv L A and Shaginyan V R 1984 *Yad. Fiz.* **39** 328–40
Artamonov S A 1985 *Yad. Fiz.* **42** 91–8
[176] Sommermann H M, Ratcliff K F and Kuo T T S 1983 *Nucl. Phys.* A **406** 109–33
[177] Nguen Dinh Thao and Stoyanov Ch 1982 *Izv. AN SSSR Ser. Fiz.* **46** 2157–62
[178] Poschenrieder P and Weigel M K 1984 *Phys. Rev.* C **29** 2355–57
Dukelsky J, Dussel G G and Sofia H M 1983 *Phys. Rev.* C **27** 2954–67
[179] Abbas A, Auerbach N, Nguen van Giai and Zamikck L 1981 *Nucl. Phys.* A **367** 189–96
[180] Vdovin A I, Soloviev V G and Stoyanov Ch 1974 *Yad. Fiz.* **20** 1131–8
Vdovin A I and Soloviev V G 1974 *Izv. AN SSSR* **38** 2598–603, 2604–9; 1975 **39** 1618–23
[181] Takuda K and Tazaki S 1979 *Prog. Theor. Phys.* B **61** 1666–81
Wenes G, van Isacker P, Waroguier M, Heyde K and van Maldeghen J 1981 *Phys. Lett.* B **98** 398–404
[182] Soloviev V G, Stoyanov Ch and Nikilaeva R 1983 *Izv. AN SSSR Ser. Fiz.* **47** 2082–8
[183] Grinberg M and Stoyanov Ch 1989 *Selected Topics in Nuclear Structure* D4–89–327 (Dubna: JINR) p 16
[184] Metzger F R 1976 *Phys. Rev.* C **14** 543–7; 1978 **18** 2138–44
Swann C P 1977 *Phys. Rev.* C **15** 1967–71
[185] Soloviev V G, Stoyanov Ch and Voronov V V 1978 *Nucl. Phys.* A **304** 503–19
[186] Voronov V V, Dao Tien Khoa and Ponomarev V Yu 1984 *Izv. AN SSSR Ser. Fiz.* **48** 1846–51
[187] Kantele J, Julin R and Luontama M, Passoja A, Poikolainen T, Bäcklin A and Jonsson N G 1979 *Z. Phys.* A **289** 157–61
Bäcklin A, Jonsson N G, Julin R, Kantele J, Luontama M, Passoja A and Poikolainen T 1981 *Nucl. Phys.* A **351** 490–508
[188] Wenes G, van Isacker P, Waroquier M, Heide K and van Maldenhem J 1981 *Phys. Rev.* C **23** 2291–2304
Wienke H, Blok H P and Blok J 1983 *Nucl. Phys.* A **405** 237–51
[189] Bijker R, Dieperink A E, Scholten O and Spanoff R 1980 *Nucl. Phys.* A **344** 207–32
[190] Casten R F, Warner D D, Brenner O S and Gill R L 1981 *Phys. Rev. Lett.* **47** 1433–6
Cill R E, Casten R F, Warner D D, Brenner O S and Walters W B 1982 *Phys. Lett.* B **118** 251–5
Wolf A *et al* 1983 *Phys. Lett.* B **123** 165–8
[191] Snelling D M and Hamilton W D 1983 *J. Phys. G: Nucl. Phys.* **9** 763
[192] Eid S A A, Hamilton W D and Elliot J P 1986 *Phys. Lett.* B **166** 267
[193] Huo Junde, Hu Dailing, Zhou Chunmei, Han Xiaoling, Hu Bachue and

Wu Yaodong 1987 *Nuclear Data Sheets* **51** 1–94
[194] Nikolaeva R, Stoyanov Ch and Vdovin A I 1989 *Europhys. Lett.* **8** 117
[195] Belyaev S T and Rumiantsev V A 1969 *Phys. Lett.* B **30** 444–7
Telitsin V B, Stoyanov Ch and Vdovin A I 1976 *Yad. Fiz.* **24** 31–9
[196] Miura K, Hiratate Y, Shoji T, Suehiro T, Yamaguchi H and Ishizaki Y 1985 *Nucl. Phys.* A **436** 221–35
Mordechai S, Fortune H T, Carchidi M and Gilman R 1984 *Phys. Rev.* C **29** 1599–702
[197] Dambasuren D, Vdovin A I and Stoyanov Ch 1975 *JINR Communication* R4–8778 (Dubna: JINR)
[198] Voinova–Eliseeva N A and Mitropolsky I A 1986 *Part. Nucl.* **50** 14–55
Takada K and Tazaki S 1983 *Nucl. Phys.* A **395** 165–81
Makishima A, Ishima M, Ohaima M, Adachi M and Taketani H 1984 *Nucl. Phys.* A **425** 1–11
Passola A, Julin R, Kantela J, Luontama M and Vergnes M 1985 *Nucl. Phys.* A **441** 261–70
[199] Aprahamian A, Brenner D S, Casten R F, Gill R L, Pionrowski A and Heide K 1984 *Phys. Lett.* B **140** 22–8
Heyde K, van Isacker, Casten R F and Wood J L 1985 *Phys. Lett.* B **155** 303–8
[200] Hamamoto I 1974 *Problems of Vibrational Nuclei* ed G Alaga, V Paar and L Sips (Amsterdam: North-Holland) pp 54–74
Paar V *Problems of Vibrational Nuclei* ed G Alaga, V Paar and L Sips (Amsterdam: North-Holland) pp 15–53
[201] Mitroshin V E 1974 *Izv. AN SSSR Ser. Fiz.* **38** 2074–6
Zvonov V S and Mitroshin V E 1985 *Izv. AN SSSR* ser. fiz. **42** 2–37
Krygin and Mitroshin V E 1985 *Part. Nucl.* **16** 927–65
[202] Alikov B A, Muminov K M, Nazmitdinov R G and Chan Zuy Khuong 1981 *Izv. AN SSSR Ser. Fiz.* **45** 2111–5
Budzinski M, Lebedev N A, Lizurej H I, Muminov T M, Nazmitdinov R G, Sarzinsky Ja, Hazratov Y and Halikov A B 1984 *Nucleonika* **29** 71–85
[203] Deryuga V A, Kratsikova T I, Finger M, Deryuga V A, Kracikova T U, Finger M, Dupak Ja, Vdovin A I and Hamilton W D 1982 *Izv. AN SSSR Ser. Fiz.* **46** 860–6, 867–73
[204] Vdovin A I, Stoyanov Ch and Andreitacheff W 1985 *Nucl. Phys.* A **440** 437–44
Vdovin A I, Rodriges O O, Andrejtscheff W and Stoyanov Ch 1985 *Izv. AN SSSR Ser. Fiz.* **49** 2173–9
[205] Levon A I, Fedotkin S N and Chan Zuy Khuong 1983 *Yad. Fiz.* **38** 577–83
Gorbachev B I, Levon A I, Nemets O F, Gorbachev B I, Levon A I, Nemets O F, Fedotkin S N and Stepanchenko V A 1984 *Zh. Eksp. Theor. Fiz.* **87** 3–13
Levon A I, Fedotkin S N and Vdovin A I 1986 *Yad. Fiz.* **43** 1416–25
[206] Vdovin A I, Voronov V V, Ponomarev V Yu and Stoyanov Ch 1979 *Yad. Fiz.* **30** 923–32
[207] Zuker A P 1974 *Proc. Int. Conf. Nuclear Structure and Spectroscopy* ed H B Blok and A E L Dieperink (Amsterdam: Scholar's Press) pp 115–26
[208] Soloviev V G 1974 *Nuclear Structure Study with Neutrons* ed Erö Szücs (Budapoest: Akadimiai Kiado) pp 85–99; 1972 *Part. Nucl.* **3** 770–831
[209] Vlasov N A, Kalinin S P, Ogloblin A A and Chuev V I 1960 *Zh. Eksp. Theor. Fiz.* **39** 1615–7

[210] Sakai M and Kubo K I 1972 *Nucl. Phys.* A **185** 217–28
Ishimatsu T, Hayahibe S, Kawamura N, Awaya T, Ohmura H, Nakajima Y and Mitarai S 1972 *Nucl. Phys.* A **185** 273–83

[211] Siemssen R H 1976 *Selected Topics in Nuclear Structure* D–9920 vol **2** (Dubna: JINR) pp 106–25

[212] Crawley G M 1980 *Proc. Int. Symp. on Highly Excited States in Nuclear Reactions* ed H Ikegami and M Muraoka (Osaka: Osaka Univ.) pp 590–611

[213] Gales S 1981 *Nucl. Phys.* A **354** 193c–234c; 1984 *Int. Symp. HESANS-83* Orsay, France pp C4-39–C4-55

[214] Mongey J 1983 *Nucl. Phys.* A **396** 39c–59c

[215] Van der Werf S Y, Kooistra B R, Hasselink W H A, Iachello F, Put L W and Siemssen R H 1974 *Phys. Rev. Lett.* **33** 712–5; 1977 *Nucl. Phys.* A **289** 141–64
Scholten O, Harakeh M N, van der Plicht, Put L W, Siemssen R H, Van der Werf S Y and Sekiguchi M 1980 *Nucl. Phys.* A **348** 301–20

[216] Gerlic F, Langevin-Joliot H, van de Wielle and Suhamel G 1975 *Phys. Lett.* B **57** 338–40
Berrier–Ronsin G, Duhamel G, Gerlic E, Kalifa J, Langevin–Joliot H, Rotbard G, Vergnes M, Vernotte J and Seth K K 1977 *Phys. Lett.* B **67** 16–8
Gales S, Hourani E, Fortier S, Laurent H, Maison J M and Schapira J P 1977 *Nucl. Phys.* A **288** 221–41

[217] Gales S, Crawley G M, Weber D and Zwieglinski B 1983 *Nucl. Phys.* **398** 19–58

[218] Langevin-Joliot H, Gerlic E, Guillot J and van de Wiele J 1984 *J. Phys. G: Nucl. Phys.* **10** 1435–47
Wagner G J, Vdovin A I, Grabmayr P, Kim T, Mairle G, Pugach V, Sigert G and Stoyanov Ch 1984 *Yad. Fiz.* **40** 1396–403
Gales S, Massolo C P, Fortier S, Schapira J P, Martin P and Comparat V 1985 *Phys. Rev.* **31** 94–110

[219] Gales S 1985 *Proc. Int. School on Nuclear Structure* D4–85–851ed V G Soloviev and Yu P Popov (Dubna: JINR) pp 51–82

[220] Stuirbrink A, Wagner G J, Knöpfle K T, Liu Ken Pao, Maide G, Riedesel H, Schindler K, Bechtold V and Friedrich L 1980 *Z. Phys.* A **297** 307–9
Crawley G M, Kasagi J, Gales S, Gerlic E, Friesel D and Bacher A 1981 *Phys. Rev.* C **23** 1818–21
Perrin G, Duhamel G, Perrin C, Gerlic E, Gales S and Comparat V 1981 *Nucl. Phys.* A **356** 61–73

[221] Gales S, Gerlic E, Duhamel G, Perrin G, Perrin C and Comparat V 1982 *Nucl. Phys.* A **381** 40–60
Gerlic E, Berrier–Rosin G, Duhamel G, Gales S, Hourani E, Langevin-Joliot H, Vergnes M and van der Wiele J 1980 *Phys. Rev.* C **21** 124–46

[222] Soloviev V G 1984 *Int. Symp. HESANS-83 (Orsay, France)* pp C4-69–C4-83
Soloviev V G 1982 *Electromagnetic Interactions of Nuclei at Low and Medium Energies* (Moscow: AN SSSR Publ.) pp 164–72

[223] Vdovin A I, Stoyanov Ch and Chan Zui Khyong 1979 *Izv. AN SSSR Ser. Fiz.* **43** 999–1005
Stoyanov Ch 1981 *Izv. AN SSSR Ser. Fiz.* **45** 1820–26

[224] Nguen Dinh Thao, Soloviev V G, Stoyanov Ch and Vdovin A I 1984 *J. Phys. G: Nucl. Phys.* **10** 517–23

Vdovin A I and Stoyanov Ch 1985 *Yad. Fiz.* **41** 1134–40
[225] Koeling T and Iachello F 1978 *Nucl. Phys.* A **295** 45–60
Doll P 1977 *Nucl. Phys.* A **292** 165–72
[226] Van Giai N 1980 *Proc. Int. Symp. on Highly Excited States* ed H Ikegami and M Muraoka (Osaka: Osaka University) pp 682–93
[227] Klevansky S P and Lemmer R H 1982 *Phys. Rev.* **25** 3137–45; 1983 **28** 1763–78
Matveev B B, Muraviev S E, Tulupov B A and Urin M G 1984 *Izv. AN SSSR* **48** 2051–3
[228] Antonov A N, Nikolaev V A and Petkov I Zh 1982 *Z. Phys.* A **304** 239–43
[229] Azaiez F, Fortier S, Gales S, Hourani E, Maison J M, Kumpulainen J and Schapira J P 1985 *Nucl. Phys.* A **444** 373–401
[230] Sakai H, Bhowmik R K, van Dijk K, Brandenburg S, Drentje A G, Harakleh M N, Iwasaki Y, Siemssen R H, van der Werf S V and van der Woude A 1985 *Nucl. Phys.* A **441** 640–75
[231] Doll P, Wagner G J, Breuer H, Knöpfle K T, Mairle G and Riedesel H 1979 *Phys. Lett.* B **82** 357–60
[232] Gales S, Massolo C P, Azaiez F, Fortier S, Gerlic E, Guillot J, Hourani E, Maison J M, Schapira J P and Crawley G M 1984 *Phys. Lett.* B **144** 323–7
[233] Gales S, Stoyanovi Ch and Vdovin A I 1988 *Phys. Rep.* **166** 125–93
[234] Harakeh M N, van Heyst B, van der Borg K and van der Woude 1979 *Nucl. Phys.* A **327** 373–96
Fujita Y, Fujivara M, Morinobu S, Katayama I, Yamazaki T, Itahashi T, Ikagami H and Hayakawa S I 1985 *Phys. Rev.* C **32** 425–30
[235] Bartholomew G A, Earle E D, Ferguson A J, Knowles J W and Lone M A 1973 *Advances Nucl. Phys.* **7** 229–324
[236] Bell Z W, Cardman L S and Axel P 1982 *Phys. Rev.* C **25** 791–803
Starr R D, Axel P and Cardman L S 1982 *Phys. Rev.* C **25** 780–90
Laszewski R M and Axel P 1979 *Phys. Rev.* C **19** 342–554
[237] Belyaev S N, Kozin A B, Nechkin A A, Semenov V A and Semenko S F 1985 *Yad. Fiz.* **42** 1050–8
[238] Schumakher M, Zurmühl U, Smend F and Nolte R 1985 *Nucl. Phys.* A **438** 499–502
[239] Crawley G M 1980 *Inst. Phys. Conf. Ser. 49* vol **1** (Bristol: Institute of Physics) pp 127–50
Crawley G M, Beneson W, Bertsch G, Gales S, Weber D and Zwieglinski B 1981 *Phys. Rev.* C **23** 589–96; 1982 *Phys. Lett.* B **109** 8–10
[240] Gerlic E, Guillot J, Langevin–Joliot H, Gales S, Sakai M, van der Wiele J, Duhamel G and Perrin G 1982 *Phys. Lett.* B **117** 20–4
Langevin–Joliot H, Gerlic E, Guillot J, Sakai M, van der Wiele J, Devaux A, Force P and Landaud G 1982 *Phys. Lett.* B **114** 103–6
[241] Nakagawa T, Tohei T, Kanozawa M, Sekine N, Yamaguchi H, Yuassa K, Iwatani K and Ishizaki Y 1982 *Nucl. Phys.* A **376** 513–32
[242] Soloviev V G, Stoyanova O and Voronov V V 1981 *Nucl. Phys.* A **370** 13–29
[243] Voronov V V 1983 *J. Phys. G: Nucl. Phys.* **9** 1273–7
Voronov V V and Zhuravlev I P 1983 *Yad. Fiz.* **38** 52–8
[244] Ipson S S, McLean K C, Booth W, Haigh J G B and Glover R N 1975 *Nucl. Phys.* A **253** 189–215
[245] Borello–Lewin T, Castro H M, Horodynski–Matsushigue L B, and Dietzsch O 1979 *Phys. Rev.* C **20** 2101–13

[246] Guillot J, van de Wiele, Langevin–Joliot H, Gerlic E, Didelez J P, Duhamel G, Perrin G, Buenerd M and Chauvin J 1980 *Phys. Rev.* C **21** 879–95
[247] Willis J E, Glagg T W and Kaitchuk 1981 *Nucl. Phys.* A **362** 8–17
[248] Lanford W A and Crawley G M 1974 *Phys. Rev.* C **9** 646–59
[249] McGrory J B and Kuo T T S 1975 *Nucl. Phys.* A **247** 283–316
Vary J and Ginocchio J N 1975 *Nucl. Phys.* A **166** 479–514
[250] Vellerli M C, Hausser O, Alford W P, Alford W P, Celler A, Frekers D, Helmer R, Henderson R, Hicks K H, Jackson K P, Jeppesen R G, Miller C A, Raywood K and Yen S 1989 *Phys. Rev.* C **40** 559
[251] Raywood K J *et al* 1990 *Phys. Rev.* C **41** 2836–51
[252] Gromov K Ya 1984 *Int. Symp. IN-BEAM Nuclear Spectroscopy* ed Dembradi and Fenyes (Budapest: Akademidi Kiado) p 269
[253] Alkhazov G D, Ganbaater N, Gromov K Ya, Isakov V I, Kalinnikov V G, Merzilev K A, Novikov Yu N, Nurmukhamedov A M, Potempa A and Tarkach F 1984 *Yad. Fiz.* **40** 554
Alkhazov G D, Artamonov S A, Isakov V I, Merzilev K A and Novikov Yu N 1987 *Phys. Lett.* B **198** 37
[254] Vogel P and Zirnbauer M R 1986 *Phys. Rev. Lett.* **57** 3148
Engel J, Vogel P and Zirnbauer M R 1988 *Phys. Rev.* C **37** 731
[255] Civitarese O, Faessler A and Tomode T 1987 *Phys. Lett.* B **194** 11
[256] Muto K and Klapdor H V 1988 *Phys. Lett.* B **201** 420
[257] Kuzmin V A and Soloviev V G 1988 *Pisma Zh. Eksp. Teor. Fiz.* **47** 68
Soloviev V G 1989 *Proc. Int. Conf. on Selected Topics in Nuclear Structure* D4, 16, 15–89–638 (Dubna: JINR) p 104
[258] Kuzmin V A and Soloviev V G 1988 *Nucl. Phys.* A **486** 118
[259] Suhonen J, Faessler A, Taigel T and Tomoda T 1988 *Phys. Lett.* B **202** 174
Suhonen J, Taigel T and Faessler A 1988 *Nucl. Phys.* A **486** 91
[260] Barden R, Kirchner R, Klepper O, Ptochocki A, Rathke G E, Roeckl E, Rykaczewski K, Schardt D and Żylicz J 1988 *Z. Phys.* A **329** 319
[261] Rykaczewski K *et al* 1985 *Z. Phys.* A **322** 263
Rykaczewski K, Ptochocki A, Grant I S, Gabelmann H, Barden R, Schardt D, Żylicz J, Nyman G and the ISOLDE Collaboration 1989 *Z. Phys.* A **332** 275
[262] Klepper D and Rykaczewski K 1991 *Proc. Predeal Int. Summer School Recent Advances in Nuclear Structure* (Singapore: World Scientific)
[263] Bloom S D, Mathews G J and Becker J A 1987 *Can. J. Phys.* **65** 684
[264] Nadjakov E 1982 *Nuclear Structure at High Spin* (Sofia: Bulgarian Academy of Science)
[265] Brianşon Ch and Mikhailov I N1982 *Part. Nucl.* **13** 245–99
[266] De Voigt, Dudek J and Szymanski Z 1983 *Rev. Mod. Phys.* **55** 949–1046
[267] Shimizu Y R, Garett J D, Broglia R A, Gallardo M and Vigezzi E 1989 *Rev. Mod. Phys.* **61** 131–68
[268] Grigoriev E P and Soloviev V G 1974 *Structure of Even Deformed Nuclei* (Moscow: Nauka) (in Russian)
[269] Ivanova S P, Komov A L, Malov L A and Soloviev V G 1976 *Part. Nucl.* **7** 450–98
[270] Malov L A and Yakovlev D G 1985 *Izv. AN SSSR Ser. Fiz.* **49** 2150–4
[271] Strutinsky V M 1966 *Yad. Fiz.* **3** 614–24
[272] Soloviev V G, Nesterenko V O and Bastrukov S I 1983 *Z. Phys.* A **309** 353–61

[273] Birbrair B L 1963 *Izv. AN SSSR* **27** 1329–37
[274] Soloviev V G 1968 *Prog. Nucl. Phys.* **10** 239–71
[275] Chasman R R, Ahmad I, Friedman A M and Erskine J F 1977 *Rev. Mod. Phys.* **49** 833–91
[276] Nesterenko V O, Soloviev V G and Sushkov A V 1986 *JINR Communication R4–86–115* (Dubna: JINR)
[277] Winter G, Sodan H, Kaun K H, Kemnits P and Funke L 1973 *Part. Nucl.* **4** 895–940
[278] Rekstad J, Engeland T Osnes E 1979 *Nucl. Phys.* A **330** 367–80
Henriguez A, Engeland T and Rekstad J 1983 *Nucl. Phys.* A **410** 1–13
[279] Dzhelepov B S 1983 *Properties of Deformed Nuclei* ed A I Muminov (Tashkent: FAN) pp 3–111
[280] Soloviev V G and Vogel P 1967 *Nucl. Phys.* A **92** 449–74
Soloviev V G, Vogel P and Yungklaussen G 1967 *Izv. AN SSSR Ser. Fiz.* **31** 518–31
Soloviev V G and Fainer U M 1972 *Izv. AN SSSR* ser. fiz. **36** 698–705
Soloviev V G and Fedotov S I 1972 *Izv. AN SSSR* ser. fiz. **36** 706–17
[281] Gareev F A, Ivanova S P, Malov L A and Soloviev B G 1971 *Nucl. Phys.* A **171** 134–64
Ivanova S P, Komov A L, Malov L A and Soloviev B G 1973 *Izv. AN SSSR Ser. Fiz.* **37** 911–21; 1975 **39** 1612–7
[282] Bunker M E and Reich C W 1971 *Rev. Mod. Phys.* **43** 348–423
Ogle W, Wahlborn S, Piepenbring R and Fredriksson S 1971 *Rev. Mod. Phys.* **43** 424–78
[283] Hoff R W, Lougheed R W, Barreau G, Börner H G, Davidson W F, Schreckenbach K, Warner D D, von Egidy T and White D H 1982 *Inst. Phys. Conf. Ser. 62* (Bristol: Institute of Physics)
Hoff R W, Davidson W F, Warner D D, Börner H G and von Egidy T 1982 *Phys. Rev.* C **25** 2232–54
Von Egidy T 1982 *Neutron Physics School* D3,4–82–704 (Dubna: JINR) pp 181–205
[284] Kvasil J, Choriev M M and Čwiok S 1985 *Czechosl. J. Phys.* **35** 949–62
[285] Bastrukov S I and Nesterenko V O 1984 *Int. Symp. on In-Beam Nuclear Spectroscopy* ed Z S Dombradi and T Fenyes (Budapest: Academiai Kiado) pp 689–99
[286] Avotina M P and Zoloavin A V 1979 *The Moments of the Ground and Excited States of Nuclei* (Moscow: Atomizdat) (in Russian)
[287] Berlovich E K, Vasilenko S S and Novikov Yu N 1972 *Lifetimes of Excited States of Atomic Nuclei* (Leningrad: Nauka) (in Russian)
[288] Begzhanov R B and Belen'kii V M 1980 *Gamma Spectroscopy of Atomic Nuclei* (Tashkent: FAN) (in Russian)
Begzhanov R B, Belen'kii V M, Zalyubovsky I N, Kuznichenko A B and Sattarov M G 1985 *Structure of Even–Even Transition Atomic Nuclei* (in Russian)
[289] Kvasil J, Mikhailov I N, Safarov R Ch and Choriev M M 1978 *Czech. J. Phys.* **28** 843–56
Kvasil J, Kracikova T I, Davaa S, Finger M and Choriev B 1983 *Czech. J. Phys.* **33** 626–41; 1985 **35** 1084–102
Kvasil J, Kracikova T I, Finger M and Choriev B 1981 *Czech. J. Phys.* **31** 1376
[290] Andrejtscheff W 1976 *Part. Nucl.* **7** 1039–79
[291] Adam I, Alikov B A, Badalov Kh N, Gonusec M, Lizúrei G I, Muminov

T M and Scharonov I A 1987 *Izv. AN SSSR Ser. Fiz.* **51** 15–23
Alikov B A, Badalov Kh N, Lizurej H I, Muminov T M and Scharonov I A 1987 *Izv. AN SSSR Ser. Fiz.* **51** 841–55
[292] Alikov B A, Badalov Kh N, Nesterenko V O, Sushkov A V and Wawryszczuk J 1988 *Z. Phys.* A **331** 265
[293] Baziat M I, Pyatov N I and Chernei M I 1973 *Part. Nucl.* **4** 941–91
Begzhanov R B, Daminov E T and Choriev M M 1981 *Izv. AN UzSSSR Ser. Fiz.-Mat.* No 3, 70–5
[294] Gnatovich V and Gromov K Ya 1966 *Yad. Fiz.* **3** 8–12
Andrejtscheff W, Manfrass P 1975 *Phys. Lett.* B **55** 159–61
[295] Vdovin A I, Malov L A, Nguen Dinh Vinh, Soloviev V G and Stoyanov Ch 1985 *Izv. AN SSSR Ser. Fiz.* **49** 834–42
Nguen Dinh Vinh 1985 *Izv. AN SSSR Ser. Fiz.* **49** 2213–7
Nguen Dinh Vinh, Malov L A and Soloviev V G 1986 *JINR Short Communications* No 16–86 (Dubna: JINR)
[296] Nguen Dinh Vinh and Soloviev V G 1986 *Yad. Fiz.* **43** 1162–8
[297] Casten R F, Breiting D, Wasson O A, Rimawi K and Chrien R E 1974 *Nucl. Phys.* A **228** 493–512
[298] Gales S, Crawley G M and Zwieglinski B 1983 *Nucl. Phys.* A **398** 19–58
Rekstad J, Lovhoiden G, Lien J R 1980 *Nucl. Phys.* A **348** 93–108
[299] Banerjee B, Mang H J and Ring P, Schmid K W and Hilton R R 1973 *Nucl. Phys.* A **215** 366–82
Faessler A, Sandhya Devi K R, Grümmer F 1976 *Nucl. Phys.* A **256** 106–26
[300] Kvasil Ya, Choriev M M, Mikhailov I N, Choriev B and Čwiok S 1984 *Izv. AN SSSR Ser. Fiz.* **48** 844–56
Kvasil Ya, Choriev M M, Tsvek S, Choriev B and Mikhailov I N 1985 *Yad. Fiz.* **42** 588–600
[301] Nesterenko V O, Soloviev V G, Sushkov A V and Shirikova N Yu 1986 *Yad. Fiz.* **44** 1443–50
[302] Soloviev V G and Shirikova N Yu 1989 *Z. Phys.* A **334** 149; 1990 *Izv. AN SSSR Ser. Fiz.* **54** 818
[303] Soloviev V G, Sushkov A V and Shirikova N Yu 1991 *Yad. Fiz.* **53** 101–12
[304] Burke D G and Elbek B 1967 *Danske Vid. Selsk. Mat. Fys. Medd.* **36** 6
[305] Hague A M *et al* 1986 *Nucl. Phys.* A **455** 231
[306] Friedman A and Kator K 1973 *Phys. Rev. Lett.* **30** 102
[307] Shihab–Eldin A A, Rasmussen J O and Stoyer M 1989 *Workshop on Microscopic Models in Nuclear Structure Physics* ed M W Guidry and J H Hamilton (Singapore: World Scientific) p 282–97
[308] Walker P M, Carvalho J L and Bernthal F M 1982 *Phys. Lett.* B **116** 393–6
Burke D G and Blezius J W 1982 *Can. J. Phys.* **60** 1751–8
[309] Bagnell D G, Tanaka Y, Sheline R K, Burke D G and Sherman J D 1977 *Phys. Lett.* B **66** 129–32
[310] Ishihara T, Sakaguohi H, Nakamura M, Noro T, Ohtani F, Sakamoto H, Ogawa H, Yosoi M, Ieiri M, Isshiki N, Takeuchi Y and Kobayashi S 1984 *Phys. Lett.* B **149** 55–8
[311] Sood P C and Sheline R K 1990 *Phys. Scr.* **42** 25
[312] Massman H, Rasmussen J O, Ward T E, Haustein P E and Bernthal F M 1974 *Phys. Rev.* C **9** 2312
[313] Soloviev V G and Sushkov A V 1990 *J. Phys. G: Nucl. Phys.* **16** L57
[314] Khoo T L, Waddington J C, O'Neil R A, Preibisz Z, Burke D G and

Johns M W 1972 *Phys. Rev. Lett.* **28** 1717
[315] Kaffrell N and Kurcewicz W 1975 *Nucl. Phys.* A **255** 339
[316] Greenwood R C, Reich C W, Baader H A, Koch H R, Breitig D, Schult O W B, Fogelberg B, Bäcklin A, Mampe W, von Egidy T and Schreckenbach K 1978 *Nucl. Phys.* A **304** 327
[317] Davidson W F, Dixon W R and Burke D G 1983 *Phys. Lett.* B **136** 161–6
Yates S W, Kleppinger E D and Kleppinger E W 1985 *J. Phys. G: Nucl. Phys.* A **11** 877–81
[318] Burke D G, Maddock B L and Davidson W F 1985 *Nucl. Phys.* A **442** 424–59
[319] Sobiczewski A, Patyk Z, Čwiok S and Rozmej P 1958 *Nucl. Phys.* A **485** 16
[320] Peker L K and Hamilton J H 1980 *Directions in Studies in Studies of Nuclei Far from Stability* ed J H Hamilton (Amsterdam: North-Holland) pp 323–35
[321] Dumitrescu T S and Hamamoto I 1982 *Nucl. Phys.* A **383** 205–23
[322] Matsuo M and Matsuyanagi K 1985 *Prog. Theor. Phys.* **74** 1227–44
[323] Čwiok S, Lojewski Z and Kvasil J 1983 *Communication JINR* E4–83–647 (Dubna: JINR)
Hayashi A and Hara K 1984 *Phys. Rev. Lett.* **53** 337–40
Sweiwert M, Mardunh J A and Hess P O 1984 *Phys. Rev.* C **30** 1779–82
[324] Börner H G, Jolie J, Robinson S J, Krusche B, Piepenbring R, Casten R F, Aprahamian A and Draager J P 1991 *Phys. Rev. Lett.* **66** 305
[325] Börner H G 1991 *Proc. VII Int. Symp. Capture Gamma-Ray Spectroscopy and Related Topics* ed R Hoff (New York: American Institute of Physics) p 112
[326] Jolie J 1991 *Proc. VII Int. Symp. Capture Gamma-Ray Spectroscopy and Related Topics* ed R Hoff (New York: American Institute of Physics) p 121
[327] Soloviev V G 1991 *Proc. VII Int. Symp. Capture Gamma-Ray Spectroscopy and Related Topics* ed R Hoff (New York: American Institute of Physics) p 16
[328] Piepenbring R and Jammari M K 1988 *Nucl. Phys.* A **481** 81; **487** 77
[329] Soloviev V G and Vogel P 1963 *Phys. Lett.* **6** 126–8
[330] Sobiczewski A 1985 *Proc. Symp. Nuclear Excited States* ed L Lason *et al* (Lodz: Lodz University) pp 79–86
Pashkevich V V 1983 *School-Seminar on Heavy Ions* D7–83–644 (Dubna: JINR) pp 405–19
[331] Jolos R W, Ivanova S P, Pedrosa R and Soloviev V G 1987 *Teor. Matem. Fiz.* **70** 154–60
[332] Soloviev V G and Shushkov A V 1989 *Phys. Lett.* B **216** 259
[333] Helmer R G 1985 *Nucl. Data Sheets* **44** 659
Shurshikov E N 1986 *Nucl. Data Sheets* **47** 433
Ignatoshkin A E, Shurshikov E N and Jabrov Yu F 1987 *Nucl. Data Sheets* **52** 365
[334] Soloviev V G 1986 *Z. Phys.* A **324** 393–401; 1984 *Pisma Zh. Eksp. Teor. Fiz.* **40** 398–401; 1988 *Yad. Fiz.* **47** 332–40
[335] Soloviev V G 1990 *Part. Nucl.* **21** 1360–404
[336] Burke D G, Davidson W F, Cizewski J A, Brown R E and Sunier J W 1985 *Nucl. Phys.* A **445** 70–92
[337] Burke D G, Cizewski J A and Davidson W F 1987 *Symmetries and Nuclear Structure* ed R A Mayer and V Paar (London: Nuclear Science

Research Conf. Series) vol **13** p 173

[338] Yoshinaga N, Akiyama Y and Arima A 1986 *Phys. Rev. Lett.* **56** 1116–9; 1988 *Phys. Rev.* C **38** 419–36

Akiyama Y, Heyde K, Arima A and Yoshinaga N 1986 *Phys. Lett.* B **173** 1–4

[339] Barfield A F, Wood J L and Barrett B R 1986 *Phys. Rev.* C **34** 2001–4

Nadjakov E C and Mikhailov I N 1987 *J. Phys. G: Nucl. Phys.* **13** 857–73

[340] Barfield A F, Barrett B R, Wood J L and Scholten O 1989 *Ann. Phys.* **182** 344

[341] Covill I M, Fulbright H W, Cline D, Weslowski E, Kotlinski B, Bäcklin A and Gridnev K 1986 *Phys. Rev.* C 793–803; 1987 **36** 1442–52

[342] Malov L A, Meliev F M and Soloviev V G 1985 *Z. Phys.* A **320** 521–7

[343] Malov L A, Nesterenko V O and Soloviev V G 1977 *J. Phys. G: Nucl. Phys.* **3** L219–L221

[344] Beene J R, Varner R L and Bertrand F E 1976 *Nucl. Phys.* A **482** 407c–420c

Moss J M, Brown D R, Youngblood D H *et al* 19 *Phys. Rev.* C **18** 741–9

[345] Van der Woude A 1988 *Nucl. Phys.* A **482** 453c–470c

[346] Beene J R, Bertrand F E, Halbert M L, Auble R L, Hensley D C, Horen D J, Robinson R L, Sayer R O and Sjoren T P 1989 *Phys. Rev.* C **39** 1307–19

Index